Map Use

MAP USE

Volume 1 · Ninth Edition
Map Reading and Design

Aileen R. Buckley

A. Jon Kimerling

Patrick J. Kennelly

Esri Press
Redlands, California

The day-night globe is by A. Jon Kimerling.

Esri Press, 380 New York Street, Redlands, California 92373-8100
Copyright © 2026 Esri
All rights reserved.
Printed in the United States of America.
30 29 28 27 26   1 2 3 4 5 6 7 8 9 10

Paperback ISBN: 9781589487758
Hardcover ISBN: 9781589488816
E-book ISBN: 9781589487765

Library of Congress Control Number: 2025943944

The information contained in this document is the exclusive property of Esri or its licensors. This work is protected under United States copyright law and other international copyright treaties and conventions. No part of this work may be used, used to train, or used to prepare derivative works, copied, reproduced, redistributed, publicly displayed or performed, or transmitted in any form or by any means, electronic or mechanical, including photocopying, digital scanning, and recording, or by any machine learning, artificial intelligence, information storage, or retrieval systems, except as expressly permitted in writing by Esri. Any unauthorized use of this work to train artificial intelligence (AI) technologies is expressly prohibited. All requests should be sent to Attention: Director, Contracts and Legal Department, Esri, 380 New York Street, Redlands, California 92373-8100, USA.

The information contained in this document is subject to change without notice.

US Government Restricted/Limited Rights: Any software, documentation, and/or data delivered hereunder is subject to the terms of the License Agreement. The commercial license rights in the License Agreement strictly govern Licensee's use, reproduction, or disclosure of the software, data, and documentation. In no event shall the US Government acquire greater than RESTRICTED/LIMITED RIGHTS. At a minimum, use, duplication, or disclosure by the US Government is subject to restrictions as set forth in FAR §52.227-14 Alternates I, II, and III (DEC 2007); FAR §52.227-19(b) (DEC 2007) and/or FAR §12.211/12.212 (Commercial Technical Data/Computer Software); and DFARS §252.227-7015 (DEC 2011) (Technical Data–Commercial Items) and/or DFARS §227.7202 (Commercial Computer Software and Commercial Computer Software Documentation), as applicable. Contractor/Manufacturer is Esri, 380 New York Street, Redlands, California 92373-8100, USA.

Esri products or services referenced in this publication are trademarks, service marks, or registered marks of Esri in the United States, the European Community, or certain other jurisdictions. To learn more about Esri marks, go to: links.esri.com/EsriProductNamingGuide. Other companies and products or services mentioned herein may be trademarks, service marks, or registered marks of their respective mark owners.

For purchasing and distribution options (both domestic and international), please visit esripress.esri.com.

Contents

Foreword — xi
Preface — xiii
Acknowledgments — xvii

Introduction — 1

- Mental maps — 1
- Cartographic maps — 3
- Basic characteristics of maps — 3
- What makes maps popular? — 4
- The map transformation process — 5
- Types of maps — 5
  - Reference maps — 6
  - Thematic maps — 7
  - Charts — 9
- Cartometric maps — 10
- Topographic maps — 10
- Other types of maps — 10
  - 3D maps — 13
  - Dynamic and interactive maps — 14
  - Real-time maps — 14
  - Story maps — 15
  - Accessible and inclusive maps — 17
  - Maps as art — 17
- Map use — 19
- Selected readings — 21

Chapter 1: The earth and earth coordinates — 23

- The earth as a sphere — 23
- The graticule — 24
  - Parallels and meridians — 24
  - Geocentric latitude and longitude — 24
  - Prime meridians — 26
- The earth as an oblate ellipsoid — 27
  - Different ellipsoids — 28
  - Geodetic latitude — 29
  - Geodetic longitude — 29
- Determining geodetic latitude and longitude — 29
- Properties of the graticule — 31
  - Circumference of the authalic sphere — 31
  - Spacing of parallels — 31
  - Converging meridians — 32
  - Quadrilaterals — 32
  - Great and small circles — 32
- Graticule appearance on maps — 33
  - Small-scale maps — 33
  - Large-scale maps — 33
- Geodetic latitude and longitude on large-scale maps — 35
  - Horizontal datums — 35
    - World Geodetic System of 1984 — 36
- The earth as a geoid — 37
  - Vertical reference datums — 37
- National Spatial Reference System — 39
  - NSRS modernization — 40
- Selected readings — 41

Chapter 2: Map scale — 43

Expressing scale — 43
    Representative fraction — 43
    Word scale — 43
    Scale bar — 45
Large- and small-scale maps — 46
    Topographic map series at different scales — 46
Multiscale maps — 50
Converting scale — 51
    Converting a word scale to an RF — 51
    Converting an RF to a word scale — 52
Determining map scale — 52
    Determining map scale from a known ground feature — 53
    Determining map scale from the equator — 53
    Determining map scale from reference material — 54
    Determining map scale from the spacing of parallels and meridians — 54
    Determining map scale from ground resolution and pixel density — 56
Selected readings — 57

Chapter 3: Map projections — 59

Globes versus flat maps — 59
The map projection process — 61
Map projection properties — 62
    Scale — 62
    Completeness — 63
    Correspondence relations — 63
    Continuity — 64
Map projection families — 64
    Map projection families based on geometric distortion — 64
        Distance — 64
        Shape — 65
        Direction — 66
        Area — 66
    Map projection families based on developable surfaces — 67
        Planar projections — 67
        Cylindrical projections — 68
        Conic projections — 69
Map projection parameters — 70
    Tangent and secant case — 70
    Aspect — 70
    Other map projection parameters — 72
Commonly used map projections — 73
    Planar projections — 73
        Orthographic — 73
        Stereographic — 73
        Gnomonic — 74
        Azimuthal equidistant — 74
        Lambert azimuthal equal area — 75
    Cylindrical projections — 76
        Equidistant cylindrical — 76
        Mercator — 76
        Web Mercator — 77
        Transverse Mercator — 78
    Conic map projections — 79
        Lambert conformal conic — 79
        Albers equal-area conic — 79
    Pseudocylindrical and other projections — 80
        Mollweide — 80
        Sinusoidal — 81
        Goode homolosine — 81
        Robinson — 83
        Natural Earth — 84
        Winkel tripel — 84
    Oblique-perspective projections — 85
Map projections on the sphere and ellipsoid — 86
Map projection selection — 87
Selected readings — 88
Notes — 89

Chapter 4: Grid coordinate systems — 91

Cartesian coordinates	91	US state grids	104
Grid coordinates	92	Grid coordinate determination on maps	105
Universal transverse Mercator system	92	Grid coordinate system appearance on maps	105
Universal polar stereographic system	94	Grid orientation	106
State plane coordinate system	95	Grid coordinate measurement	107
SPCS 27 and 83	95	Grid cell location systems	109
SPCS2022	97	Military Grid Reference System	109
Zones	98	US National Grid	112
Minimizing distortion	99	Reference grids	113
Low-distortion projections	101	Selected readings	114

Chapter 5: Land partitioning — 115

Irregular systems	115	Government lots and platted lots	125
Metes and bounds	115	Survey irregularities	125
French long lots	117	PLSS appearance on maps	127
Spanish and Mexican land grants	119	Land records	128
Donation land claims	119	Subdivision plats	128
Regular systems	121	The cadastre	129
US Public Land Survey System	121	Land information systems	129
Townships and sections	123	Extensions of the cadastre	129
Fractional divisions	124	Selected readings	132

Chapter 6: Map design — 133

Purpose, audience, and other considerations	133	Graphic marks	150
Data for map compilation	135	Levels of measurement	150
Map scale	135	Visual variables	153
Map projection	135	Qualitative visual variables	153
Cartographic abstraction	137	Quantitative visual variables	154
Selection	137	Bivariate and multivariate symbols	154
Generalization	139	Color	158
Classification	141	Properties of color	158
Qualitative data classification	143	Color systems	158
Quantitative data classification	144	Common color models	159
Number of classes	144	Color schemes	161
Classing methods	147	Type	162
Adjusting class breaks	150	Properties of type	163
Symbols	150	Selected readings	164

Chapter 7: Map organization and layout — 167

Basic design concepts	167
Figure-ground organization	167
Visual hierarchy	167
Visual contrast	169
Color contrast	169
Legibility	171
Map elements	174
Common map elements	174
Layout design	176
Golden ratio	176
Proportion	178
Balance	178
Harmony	179
Web maps	179
Web map organization	180
Web map design	180
Digital twins	181
Selected readings	182

Chapter 8: Qualitative thematic maps — 183

Qualitative information	183
Point features	183
Line features	185
Area features	186
Census data	186
Data accuracy	187
Homogeneity	188
Single-theme maps	189
Point feature maps	189
Line feature maps	191
Area feature maps	191
Multivariate maps	193
Multivariate points	193
Multivariate lines	193
Multivariate areas	195
Mapping change	195
Change in location	195
Point feature locations	195
Line feature locations	195
Area feature locations	195
Change in attributes	197
Change maps	197
Small multiples	197
Dynamic maps	198
Selected readings	201

Chapter 9: Quantitative thematic maps — 203

Quantitative information	203
Spatial samples	203
Point features	205
Line features	206
Area features	206
Continuous surfaces	209
Interpolation	209
Data normalization	210
Single-theme maps	210
Point feature maps	210
Graduated and proportional symbol maps	210
Line feature maps	212
Flow maps	212
Vector flow maps	213
Area feature maps	213
Choropleth maps	215
Dot maps	215
Dasymetric maps	215
Cartograms	215
Prism maps	220
Binned maps	220
Continuous surface maps	220
Isoline maps	220
Heat maps	222
Dot density maps	223
3D surface maps	225
Multivariate maps	225
Multivariate symbols	225
Composite variable maps	227
Mapping change	227
Time series maps	228
Space-time cubes	228
Dynamic maps	230
Selected readings	231

Chapter 10: Relief portrayal — 233

Elevation data — 233
  Spatial resolution — 233
  DEM data collection — 235
    Lidar — 235
  Vertical exaggeration — 236
Map perspective — 237
Absolute-relief maps — 237
  Spot elevations, benchmarks, and soundings — 237
  Contours — 239
    Types of contours — 240
  Isobaths — 241
  Hypsometric tints — 241
Relative-relief maps — 243
  Variable perspective maps — 243
    Raised-relief globes — 244
    Raised-relief maps — 245
    Terrain models — 245
Plan perspective maps — 246
  Relief shading — 246
  Multidirectional relief shading — 246
  Relief shading and shadowing — 247
  Relief shading and hypsometric tints — 249
  Relief shading and contours — 250
Oblique-perspective maps — 250
  Hillsigns and hachures — 251
  Block diagrams — 251
  Diorama maps — 252
  Landscape drawings — 252
True 3D — 252
  Stereoscopic views — 252
  Anaglyphs — 254
  ChromaDepth maps — 254
Dynamic maps — 254
Selected readings — 256

Chapter 11: Imagery and maps — 259

Image quality — 261
  Spectral sensitivity — 261
  Spatial resolution — 262
  Sensor height — 262
  Relief displacement — 263
Aerial imagery — 264
  Panchromatic imagery — 264
  Natural-color imagery — 264
  Color-infrared imagery — 266
  Aerial imagery capture — 266
  Drone imagery — 267
Orthophotos and orthophotomaps — 269
  Orthophotos — 269
  Orthophotomaps — 269
  Digital orthophotoquads — 269
  Annotated orthophotomaps — 271
Satellite imagery — 272
  Weather satellites — 272
  Earth resources satellites — 275
    Landsat — 276
    Sentinel-2 — 278
    MODIS — 278
  Submeter-resolution systems — 281
  Hyperspectral systems — 282
  Map examples — 283
Dynamic maps — 284
Selected readings — 286

Chapter 12: Maps and reality — 287

The map as reality	287	Reality as a map	295
Cartographic artifacts	288	Survival	296
The missing essence	289	We all are human	296
Imaginative map use	289	Living with map limitations	296
Map misuses	292	The path ahead	297
A degraded environmental image	294	Selected readings	297
Abstract decision-making	294		

Appendix — 299

Table A.1. Prime meridians used previously on maps published in different countries and their longitudinal distances from the Greenwich meridian (in degrees, minutes, seconds) — 300

Table A.2. Metric and Imperial distance and area equivalents — 301

Table A.3. Variation in the length of a degree of latitude — 302

Table A.4. Areas of quadrilaterals of 1° extent — 303

Table A.5. Web map scales (at the equator) associated with zoom levels — 304

Table A.6. Variation in the length of a degree of longitude at different latitudes — 308

Table A.7. Common Landsat 8 and 9 band combinations — 309

Index — 311

Foreword

Today, the world of creating maps is accelerating, thanks to advances in digital cartography, geographic information systems (GIS), and web mapping. These technologies have become powerful engines for streamlining the design and production of maps. They have also expanded the range of topics being mapped and the actual number of maps being made. More important, they have broadened the community of people making and using maps.

But have the maps themselves changed?

Once primarily used as a tool to help people navigate their environment, maps today are increasingly used to help communicate and address the world's important challenges.

I see maps as a new kind of language that facilitates improved communication and collaboration in every aspect of society.

Mastering this language will become an essential skill set for 21st-century living.

Maps are abstractions of the earth—representations of the complexities of the world and useful for understanding the world around us. They distill and summarize information in a special type of graphic display. Creating these abstractions so that they provide valuable information that reflects and represents the complexities of the earth is the real challenge for those who produce today's maps.

Map Use, now in its ninth edition, is a comprehensive book for using maps effectively. By providing the keys to unlock the codes that cartographers use to represent the world around us, this book is an invaluable companion for college students and instructors, for professionals in a variety of fields where maps are important, and for casual map users.

Jack Dangermond
Cofounder and president, Esri®

Preface

Many books are written on the details of cartography, but because map use is not simply the offshoot of cartography, most of these books are of limited value to you as a map user. By contrast, this book is written for those who need to know how to use and make their own maps to build or enhance their spatial understanding of the world. Fully revised to better provide the context and demonstration of skills for reading and properly analyzing this complex form of communication, *Map Use: Map Reading and Design*, Volume 1 is a definitive resource for an introductory map use, cartography, or GIS course.

Map Use is also a valuable tool for people whose vocation or recreation requires knowing how to read and use maps because it takes readers beyond visual representations and into the decision-making processes of cartographers.

Academics tend to treat maps as indoor objects, rarely including in their textbooks the fact that one of the most exciting ways to use maps is in the field. Conversely, manuals and field guides on map and compass use focus narrowly on wayfinding, geocaching, and military map reading, virtually ignoring the role that maps play in how we think about and communicate environmental information. We depart from this tradition by bridging the gap between these two extremes, carefully weaving information from many fields into a coherent view of map use. This book offers readers a comprehensive, philosophical, and practical treatment of map use in several primary ways.

First, we define a map as a graphic representation of an environment that shows relations between geographic features. This encompassing definition lets us include a variety of important map forms that are otherwise awkward to categorize, such as mental maps, which are discussed in the introduction, or web maps, which may exist only ephemerally. Our definition should also accommodate new cartographic forms developed in the future. This book integrates discussions of a variety of map forms, including standard planimetric maps, perspective diagrams, aerial and satellite images, web maps, 3D maps, and others, rather than partitioning each type of map into a separate category. This integrative approach, focused on commonality rather than uniqueness, showcases the enhanced insight into the environment that the variety of mapping perspectives can inspire. It helps readers understand that the definition of a map is fluid. This fluidity lets us view our environment from countless perspectives to glean insights that would otherwise be lost.

Second, we make a clear distinction between the tangible cartographic map and the mental or cognitive map of the environment that we hold in our heads. Ultimately, it is the map in our minds, not the map in front of our eyes, that we use to make decisions. Throughout the text, we stress that cartographic maps are valuable aids for developing better mental maps. We should strive to become familiar enough with the environment that we can move through it freely, interact with it, and view it remotely, in both a physical and a mental sense. Ideally, our cartographic and mental maps should merge so that our spatial understanding, communication, and behavior have the greatest chance of being tuned to the reality of the environment.

Third, where appropriate, we reference commercial products and services of special interest to the map user. A few decades ago, discussing these commercial products would have seemed strange since most mapping was done by large government agencies, but times have changed. The strong recent trend toward commercialization of all things cartographic now makes these products indispensable to people who are intent on using maps efficiently and effectively. GIS, web apps, and digital data for mapping on home computers and mobile devices are developed and made accessible not only by government agencies at all levels but also by private companies, academics, and NGOs. What you do with maps in the future will be strongly influenced by the nature of these commercial products and services.

Finally, we show how map use is relevant to daily life. Whenever possible, we use examples and illustrations from popular sources and common practices. Maps touch so many aspects of our daily lives that it is simple and natural to make points and reinforce ideas with illustrations from topics of everyday interest. These illustrations are included in the text to demonstrate and reinforce basic map use principles and illustrate the universality of our relationship with maps.

Many other books treat the using and making of maps primarily in an engineering and technical context. In contrast, *Map Use* takes a unique approach in the depth of knowledge it offers in a clear and readable style. We believe that learning to understand the intentions or goals of a map leads to making better choices in your own map use and mapmaking.

Learning to use a map is a relatively easy and painless process, with an immense payoff—maps offer a look at the virtually invisible. They let us see the environment from vantage points that are distant from us in time and space. Maps also allow us to visualize aspects of our environment that are intangible, imperceptible, or purely conceptual. Most important, maps focus our attention on selected features by keeping the display free of distracting detail. Maps free us from our natural limitations, transcend our senses, and let us see the world anew.

New to this edition

The ninth edition of *Map Use* includes many updates to help previous readers stay up-to-date on advancements in our field and to give new readers a clear understanding of the current state of the art and science of cartography. Many of the more than 300 full-color illustrations in this edition are new or updated. An important change from previous editions is that we have focused our efforts and this book on the most critical concepts that map users and introductory mapmakers need to know to get started in this exciting field. A substantial glossary of key terms is available online at links.esri.com/MapUseGlossary. These glossary terms are indicated throughout the book in **bold font**. The index has been fully revised for this edition. *Map Use* is available as an ebook as well as in its traditional print format.

When *Map Use* was first published in 1978, very little mapping was done in a computer environment. Today, not only is most mapping done with the aid of computers, but the map user is often the one who guides the mapping process. Especially with the aid of GIS software and applications, the map user is increasingly the mapmaker. Even more significant, the map user can establish insightful dialogues with maps by manipulating the digital data in various ways. To a large degree, GIS is responsible for this exciting interchange between people, maps, and environmental data, and has changed the essential nature of map use to our immense benefit. At the same time, maps contribute greatly to GIS by providing a familiar visual interface through which GIS technology's powerful computing resources can be fully realized. Now that GIS technology is so widely available, a broader population is using it in a variety of ways. Clearly, the ability to think and communicate visually through the medium of maps is more important than ever, and this book can enhance this ability in a broad range of users.

Yet for all the technological advances in map use and mapmaking, the philosophy underlying *Map Use* remains the same. As in earlier editions, we stress that a good map user must understand, at a basic level, what goes into the making of a map. From mapmakers, we ask for little less than a miracle. We want the overwhelming detail, complexity, and size of our surroundings reduced to a simple representation that is convenient to access. We also want abstract maps to provide us with a meaningful basis for relating to the real environment. In return, we must make a corresponding effort to become educated map users.

How this book is organized

Map Use is specifically designed for use in a three-credit semester course at the college freshman level. Presentation of the material is geared to upper-division high school and intermediate college level. The book is aimed at both specialized and general map users. You can use it with equal effectiveness as a basic reference work or as the textbook for a beginning map use or cartography course. We intentionally avoid the confusing terms and details that characterize so many cartographic texts. Here, terms are defined precisely, and the relationship of each one to the other is clear—these definitions are also presented in the book's expanded online glossary. It includes terms in the book that are newly introduced or of special importance to map use and mapmaking. An appendix with relevant mathematical tables complements material presented in the text.

Map Use helps readers develop an appreciation of how mapmakers represent the environment in the reduced, abstract form of a map. In map reading, you must mentally "undo" the mapping process. We discuss the geographic data that underpins a map, the process required to transform that information through mapping techniques, and the basic principles of map design and layout. The topics covered in these chapters also provide the foundation for effective mapmaking, and we stress their close interconnection throughout the text. The emphasis in the final chapter

("Maps and Reality") is on environmental comprehension and understanding, for it is our surroundings, not the map, that are the real subject of map use.

Once readers grasp the degree to which cartographic procedures can influence the appearance and form of a map, they are in a position to use maps to analyze spatial patterns and interpret relationships in the mapped environment. Although a systematic development of subject matter is followed throughout the book, each part and chapter can be used independently and are cross-referenced to the rest of the material. This flexibility of design makes *Map Use* a versatile text, useful for classroom instruction and as a reference for practicing users. The book's organization provides a logical development of concepts and a progressive building of understanding and skills, from beginning to end. More experienced map users may focus initially on sections or chapters of special interest and then refer to other parts of the book to refresh their memories or clarify terms, concepts, and methods.

This book will serve its purpose if you finish it with a greater appreciation of maps than when you started. In even the simplest map, there is much to respect. Mapmakers have managed to shape the jumble of reality into a compact, usable form. They have done a commendable job. Now, map user, the rest is up to you.

Authors' note

Readers of earlier editions of *Map Use* will notice major authorship changes in this edition. The biggest change is that the torch of lead author has passed to Aileen R. Buckley, coauthor of the sixth, seventh, and eighth editions. Aileen studied with Jon Kimerling at Oregon State University as a graduate student. After receiving her PhD, she went on to several years of teaching and research in cartography at the University of Oregon, and then to a challenging career as a research cartographer with Esri. There she has made outstanding contributions to GIS and cartography. She is recognized internationally as a top research scholar in private industry.

Jon was lead author of the fifth through eighth editions of *Map Use*, inheriting the role from Phil Muehrcke in 2005. Jon is professor emeritus of geosciences at Oregon State University, where he taught maps, GIS, and geography and conducted research in map creation, design, and understanding. He coauthored *Elements of Cartography*, sixth edition, and four editions of *Atlas of the Pacific Northwest*.

Patrick J. Kennelly has also joined the author team for this edition. Patrick also studied with Jon Kimerling at Oregon State University as a graduate student. After receiving his PhD, he went on to several years of research and teaching graduate and undergraduate courses in GIS at Long Island University. He is recognized as one of the world's foremost scholars with respect to terrain representation. He currently serves as director of the GIS program at Central Oregon Community College.

On a final note, Phillip and Juliana Muehrcke, who conceived of *Map Use* in the 1970s and privately published its first five editions, happily retired from authorship of the eighth edition. Their creativity and wonderful writing showing their love of map use still fill this text, and we will always remember them fondly when reading these chapters.

Aileen R. Buckley
A. Jon Kimerling
Patrick J. Kennelly

Acknowledgments

In previous editions of *Map Use*, we acknowledged the importance of contributions made by special teachers and colleagues, teaching assistants, and students. We continue to be grateful for the inspiration and assistance provided by these people. To our delight, some of these individuals have taken the time to comment on earlier editions of this book. All these responses are gratefully received, and many were useful in crafting this improved edition.

To these students and colleagues, we add the valuable contributions made by the many people, agencies, companies, and communities that lent us the use of their work to illustrate this edition of *Map Use*. Their maps and graphics bring the text to life with real-world examples of the art and craft of cartography. Our journey to discover the hundreds of maps, diagrams, and images in this book led us to reconnect with old friends and make new ones. A special joy was the opportunity to learn about maps and mapmakers unknown to us. We are honored to showcase their work in our book.

We also acknowledge the contributions of the people who lent us their data, advice, and time. Although only three authors are listed on the cover, many others had a hand in bringing this book up to date for the ninth edition. We are grateful to the folks at Esri who contributed to this book. Inside, you will find maps by Aileen's colleagues on the ArcGIS® Living Atlas of the World team, including John Nelson, Emily Meriam, Andrew Skinner, Craig McCabe, Jinnan Zhang, and Keith VanGraafeilan. Works by the Esri ArcGIS StoryMaps team, ArcGIS Defense Mapping Extension team, ArcGIS Parcel Fabric team, Esri R&D Center Zurich, and Esri UK are also included. In addition, Alex Posen, Andrew Green, Bojan Šavrič, Carsten Bjornsson, Chris Wesson, Christian Harder, Christopher Patterson, Clint Brown, Cody Benkelman, Damian Spangrud, David Asbury, Heather Smith, Helen Thompson, John Grammer, Ken Field, Kristian Ekenes, Melita Kennedy, Peter Becker, Robert Waterman, and Sean Breyer graciously provided us with maps, data, or expertise. A special thanks goes to Dawn Wright, Esri Chief Scientist, for her bike map and her unflagging support of our efforts on this edition.

We gratefully acknowledge the hard work and dedication of the Esri Press and Esri Legal staff who helped make this new edition possible. Stacy Krieg was our acquisitions manager, David Oberman edited our text, Carolyn Schatz also helped with editing, Catherine Ortiz provided top-down guidance, and Jon Carter kept us on track. Victoria Roberts deserves special thanks for not only laying out the book but also updating graphics and creating new ones. Reflected throughout this edition of *Map Use* are many maps from the Esri Press publication *The Power of Where: A Geographic Approach to the World's Greatest Challenges*, by Jack Dangermond, Esri, and the GIS Community. Acknowledgments must also go to Pete Schreiber, Roanan Harris, Tristan Cassel, and others who helped ensure that we stayed between the lines.

This work reflects our deep love of maps and a desire to help others bring maps into their lives. We alone, of course, bear full responsibility for errors in the text or illustrations and for any controversial statements.

Acknowledgments would be incomplete without expressing gratitude to our families and many friends who helped us in so many ways. Particular thanks are given to David Sandoval, Ann Kimerling, and Sharon Maier-Kennelly for their unending encouragement and patience during the creation of this edition.

Aileen R. Buckley
Redlands, California

A. Jon Kimerling
Corvallis, Oregon

Patrick J. Kennelly
Bend, Oregon

Introduction

It should be easier to use a map than to read this book. After all, we know that a picture is worth a thousand words. Everyone, from poets to politicians, works from the assumption that maps are easy to understand and follow. The very term *map* is ingrained in our thinking. We use it to suggest clarification, as in "Map out your plan" or "Do I have to draw you a map?" How ironic, then, to write a book using language that is, supposedly, more complicated than the thing we are trying to explain!

The problem is that maps are not nearly as simple and straightforward as they seem. Using a map to represent our detailed, complex, and interrelated surroundings can be deceptive. This is not to say that the map itself is lying. But it is the environment, not the map, you are trying to understand. A map lets you view the environment as if it were less complicated. There are advantages to such a simplified picture, but there is also the danger that you will end up with an unrealistic understanding.

In this book, we define a **map** as a spatial representation of the environment that is presented graphically. By **representation**, we mean something that stands for the environment, portrays it, and is both a likeness and a simplified model of it. This definition encompasses such diverse maps as those on the wall, those on phones or computers, those that appear ephemerally and then are gone, and those held solely in the mind's eye. You may envision the environment by using **cartographic maps**, which are what most people think of as traditional maps displayed graphically on a physical medium such as paper or a computer screen. Or you may use **mental maps**, or **cognitive maps**, that exist only in your mind. These kinds of maps are often slighted although they are ultimately the maps that you use to make decisions about the environment.

Mental maps

As a child, your mental map was probably based on direct experience—for example, connected pathways, such as the routes from your home to your school or the park. You had a self-centered view of the world in which you related everything to your own position. The cartoon in figure I.1 graphically portrays this type of mental map. As an adult, you can appreciate this cartoon because you see how inefficient the child's mental map is. But the truth is, you will often resort to this way of visualizing the environment when thrown into unfamiliar surroundings. If you go for a walk in a strange city, you may remember how to get back to your hotel by visualizing a pathway like that in the cartoon. Landmarks will be strung like beads along your mental path as you head back. Even if you might be able to guess at a more direct route back, you may feel more comfortable following the string of landmarks you remember from your original journey to make sure that you do not get lost.

Most of your mental maps are more detailed than this path, however. You take advantage of indirect as well as direct experience to build the map. You acquire information through television, photographs, books and magazines, the internet, and other indirect sources. You can transcend your physical surroundings and visualize distant environments, even those you have never visited. Your mental maps become increasingly complex as they expand to encompass places and times that you have never seen and may never experience personally.

As you grow older, your self-centered view of the world is replaced by a **geocentric view**. Rather than relating everything to your own location, you learn to mentally orient yourself with respect to the external environment. Once you learn to separate yourself from your environment, you do not have to structure your mental map in terms of connected pathways and landmarks. You can visualize how

Figure I.1. The geometry of a child's mental map is based on direct experience and connected pathways. With permission of Jeff Keane.

to get from one place to another as the crow flies—the way you would if you were not restricted to roads and other connected routes. It is your adult ability to visualize the big picture that makes the cartoon path amusing.

Sharing your mental map with others, either in conversation or on maps that you draw, is much easier when you use a **geometric reference framework**, or a framework in which you can easily describe and determine locations, distances, directions, and other geographic relationships. The system of **cardinal directions** (north, south, east, and west) is such a framework. You can pinpoint the location of something by stating its cardinal direction and distance from a starting point. You can say, for example, that the post office is two miles north of your workplace or the police station is 200 meters west of the courthouse.

This visualization of space is based on **Euclidean geometry**—the geometry you first learned in school. It is the geometry that says parallel lines never cross, the shortest distance is a straight line, space is three-dimensional, and so on. The ability to visualize the environment in terms of Euclidean geometry is an essential part of developing a geocentric mental map. But even if you develop mental maps based on Euclidean geometry, they will only be correct over small areas. This is because the earth is spherical, and the spherical geometry of the earth's surface is inherently non-Euclidean. Very few of us have well-developed spherical mental maps simply because the small portion of the environment that we typically experience ourselves appears flat from our ground perspective.

Even if you can visualize the world geocentrically in terms of Euclidean or spherical geometry, it is hard for most people to transform their mental map to a cartographic map in a geometrically accurate manner. Try drawing from memory a map of the area in which you live. The hand-drawn map will tell you a great deal about the geometric accuracy of your mental map. Not only will you probably draw the places you know best with the greatest detail and spatial accuracy, but you will also probably draw the things that are important to your life and leave out those you do not care about. Figure I.2 shows an illustration that depicts a distorted mental image that a person from Santa Cruz, California, might have of the world to the east. Tongue in cheek as this map may be, it captures the fact that people visualize their own surroundings as far more important than the rest of the world. In the same way, your mental map emphasizes your own familiar part of the world over more distant places.

It is important to recognize these biases and how they affect the accuracy of your mental map. The quality of your mental map is crucial because your behavior in the environment largely depends on it. You relate to your surroundings as you visualize them mentally, not necessarily as they really are. If the discrepancy between your mental map and the real world is great, you may act in self-defeating or even disastrous ways. But you do not have to rely solely on your mental map because cartographic maps are available to you.

Figure I.2. A view of the world to the east seen through the eyes of a resident of Santa Cruz. © 2012 *The Cruz* poster, an interpretive map of Santa Cruz, California, by Kirby Scudder.

Cartographic maps

A cartographic map is a graphic representation of the environment. By graphic, we mean that a cartographic map is something you can see or touch. Cartographic databases or digital images are not in themselves maps, but they are essential to the methods we use to make maps today.

Cartographic maps come in many forms. Globes, physical landscape models, and tactile maps for the visually impaired are truly three-dimensional objects, but most maps are two-dimensional representations that have been cartographically enhanced. Cartographic maps have been carved, painted, or drawn on a variety of media for thousands of years, and print maps have been produced for the last five centuries. Today, maps displayed on digital devices are probably the most seen and used maps.

Basic characteristics of maps

What gives a graphic representation of the environment its "mapness"? First, maps are vertical or oblique views of the environment, not profile views such as a drawing or photograph of a building taken from a street view. Second, maps are a scaled-down representation of the environment, as you will see in chapter 2. Third, except for 3D globes and landscape models that faithfully represent the earth's curvature, maps are made using a map projection, which is a

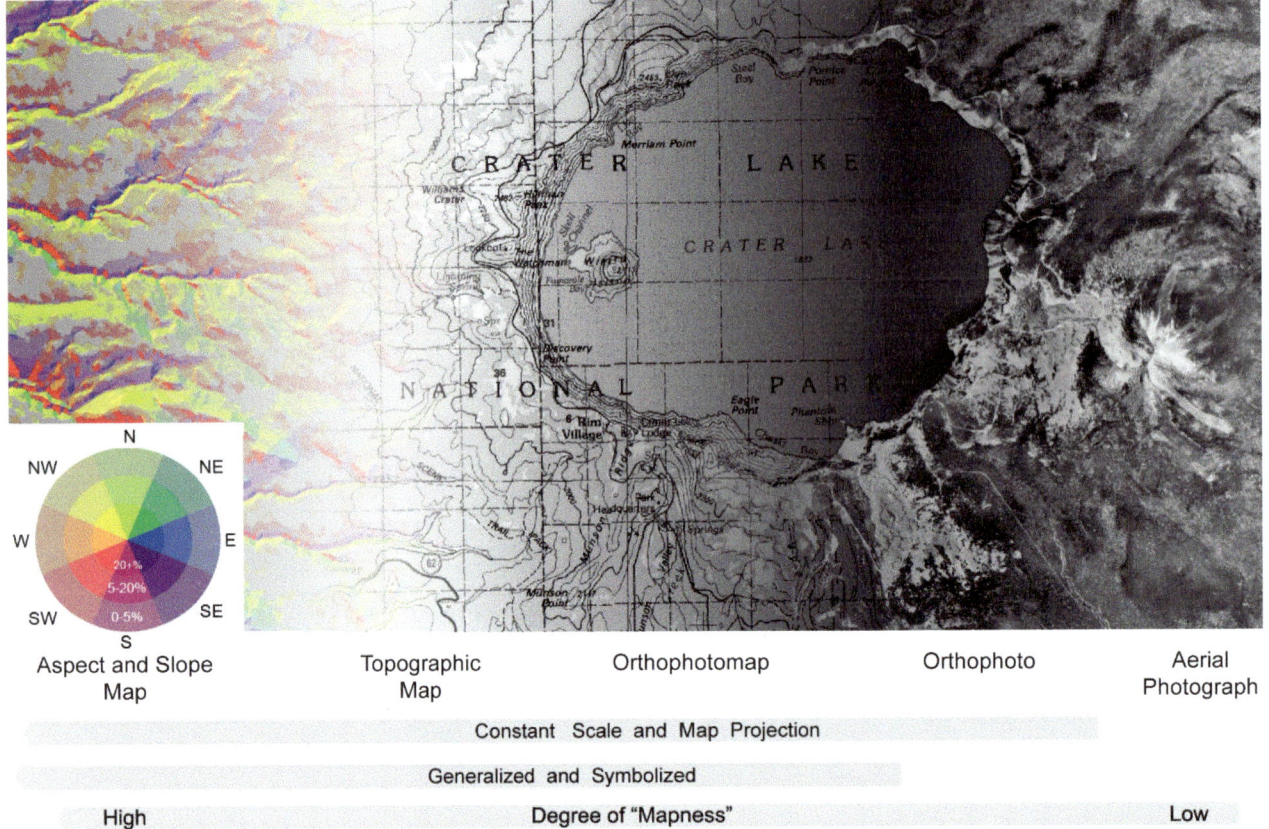

Figure I.3. Different types of maps lie at different positions along a mapness continuum: (*from left*) aspect and slope map, topographic map, orthophotomap, orthophoto, and aerial photograph. Their position on the continuum is defined by how many characteristics of maps they possess. The degree of mapness decreases in the examples as you move to the right. Data courtesy of the US Geological Survey.

mathematically defined transformation of locations on the spherical earth to a map surface, as we explain in chapter 3. Fourth, maps are generalized and symbolized representations of the environment, as we discuss in chapter 6.

A cartographic map does not need all four of these characteristics, but it should have at least one. Different types of maps fall at different points along a mapness continuum, defined by the degree to which they exhibit these four characteristics. This continuum is illustrated in figure I.3 for a collage of map types.

All the Crater Lake maps are vertical views and so possess at least one of the mapness characteristics. The **topographic map** (which you will learn about later in this introduction and in chapter 2) and the aspect and slope map strongly reflect all four characteristics and are good examples of what most people think of as a map. The **orthophotomap** in the center of the figure also has all four characteristics because generalized topographic map symbols were printed over a geometrically corrected aerial photograph called an **orthophoto**. Finally, the **aerial photo** from which the orthophoto is made is not on a map projection surface and varies in scale. Although the aerial photo has only the characteristic of being a vertical view of the environment, it is still a form of cartographic map.

What makes maps popular?

Maps are ubiquitous in today's world. The obvious question is, what accounts for their widespread popularity? First, maps are convenient to use. Paper maps are usually small and flat for ease of storage and handling. Maps on digital devices are easily accessible and often shown in apps for uses such as navigation or tracking. Thus, maps bring reality into a less unwieldy proportion and a more accessible format. Second, maps simplify our surroundings. Without them, our world often seems like a chaos of unrelated phenomena. Maps make the world intelligible. Third, maps are credible. The relationship between the map and reality seems so straightforward that we are comfortable letting maps stand for the environment. Maps, even more than the printed word, impress people as authentic and authoritative. It is axiomatic that seeing is believing. What we see on a map informs our mental map, giving the cartographic map a sense of credibility. We tend to accept the information on maps without question. This blind acceptance is potentially disastrous when using maps without thinking about it. For example, using GPS to navigate without checking other information, such as road signs or even your surroundings, may land you in a dangerous position. Fourth, maps have a strong visual impact. Maps create a direct and lasting impression of the environment. Their graphic form appeals to our visual sense and allows us to take advantage of our most powerful information-processing system—our vision.

These characteristics combine to make maps appealing and useful. Yet these same four things, when viewed from a different perspective, can be seen as limitations. Take convenience. It is what makes fast food popular. When we buy processed foods, we trade quality for ease of preparation. Few would argue that the result tastes like the same thing made from fresh ingredients. The same is true of maps. We gain ease of handling and storage by creating a prepared image of the environment. This representation of reality is bound to make maps imperfect in many ways.

Simplicity, too, can be seen as a liability as well as an asset. Simplification of the environment through mapping appeals to our limited ability to process information, while at the same time it reduces the complexity and, potentially, the intricacies that we need to understand our environment. With maps, the overwhelming and confusing natural state of reality is reduced. But the environment remains unchanged—it is just your view of it that lacks detail and complexity.

You should also question the credibility of maps. Maps reflect the capabilities of their maker, and mapmakers' decisions may not be as reliable or rational as you assume. Maps are like statistics—people can use them to show whatever they want. Some map features are distortions, others are errors, and still others are omitted through oversight or design. So many perversions of reality are inherent in mapping that the result is best viewed as an intricate, controlled representation. A map is a snapshot of a place at a point in time, but the world keeps changing. For all these reasons, a map's credibility is open to debate.

Also, be careful not to confuse a map's visual impact with proof or explanation. Just because a map leaves a powerful visual impression does not make it meaningful or insightful. From the single view of the map, it is sometimes difficult or impossible to understand the processes that caused the patterns we see. For explanations, you must look beyond maps and examine the real world.

The map transformation process

Mapping is an art and a science. Maps are visual representations that convey complex themes and can express emotions and unique perspectives. Unlike an artist's representation of the environment, in which geometric liberties are taken to convey an idea or emotion, a map is expected to be true to the location and nature of our surroundings. Indeed, our willingness to let maps stand for the environment is because of this expected correspondence to reality. Figure I.4 illustrates the **map transformation process** used to transform data collected about the geographic environment into a map that, in the end, is used by readers. By maintaining the highest possible fidelity in this transformation process, the mapmaker assures accurate correspondence to reality. You, the map user, also play a part through reading, analysis, and interpretation. You are responsible for checking the map against reality, to assure that mapmakers have met their responsibility to maintain the map's accuracy. Although the map transformation process used to be a time-consuming effort left to professional cartographers, geographic information systems (GIS) and other geospatial technologies (discussed throughout this book) have simplified the process and allowed anyone interested in maps to make them.

Types of maps

The sheer number and variety of maps have increased greatly over the past few decades, until map use is now a part of your daily life. You may not have given much thought to the types of maps you use, but cartographers often think of maps as falling into one of three categories: reference maps, thematic maps, and charts.

You can first think of maps that serve as a reference library of geographic information. These types of **reference maps** are an efficient form of geographic reference material that catalogs feature types and records their geographic locations. Reference maps allow you to instantly see the relative size and position of features. Explaining these

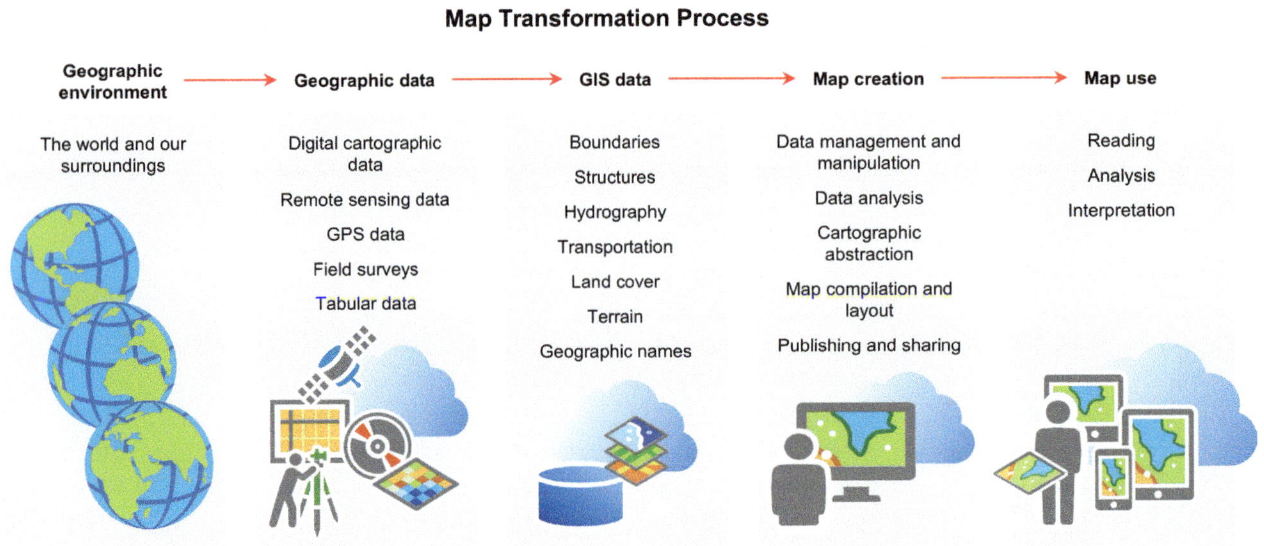

Map Transformation Process

Figure I.4. The map transformation process (after Tobler 1979) begins with the collection of data about our geographic environment. The data is compiled into a GIS database and is then used to create the map. The map user completes the process when reading, analyzing, or interpreting the information on the map.

spatial relationships among features in writing would be a laborious task.

Maps can also focus on a specific topic, much like an essay. Like a well-written theme, a map can introduce a subject in a way that not only captures the reader's attention but also informs in a way that the reader understands, has good structure and organization, and has clarity and focus. These types of **thematic maps** help you focus on geographic patterns and relationships that you might have a hard time visualizing from a written description.

Charts are a more specialized type of map designed for plotting and planning routes, determining distances and directions, and making accurate measurements and estimations. Charts help navigators find their way over land or sea. We will look at each of these in detail.

Reference maps

The earliest known maps, from several thousand years ago, were reference maps. Symbols on reference maps are used to locate and identify prominent landmarks and other pertinent features. Mapmakers try to be as detailed and spatially truthful as possible so that the information on the map can be used with confidence. These maps have a basic "Here is found…" characteristic and are useful for finding the location of geographic features that readers are interested in. On reference maps, no feature is emphasized over the others. As much as possible, all features are given equal visual prominence.

Globes, like the one in figure I.5A, are reference maps that show geographic features in a more generalized form than topographic maps because of their scale (discussed in chapter 2). Maps produced for counties, states, nations, and other geographic areas are used for reference, such as the wall map of Japan in figure I.5B. The **basemap** layer with geographic information that serves as the background for web maps is also an example of a reference map (figure I.5C).

The geographic information on reference maps makes them useful for professional and recreational map users alike. Because of their versatility, they are also called **general purpose maps**. These maps are used in engineering, energy exploration, natural resource conservation, environmental management, public works design, commercial and residential planning, and more. Maps for special purposes, such as biking, camping, and fishing, are often variations of reference maps with additional features, such as bike trails, campsites, and fishing spots, added to them. The **special purpose map** in figure 1.5D is an example of a bike map with route types classified by the level of comfort (shown with the different colored lines) and relevant points of interest, such as metro bike kiosks (shown with the symbol *B*).

Figure I.5. Reference map examples include a world globe of physical features (*A*), a 1:1,800,000 general reference map of Japan (*B*), an Esri basemap displayed on a smartphone (*C*), and an online bike map for the city of Austin, Texas (*D*). Courtesy of RäthGloben, Germany (*A*), Benchmark Maps, an imprint of East View Information Services (*B*), and City of Austin (*D*).

Thematic maps

Unlike reference maps, which show many types of features but are designed to emphasize none over the other, thematic maps focus on a single type of feature or phenomenon that is the theme of the map. Reference maps focus on the locations of different features, whereas thematic maps emphasize the geographic distribution of the features or phenomena. The map of air quality in Europe in figure I.6A is a good example. Although the points on the map represent the locations of monitoring stations, it is the pattern of air quality across the region that is of primary importance. We can see that air quality in Poland, for example, is worse than for Germany. Although this map does not tell us why that is the case, it does help us see the geographic pattern.

The map of US marine vessel traffic in figure I.6B shows us two years of shipping tracks for cargo ships off the US coast. The track of each individual ship is displayed, but it is the composite display of the tracks that is the theme. The geologic map in figure I.6C shows the composition and

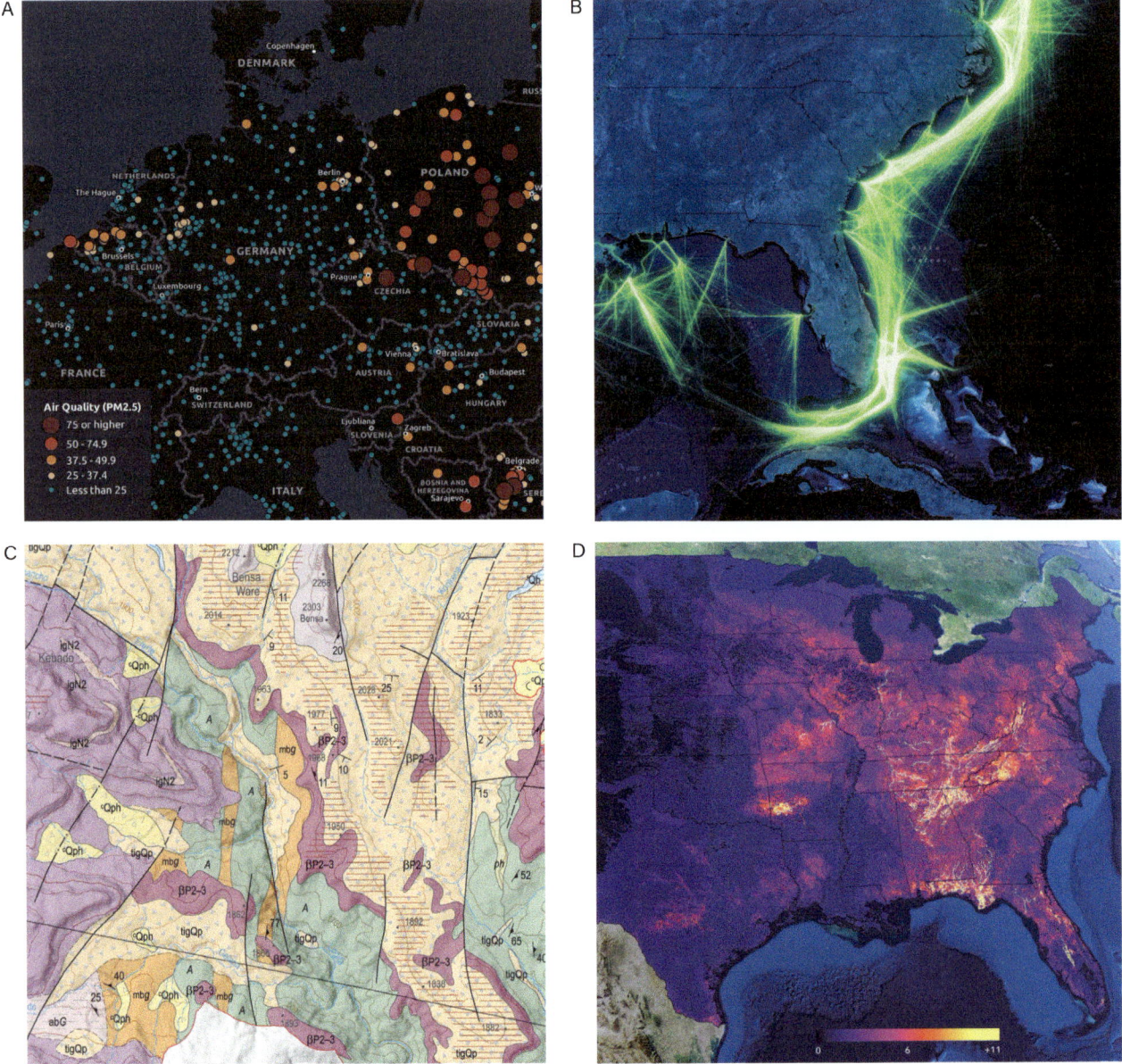

Figure I.6. Thematic map examples include maps of air quality in Europe (*A*), US cargo ship traffic (*B*), geology in the Sidama Region of Ethiopia (*C*), and biodiversity importance (*D*). Courtesy of OpenAQ and Esri (*A*), US Coast Guard, National Oceanic and Atmospheric Administration (NOAA), and US Bureau of Ocean Energy Management (*B*), courtesy of the Czech Geological Survey (*C*), and NatureServe (*D*).

structure of geologic materials at the earth's surface and at depth. This type of thematic map is more complicated than the ship traffic map, as you can see from the greater number of features, symbols, and labels on the map. Thematic maps like this are designed for a specialized audience. The theme of the map in figure I.6D is the number of plant and animal species that are endangered or at risk. Its objective is to highlight the areas of highest biodiversity importance to guide effective conservation decision-making. Scientific maps like these need to be clear and concise to be used by policymakers.

Whereas a geologic map shows the geographic distribution of features that can be found on (or under) the ground, maps like the biodiversity example show the geographic distribution of phenomena that do not physically exist. These maps often focus on statistical measures. Although

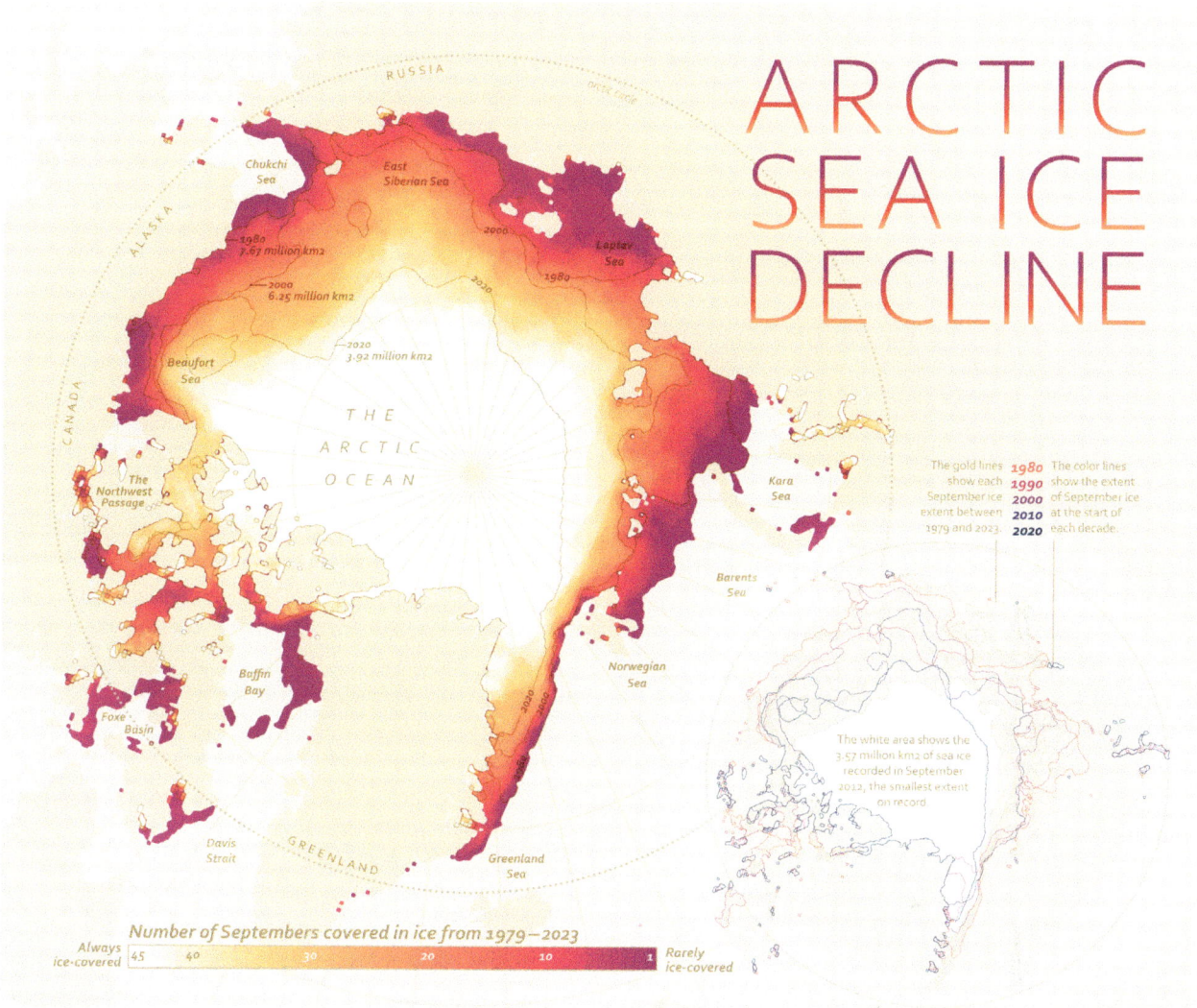

Figure I.7. This thematic map shows the diminishing extent of sea ice in the Arctic from 1979 to 2023.

you cannot see areas that have a value of 11 or more in the physical environment, you may see evidence of places that support more biodiversity, such as denser forests or protected areas with many plant species.

Thematic maps ask, "What if you want to look at the spatial distribution of some aspect of the world in this particular way?" The map in figure I.7, for example, asks, "What if you want to look at the Arctic to see whether the extent of sea ice (the area of the ocean that is frozen) has diminished since the 1980s?" A map like this gives you a way to look at a complex and challenging problem so that you will understand it better.

In figure I.7, note how little reference information is shown on the map. You will find only country boundaries, oceans and other major water bodies, and place names for the countries and Alaska. Most thematic maps have similar limited reference information, as they often show only enough to give geographic context. Be careful not to use contextual reference information to make precise measurements—that is not the intent of thematic maps. When using thematic maps, focus on their function of showing the geographic distribution of the theme.

Charts

Charts are a special type of map designed to help mariners and aviators navigate over water and land. **Nautical charts** aid the navigation of watercraft through large bodies of water, such as oceans and seas, as well as other navigable waters, such as large rivers and bays. **Aeronautical charts** for flight navigation have been created for all land masses, and charts of ocean areas are used to plan and monitor transoceanic travel.

Nautical charts, created specifically for **water** or **marine navigation**, facilitate safe maritime navigation. Marine navigators use these maps to determine the position of their ship or another vessel and to plan and follow a route over or through water. Recreational and commercial vessel navigators alike use detailed information about unseen hazards beneath the water's surface, aids to navigation, shorelines, landmarks, and other information critical to safe navigation to plan and follow routes between ports or anchorages. Figure I.8 maps these as well as berth numbers, fish havens, and other potentially useful features. Each nautical chart is made using a special map projection that allows navigators to quickly and easily measure the distance and direction of each route.

Aeronautical charts are maps designed for the air navigator. **Air navigation** involves using maps to determine the position of a plane or other aircraft and to plan and follow a route by air. Pilots use the information on the charts to find distances, directions, and travel times between destinations. Look at figure I.9, which is part of an aeronautical chart of Kauai in the Hawaiian Islands. The chart is filled with information important for safe flying, such as visual checkpoints used during flight (such as cities, roads, and rivers), controlled airspace (shown with purple bands), restricted areas (shown with blue hatched bands), visual aids to navigation (such as airport beacons), and radio aids to navigation (shown in the box to the right of the chart segment). Air navigation also involves maintaining a safe altitude above the ground, so you can also see information about towers, antennas, smokestacks, and other obstructions to navigation, as well as special ground elevation symbols that help air navigators quickly determine the minimum safe in-flight altitude.

Historically, nautical and aeronautical charts were available only on paper, but like all information products, their methods of creation and use have grown and improved in the digital age. Charts today are also integrated with digital navigation systems so aviators and sailors can easily access the information they need about routes and destinations, as well as stay informed of changing conditions, such as storms, strong winds or currents, or updates about restrictions.

Cartometric maps

A chart is one example of a **cartometric map**. **Cartometrics**, or **cartometric analysis**, refers to mathematical operations, such as counting, measuring, and estimating. Another example of a cartometric map is a topographic map, which we look at next. Like charts, topographic maps can be used cartometrically to determine positions, distances, directions, and more.

Topographic maps

Topographic maps show the **topography** of an area by depicting the shape of and features on the earth's surface. Topographic maps, also called **topos**, show and label **natural features** that are found in the physical environment, such as mountains, valleys, plains, lakes, rivers, and vegetation (figure I.10). They also show **human-made features**, such as roads, buildings, and boundaries. You can identify a topographic map by its depiction of the land surface form through symbols such as **contour lines**, which connect points of equal elevation (we describe contours in chapter 10). These are shown in brown on the map in figure I.10.

Topographic maps are sometimes created by independent cartographers or commercial map companies for recreational uses, such as camping, hiking, or biking. The map in figure I.11 is an example.

A topographic map is not only a valuable reference map but also an important **land navigation** tool that you can use to find your way across the land. A topographic map shows you ground features, such as roads, trails, lakes, and streams, that are both landmarks and obstacles. They allow you to determine elevation changes and estimate the slopes you will encounter along a route. This information helps you estimate the time and effort it will take to complete your trip. And topographic maps are drawn on map projections that allow you to measure distances and directions between locations along your route. Because of all these features, topographic maps are used by hikers, off-road vehicle enthusiasts, and other recreationists, as well as foresters, park rangers, and other land resource managers.

Other types of maps

We have said that geospatial technology has simplified and streamlined the map transformation process. This technology is becoming easily accessible and cheap, or even free, and the skill levels to use it are decreasing. The data to

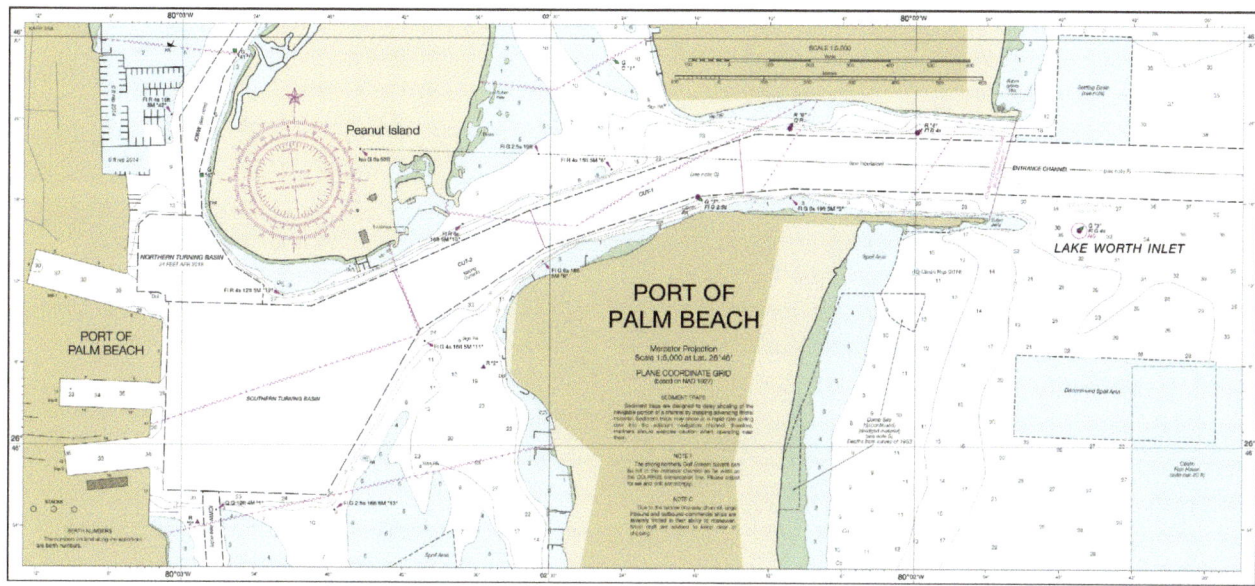

Figure I.8. A 1:5,000-scale nautical chart of the Port of Palm Beach. Courtesy of the US Office of Coast Survey.

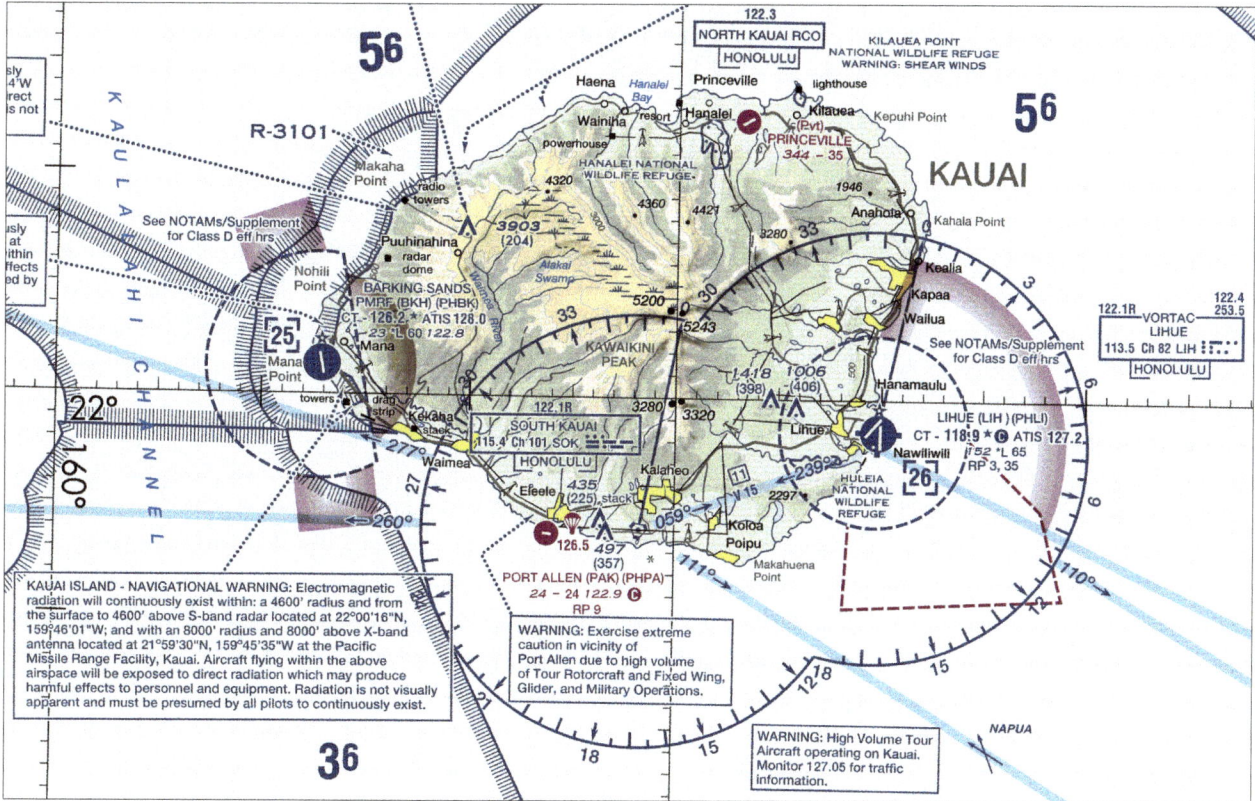

Figure I.9. A segment of an IFR Enroute Low-Altitude Chart, which provides aeronautical information for navigation under instrument flight rules (IFR) below 18,000 feet. Courtesy of the Federal Aviation Administration.

12 Map Use: Map Reading and Design

Figure I.10. A segment of a 1:100,000-scale national topographic map of Switzerland. © swisstopo.

Figure I.11. A segment of a 1:50,000-scale hiking map of an area near Walenstadt, Switzerland. © swisstopo.

make the maps is easier to find and often free, and the steps to share the maps are straightforward. Because the maps are shared online, the audiences for the maps are growing and becoming more diverse. As a result of these technical advances, we are seeing a greater variety of maps now than ever before. Although most maps can still be classified using the three-category system, many are hybrids, composed of or derived from primarily thematic and reference maps. Rather than taking an exhaustive look at these map forms, we focus next on other map types that deserve special mention.

3D maps

3D maps add height, or a **z-value**, to the x,y horizonal dimensions of 2D maps, allowing users to visualize, analyze, and manage geographic information in three dimensions. 3D maps can depict our world more realistically because they are more like what we see when we look around us.

Increasingly, data is collected and managed so that it can be visualized in 3D. High-resolution elevation data can be used to create a digital representation of the 3D geometry of an area, such as Prague, Czechia, in figure I.12A. When integrated with photo-realistic photographic

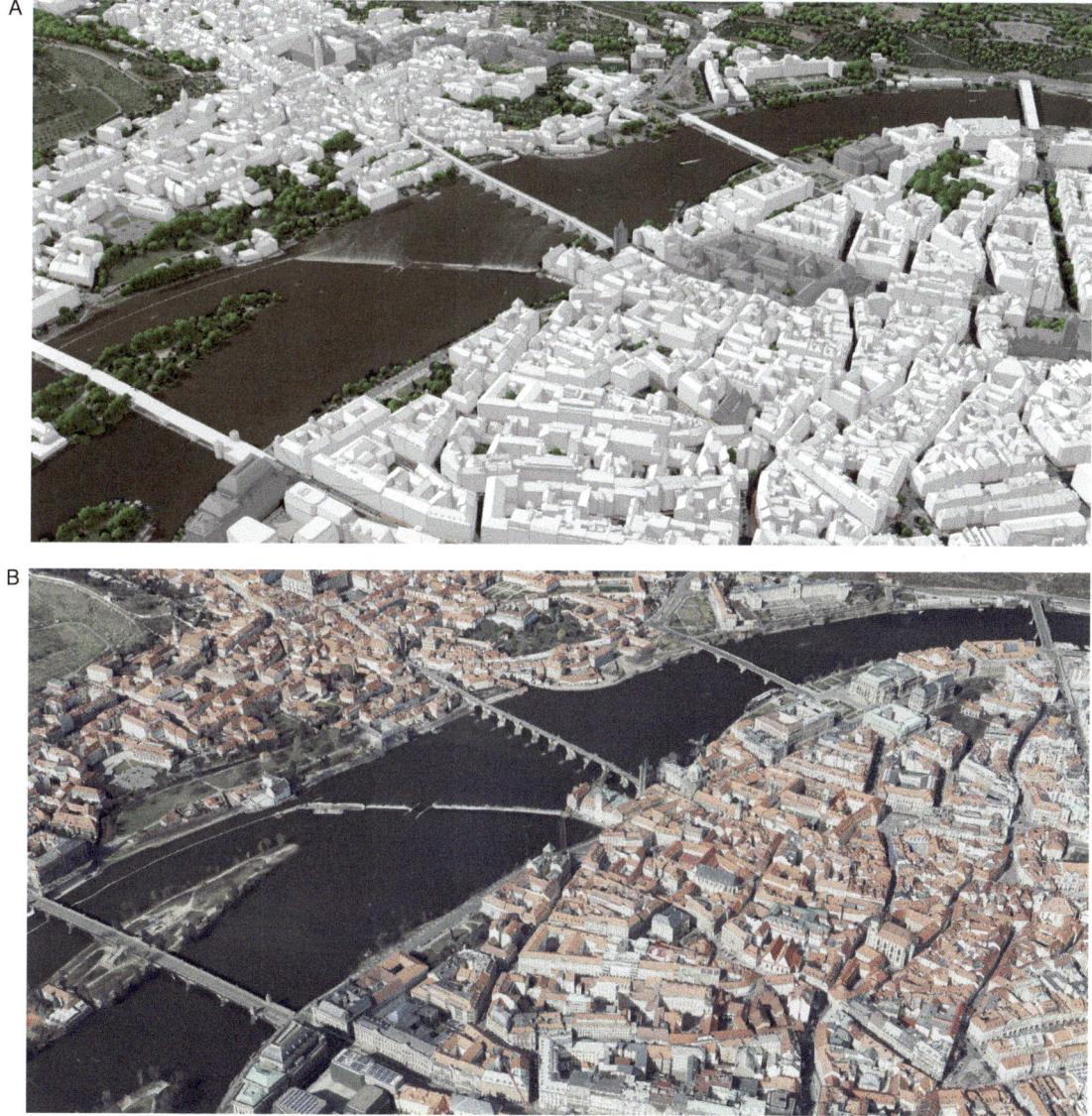

Figure I.12. A high-resolution 3D model of Prague, Czechia (*A*), can be combined with aerial photography to create a photo-realistic digital twin (*B*).

Figure I.13. Dynamic maps can provide views from global (*A*) to regional (*B*) to local (*C*) to site scales (*D*).

images (figure I.12B), the result is a visually stunning and highly accurate 3D map that helps us better understand unique aspects of the environment that are not obvious from a top-down perspective. Interacting with 3D maps in realistic and immersive experiences helps users gain clarity, evaluate decisions, and improve awareness.

Dynamic and interactive maps

The maps you will see in this book are **static maps** printed on paper or displayed on digital devices. The features on static maps lack movement, action, or change—they cannot shuffle across the paper, appear or disappear on the page, or even shimmy in place. **Dynamic maps**, on the other hand, show change, activity, or progress. The earliest dynamic maps were created as animations, like motion pictures shown frame by frame with small changes in positions of features between frames. Today, dynamic maps allow users to pan and zoom to view the environment from different vantage points. **Animated maps** give a sense that things are changing over time, through space, or both. Early examples appeared in movies, mostly World War II newsclips. The development of broadcast television followed by computer displays heralded an explosion in the type and number of animated maps produced. Today, these kinds of maps are ubiquitous. For example, you can see animated weather maps on TV weather reports or websites. Satellite images of clouds that move across a static basemap of your local area are shown along with moving map symbols for weather fronts, rain or snow showers, and developed storms.

Many maps found online or in web apps today are also interactive. **Interactive maps** give you control over what is shown on the map and how it is shown. Aside from the common ability to click the web map to see information, charts, and images in a pop-up window, you may be able to change the basemap, turn layers on and off, and even add your own layers. A good example is the MapMaker web app developed by National Geographic and Esri (figure I.14), which can be used to explore the world in 2D or 3D, compare areas with a swipe tool, make measurements, and more.

An interactive map may also interact with or respond to you—for example, when your vehicle navigation system gives you driving instructions while you track your changing position on the displayed map.

Real-time maps

Some web maps use **real-time data** to provide a continuous stream of up-to-date information, allowing for more accurate and timely maps. These constantly updated maps reflect the current state of the world. Real-time maps have a range of applications—for example, to give accurate weather forecasts and provide live traffic updates. Real-time data can be used to track natural and other phenomena that pose threats to life and property, such as unhealthy air quality, storms, tornadoes, hurricanes and cyclones, drought, flooding, earthquakes, wildfires, and more (figure I.15). In emergency response and disaster management, real-time maps can play a crucial role by providing immediate and accurate information on incidents to enable faster response times and more effective resource allocation.

Story maps

Story maps are visual narratives of a place, event, issue, trend, or pattern in a geographic context. Often shared online, interactive and engaging story maps combine text, charts, and images with maps, videos, audio, and other multimedia elements. Story maps can deepen our understanding of a topic, be used for educational purposes, and showcase projects or initiatives. The story builder is accessible to a wide audience, allowing thousands of users to tell interactive and visually compelling narratives and briefings that combine maps, 3D scenes, and multimedia.

An example is "Living in the Age of Humans"—a collection of story maps about the impacts of human activity on the planet. The story maps (figure I.16) take you on a

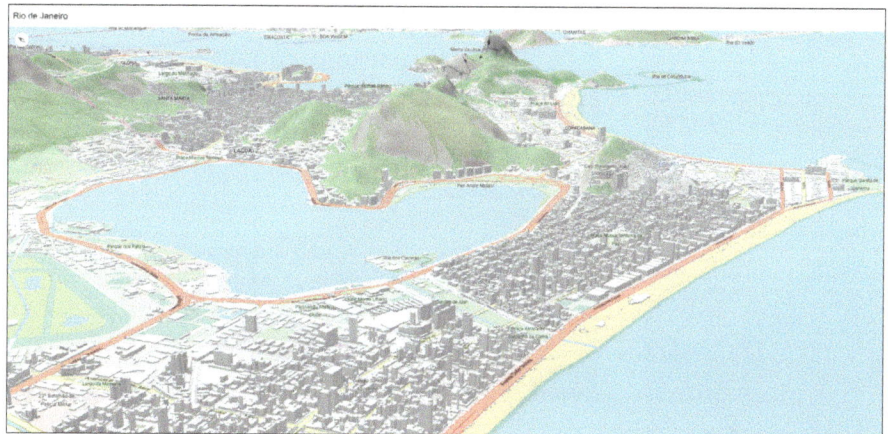

Figure I.14. The MapMaker web app was designed to make teaching and learning with interactive maps fun and easy. The app provides various navigational and zoom tools and the ability to search for a location, choose basemaps, turn layers on and off, view a legend, and take measurements, as well as switch between 2D and 3D views. Courtesy of National Geographic and Esri.

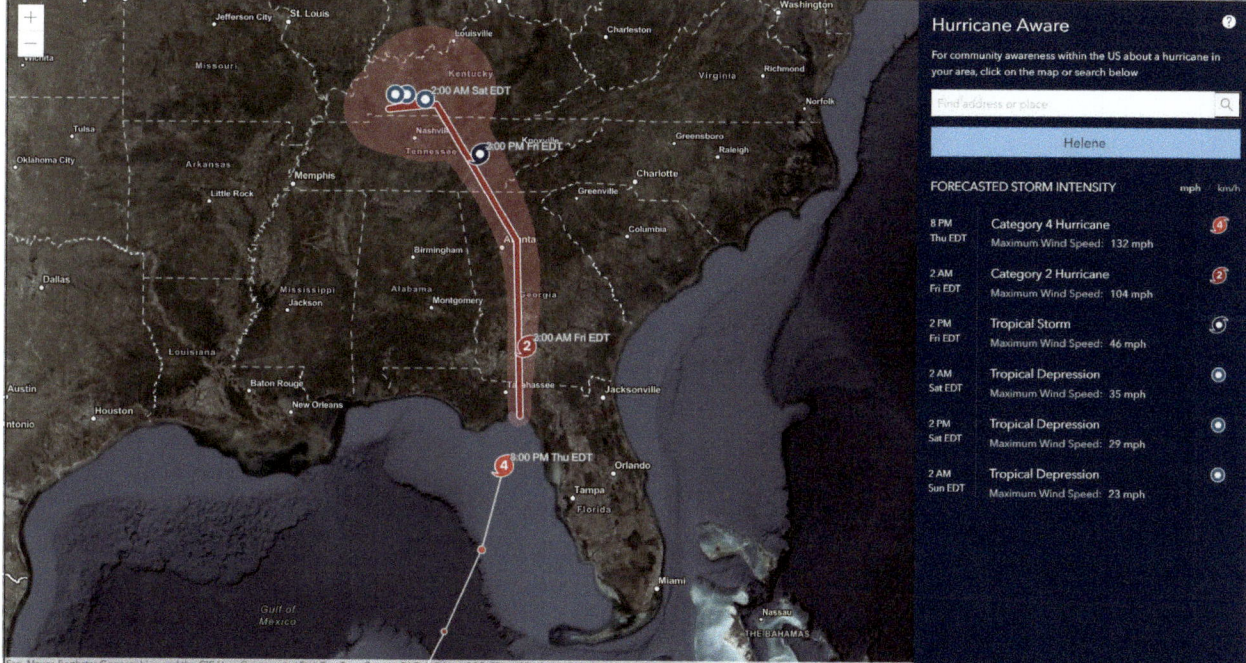

Figure I.15. The Hurricane Aware web app provides information about the potential impacts of tropical storms in the conterminous United States, with storm track information for the rest of the world. Real-time data from the National Weather Service and Joint Typhoon Warning Center forecasts hurricanes, precipitation, and wind gusts, and US census data is used to forecast demographic impacts. Courtesy of Esri, National Oceanic and Atmospheric Administration, and Joint Typhoon Warning Center.

16 Map Use: Map Reading and Design

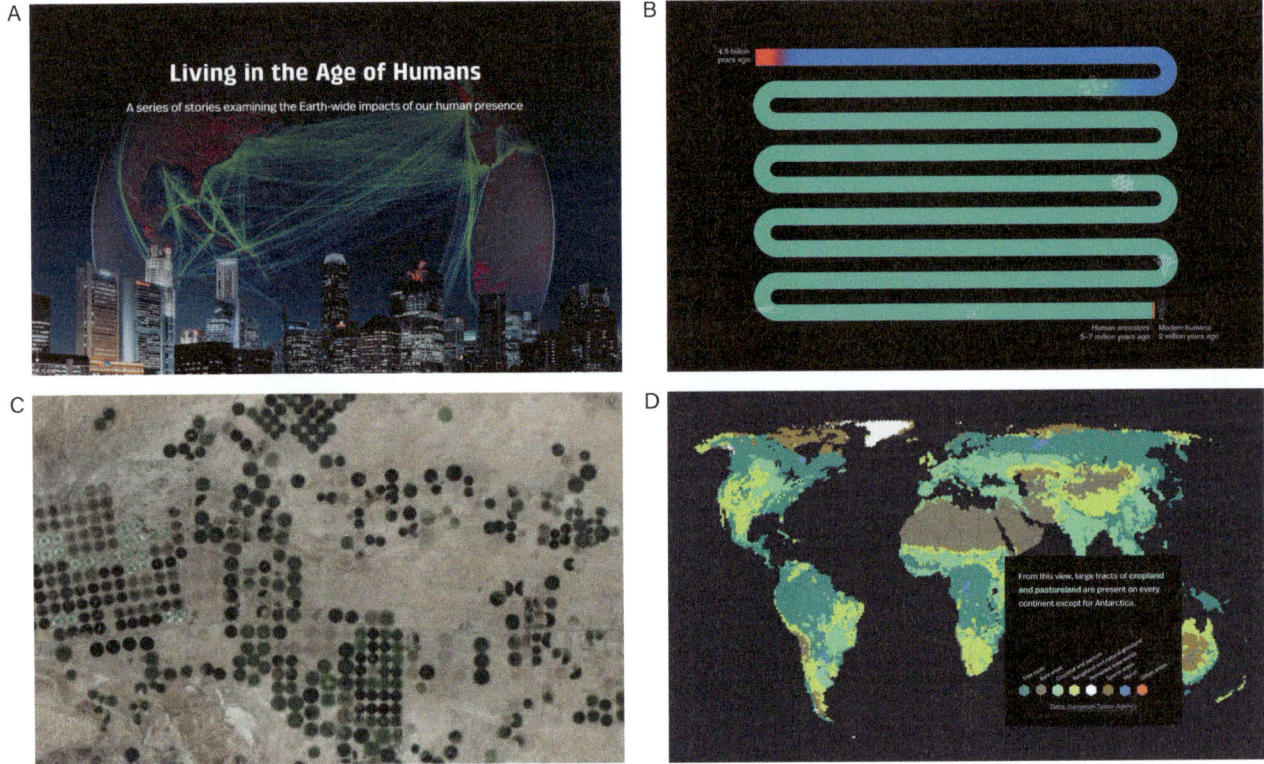

Figure I.16. The "Living in the Age of Humans" story map collection includes a cover page (*A*), graphic charts with explanatory text (*B*), satellite images (*C*), and maps with related descriptions and photos (*D*).

Figure I.17. Interactive globes in the "Living in the Age of Humans" story map allow readers to explore the entire earth, with content tabs that focus on density, urbanization, the earth at night, networks, and the human footprint. The image shows regions of higher population density (higher symbols) and lower population density (lower symbols).

guided tour through a written narrative of earth's history, enhanced with charts and pictures. Satellite images illustrate the impact of human activity on landscapes across the globe.

The story maps also include elements such as videos and interactive globes (figure I.17). Story maps can include swipe functions to compare two images or maps side by side, time lines to lead readers through a sequence of events, and embedded content from other online sources such as social media. Links to websites, online tutorials, and related articles can be used to provide further details. Because story maps are shared online, they support accessibility by following the World Wide Web Consortium's *Web Content Accessibility Guidelines*.

Accessible and inclusive maps

In today's digital age, when maps are accessed through websites, apps, and other digital platforms, there are requirements to meet standards for accessibility. **Accessibility** refers to the design and implementation of products, services, environments, and digital content that can be accessed and used by people with disabilities. It involves removing barriers and providing accommodations to ensure equal access and opportunities for people with disabilities. **Inclusive maps** consider diverse communities and perspectives by considering the needs and views of different groups of people, including those from different cultural backgrounds, ethnicities, and socioeconomic statuses, to ensure that all map users feel represented and included.

In the context of maps, accessibility refers to the ability to access and use maps effectively by designing those maps to accommodate the needs of people with various disabilities, such as visual impairments, hearing impairments, cognitive limitations, or mobility limitations. For people with hearing impairments, accessibility can be enhanced by providing visual cues and captions in map-related content, such as videos or interactive elements. For those with mobility limitations, accessibility can be improved by ensuring that maps are navigable using alternative input methods, such as keyboard controls or voice commands.

Accessibility for the visually impaired can be improved with the use of colorblind-safe colors for those with **color vision deficiency**—a form of visual deficiency that prevents someone from identifying certain colors (or, in some cases, all colors). Access to map information can also be improved through screen readers—machines that vocalize map contents—or **tactile maps**, in which the physical representation of the map can be explored through touch. An

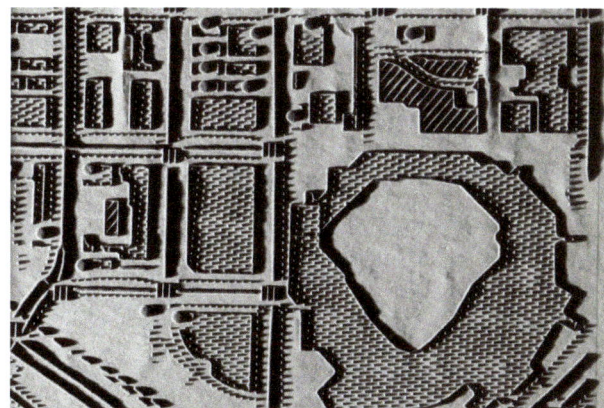

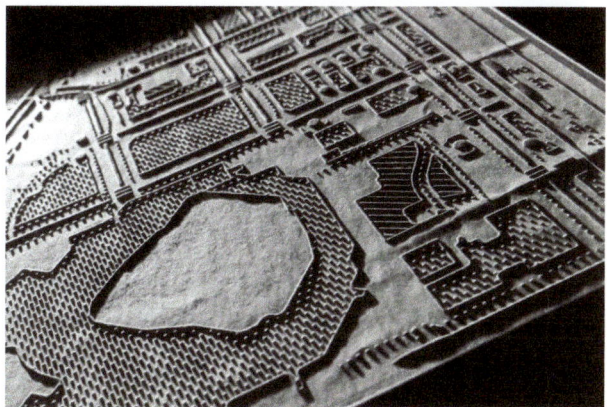

Figure I.18. Tactile maps help people who are totally or partially blind by providing geographic context for an area that they may or will have navigated with mobility aids, such as voice navigation on smart devices. Because the tactile maps are made to be touched rather than seen, the symbols are raised and text is in Braille. Map courtesy of the Dutch Tactile Mapping Project. Photos by Esri.

example of a tactile map for an area around Petco Park in San Diego, California, is shown in figure I.18. This map was compiled using GIS and printed on special paper. When the paper is heated properly, a chemical reaction occurs that causes the black ink on the paper to swell, thus embossing the features and labels.

Maps as art

With the growing ubiquity and diversity of maps comes a flourishing of map art. Think about how you would describe the look of the maps in figure I.19. Do they have a certain beauty that makes them a work of art as well as a spatial representation of the environment? Some artists are not interested in making maps to show us how to navigate from place to place or in helping us visualize the spatial distribution of quantitative data. These artists use maps as the medium for their art, but the maps are more a way to explore the artistic relationships between space, time, and our emotions. Map art can be found in galleries, museums, public spaces, and private homes. Map art has gained popularity as a form of artistic expression in recent years because it offers a creative and visually engaging way to appreciate and interpret maps beyond their traditional use as navigational tools.

Figure I.19. Map art examples include Matthew Cusick's "Empire Revisited" (2011) map collage (*A*), a topographically accurate 3D model of Cape Cod created by Treeline Terrains as wood-carved wall art (*B*), the intricate channels of the Willamette River in Oregon captured in a colorized rendering of lidar data by Daniel Coe (*C*), and a detailed illustrated map hand-drawn by Anton Thomas with colored pencil and pen (*D*). Courtesy of Matthew Cusick and the Pavel Zoubok Gallery (*A*), Treeline Terrains (*B*), Daniel Coe/Meander & Flow Design (*C*), and © 2023 Anton Thomas Art (*D*).

Map use

We wrap up this introduction by considering **map use**—the process of obtaining useful information from one or more maps to help you understand the environment and improve your mental map. Map use involves three main activities: reading, analysis, and interpretation. And mapmaking gives insight into the conditions and decisions that affect the appearance of the map, thus facilitating map use activities.

In **map reading**, you determine what the mapmaker has depicted and how it has been depicted. If you read the maps in figure I.20, for example, they give you an idea of where some of the food for a typical American Thanksgiving dinner comes from. Looking at the distribution of symbols in the map at the top, you can determine that the data relates to US counties whose boundaries are shown in light-brown lines within the states. From the spread of the symbols over the map, you can see that turkeys and green beans come from many regions of the United States, and sweet potatoes are primarily grown in the central and eastern states and California. Looking at the circle symbols, you can see the regions where most of the turkeys and selected vegetables come from because the symbols in these areas are larger and overlap more. When you look at the map legends, you learn that the turkey map at the top shows the number of turkeys sold divided by land area. This is a different measure from that used for the other three maps, which show acres harvested divided by land area. All the maps show a density in the land area, so they can be compared for relative similarities and differences. But reading the legends carefully, you learn that the values are not the same for each symbol size class, so a one-to-one comparison is not possible.

Map design is the process of compiling the map, all the while making decisions that align the look of the map with its intended audience and purpose. Familiarity with map design decisions will help you better understand the color choices, symbol sizes, class ranges and breaks, and other design considerations for the maps in figure I.20. Understanding these decisions will, therefore, lead you to better read the map.

Learning to read the information on maps is only the first step. Your curiosity or work may lead you to go further and analyze the information in one or more maps. With **map analysis**, you make measurements and look for evidence of spatial structure (patterns, associations, and correspondence). Both reference and thematic maps are tools used to understand spatial patterns. Analyzing the spatial patterns on the maps in figure I.20, for example, you may see clusters and outliers on the maps, but another map user may not see things the same way. Quantitative measures of spatial patterns on a map and associations among patterns on two or more maps add rigor and repeatability to your map analysis.

After analyzing the maps in figure I.20 and finding spatial patterns of wide and sparse distributions, high and low densities, and similarity on two or more maps, your curiosity may now be aroused still further. You may wonder why the geographic phenomena are found where they are, and you may want to learn more about their spatial relations. Finding such explanations takes you into the realm of **map interpretation**. To understand why things are related spatially, you must search beyond the maps. To do so, you may draw on your personal knowledge, fieldwork, written documents, interviews with experts, or other maps and images.

In your search, you will probably find that crop production is associated with many environmental factors, including climate and weather conditions (temperature, precipitation, and sunlight), soil quality (nutrients, soil structure, and pH level), topography (elevation, aspect, and slope), and water availability. You will find that some crops, such as green beans, are versatile and can be grown in different climates and soil types, and others, such as sweet potatoes, require a unique growing environment, such as four months of warm weather and warm soil. Human factors can also affect crop production, so you may look at factors relating to irrigation, fertilization, weed control, planting decisions, and the use of machinery or technology.

Because maps are used to represent a variety of environmental subjects, map use is intertwined with many disciplines. It is impossible to appreciate them in isolation. The more fields, particularly environmental, that you study, the better you will be at using maps. Map interpretation grows naturally out of an appreciation of a variety of subjects. The reverse is also true. An appreciation of maps leads to a better understanding of the world around you—for the subject of maps, after all, is the world itself.

This brings us to a final, important point. As you gain an understanding of map use, be careful not to confuse the mapped world with the real world. Remember, the reason you are using maps is to understand the physical and human environment. The aim of map use is to stimulate you to interact with your environment and to learn and experience more while you do.

20   Map Use: Map Reading and Design

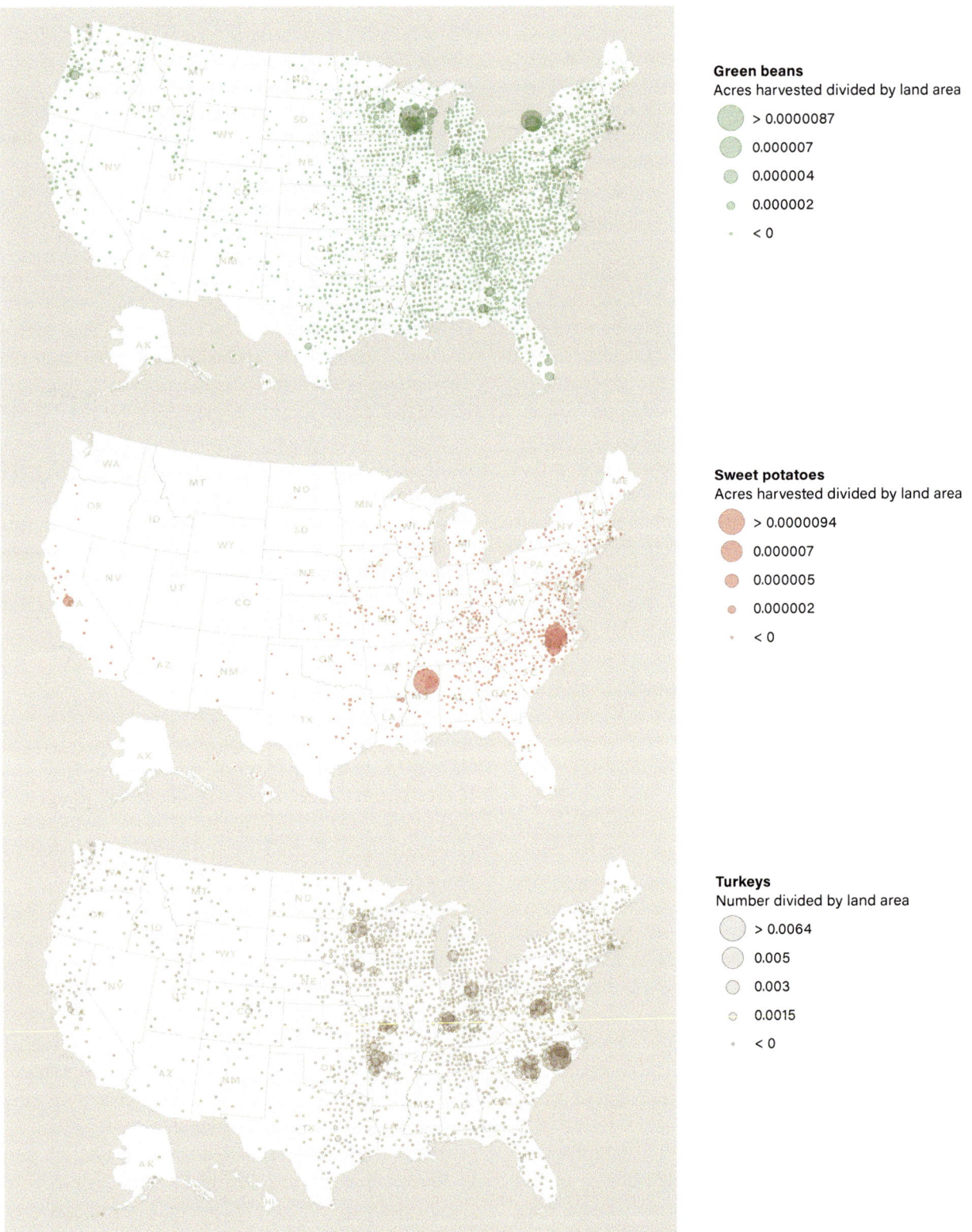

Figure I.20. Circle symbols on these maps show where some of the foods for a typical Thanksgiving dinner come from. The larger the circle on the map, the higher the value—number of turkeys or acres harvested.

Selected readings

Caquard, S. 2013. "Cartography I: Mapping Narrative Cartography." *Progress in Human Geography* 37 (1): 135–44.

Caquard, S., and W. Cartwright. 2014. "Narrative Cartography: From Mapping Stories to the Narrative of Maps and Mapping." *The Cartographic Journal* 51 (2): 101–6.

Cosgrove, D. 2005. "Maps, Mapping, Modernity: Art and Cartography in the Twentieth Century." *Imago Mundi* 57 (1): 35–54. https://doi.org/10.1080/0308569042000289824.

Cusick, M. 2014. "Map Works." *Cartographic Perspectives* 78: 82–85.

Dent, B. D., J. S. Torguson, and T. W. Hodler. 2009. "Introduction to Thematic Mapping." Chapter 1 in *Cartography: Thematic Map Design*, 6th edition. McGraw-Hill Education.

Esri. 2023. "Tactile Maps Built with GIS Help People Who Are Blind Gain Spatial Awareness." *ArcNews* Summer 2023. www.esri.com/about/newsroom/arcnews/tactile-maps-built-with-gis-help-people-who-are-blind-gain-spatial-awareness.

Federal Aviation Administration. 2024. *Aeronautical Chart Users' Guide.* https://www.faa.gov/air_traffic/flight_info/aeronav/digital_products/aero_guide/.

Kelly, M. 2022. "Narrative and Storytelling." *The Geographic Information Science & Technology Body of Knowledge*, 2022 edition. https://gistbok-topics.ucgis.org/CV-04-033.

Kent, A. J. 2020. "All That Glitters: Art, Fire and Post-Cartographic Design." *The Cartographic Journal* 57 (1): 1–5.

Miller, G. 2017. "A Tactile Atlas Helps the Blind 'See' Maps." *National Geographic Newsletter.*

Mocnik, F. 2023. "Why We Can Read Maps." *Cartography and Geographic Information Science* 50 (1): 1–19.

Monmonier, M. 1991. *How to Lie with Maps.* University of Chicago Press.

National Oceanic and Atmospheric Administration. "*Nautical Cartography: The Making of a NOAA Nautical Chart.*" www.nauticalcharts.noaa.gov/learn/nautical-cartography.html.

Ng-Chan, T., N. Lally, and S. Hayashi. 2024. "Art and Cartography: Views from Somewhere." *International Journal of Cartography* 10 (2): 139–43. https://doi.org/10.1080/23729333.2024.2362033.

Roth, R. E. 2013. "Interactive Maps: What We Know and What We Need to Know." *Journal of Spatial Information Science* 6: 59–115. http://dx.doi.org/10.5311/JOSIS.2013.6.105.

Slocum, T. A., R. B. McMaster, F. C. Kessler, and H. H. Howard. 2022. *Thematic Cartography and Geovisualization*, 4th edition. CRC Press. https://doi.org/10.1201/9781003150527.

Tobler, Waldo. 1979. "A Transformational View of Cartography." *The American Cartographer* 1 (2): 101–6.

Vision Loss Expert Group of the Global Burden of Disease Study. 2024. GBD 2019 Blindness and Vision Impairment Collaborators. "Global Estimates on the Number of People Blind or Visually Impaired by Cataract: A Meta-Analysis from 2000 to 2020." *Eye* 38:2156–72. https://doi.org/10.1038/s41433-024-02961-1.

W3C Web Accessibility Initiative. 2024. *Web Content Accessibility Guidelines: WCAG 2 Overview.*

Wood, D., and J. Fels. 1992. *The Power of Maps.* Guilford.

CHAPTER 1
The earth and earth coordinates

Of all the jobs that maps do for you, one stands out—they tell you where things are and allow you to communicate this information efficiently to others. This, more than any other factor, accounts for the widespread use of maps. Maps give you a superb **positional reference system**—a way to pinpoint the locations of things in space.

There are many ways to determine the position of a feature shown on a map. All begin with defining a geometric figure that approximates the true shape and size of the earth. This figure is either a **sphere** (a three-dimensional solid in which all points on the surface are the same distance from the center) or an **oblate ellipsoid**, or **ellipsoid** for short, (a slightly flattened sphere) of precisely known dimensions. Once the dimensions of the sphere or ellipsoid are defined, a **graticule** of east–west lines called **parallels** and north–south lines called **meridians** is draped over the sphere or ellipsoid. The angular distance from the zero lines of latitude and longitude, the equator and **prime meridian** respectively, gives us latitude and longitude coordinates of a feature. The locations of elevations measured relative to an average gravity or sea level surface called the **geoid** and to a second geometric surface that closely fits the geoid called the **reference ellipsoid** can then be defined by 3D (latitude, longitude, elevation) coordinates. Next, we look at these concepts in more detail.

The earth as a sphere

We have known for more than 2,000 years that the earth is spherical in shape. We owe this knowledge to several ancient Greek philosophers—in particular, **Aristotle** (fourth century BC), who believed that the earth's sphericity could be proved by careful visual observation. Aristotle noticed that as he moved to north from south, the stars were not stationary—new stars appeared on the northern horizon while familiar stars disappeared to the south. He reasoned that this could occur only if the earth was curved north to south. He also observed that departing sailing ships, regardless of their direction of travel, always disappeared from view first by their hulls just above the waterline. If the earth was flat, the ships would simply get smaller as they sailed away. Only on a sphere would the hulls always disappear first. His third observation was that a circular shadow is always cast by the earth on the moon during a lunar eclipse, something that occurs only if the earth is spherical. These arguments entered the Greek literature and persuaded scholars over the succeeding centuries that the earth must be spherical in shape.

Determining the size of our spherical earth was a daunting task for our early scholars. Around 250 BC, the Greek scholar **Eratosthenes**, head of the famous library and museum in Alexandria, Egypt, made the first scientifically based estimate of the earth's circumference. The story that has come down to us is of Eratosthenes reading an account of a deep well at Syene near modern Aswan, about 500 miles (800 kilometers) south of Alexandria. The well's bottom was illuminated by the sun only on June 21, the day of the summer solstice. He concluded that the sun must be directly overhead on this day, with its rays perpendicular to the level ground (figure 1.1). Then he reasoned brilliantly that if the sun's rays are parallel and the earth is spherical, a vertical column such as an obelisk should cast a shadow in Alexandria on the same day. Knowing the angle of the shadow would allow the earth's circumference to be measured if the north–south distance from Alexandria to Syene could be determined. The simple geometry involved here is that if two parallel lines are intersected by a third line, the alternate interior angles are equal. From this supposition, he reasoned that the shadow angle at Alexandria equals the angular difference (7°12′) at the earth's center between the two places.

The story continues that on the next summer solstice,

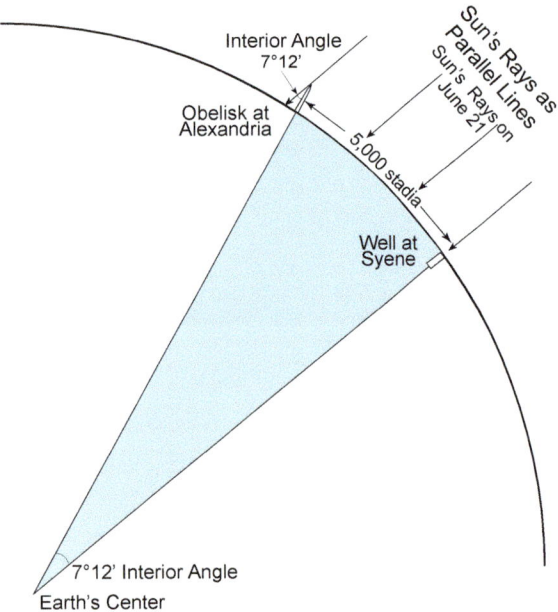

Figure 1.1. Eratosthenes's method for measuring the earth's circumference.

Eratosthenes measured the shadow angle from an obelisk in Alexandria, finding it to be 7°12′, or 1/50th of a circle. Hence, the distance between Alexandria and Syene is 1/50th of the earth's circumference. He was told that Syene must be about 5,000 stadia south of Alexandria because camel caravans traveling at 100 stadia per day took 50 days to make the trip between the two cities. From this distance estimate, he computed the earth's circumference as 50 × 5,000 stadia, or 250,000 stadia. A **stadion** is an ancient Greek unit of measurement based on the length of a sports stadium at the time. A stadion varied from 200 to 210 modern yards (182 to 192 meters), so his computed circumference was somewhere between 28,400 and 29,800 modern statute miles (45,700 and 47,960 kilometers), 14 to 19 percent greater than the currently accepted circumference distance of 24,874 statute miles (40,030 kilometers).

We now know that Eratosthenes' error was because of an underestimate of the distance between Alexandria and Syene and because the two cities are not exactly north–south of each other. However, his method was sound mathematically, and it was the best circumference measurement until the 1600s. Equally important, Eratosthenes had the idea that careful observations of the sun allowed him to determine angular differences between places on earth, an idea that was later expanded to other planets and stars to find positions on the earth. For example, you will see later in this chapter that ocean navigators use special instruments called sextants to observe the angular distance of the sun and other stars above the horizon, from which they determine their position at sea.

The graticule

Once the shape and size of the earth were known, mapmakers required a system for defining locations on the earth's surface. We are again indebted to ancient Greek scholars for devising a means for placing imaginary reference lines on the spherical earth from which we can determine precise locations.

Parallels and meridians

Astronomers before Eratosthenes placed horizontal lines on maps to mark the **equator** (forming the imaginary circle around the earth that is equidistant from the North and South Poles) and the **Tropics of Cancer** and **Capricorn** (lines of latitude that mark the northernmost and southernmost positions in which the sun is directly overhead on the summer and winter solstices, respectively. Syene, mentioned earlier, is located almost atop the Tropic of Cancer.) These lines were important, but they did not provide enough information to locate features on the earth's surface.

This came later when the Greek astronomer and mathematician **Hipparchus** (190–125 BC) proposed that a set of equally spaced east–west lines called parallels be drawn on maps (figure 1.2). He then added a set of north–south lines called meridians that are equally spaced at the equator and converge at the North and South Poles. We now call this arrangement of parallels and meridians the graticule, from the Latin for *grill* or *grating*. Hipparchus's numbering system for parallels and meridians was, and still is, called latitude and longitude, from the Latin for *breadth* and *length*, respectively. We now look at latitude and longitude on a sphere, sometimes called geocentric latitude and longitude. Later in this chapter, we explain how geocentric latitude and longitude differ from geodetic latitude and longitude on an oblate ellipsoid.

Geocentric latitude and longitude

Geocentric latitude on the spherical earth is the north–south angular distance from the equator to the place of interest (figure 1.3). The numerical range of latitude is from 0° at the equator to 90° at the poles. The letters *N* and *S*, such as 45° N for Fossil, Oregon, are used to

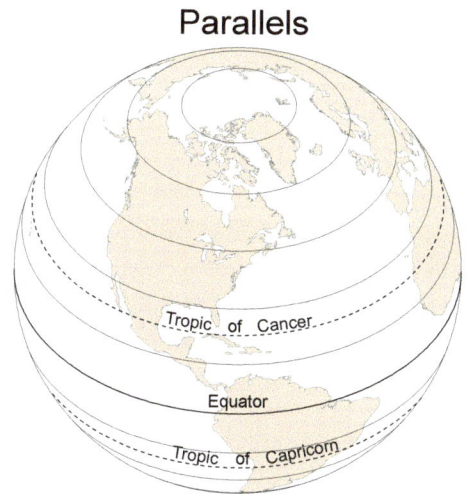
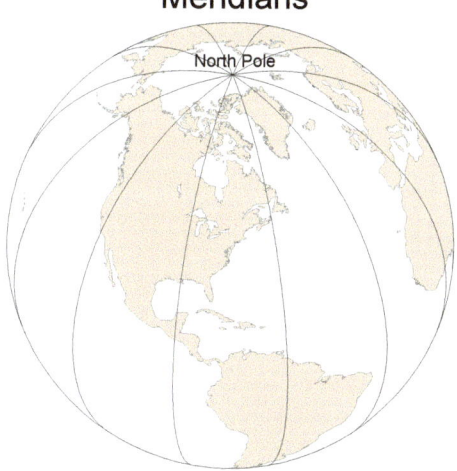

Figure 1.2. Parallels and meridians.

indicate north and south latitudes. Instead of the letter *S*, you may see a minus sign (–) for south latitudes; however, a plus sign (+) is not used for north latitudes.

Longitude is the angle, measured along the equator, between the intersection of the reference meridian, called the prime meridian, and the point at which the meridian for the feature of interest intersects the equator. The numeric range of longitude is from 0° to 180° east and west of the prime meridian, twice as long as for the parallels. East and west longitudes are labeled *E* and *W*, so that Fossil, Oregon, has a longitude of 120° W. As with south latitudes, west longitudes may also be indicated by a minus sign, but, as for north latitudes, a plus sign is not used for east longitudes.

Putting latitude and longitude together into what is called a **geographic coordinate** (such as 45° N, 120° W or 45°, –120°) pinpoints a place on the earth's surface. There are several ways to write latitude and longitude values. The oldest is the **Babylonian sexagesimal system** of **degrees (°), minutes (´), and seconds (˝)**, commonly referred to as **DMS**, where degrees range from 0 to 360 and there are 60 minutes in a degree and 60 seconds in a minute. For example, the latitude and longitude of the capitol dome in Salem, Oregon, is 44°56´18˝ N, 123°01´47˝ W (or 44°56´18˝, –123°01´47˝).

Latitude and longitude can also be expressed in **decimal degrees (DD)** through equation (1.1).

$$\text{DD} = dd + \frac{mm}{60} + \frac{ss}{3600} \quad (1.1)$$

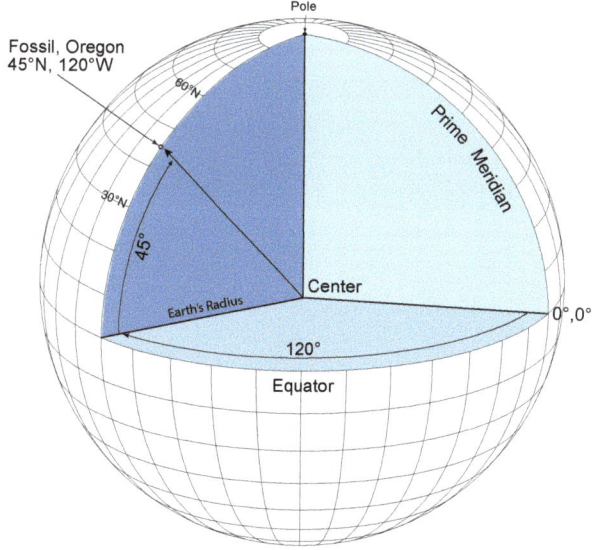

Figure 1.3. Latitude and longitude on the sphere allow explicit identification of the positions of features on the earth.

Here, *dd* is the number of whole degrees, *mm* is the number of minutes, and *ss* is the number of seconds. An example is shown in equation (1.2).

$$44°56´18˝ = 44 + \frac{56}{60} + \frac{18}{3600} = 44.9381° \quad (1.2)$$

Decimal degrees are often rounded to two decimal places, so the location of the Oregon state capitol dome is written in decimal degrees as 44.94, –123.03. If you can define latitude to a precision of two decimal places (see table 1.1), you can specify its north–south location on the earth to an accuracy of within 7/10th of a mile (about 1.1 kilometers).

Because the earth is spherical, the east–west distance between equally spaced meridians decreases from the equator to the North or South Pole, as seen in figure 1.3. Notice in table 1.2 that the accuracy of the longitude part of the geographic coordinate increases the farther you get north or south from the equator. For example, longitude values with two–decimal place precision allow you to specify east–west location on the earth to an accuracy of within 6/10th of a mile (about 1 kilometer) at 30° N and at a higher accuracy of within 0.34 mile (0.56 kilometers) at 60° N. At the equator, the east–west accuracy is the same as the 7/10th of a mile north–south accuracy for parallels shown in table 1.1.

Prime meridians

The choice of prime meridian (the 0° meridian used as the reference from which longitude east and west are measured) is entirely arbitrary because there is no physically definable starting point such as the equator for latitude. In the fourth century BC, Eratosthenes selected Alexandria, Egypt, as the starting meridian for longitude. In European medieval times, the Canary Islands off the west coast of Africa were used because they were then the westernmost outpost of Western civilization. In the 18th and 19th centuries, many countries used their capital city, their country's most prominent maritime port or city (such as Philadelphia in the United States), a respected national observatory (for example, the Royal Observatory in Greenwich, England, or Paris Observatory in France), or a religious site (such as Jerusalem or Saint Petersburg) as the prime meridian for their national maps (see appendix table A.1 for a listing of many historical prime meridians).

You can imagine the confusion that must have existed when trying to locate places on maps using different lines of 0° meridian during these times. The problem was largely eliminated in 1884, when the **International Meridian Conference**, held in Washington, DC, selected as the international standard the British prime meridian, defined by the north–south optical axis of a telescope at the Royal Observatory in Greenwich, a suburb of London. This standard is called the **Greenwich meridian**. This choice was largely based on the widespread use of British nautical charts with Greenwich as the prime meridian due to the country's colonial and seafaring dominance in the 18th and

Decimal places	DD Precision	DMS Precision	Distance Accuracy (Imperial)	(metric)
0	1	1°0'0"	68.97 mi	111 km
1	0.1	0°6'0"	6.9 mi	11.1 km
2	0.01	0°0'36"	0.69 mi	1.11 km
3	0.001	0°0'3.6"	364.17 ft	111 m
4	0.0001	0°0'.36"	36.42 ft	11.1 m
5	0.00001	0°0'.36"	3.64 ft	1.11 m

Table 1.1. North–south distance accuracy for latitude precisions in decimal degrees and degrees-minutes-seconds

Table 1.2. East-west distance accuracy for longitude precisions in decimal degrees and degrees-minutes-seconds at 30° N and 60° N

Decimal places	DD Precision	DMS Precision	30° N Distance Accuracy		60° N Distance Accuracy	
			(Imperial)	(metric)	(Imperial)	(metric)
0	1	1°0'0"	59.7 mi	96 km	34.5 mi	55.5 km
1	0.1	0°6'0"	6.0 mi	9.6 km	3.4 mi	5.6 km
2	0.01	0°0'36"	0.6 mi	0.96 km	0.34 mi	0.56 km
3	0.001	0°0'3.6"	315 ft	96 m	182 ft	56 m
4	0.0001	0°0'.36"	31.5 ft	9.6 m	18.2 ft	5.6 m
5	0.00001	0°0'.036"	3.15 ft	0.96 m	1.82 ft	0.56 m

19th centuries. However, France abstained from the 1884 vote and continued to use the Paris meridian until 1914 for its navigational maps.

You may occasionally come across a historical map that uses one of the prime meridians listed in the appendix. When you do, knowing the angular difference between the prime meridian used on the map and the Greenwich meridian becomes useful information. As an example, on a map in an old French atlas that uses the Paris meridian, the longitude of Constantinople, Türkiye (modern-day Istanbul) is 26°37′ E (based on the Paris meridian), and you know from appendix table A.1 that the Greenwich longitude of the Paris meridian is 2°20′ E. You can determine the Greenwich-based longitude of Istanbul through addition, in equation (1.3).

$$26°37′ \text{ E} + 2°20′ \text{ E} = 28°57′ \text{ E} \quad (1.3)$$

The computation is done more easily in decimal degrees, as described earlier, in equation (1.4).

$$26.63° \text{ E} + 2.33° \text{ E} = 28.95° \text{ E} \quad (1.4)$$

The earth as an oblate ellipsoid

Scholars assumed that the earth was a perfect sphere until the 1660s, when the English physicist and mathematician **Sir Isaac Newton** developed the theory of gravity. Newton thought that gravity should produce a perfectly spherical earth if it was not rotating about its **polar axis**—the axis from the center of the earth to the poles. The earth's 24-hour rotation, however, introduces outward **centrifugal forces** that are perpendicular to the axis of rotation. The amount of force varies from zero at each pole to a maximum at the equator.

Newton noted that these outward centrifugal forces counteract the inward pull of gravity, so the net inward force decreases progressively from the pole to the equator causing the earth to be slightly flattened. Slicing the earth in half from pole to pole would then reveal an ellipse with a slightly longer equatorial radius and slightly shorter polar radius; we call these radii the **semi-major** and **semi-minor axes**, respectively (figure 1.4). If we rotate this ellipse 180° about its polar axis, we obtain a 3D solid called an **oblate ellipsoid**, or **ellipsoid**.

The oblate ellipsoid is important because parallels are not spaced equally as they are on a sphere but decrease slightly in spacing from the pole to the equator. This

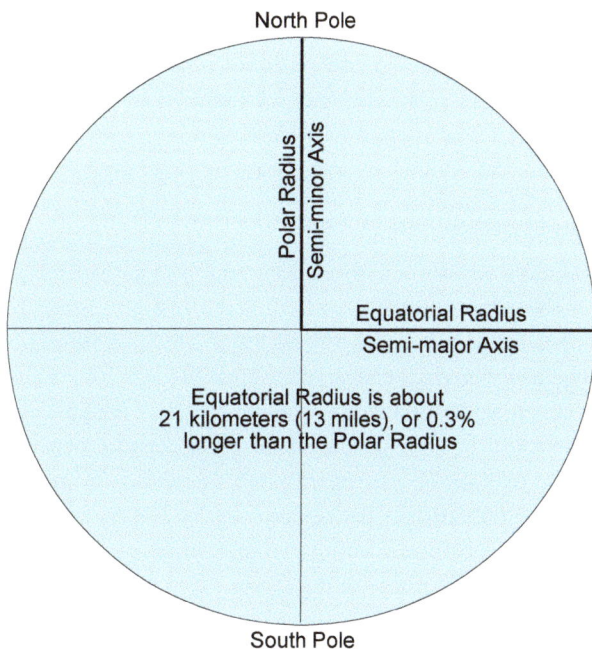

Figure 1.4. The form of the oblate ellipsoid was determined by measurements of degrees of longitude at different latitudes, beginning in the 1730s. Its equatorial radius was about 13 miles (21 kilometers) longer than its polar radius. This figure is true to scale, but our eye cannot see the flattening because the difference between the north–south and east–west axis is so small.

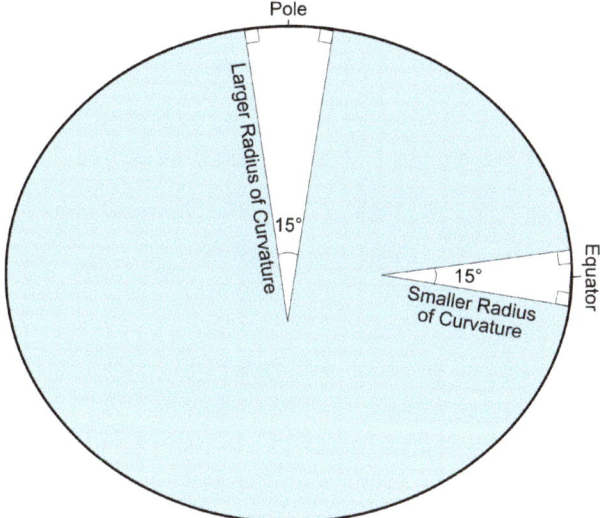

Figure 1.5. This north–south cross section through the center of a greatly flattened oblate ellipsoid shows that a larger radius of curvature at the pole results in a larger ground distance per degree of latitude relative to the equator.

variation in the spacing of parallels is shown in figure 1.5, a cross section of a greatly flattened oblate ellipsoid. We say that on an oblate ellipsoid, the **radius of curvature** (the measure of how curved the surface is) is largest at the pole and smallest at the equator. Near the pole, the ellipse is flatter than near the equator. You can see that there is less curvature on the ellipse with a larger radius of curvature and more curvature with a smaller radius of curvature.

The north–south distance between two points on the ellipsoidal surface equals the radius of curvature times the angular difference between the points. Because the radius of curvature on an oblate ellipsoid is largest at the pole and smallest at the equator, the north–south distance between points that are a degree apart in latitude should be greater near the pole than at the equator. In the 1730s, scientific expeditions to Ecuador and Finland measured the length of a degree of latitude at the equator and near the Arctic Circle, proving Newton correct. These and additional measurements in the following decades for other parts of the world allowed the semi-major and semi-minor axes of the oblate ellipsoid to be computed by the early 1800s, giving about a 13-mile (21-kilometer) difference between the two, only one-third of 1 percent.

Different ellipsoids

During the 19th century, better surveying equipment was used to measure the length of a degree of latitude on different continents. From these measurements, slightly different oblate ellipsoids, varying by only a few hundred meters in axis length, best fit the measurements. Table 1.3 is a list of these ellipsoids, along with their areas of use. Note the changes in ellipsoid use over time. For example, the **Clarke 1866 ellipsoid** was the best fit for North America in the 19th century and was used as the basis for latitude and longitude on topographic and other maps produced in Canada, Mexico, and the United States from the late 1800s to about the late 1970s. By the 1980s, vastly superior surveying equipment, coupled with millions of observations of satellite orbits, allowed us to determine oblate ellipsoids that are excellent average fits for the entire earth. Satellite data is important because the elliptical shape of each orbit monitored at ground receiving stations mirrors the earth's shape. The Geodetic Reference System of 1980 (GRS80) replaced the Clarke 1866 ellipsoid in North America and is used as the basis for latitude and longitude on maps throughout the world. The most recent of these ellipsoids, the **WGS84**, is virtually identical to the GRS80 and is a key part of the **World Geodetic System of 1984** discussed later in this chapter. In table 1.3, the WGS84 ellipsoid has an equatorial radius of 6,378.137 kilometers (3,963.191 miles) and a polar radius of 6,356.752 kilometers (3,949.903 miles). On this ellipsoid, the distance between two points that are one degree apart in latitude between 0° and 1° at the equator is 110.567 kilometers (68.703 miles), shorter than the 111.699-kilometer (69.407-mile) distance between two points at 89° and 90° north latitude. Hence, on equatorial- and polar-area maps of the same scale, the spacing between parallels is not the same, but the difference is small.

Table 1.3. Historical and current oblate ellipsoids

Name	Date	Equatorial Radius (km)	Polar Radius (km)	Areas of Use
WGS84	1984	6,378.137	6,356.75231	Worldwide
GRS80	1980	6,378.137	6,356.7523	Worldwide (NAD83)
Australian	1965	6,378.160	6,356.7747	Australia
Krasovsky	1940	6,378.245	6,356.863	Soviet Union
International	1924	6,378.388	6,356.9119	Remainder of the world not covered by older ellipsoids (European Datum 1950)
Clarke	1880	6,378.2491	6,356.5149	France; most of Africa
Clarke	1866	6,378.2064	6,356.5838	North America (NAD27)
Bessel	1841	6,377.3972	6,356.079	Central Europe, Sweden, Chile, Switzerland, Indonesia
Airy	1830	6,377.5634	6,356.2569	Great Britain, Ireland
Everest	1830	6,377.2763	6,356.0754	India and the rest of South Asia

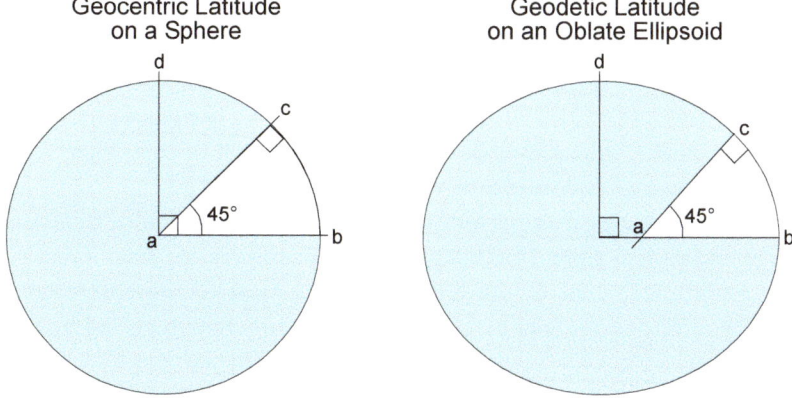

Figure 1.6. Geocentric and geodetic latitudes of 45°. On a sphere, circular arc distance b–c is the same as circular arc distance c–d. On the greatly flattened oblate ellipsoid, elliptical arc distance b–c is less than elliptical arc distance c–d. On the WGS84 oblate ellipsoid, arc distance b–c is 4,984.94 kilometers (3,097.50 miles) and arc distance c–d is 5,017.02 kilometers (3,117.43 miles), a difference of about 32 kilometers (20 miles).

Geodetic latitude

Geodetic latitude is defined as the angle made by the equatorial plane and a line perpendicular to the ellipsoidal surface at the parallel of interest (figure 1.6). Geodetic latitude differs from latitude on a sphere because of the unequal spacing of parallels on the ellipsoid. Lines perpendicular to the ellipsoidal surface pass through the center of the earth only at the poles and the equator, but all lines perpendicular to the surface of a sphere pass through its center. This centricity is why the latitude defined by these lines on a sphere is called **geocentric latitude**, introduced earlier in this chapter.

Geodetic longitude

There is no need to make a distinction between **geocentric longitude** (described earlier in this chapter) on the sphere and **geodetic longitude** on the ellipsoid. Both types of longitude are defined as the angular distance between the prime meridian and another meridian passing through a point on the earth's surface, with the center of the earth as the vertex of the angle. Because of the geometric nature of the ellipsoid, the angular distance turns out to be the same as for geocentric longitude on the sphere.

Determining geodetic latitude and longitude

The oldest way to determine geodetic latitude is with instruments for observing the positions of celestial bodies. The essence of the technique is to establish celestial lines of position (east–west and north–south) by comparing the predicted positions of celestial bodies with their observed positions. A handheld instrument, called a **sextant**, was

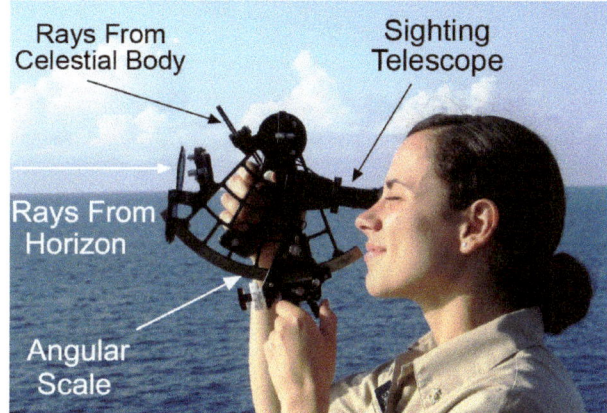

Figure 1.7. A sextant is used at sea to find latitude from the vertical angle between the horizon and a celestial body, such as the sun, a planet, or a star. Courtesy of Dr. Bernie Bernard, TDI-Brooks International Inc.

the tool historically used to measure the angle (or altitude) of a celestial body above the earth's horizon (figure 1.7). Before the advent of the **Global Navigation Satellite System (GNSS)**, of which the US **Global Positioning System (GPS)** is a key component, it was the tool that nautical navigators used to find their way using the moon, planets, and stars, including the sun.

Astronomers study and tabulate information on the actual motion of the celestial bodies that help pinpoint latitude. Because the earth rotates on an axis defined by the North and South Poles, stars in the Northern Hemisphere's night sky appear to move slowly in a circle centered on **Polaris** (the **North Star**), which lies almost directly above the North Pole. A navigator needs only to locate Polaris to find the approximate North Polar axis because the star is less than one degree away from the north celestial pole.

In addition, because the star is so far away from the earth, the angle from the horizon to Polaris is approximately the same as the latitude (figure 1.8).

However, Polaris is, in actuality, 0.75 degrees from directly above the North Pole, circling the pole in a small circle that is 1.5 degrees in diameter. Precisely determining the spot near Polaris that is directly above the North Pole requires accurately knowing the time and using Polaris position data from the **Astronomical Almanac**, published jointly by the US Naval Observatory and HM Nautical Almanac Office in the United Kingdom, which contains data about the position and brightness of the sun, planets, and stars. Errors in determining latitude can otherwise approach three-quarters of a degree, which translates into a ground position error of about 50 miles (80 kilometers).

In the Southern Hemisphere, latitude is harder to determine by celestial measurement because there is no equivalent to Polaris directly above the South Pole. Navigators instead use a small constellation called **Crux Australis** (the **Southern Cross**) to serve the same function (figure 1.9). Finding south is more complicated because the Southern Cross is a collection of five stars that are part of the constellation **Centaurus**. The four outer stars form a cross, whereas the fifth, much dimmer star (**Epsilon**) is offset about 30 degrees below the center of the cross.

Longitude can be determined on land or sea by careful observation of time. In previous centuries, accurately determining longitude was a major problem in both sea navigation and mapmaking. It was not until 1762 that a clock that was accurate and portable enough for longitude finding was invented by Englishman **John Harrison**. This clock, called a **chronometer**, was set to the time at Greenwich, England, later called **Greenwich Mean Time (GMT)**, before departing on a long voyage. (Coordinated Universal Time [UTC] is the current global standard.) As you saw earlier, the prime meridian at 0° longitude passes through Greenwich, England. Therefore, each hour difference between your time and that of Greenwich is equivalent to 15 degrees of longitude from Greenwich. The longitude of a distant locale was found by noting the GMT at local noon (the highest point of the sun in the sky, found with a sextant). The time difference in hours was simply multiplied by 15 to find the longitude.

Today, you can find your longitude in the field simply by looking at the value computed by your GNSS receiver or GNSS unit in your smartphone. The details of how GNSS receivers find your latitude and longitude are found

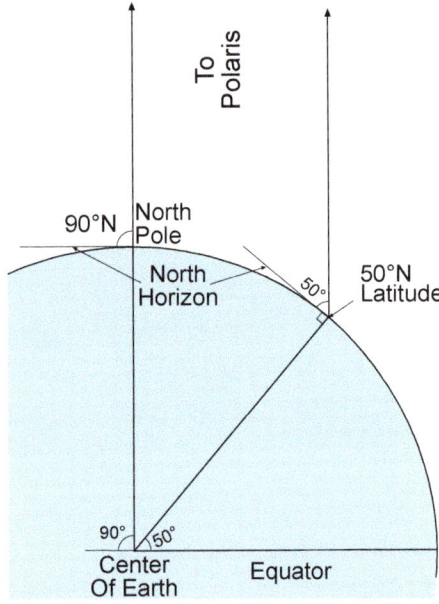

Figure 1.8. In the Northern Hemisphere, it is easy to determine your approximate latitude by observing the height of Polaris above your northern horizon. For example, at 50° N latitude, you will see Polaris at 50 degrees above the north horizon.

Figure 1.9. The Southern Cross is used for navigation in the Southern Hemisphere. To approximately locate the point in the sky directly above the South Pole, extend the long (Gacrux to Acrux) axis of the Southern Cross 4.5 times the length of the axis and then draw in the sky a line perpendicular from the end of this line to the left that is the length of the short (Becrux to Delta) axis.

in their reference manuals, but you can rest assured that the GNSS coordinates are more accurate than the most carefully measured time differences made in the past using Harrison's invention and similar chronometers.

Properties of the graticule

Circumference of the authalic sphere

When determining latitude and longitude, we sometimes use the spherical approximation to the earth's shape rather than the oblate ellipsoid. Using a sphere leads to simpler calculations, especially when working with small-scale maps of countries, continents, or the entire earth (see chapter 2 for more on small-scale maps). On these maps, differences between locations on the sphere and the ellipsoid are negligible. Cartographers use a value of the earth's spherical circumference called an **authalic sphere** (meaning "area preserving sphere"). The authalic sphere is a sphere that has the same surface area as the oblate ellipsoid being used. The equatorial and polar radii of the WGS84 ellipsoid (used in the World Geodetic System of 1984 described later in this chapter) are what we use to calculate the radius and circumference of the authalic sphere that is equal to the surface area of the WGS84 ellipsoid. The computations involved are moderately complex and best left to a short computer program, but the result is a sphere of radius of 3,958.76 miles (6,371.017 kilometers) and circumference of 24,873.62 miles (40,030.22 kilometers).

Spacing of parallels

As you saw earlier, on a spherical earth, the north–south ground distance between equal increments of latitude does not vary. However, it is important to know how you want to define that distance. As you will see next, there are different definitions for terms that you may take for granted, such as a "mile."

Using the authalic sphere circumference based on the WGS84 ellipsoid, latitude spacing is always 24,873.62 miles ÷ 360°, or 69.09 statute miles per degree. **Statute miles**, sometimes called **land miles**, of 5,280 feet in length are used for land distances in the United States, whereas **nautical miles** are used around the world for maritime and aviation purposes. The original nautical mile was defined as one minute of latitude measured north–south along a meridian. The current international standard for a nautical mile is 6,076.1 feet (about 1.15 statute miles) so that there are 60.04 nautical miles per degree of latitude on the authalic sphere.

The **metric system** is used to express distances in all countries other than the United States, Liberia, and Myanmar, which continue to use **Imperial units**. The **Imperial measurement system** (also called the **English system**) is sometimes referred to as foot-pound-second, after the units of length, weight, and time. **Metric units**, like the nautical mile, are closely tied to distances along meridians, because the **meter** (equal to 39.37 inches) was initially defined as one 10-millionth of the distance along a meridian from the equator to the North or South Pole. There are 1,609.344 meters in a statute mile and 1,852 meters in a nautical mile (see appendix table A.2 for these and additional metric and Imperial unit distance equivalents). Expressed in metric units, there are 111.20 kilometers per degree of latitude on the authalic sphere.

You have seen that parallels on an oblate ellipsoid are not spaced equally as they are on the authalic sphere but instead decrease slightly from the pole to the equator. You can see in appendix table A.3 the variation in the length of a degree of latitude for the WGS84 ellipsoid, measured along a meridian at one-degree increments from the equator to the pole. The distance per degree of latitude ranges from 69.407 statute miles (111.699 kilometers) at the pole to 68.703 statute miles (110.567 kilometers) at the equator. The graph in figure 1.10 shows how these distances differ from the constant value of 69.09 statute miles (111.20 kilometers) per degree for the authalic sphere. The WGS84 ellipsoid distances per degree are about 0.3 miles (0.48 kilometers) greater than the sphere at the pole and 0.4 miles (0.64 kilometers) less at the equator. The ellipsoidal and spherical distances are almost the same in the mid-latitudes, somewhere between 45 and 50 degrees.

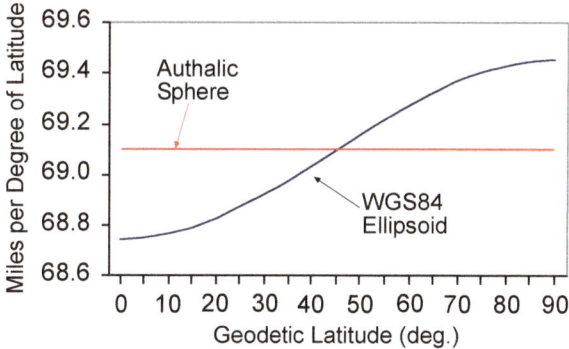

Figure 1.10. Distances along the meridian for one-degree increments of latitude from the equator to the pole on the WGS84 ellipsoid and authalic sphere.

Converging meridians

A quick glance at any world globe (or figure 1.2) shows **converging meridians**—that is, the meridians progressively converge from the equator to a point at the North and South Poles. You will also see that the length of a degree of longitude, measured east–west along parallels, decreases from the equator to the pole. The spacing of meridians on the authalic sphere at a given latitude is found by using equation (1.5).

$$69.09 \text{ miles (or } 111.20 \text{ kilometers)} \times \cos(\text{latitude}) \quad (1.5)$$

For example, at 45 degrees north or south of the equator, cosine(45°) = 0.7071. Therefore, the length of a degree of longitude is found in equation (1.6).

$$69.09 \times 0.7071, \text{ or } 48.85 \text{ statute miles}$$
$$(111.20 \times 0.7071, \text{ or } 78.63 \text{ kilometers}) \quad (1.6)$$

This spacing of meridians is roughly 20 miles (32 kilometers) shorter than the 69.09-mile (111.20-kilometer) spacing at the equator.

Quadrilaterals

Many navigational maps cover **quadrilaterals**, which are areas bounded by increments of latitude and longitude, although the increments for latitude need not be the same as for longitude. Because meridians converge toward the poles, the shapes of the quadrilaterals vary from a square on the sphere at the equator to a narrow spherical triangle at the pole, such as the two 15° × 15° quadrilaterals from 75° N to the North Pole in figure 1.11. The equation cosine of mid-latitude, or cos(mid-latitude), gives the **aspect ratio** (width to height ratio) of a quadrilateral. A quadrilateral centered at the equator has an aspect ratio of 1.0, whereas a quadrilateral centered at 60° N has an aspect ratio of 0.5. A map that covers 1° × 1° or any other equal latitude and longitude extent quadrilateral looks long and narrow at this high latitude, whereas an equal-extent quadrilateral map at the equator is essentially square in shape.

Great and small circles

A **great circle** is the largest possible circle that can be drawn on the surface of the spherical earth, and it is also the shortest distance between two points on the earth's surface. Its circumference is that of the sphere, and its center is the center of the earth, so that all great circles divide the

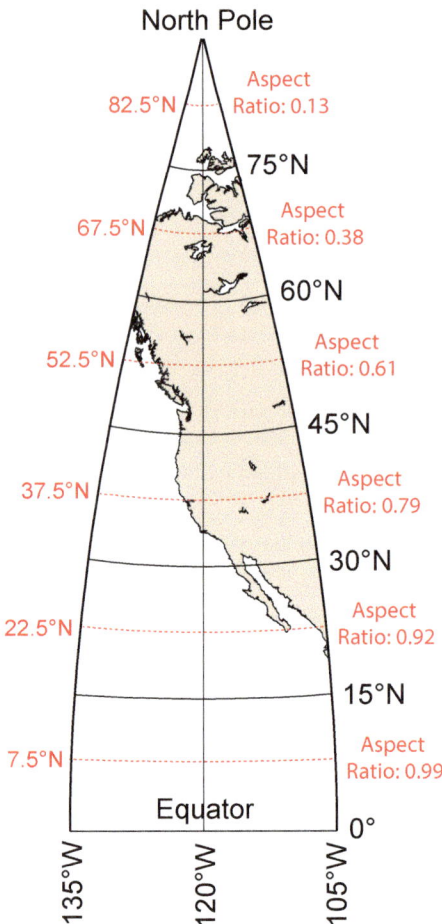

Figure 1.11. Aspect ratios for 15° × 15° quadrilaterals on the earth from the equator to the North Pole. The cosine of the mid-latitude for each quadrilateral gives its aspect ratio.

earth into halves. Notice in figure 1.3 that the equator is a great circle that divides the earth into the Northern and Southern Hemispheres. Similarly, the prime meridian and its **antipodal meridian**, situated at the opposite side of the earth at 180°, form a great circle that divides the earth into the Eastern and Western Hemispheres. All other pairs of meridians and their antipodal meridians are also great circles. Because a great circle is the shortest route between any two points on the earth, great circle routes are fundamental to long-distance navigation.

Any circle on the earth's surface that intersects the interior of the sphere at any location other than the center is called a **small circle**, and its circumference is smaller than a great circle. You can see in figure 1.3 that all parallels other than the equator are small circles. The circumference of a particular parallel is given by equation (1.7).

$$24{,}874 \text{ miles (or } 40{,}030 \text{ kilometers)} \times$$
$$\text{cosine (latitude)} \quad (1.7)$$

For example, the circumference of the 45th parallel is 24,874 × 0.7071, or 17,588 statute miles (40,030 × 0.7071, or 28,305 kilometers).

Graticule appearance on maps

Small-scale maps

A **small-scale map** is a map that represents a large area on the earth in a small space on the map. Small-scale world or continental maps such as globes and world atlas sheets normally use geocentric latitude-longitude coordinates based on an authalic sphere. There are several reasons for using the authalic sphere. Before using digital computers to make these types of maps numerically, it was much easier to construct them from spherical coordinates. Equally important, the differences in the plotted positions of geocentric and corresponding geodetic parallels become negligible on maps that cover so much area on a small page.

For example, mathematicians have computed the maximum difference between geocentric and geodetic latitude as 0.128 degrees at the 45th parallel. If you draw parallels at 45° and 45.128° on a map scaled at one inch per degree of latitude, the two parallels are drawn a very noticeable 0.128 inches (0.325 centimeters) apart. Now imagine drawing parallels on a map scaled at one inch per 10 degrees of latitude, a scale that corresponds to a world wall map approximately 18 inches (46 centimeters) high and 36 inches (92 centimeters) wide. The two parallels are now drawn 0.013 inches (0.033 centimeters) apart, a difference that is not even noticeable considering the width of a line on a piece of paper.

Large-scale maps

A **large-scale map** shows a small area on the earth in a high amount of detail. Parallels and meridians are shown in different ways on large-scale maps that show small areas in great detail (see chapter 2 for more on large-scale maps). Large-scale topographic maps, which show elevations and landforms as well as other ground features, use tick marks to show the location of the graticule. In the United States, the agency responsible for the nation's topographic maps is the **US Geological Survey (USGS)**. All USGS **7.5-minute topographic maps** (quadrilateral maps that cover 7.5 minutes of one degree of longitude and latitude) have graticule

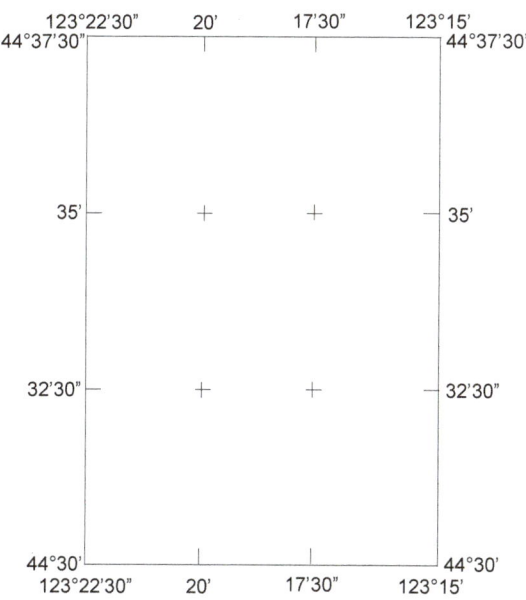

Figure 1.12. Graticule ticks on the 7.5-minute topographic map for Corvallis, Oregon.

ticks at 2.5-minute intervals (figure 1.12). The latitude and longitude coordinates are printed in each corner, but only the minutes and seconds of the intermediate edge ticks are shown. Note the four plus sign symbols used for the interior 2.5-minute graticule ticks in the figure. Also note that the quadrilateral is rectangular with greater height than width, as it is located almost 45 degrees north of the equator, having an aspect ratio of 0.7 (cos (44.5 degrees)).

The graticule is shown in a different way on nautical charts for marine navigation. The segment of the 1:250,000-scale (at latitude 20°30′) nautical chart for the Island of Maui in figure 1.13 shows that alternating white and dark bars spaced at the same half-minute increment of latitude and longitude line the edge of the chart, along with external ticks at one-minute increments. More closely spaced ticks closer to the mapped area are used to find the latitude and longitude of mapped features to within a fraction of a minute. Because of the convergence of meridians, the spacing between ticks, which shows equal increments of latitude, is longer within the vertical bars on the left and right edges of the chart than the spacing between ticks within the horizontal bars at the top and bottom. The one exception is charts of areas along the equator, in which the spacing between ticks is the same on all edges.

Aeronautical charts display the graticule in yet another way. The segment of the 1:500,000-scale sectional aeronautical chart for Albuquerque, New Mexico (figure 1.14),

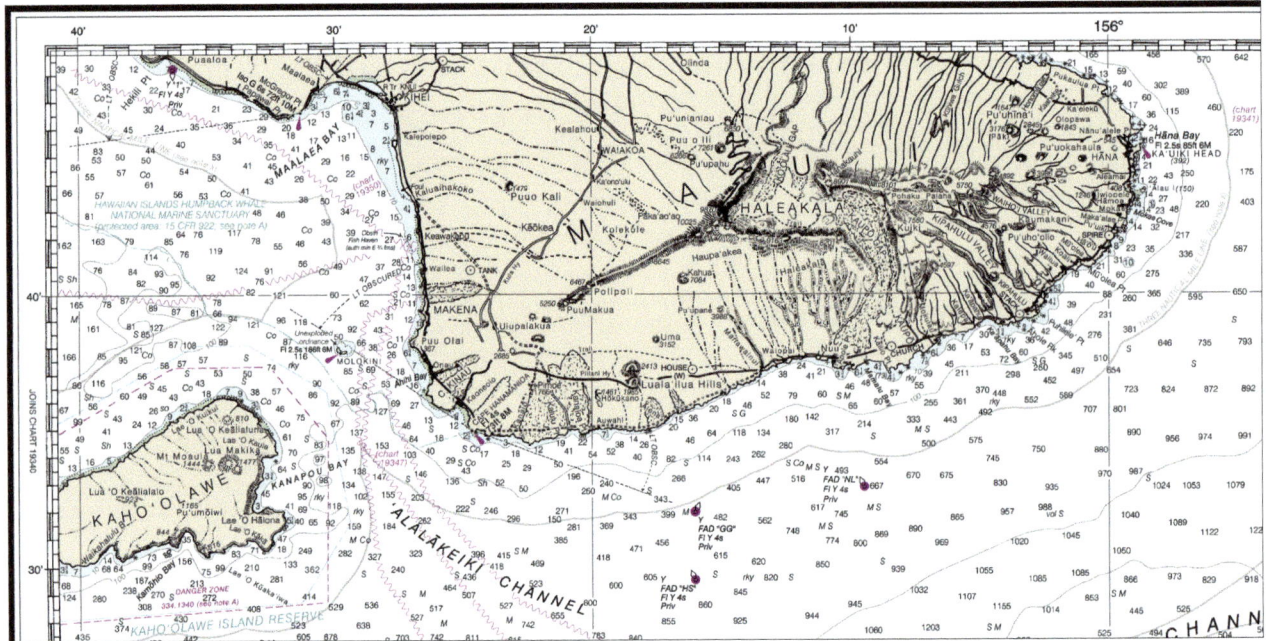

Figure 1.13. Graticule bars with ticks on the edges of a small segment of a paper nautical chart for the Island of Maui. Courtesy of the National Oceanic and Atmospheric Administration.

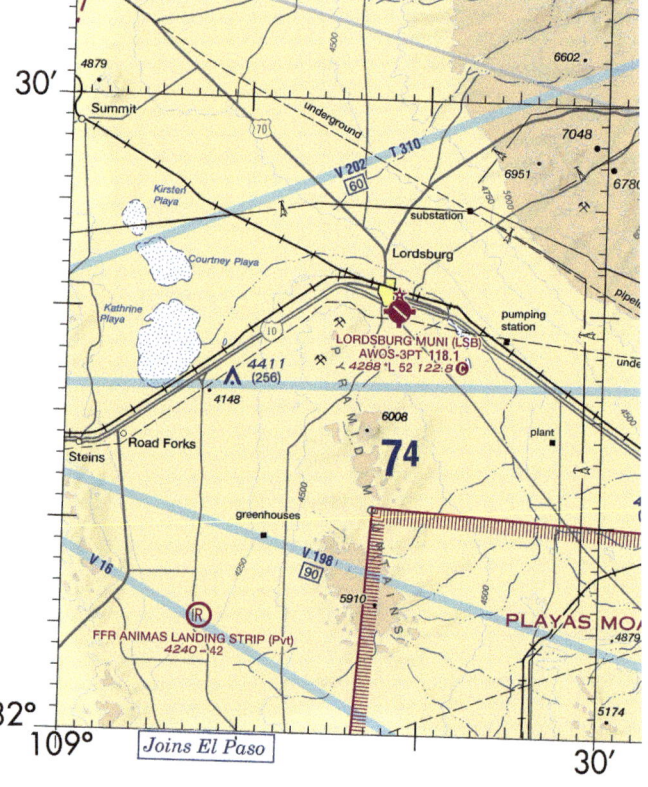

Figure 1.14. Graticule ticks on a small segment of the Albuquerque Sectional Aeronautical Chart. Scale bars for the map in nautical miles, statute miles, and kilometers are also shown. Courtesy of the Federal Aviation Administration.

shows that parallels and meridians are drawn at 30-minute latitude and longitude intervals. Ticks are placed at one-minute increments along each graticule line, allowing features to be located easily to within a fraction of a minute. Notice that the one–nautical mile increments on the nautical mile scale bar are the same width as the one-minute latitude ticks on the left edge of the chart, as they should be because one nautical mile was initially defined as one minute of latitude.

Geodetic latitude and longitude on large-scale maps

You will often find parallels and meridians of geodetic latitude and longitude on detailed maps of small areas. Geodetic coordinates are used to make the map a close approximation to the size and shape of the part of the ellipsoidal earth it represents. To see the perils of not doing this step, you need only examine one-degree quadrilaterals at the equator (from 0° to 1° in latitude) and at the North or South Poles (from 89° to 90° in latitude).

You can see in appendix table A.3 that the ground distance between these pairs of parallels on the WGS84 ellipsoid is 68.703 and 69.407 miles (110.567 and 111.699 kilometers), respectively. If the equatorial quadrilateral is mapped at a scale so that it is 100 inches (254 centimeters) high, the polar quadrilateral mapped at the same scale is 101 inches (256.5 centimeters) high. If both quadrilaterals are mapped using the authalic sphere with 69.09 miles (111.20 kilometers) per degree, both quadrilaterals are 100.6 inches (255.5 centimeters) high. Having both maps several 10ths of an inch (or around a centimeter) longer or shorter than they should be seems a small error, but it can be an unacceptably large error for maps that are used to make accurate measurements of distance, direction, or area. Yet you can see in figure 1.10 that the height differences at the equator and the poles are the extremes and that there is little difference at mid-latitudes and no difference at 45°.

Horizontal datums

To further understand the use of different types of coordinates on large-scale maps of smaller extents, you must first look at **horizontal datums** (or **horizontal reference frames**) that define the earth's shape and provide a coordinate system that serves as a basis for defining latitude and longitude coordinates on the earth's surface. The coordinates in a reference frame are defined by its **datum**, which consists of a collection of highly accurate **control points** (points with known positional accuracy), an ellipsoid model representing the earth's shape, and parameters defining the orientation and scale of the ellipsoid.

Surveyors determine the precise geodetic latitude and longitude of horizontal control points in a datum. You may have seen a horizontal control point **monument** (also called a **survey mark**, survey marker, or sometimes a geodetic mark), which is a fixed object established by surveyors when they determine the exact location of a point. A **benchmark** is a type of survey marker that indicates elevation. Monuments are often brass, bronze, or aluminum disks placed on top of rods sunk in the ground (as in figure 1.15), set in rock, or embedded in the tops of pillars, water towers, or even church spires so that they are permanent. Disturbing them is often against federal and state law.

From the 1920s to the early 1980s, these control points were surveyed relative to the Clarke 1866 ellipsoid, together forming the **North American Datum of 1927 (NAD27)**. Topographic maps, nautical and aeronautical charts, and many other large-scale maps of this period have graticule lines or ticks based on this datum. For example, the southeast corner of the Corvallis, Oregon, topographic map first published in 1969 (figure 1.16) has a NAD27 latitude and longitude of 44°30′ N, 123°15′ W.

By the early 1980s, better knowledge of the earth's shape and size and far better surveying methods led to the creation of a new horizontal reference frame, the **North American Datum of 1983 (NAD83)**. The NAD27 control points were corrected for surveying errors where required

Figure 1.15. Horizontal control point marker cemented in the ground. Courtesy of the US Geological Survey.

and then were added to thousands of more recently acquired points. The geodetic latitudes and longitudes of all these points were determined relative to the **Geodetic Reference System of 1980 (GRS80)** ellipsoid, which is essentially identical to the newer WGS84 ellipsoid.

Changing the horizontal reference frame from NAD27 to NAD83 meant that the geodetic latitude-longitude coordinates for control points across the continent changed slightly in 1983, and this change had to be shown on large-scale maps published earlier but still in use. On topographic maps, the NAD83 position of the map corner is shown by a dashed plus sign, as in figure 1.16. Often, the shift is in the 100-meter range and must be accounted for when plotting on older maps the geodetic latitudes and longitudes obtained from GNSS receivers and other modern position-finding devices.

There are three versions of NAD83, each serving specific purposes in geodetic surveying and mapping. For mapping and surveying applications, especially in North America, NAD83 (2011) is most used.

World Geodetic System of 1984

The **World Geodetic System of 1984 (WGS84)** was created by the US Department of Defense as a high-precision geodetic coordinate system for worldwide feature positioning, mapping, and navigation using GNSS technology. In figure 1.17, what is called the **WGS84 datum** is based on the WGS84 ellipsoid (refer to values of the equatorial and polar radii in table 1.3). The ellipsoid origin at the earth's center of mass is also the origin of a three-axis (x,y,z) Cartesian coordinate system that extends to the surface of the ellipsoid and beyond into outer space. This coordinate system allows you to determine the highly accurate (x,y,z) coordinates for the positions of features to be mapped in the earth's interior or underwater, on the land or water surface, and in the atmosphere. Geodesists then use an iterative mathematical procedure (the subject is beyond the scope of this book) to transform (x,y,z) coordinates on the surface of the WGS84 ellipsoid into geodetic latitude and longitude coordinates.

You may wonder how the geodetic latitude and longitude of a feature can be determined precisely when the earth's surface comprises several continental and oceanic tectonic plates that are slowly drifting apart, sliding past each other or coming together over time. Because the WGS84 geographic coordinate system is designed for worldwide positioning and mapping from GNSS satellite data, it could not be "locked" in place on a particular

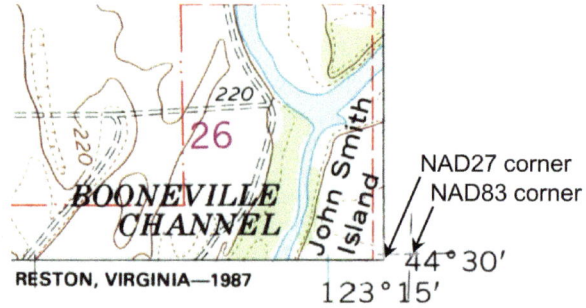

Figure 1.16. Bottom-right corner of a historical 1:24,000-scale USGS topographic map for Corvallis, Oregon, showing the difference between its NAD27 and NAD83 positions. Map courtesy of the US Geological Survey.

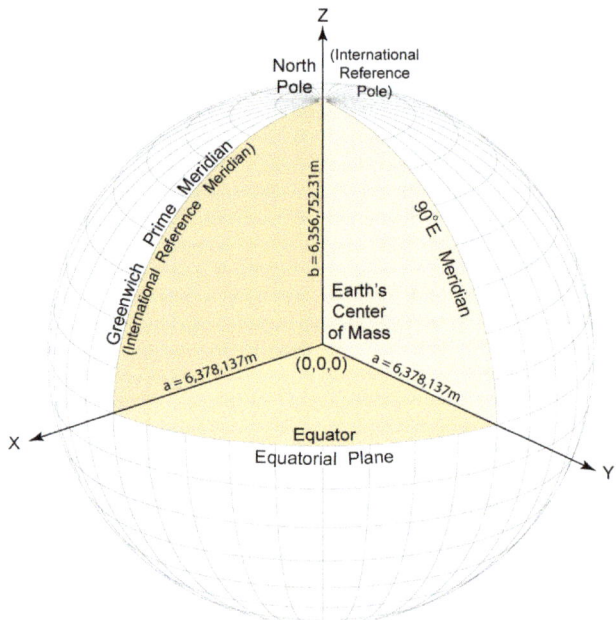

Figure 1.17. Definition of the coordinate system based on the WGS84 (x,y,z) datum.

tectonic plate. Geodesists analyze the relative rate of movement of the tectonic plates on the earth's surface and define an **international reference pole** (z-axis), **international reference meridian** (x-axis), and y-axis (figure 1.17) that are close to the geographic North Pole, the Greenwich meridian, and the 90° E meridian. We only show half of each axis in figure 1.17—to cover the entire earth, the z-axis line extends from the South to North Pole, the x-axis extends from the earth's center of mass outward to the 180° meridian, and the y-axis extends outward to the 90° W antipodal meridian. You can think of these three axis lines as defining a geodetic latitude and longitude graticule fixed in space, over which features on land and in the ocean slowly drift, changing their WGS84 latitude and longitude position slightly each year.

To give you an idea of how tectonic plates are moving relative to the international reference meridian and equator, in England near the international reference meridian all features were moving at a rate of about 2.5 cm (one inch) per year in a northeasterly direction. This movement translates into WGS84 geodetic latitudes and longitudes increasing by about 0.0005 and 0.001 seconds per year, respectively. This change in geodetic coordinates is extremely small, but in places such as Hawaii and Australia the rate of movement is about one meter per decade. Changes of this magnitude in geodetic latitude and longitude must be adjusted for in the WGS84 system because survey-grade GNSS receivers and other modern positioning technology can locate features to centimeter accuracy. Every few years, worldwide adjustments due to tectonic plate motion are used to update the **International Terrestrial Reference Frame (ITRF)** and thus the WGS84 geodetic latitudes and longitudes.

The ITRF consists of nearly 300 sites throughout the world. Highly accurate geodetic latitude and longitude coordinates are constantly determined at these sites from GNSS data and other high-precision global positioning information. Changes in these coordinates over time allow the velocity (amount and direction) of tectonic plate movement to be determined for each site. These velocities are used by geodesists to compute the change in latitude and longitude per year at the site so that the geographic coordinates of these sites can be updated annually. We currently use ITRF2020, with latitude and longitude coordinates as of 2020.

These tiny shifts in location over time throughout the earth are unlikely to affect your use or making of maps for all but the very largest map scales where extremely accurate map positions are important. However, it is still important to understand that the latitude and longitude of locations on the earth are not unchanging numerical values as they change ever so slightly over time.

The earth as a geoid

When the earth is treated as an authalic sphere or oblate ellipsoid, mountain ranges, ocean trenches, and other surface features that have vertical relief are neglected. There is justification for this treatment because the earth's surface is truly smooth when you compare the surface undulations with the 7,918-mile (12,742-kilometer) diameter of the earth based on the authalic sphere. The greatest relief variation is the approximately 12.3-mile (19.8-kilometer) difference between the summit of Mount Everest (29,035 feet or 8,852 meters) and the deepest point in the Mariana Trench (36,192 feet or 11,034 meters). This vertical difference is immense on the human scale, but it is only 1/640th of the earth's diameter. If you look at the difference between the earth's average land height (2,755 feet or 840 meters) and ocean depth (12,450 feet or 3,795 meters), the average roughness is only 1/2,750th of the diameter. In fact, if the earth were reduced to the diameter of a bowling ball, it would be smoother than the bowling ball.

The earth's global-scale smoothness aside, knowing the elevations and depths of features is important to us. Defining locations by their geodetic latitude, longitude, and elevation gives you a simple way to collect elevation data and display this information on maps. The top of Mount Everest, for example, is located at 27°59′ N, 86°56′ E, 29,035 feet (8,852 meters), but to what is this elevation relative? This question leads us to another approximation of the earth called the geoid, which is a surface of equal gravity used as the reference datum for elevations.

Vertical reference datums

Elevations and depths are measured relative to what is called a **vertical reference datum** (or **vertical datum**), an arbitrary surface with an elevation of zero. The traditional datum used for land elevations at individual survey points on aeronautical charts, topographic maps, engineering plans, and other large-scale maps is **mean sea level** (**MSL**, sometimes shortened to **sea level**) (see chapter 10 for more on mean sea level). Surveyors define mean sea level as the average of all low and high tides at a particular starting location over a **Metonic cycle** (a period of approximately 19 years or 235 lunar months, at the end of which the phases of the moon begin to occur in the same order and on the same days as in the previous cycle). Early surveyors chose this datum because of the measurement technology of the day. Surveyors first used the method of **leveling**, in which elevations are determined relative to the point at which mean sea level is defined, using horizontally aligned telescopes and vertically aligned leveling rods. A small circular monument was placed in the ground at each surveyed benchmark elevation point. Later, surveyors could determine elevation by making gravity measurements at different locations on the ground and relating them to the strength of gravity at the point used to define mean sea level. Gravity differences translate into elevation differences because the strength of gravity changes with elevation.

Mean sea level is easy to determine along coastlines, but what about inland locations? It requires extending MSL across the land. Imagine that MSL is extended across the continental land masses on the ellipsoidal surface, which is the same thing as extending a surface that has the same strength of gravity as MSL (figure 1.18). However, this imaginary equal-gravity surface does not form a perfect ellipsoid because differences in topography and varying densities of earth materials affect gravity's pull at different locations and thus the shape of the surface.

The slightly undulating, nearly ellipsoidal surface that best fits MSL for all the earth's oceans is called a global geoid. The **global geoid** varies approximately 100 meters (about 328 feet) above and below the oblate ellipsoid surface in an irregular fashion. World maps that show land **topography** and ocean **bathymetry** use land heights and water depths relative to the global geoid. In the conterminous United States, geoid heights range from a low of −39.1 meters on the Atlantic Seaboard to a high of −7.2 meters in the Rocky Mountains (figure 1.19). Worldwide, geoid heights vary from −105 meters (about 345 feet) just south of Sri Lanka to 85 meters (about 279 feet) in Indonesia.

The MSL datum based on the geoid is so convenient that it is used to determine elevations around the world, so it is the basis for the elevation data found on nearly all topographic maps and nautical charts. But the **local geoid** used in your area is probably slightly above or below (usually within two meters of) the global geoid used for world maps. This difference is due to the MSL at one or more nearby locations being used as the vertical reference datum for your nation or continent, rather than the average sea level for all the earth's oceans.

In the United States, for example, you may see elevations relative to the **National Geodetic Vertical Datum of 1929 (NGVD29)** on older topographic maps. This datum was defined by the observed heights of MSL at 26 tide gauges, 21 in the United States and 5 in Canada. It also was defined by the set of elevations of all benchmarks resulting from over 60,000 miles (96,560 kilometers) of leveling across the continent, totaling more than 500,000 vertical control points. In the late 1980s, surveyors adjusted the 1929 datum with new data to create the **North American Vertical Datum of 1988 (NAVD88)**. Topographic maps,

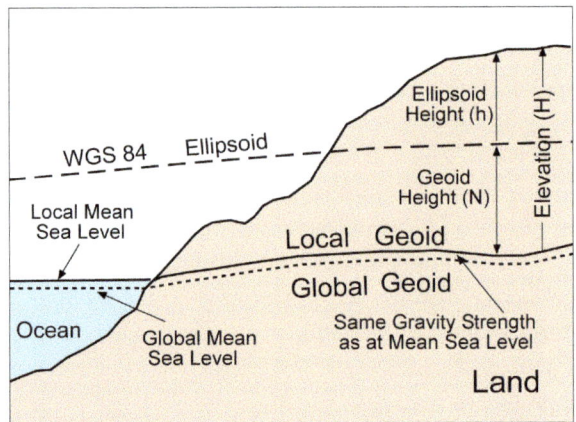

Figure 1.18. The geoid is the surface at which gravity is the same as at mean sea level. Elevations on maps are measured relative to the global geoid, but modern GNSS-determined heights are relative to the WGS84 ellipsoid.

nautical charts, and other cartographic products for the United States, Canada, Mexico, and Central America made from this time forward use elevations based on NAVD88. MSL for the continent was defined at one tidal station on the Saint Lawrence River at Rimouski, Quebec, Canada. NAVD88 was a necessary update of the 1929 vertical datum because about 400,000 miles (650,000 kilometers) of leveling was added to the NGVD since 1929. Additionally, numerous benchmarks were lost over the decades, and the elevations at others were affected by vertical changes because of rising land elevations that date from the retreat of glaciers at the end of the last ice age or subsidence from the extraction of natural resources such as oil and water.

GNSS has created a second option for measuring elevation rather than basing elevation on a vertical datum. Height measurements in GNSS receivers are often displayed as **height above the ellipsoid (HAE)**, or **ellipsoidal height** (h in figure 1.18), which is the height above or below a reference ellipsoid. The ellipsoids used by GNSS devices vary, but most currently use the WGS84 ellipsoid. HAE is the distance from a point on the earth's surface to the ellipsoid along a line perpendicular to the ellipsoid (see figure 1.18). The reference ellipsoid (WGS84) may be above or below the surface of the earth at a particular place. If the ellipsoid's surface is below the surface of the earth at the point, the ellipsoidal height has a positive sign; if the ellipsoid's surface is above the surface of the earth at the point, the ellipsoidal height has a negative sign.

Because ellipsoidal height is not a particularly useful measure, it is converted into **orthometric height** so that

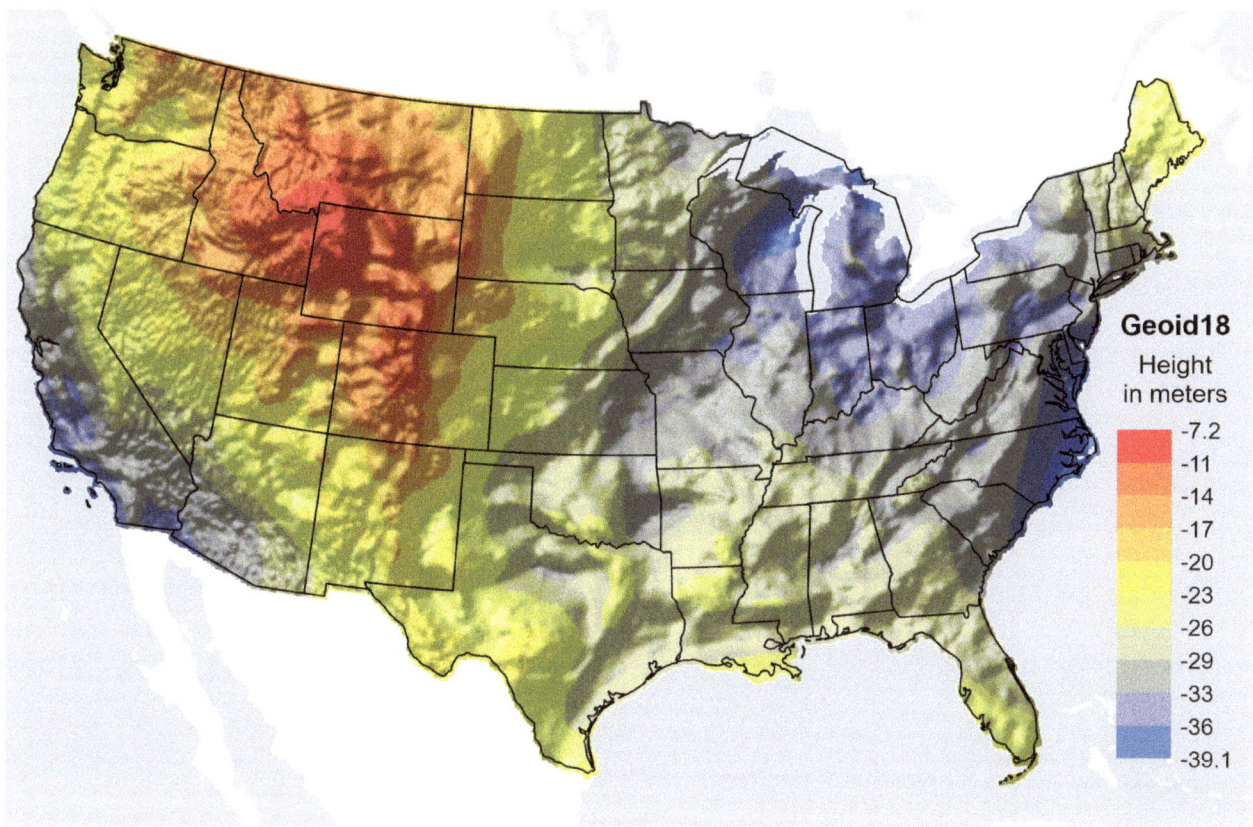

Figure 1.19. Geoid heights in the United States. Data courtesy of the National Oceanic and Atmospheric Administration, National Ocean Service, and National Geodetic Survey.

you can tell the actual elevation of a point. This is done using the geoid as the baseline to convert HAE into more user-friendly height measurements. First, the geoid height (N) is determined—this is the vertical distance from the ellipsoid to the geoid. Then the **geoid height** N is added to the ellipsoidal height h using equation (1.8) to get the orthometric height H:

$$H = h + N \quad (1.8)$$

This is what the GNSS device displays. Orthometric height is the type of elevation data that surveyors, engineers, hydrologists, and others who need precise elevation data use to work practically and accurately.

National Spatial Reference System

The **National Spatial Reference System (NSRS)** supplies defined latitude, longitude, height, and other authoritative geospatial values throughout the United States. Although the average person would likely not notice, the NSRS provides the foundation for nearly every aspect of our lives in which geographic position matters, such as navigation, emergency response, floodplain mapping, conservation, construction, agriculture, water resource management, transportation, and much more. The NSRS ensures consistent geographic coordinates across all locational applications in the country.

The NSRS is maintained by NOAA's **National Geodetic Survey (NGS)**, which provides the framework for all positioning activities in the nation with foundational latitude, longitude, elevation, and shoreline information. For more than two centuries, the NGS and its predecessor agencies have worked with private and public surveyors to place more than one million geodetic survey marks, such as the horizontal control point marker in figure 1.15, throughout and nearby the conterminous United States. Each survey mark is published with horizontal or vertical information for latitude, longitude, and elevation. These survey marks are at the heart of the NAD83 reference frame and NAVD88 vertical reference datum, which are the key components of the current NSRS.

Precise and highly accurate GNSS data from a permanent network of several thousand **continuously operating reference stations (CORS)** in the conterminous United States and adjacent areas augments the survey marks in the NSRS (figure 1.20). Real-time data is collected in intervals ranging from 1 to 30 seconds at stations that record either GPS data from the US-developed satellite navigation system or GNSS data from various global and regional satellite navigation systems.

Currently, the NSRS relies on a combination of more than 1,000 continuously operating reference stations and an aging network of survey marks. Although the reference stations are tracked daily for movement, the survey marks may become deteriorated, destroyed, or unmaintained, leading to outdated and inaccurate height and position data. This in turn degrades the accuracy of the NSRS. To modernize the NSRS, the existing NAV83 reference frames and NAVD88 vertical reference datum are being replaced with new versions that rely completely on GNSS reference station data. This modernization helps lessen problems due to deteriorating survey marks, in addition to increasing the precision and accuracy of newly surveyed positions.

NSRS modernization

To modernize the NSRS, the deficiencies of NAD83 and NAVD88—the horizontal and vertical datums of the NSRS—are best addressed by defining new horizontal and vertical datums. Specifically, NAD83 is misaligned to the earth's center by about 2.2 meters, and NAVD88 is both biased (by about one-half meter) and tilted (about one meter coast to coast) relative to the best global geoid models available today. By correcting these two issues, every latitude, longitude, ellipsoid height, and orthometric height reported in the current NSRS will change by as much as four meters in the modernized NSRS.

The effort to modernize the NSRS (**NSRS modernization**) requires that the NGS replace the three current NAD83 reference frames that are in use and all current

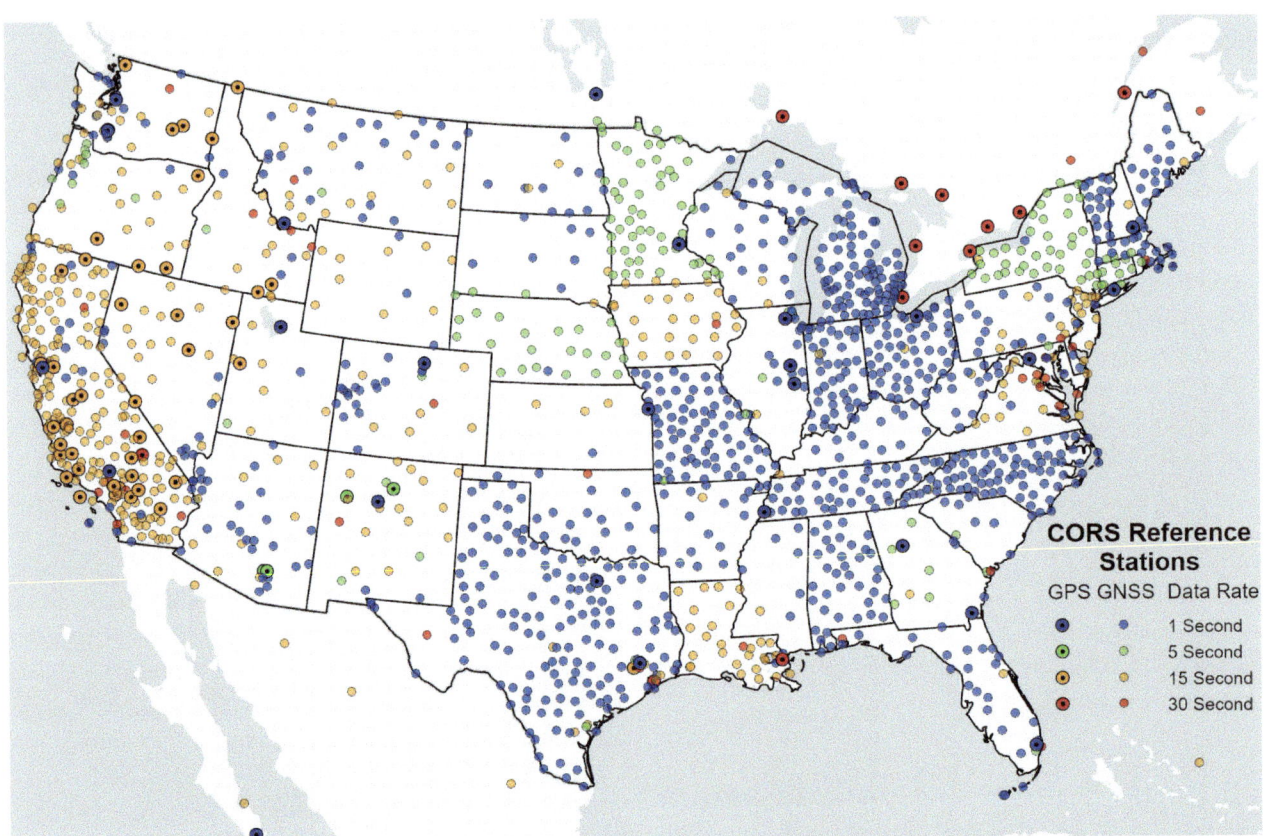

Figure 1.20. 2023 CORS reference stations for the conterminous United States. GPS or other GNSS station colors show the station's data collection rate from each second to every 30 seconds. Data courtesy of the National Oceanic and Atmospheric Administration, National Ocean Service, and National Geodetic Survey.

vertical datums, including NAVD88, with four new terrestrial reference frames and a new vertical datum. The modernized NSRS will align with both international standards as well as all GNSS and will offer improved horizontal and vertical accuracy, easier maintenance and updates, better response to events such as earthquakes and cyclones, and better planning for sea level rise and subsidence from oil and gas withdrawal.

Selected readings

Altamimi, Z., P. Rebischung, X. Collilieux, L. Métivier, and K. Chanard. 2023. "ITRF2020: An Augmented Reference Frame Refining the Modeling of Nonlinear Station Motions." *Journal of Geodesy* 97 (47). DOI: https://doi.org/10.1007/s00190-023-01738-w.

Amin, H., L. E. Sjöberg, and M. Bagherbandi. 2019. "A Global Vertical Datum Defined by the Conventional Geoid Potential and the Earth Ellipsoid Parameters." *Journal of Geodesy* 93: 1943–61.

Greenberg, J. 1995. "Isaac Newton and the Problem of the Earth's Shape." *Archive for History of Exact Sciences* 49 (4): 371–91.

Hoare, M. R. 2005. *The Quest for the True Figure of the Earth*. Routledge.

Iliffe, J. C. 2000. *Datums and Map Projections for Remote Sensing, GIS and Surveying*. Whittles.

Kessler, F. 2006. "Working with Projections and Datum Transformations in ArcGIS: Theory and Practical Examples by Werner Flacke and Birgit Kraus." *Cartography and Geographic Information Science* 33 (3): 233–37. https://doi.org/10.1559/152304006779077282

Kessler, F. 2022. "Vertical (Geopotential) Datums." In *The Geographic Information Science and Technology Body of Knowledge*, edited by John P. Wilson. Association of American Geographers. https://gistbok-ltb.ucgis.org/30/concept/9401

Maling, D. H. 2013. *Coordinate Systems and Map Projections*. Pergamon Press.

Maupertuis, P. L. 1738. *The Figure of the Earth Determined from Observations Made by Order of the French King at the Polar Circle*. Translated from French. Gallica Digital Library.

Meyer, T. 2021. "Earth's Shape, Sea Level, and the Geoid." In *The Geographic Information Science & Technology Body of Knowledge*, edited by John P. Wilson. https://gistbok-topics.ucgis.org/DM-05-044.

Mugnier, C. J. 2022. *Coordinate Systems of the World: Datums and Grids*. CRC Press.

National Oceanic and Atmospheric Administration (NOAA). 2024. "What Is the National Spatial Reference System?" https://oceanservice.noaa.gov/facts/nsrs.html.

National Oceanic and Atmospheric Administration (NOAA). 2000. *Tidal Datums and Their Applications*. NOAA Special Publication NOS CO-OPS 1. https://tidesandcurrents.noaa.gov/publications/tidal_datums_and_their_applications.pdf.

Pearson, C. 2005. "National Spatial Reference System Readjustment of NAD 83." National Oceanic and Atmospheric Administration (NOAA). www.ngs.noaa.gov/NationalReadjustment/Items/CFP.The%20national%20readjustment-1.html.

Smith, D. A., D. R. Roman, and V. A. Childers. 2015. "Modernizing the Datums of the National Spatial Reference System." *Marine Technology Society Journal* 49 (2): 151–58.

Torge, W., and J. Muller. 2012. *Geodesy*, 4th ed. De Gruyter.

Van Sickle, J. 2010. *Basic GIS Coordinates*, 3rd ed. CRC Press.

CHAPTER 2
Map scale

Maps are always smaller in size than the environment they represent. The amount of size reduction is known as the **map scale**, which tells you the relationship between distances on the map and their corresponding ground distances. To use maps effectively, you must convert measurements from map units to ground units (inches to miles, for example). As you might expect, an understanding of map scale is central to performing this task. In this chapter, we explore what you need to know about map scale to become a skilled map user and make your own maps.

Expressing scale

Map scale is almost always given in the form: "This distance on the map represents this distance on the earth's surface." The relationship between map and ground distance can be expressed in a number of ways—most commonly as a representative fraction, word scale, or scale bar.

Representative fraction

A common way to describe scale is to use a **representative fraction (RF)**, which is the ratio between map and ground distance. The RF simplifies calculations involving scale. An RF is written as either $1/x$ or $1:x$. The numerator is always 1 and represents map distance, and the denominator x indicates distance on the ground in the same unit of measurement as the distance on the map. Consider equations (2.1) and (2.2):

$$\frac{1}{x} = \frac{\text{map distance}}{\text{ground distance}} \quad (2.1)$$

$$\text{scale denominator } x = \frac{\text{ground distance}}{\text{map distance}} \quad (2.2)$$

The advantage of having identical units on the top and bottom of the fraction is that map measurements can be made in centimeters, inches, or whatever distance unit you choose. For example, an RF of 1/24,000 or 1:24,000 means that one inch on the map represents 24,000 inches on the ground, one centimeter on the map represents 24,000 centimeters on the ground, and so on. You can also say that the scale reduction is 24,000 to 1. If you have ever collected components of HO-gauge model railways, the tracks, engines, and boxcars are all scaled so that they are a size that is 1/87th of real-world components.

Sometimes a map is described by the denominator of the RF. For example, you might hear someone refer to a 24,000-scale map. This is sometimes shortened to a 24K map, where K stands for 1,000.

Word scale

Another familiar way to express scale is to use a descriptive **word scale**. The word scale expresses large RF denominator values in more familiar (larger) units of measurement. We express the scale in words as so many "inches to one (statute) mile" (see chapter 1 for more on statute miles) or "centimeters to one kilometer." An RF of 1:24,000 can be expressed using a word scale as "1 inch to 2,000 feet" because 2,000 feet equals 24,000 inches, and an RF of 1:100,000 can be expressed as "one centimeter to one kilometer" because there are 100,000 centimeters in a kilometer. These expressions are also sometimes seen as "inches to the mile" or "centimeters to the kilometer" statements. Word scales can also be expressed as so many "miles to one inch" or "kilometers to one centimeter." Word scales for commonly used map scales are given in table 2.1, along with examples of maps published at each scale.

At first, it may be confusing to find that one map indicates scale as "one centimeter to one kilometer" and another as "four miles to one inch." This lack of consistency

Table 2.1. Common ways of expressing map scale for map series for three US national mapping agencies: The US Geological Survey (USGS), National Oceanic and Atmospheric Administration (NOAA), and Federal Aviation Administration (FAA)

RF	Word scale	US government map series at this scale
1:2,500	1 centimeter to 25 meters	NOAA Electronic Navigational Charts (ENC) berthing charts
1:5,000	1 centimeter to 50 meters	NOAA ENC berthing charts
1:10,000	10 centimeters to 1 kilometer; ~6¼ inches to 1 mile	NOAA ENC harbor charts
1:20,000	5 centimeters to 1 kilometer; ~3 inches to 1 mile	Current USGS topographic (topo) maps for Puerto Rico; NOAA ENC harbor charts
1:24,000	1 inch to 2,000 feet	Current USGS topo maps; USGS OnDemand topo maps; USGS Historic Topographic Map Collection (HTMC) maps
1:25,000	~2½ inches to 1 mile; 4 centimeters to 1 kilometer	Current USGS topo maps for Alaska; USGS HTMC maps for Alaska
1:40,000	1 centimeter to 400 meters	NOAA ENC approach charts
1:48,000	1 inch to 4,000 feet	USGS HTMC maps
1:62,500	~1 inch to 1 mile	USGS HTMC maps
1:63,360	1 inch to 1 mile	USGS HTMC maps
1:80,000	1 centimeter to 800 meters	NOAA ENC approach charts
1:100,000	~1 inch to 1½ miles; 1 centimeter to 1 kilometer	USGS OnDemand topo maps; USGS HTMC maps
1:160,000	~1 inch to 2½ miles; 1 centimeter to 1.6 kilometers	NOAA ENC coastal charts
1:250,000	~1 inch to 4 miles; 1 centimeter to 2.5 kilometers	USGS HTMC maps; FAA terminal area charts
1:320,000	~1 inch to 5 miles; 1 centimeter to 3.2 kilometers	NOAA ENC coastal charts
1:500,000	1 centimeter to 5 kilometers	FAA sectional aeronautical charts
1:640,000	~1 inch to 10 miles; 1 centimeter to 6.4 kilometers	NOAA ENC general charts
1:1,000,000	~16 miles to 1 inch; 1 centimeter to 10 kilometers	FAA world aeronautical charts
1:1,280,000	~20 miles to 1 inch; 1 centimeter to 12.8 kilometers	NOAA ENC general charts
1:2,560,000	~40 miles to 1 inch; 1 centimeter to 25.6 kilometers	NOAA ENC overview charts
1:5,120,000	~80 miles to 1 inch; 1 centimeter to 51.2 kilometers	NOAA ENC overview charts

should cause little trouble, however, because the smaller unit of measurement (inches or centimeters) always refers to the map whereas the larger measurement unit refers to the ground.

Many word scales are only close approximations to the RF (shown by the ~ sign in table 2.1) and should not be used in exact mathematical calculations. An RF of 1:250,000 is really 3.9457 miles to 1 inch, but the word scale is much easier to remember if it is rounded to four miles. The reason the values can be rounded is because of the way they are used. Word scales are often used so map readers with rulers can determine the length of a feature. Because the divisions of a ruler are fairly large, to state the map scale using multiple decimal places is no more useful than stating the rounded values. However, if you want to indicate more precision, you can use more decimal places in the word scale.

Scale bar

A third way to show map scale is to use a **scale bar** (also called a **bar scale**). A scale bar looks like a small ruler that can be placed on the map to assist with map distance measurements. You will usually read the scale bar from left to right, beginning at zero. Sometimes the scale bar is extended to the left of the zero point, using smaller divisions than on the right (figure 2.1). This **scale bar extension** allows you to determine distance not only in whole units but also in fractions of units, such as 10ths of a mile or kilometer.

The divisions of the scale bar are arranged to provide whole numbers or fractions of miles or kilometers of ground distance. This method of division means that the divisions do not necessarily represent whole numbers of centimeters or inches—some fraction will almost always be left over. In other words, although the scale bar looks like a ruler, its divisions do not coincide with those on your ruler. Rare exceptions are a scale of 1:100,000, because at this scale one centimeter on the map is exactly one kilometer on the ground, and 1:63,360, because one inch on the map is exactly 63,360 inches, or one mile, on the ground.

The scale bar has three features that make it especially useful. First, if the map is enlarged or reduced because of photocopying or screen display, the scale bar changes size in direct proportion to the physical size of the map. However, the word scale and representative fraction will be incorrect when the map size changes. Second, both kilometers and miles (or other units) can be shown conveniently on the same scale bar (see the bottom example in figure 2.1). This is called a **stacked scale bar** because the two scale bars using different units are stacked on top of each other. And finally, the scale bar is easy to use when figuring distances along a straight or curved line on a map.

On one special map, the standard scale bar is sometimes replaced with a **variable scale bar**. Figure 2.2 shows an example of this type of scale bar. It is only appropriate to use a variable scale bar on a map in which scale varies systematically in the north–south direction (along the parallels) and the projection is conformal so that distortion in

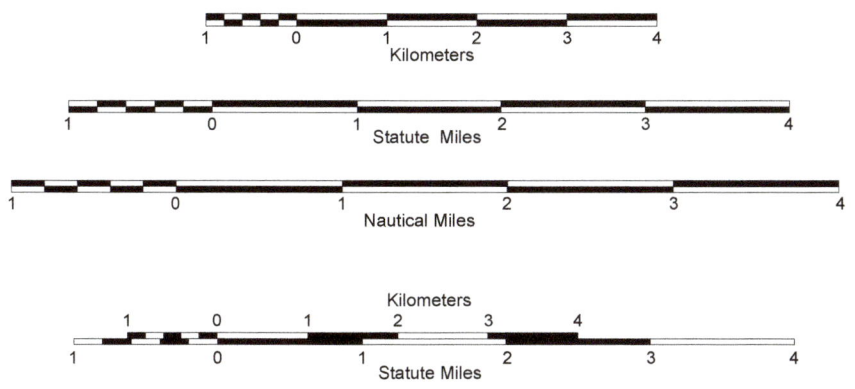

Figure 2.1. Scale bars in kilometers, statute miles, and nautical miles taken from maps identical in scale, as well as a stacked scale bar that shows both kilometers and miles. The scale bar extensions to the left of the zero point are used when making more precise distance measurements. Statute and nautical miles are discussed in chapter 1.

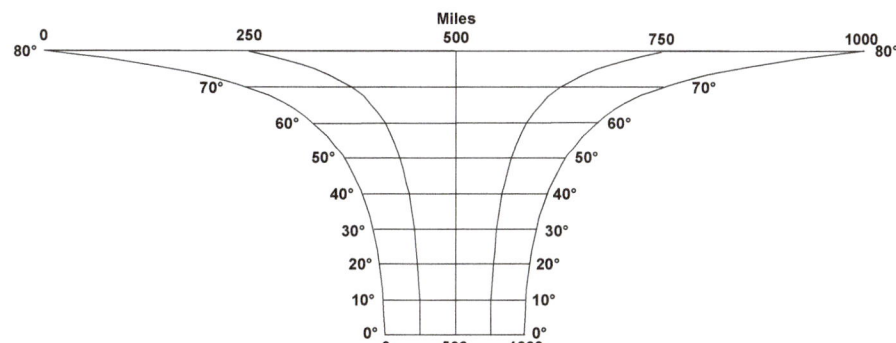

Figure 2.2. A variable scale bar might be seen correctly on a map using the Mercator projection.

any direction is the same at a given latitude (see chapter 3 for more on map projections). The only map that meets these qualifications is a map made using the Mercator projection. If you have seen a variable scale bar on a map with any other type of map projection, it was added to the map incorrectly.

When you use a variable scale bar, you are working with a scale bar that is stretched to match the local map scale at different latitudes. First, decide at what latitude you want to use the variable scale bar, and then find the scale bar for that latitude on the variable scale bar. You can do this for latitudes north or south of the equator. If the latitude falls between two of the scale bars, you can visualize or physically add a scale bar for the latitude as a horizontal line that you place in the correct vertical position on the variable scale bar. For example, you can draw the scale bar for 45° latitude as a horizontal line halfway between the 40° and 50° scale bars. You then use the scale bar as a ruler to measure in the east–west direction.

Large- and small-scale maps

Because map distance is always stated in the numerator of the RF as 1, it follows that the smaller the denominator, the closer to a 1:1 ratio and, consequently, the larger the scale will be. Thus, a map scale of 1:20,000 is twice as large as a scale of 1:40,000. If that sounds backward, it will help to remember that the terms *large scale* and *small scale* come from the numeric value of the representative fraction 1/*x*. The number 1/1,200 is much larger than 1/100,000,000. Thus, a USGS topographic map (described below) at a scale of 1:24,000 is a **large-scale map**, in which small areas are shown in great detail, and a world map at a scale of 1:100,000,000 is a **small-scale map**.

Although it is common to classify maps by their scales, there is no general agreement on where the class limits should be set for large-, medium-, and small-scale maps. One reason for dividing maps into large and small scale is that you can use large-scale maps to make accurate distance and area measurements, but you cannot do so on small-scale maps. Thus, if you sort maps into two groups—large- and small-scale—1:500,000 is a feasible dividing point. World atlas and wall maps of continental coverage then fall into the small-scale group, whereas topographic maps, city street guides, and other detailed maps of small areas are in the large-scale class.

If a more detailed three-way grouping is used, maps with scales of 1:1,000,000 and smaller can be classed as small-scale, and maps with scales of 1:250,000 and larger can fall in the large-scale class. Maps that range in scale between these extremes can then be referred to as **medium-scale maps**. Any such classification of maps by scale, of course, is arbitrary and should not be given meaning beyond the organizational convenience that it provides.

When you use a map, be sure to note whether it is a large-, medium-, or small-scale product. Check that the features of interest are displayed at the correct scale for your purposes. If you want to study a small ground area in great detail with little generalization of features (see chapter 6 for more on generalization), you need a large-scale map. When you are required to make accurate distance, direction, and area measurements, you should use only large-scale maps. The change in scale across the surface of a large-scale map is negligible, so you can trust the map as a geometrically exact representation of the small piece of the earth it covers. If you are more interested in a generalized presentation of a large area, such as a state, country, continent, or the entire earth, choose a small-scale map. The scale changes continuously across small-scale maps, so the RF or scale bar printed on these maps gives the scale at a particular point or along a given line but not for the entire map (unless, as noted, a variable scale bar appears on a map that uses the Mercator projection).

Topographic map series at different scales

To better understand map scale, we can look at maps that have been designed for a series of scales. The creation of such map series is often carried out by **national mapping agencies (NMAs)** that produce maps and geographic information for their country. Table 2.1 shows standard map series produced by three NMAs in the United States. Many NMAs in other countries have produced similar series.

In this section, we focus on one of the map series of the NMAs included in table 2.1—the topographic map series produced by the **USGS**, whose mission is to map the nation and survey its resources. Topographic maps, or **topo maps** for short, use contours to show the elevation of the earth's surface, revealing its shape (see the introduction). The USGS historically produced three standard national topographic map series that covered the **conterminous United States** (the Lower 48). Most of these historical maps can be found in the **Historical Topographic Map Collection (HTMC)**. The HTMC includes all editions of the 1:250,000 scale and larger topographic quadrangle maps originally published as paper maps in the period 1884–2006. The

Chapter 2: Map scale   47

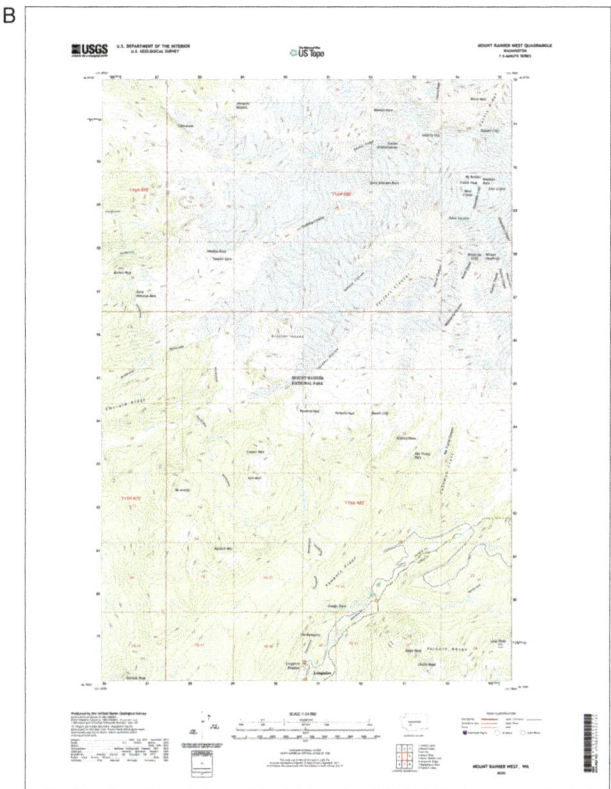

Figure 2.3. The current USGS topo map (*B*) has the same familiar look and map extent as a historical paper map (*A*), and there is also a version with an orthoimage (*C*). Courtesy of the US Geological Survey.

scanned maps in this collection can be viewed online or downloaded as digital georeferenced images.

The current USGS topographic map series for the nation, called **US Topo (UST)**, contains maps at a scale of 1:24,000 for the conterminous US, 1:25,000 for Alaska, and 1:20,000 for Puerto Rico. These maps, which date from 2009 and later, have the familiar look and map extents of the historical paper maps, but they also include a version with high-resolution color imagery in the background (figure 2.3). The production cycle is roughly every three years, so the map content is continuously updated and enriched. These topo maps are available in paper or digital form directly from the USGS and from private vendors.

With the USGS TopoBuilder web app, you can customize the map to show the features, as well as the backgrounds (such as color imagery or shaded relief, described in chapter 10), that you want on the map. You can also customize the area shown by changing the center of the map. This solves the age-old problem of your area of interest falling at the corner of four map sheets. The on-demand topo maps are available for digital download at scales of 1:24,000 (figure 2.4A) and 1:100,000 (figure 2.4B).

In the USGS topo map series (figure 2.5), the 1:250,000 scale topographic series is the smallest in scale, at about one inch to four miles (1 centimeter to 2.5 kilometers). Each of the 489 maps in the series is a 1° × 2° quadrilateral, called a **quadrangle**, that is given the name of a prominent town or physical feature in the mapped area. Washington state, for example, is covered by 18 quadrangles (figure 2.5, *top left*) placed at one-degree latitude increments (46° N, 47° N, and so on) and two-degree longitude increments of (118° W, 120° W, and so on). For example, the Yakima map sheet (figure 2.5, *top right*) covers an area from 46° to 47° N and 120° to 122° W.

The second USGS topographic map series is at the 1:100,000 scale (approximately 1 inch to 1.5 miles or exactly 1 centimeter to 1 kilometer). The nearly 2,000 maps in this series are quadrangles with parallels of latitude spaced 30 minutes apart and meridians of longitude spaced at 60-minute (one-degree) increments. The 52 map sheets that cover the state of Washington are shown in the middle-left index map in figure 2.5. Each map is named in a similar manner as the 1:250,000-scale series on the basis of a prominent geographic feature. Four 1:100,000-scale maps nest in each quadrangle of the 1:250,000-scale map series (for example, Mount Rainier, Yakima, Mount Adams, and Toppenish are quarters of the Yakima 1:250,000-scale map). Also, the Mount Rainier 1:100,000-scale map sheet centered on the peak is a more detailed representation of

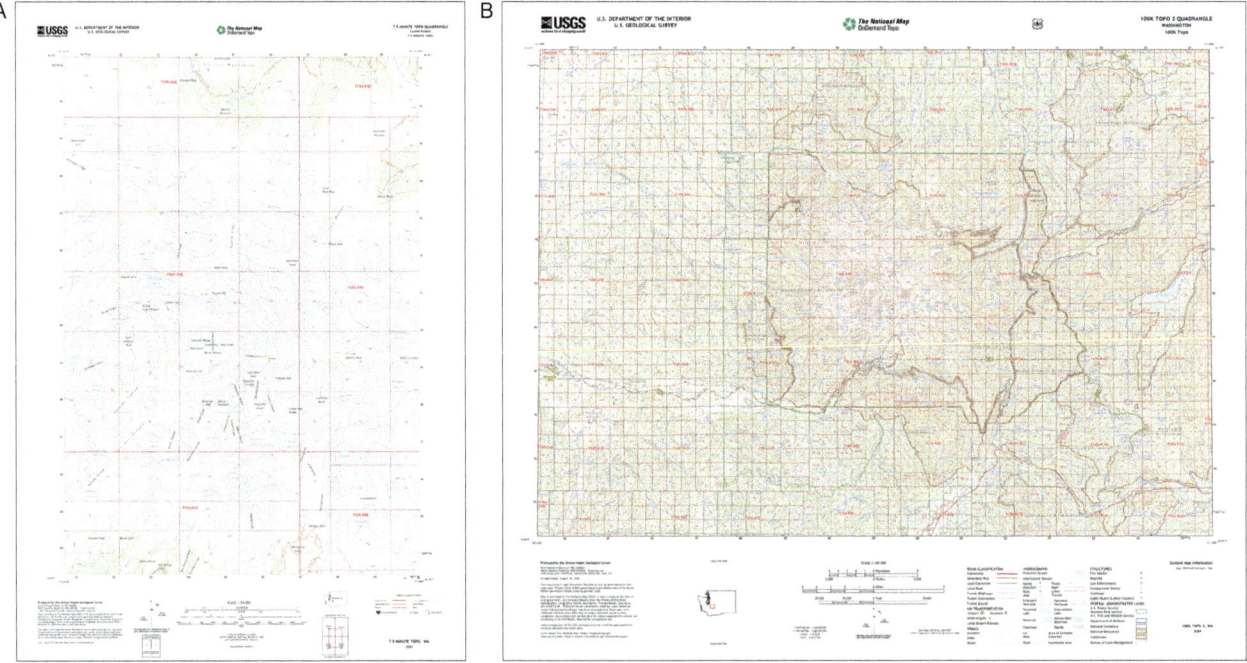

Figure 2.4. On-demand topographic maps at scales of 1:24,000 (*A*) and 1:100,000 (*B*) can be customized to show the map area, features, and background you want. Courtesy of the US Geological Survey.

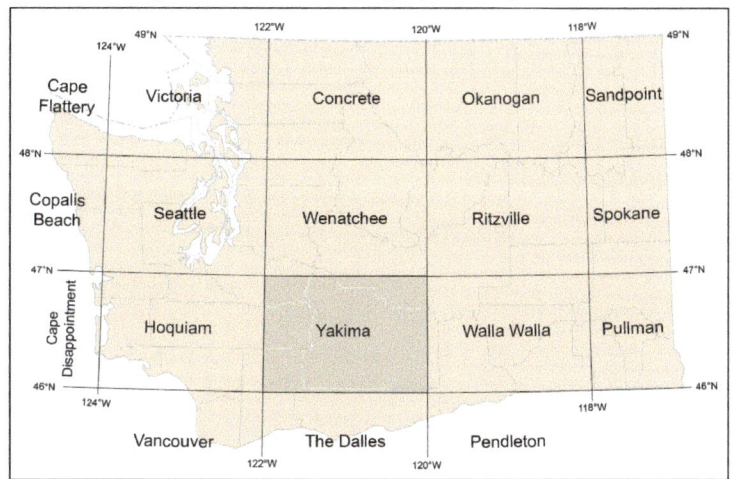
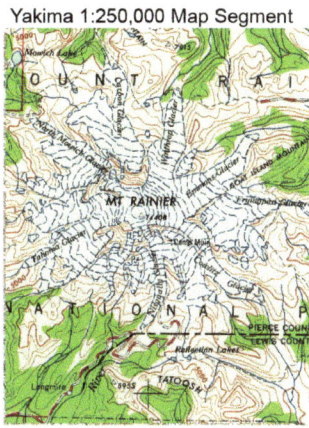
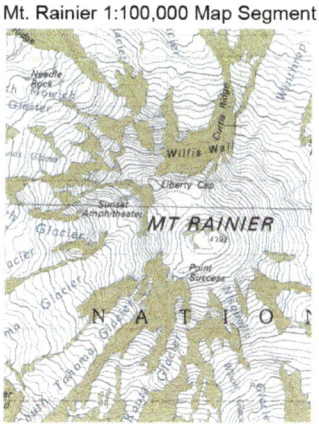

Figure 2.5. Index maps showing the 1:250,000-scale, 1:100,000-scale, and part of the 1:24,000-scale USGS topographic maps that cover the state of Washington. Representative map segments showing the summit of Mount Rainier at the three scales are on the right of each index map. Topographic map segments courtesy of the US Geological Survey.

the area than the 1:250,000-scale map, showing the names and locations of glaciers and other physical features on the mountain.

The largest scale and best-known USGS map series is at the 1:24,000 scale (1 inch to 2,000 feet or 1 centimeter to 240 meters). Nearly 57,000 topographic maps cover the conterminous United States, Hawaii, and parts of Alaska, whereas the rest of Alaska is mapped at 1:25,000 scale (~2.5 inches to 1 mile or 4 centimeters to 1 kilometer). Each 1:24,000 scale map is a quadrangle bounded by parallels of latitude and meridians of longitude that are spaced 7.5 minutes apart (1/64th of a 1° × 1° quadrilateral). As such, these are often called **7.5-minute maps** or **quads**. (In common use, all these terms mean the same thing: 7.5-minute quadrangle or quad or map; 1:24,000 quadrangle or quad or map; and 24K quadrangle or quad or map). Each map is named in a similar manner to the smaller scale map series. The bottom-left index map in figure 2.5 shows that 32 1:24,000-scale topographic maps are nested within a 1:100,000-scale map sheet. The Mount Rainier West quadrangle, a portion of which is shown in figure 2.5, shows a highly detailed representation of Mount Rainier's peak.

You can use the areas of quadrilaterals in appendix table A.4 to find the number of square miles or kilometers covered by topographic maps at different latitudes. For example, at the southern tip of Florida, near 25° N, a 7.5-minute quadrangle covers an area of about 67 square miles (174 square kilometers). Each quadrangle covers 49 square miles (127 square kilometers) at the Canadian border along the 49° N parallel. The Mount Rainier West quadrangle covers an area of approximately 51 square miles (131 square kilometers).

Multiscale maps

The maps described in the previous section were designed and printed for a single scale. If you see them online, you can zoom in to see an enlargement, but the map itself does not change. This is different from the **"slippy" web maps** that allow you to pan and zoom. These **multiscale maps** have varying **levels of detail (LODs)** at a range of scales. They are designed to be visually continuous as you zoom in and out. Subtle modifications between scales assure that you are not distracted by abrupt changes in the display as you zoom. Most multiscale maps are composed of up to 23 layers, each with a different LOD. The LOD relates to the map scale at the equator for each **zoom level**, from level 0 at the smallest map scale of 1:591,657,527.59 to level 23 at a map scale of 1:70.53 for raster data served as image tile layers. The zoom level for **vector tile layers** (with vector representation of data as points, lines, and polygons) is one less than for **image tile layers** (with raster representation of data as grids of cells, also known as pixels). For example, the map scale for level 0 is 1:295,828,763.80, and for level 23 it is 1:35.27. These map scales are associated with

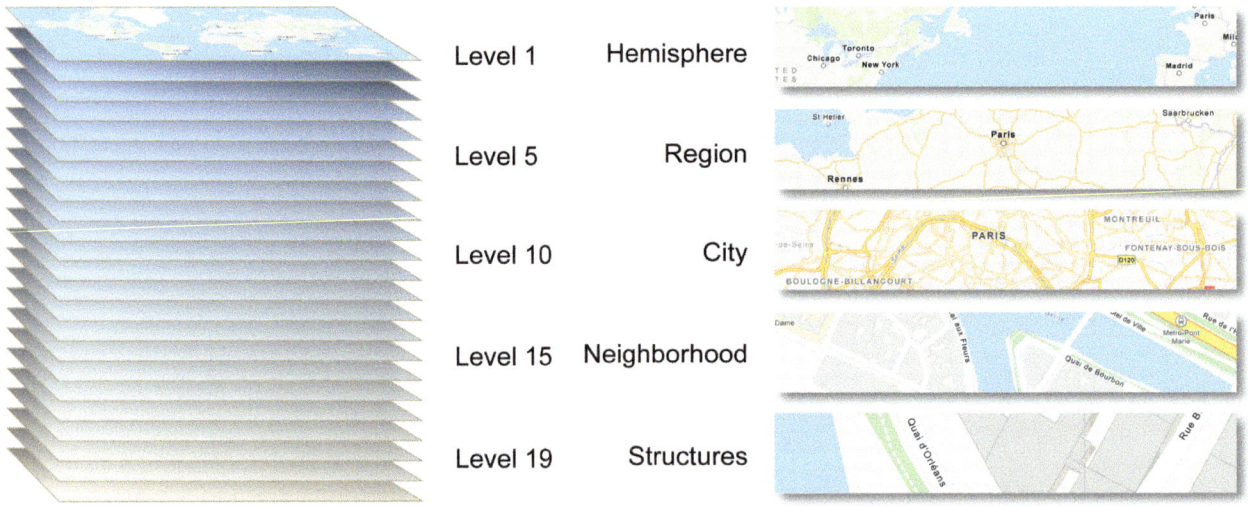

Figure 2.6. Multiscale maps can consist of up to 23 separate layers, with the scale doubling at each larger level, allowing a single web map to be zoomed from a world view to individual buildings.

the zoom levels for web maps that use the standard web Mercator projection and tiling scheme (see appendix table A.5), and they support mapping from the global scale for the entire earth to a very local scale for the area of a single structure or property (figure 2.6).

It is not only the creation of web maps that differs from paper maps. You must also consider their use and misuse: for example, to measure areas and distances. First, determining the map scale of the level displayed on a web map is often difficult because the zoom level and map scale are not usually shown. In addition, web maps often use the web Mercator projection (discussed in detail in chapter 3). As a Mercator projection, scale varies systematically from north to south, as mentioned earlier in this chapter, so that variations in measured distances and areas must also be considered. Although these limitations make measurements difficult on web maps, the apps in which they appear often provide tools to correctly measure distances and areas.

Converting scale

The scale depicted on the map you are using may be in the wrong form to best serve your purpose. Therefore, you may need to make conversions between a word scale, RF, and scale bar. **Scale conversions** are based on applying common distance equivalents such as 1 foot = 12 inches, 1 statute mile = 5,280 feet, 1 statute mile = 63,360 inches, or 1 kilometer = 100,000 centimeters. Appendix table A.1 provides a convenient list of Imperial and metric distance equivalents.

Converting a word scale to an RF

You may have a map with a word scale and want to know the RF. The first thing to remember with any scale conversion is that the ratio $1/x$ is always map distance (numerator) to ground distance (denominator). Suppose that the word scale is "3 inches to 10 miles." In converting to an RF, the ratio is shown in equation (2.3).

$$\frac{1}{x} = \frac{3 \text{ inches}}{10 \text{ miles}} \quad (2.3)$$

Because you cannot have an RF with different units in the numerator and denominator, you must convert miles to inches—the math is no problem if you remember that there are 63,360 inches in a statute mile. So, you have equation (2.4).

$$\frac{1}{x} = \frac{3 \text{ inches}}{10 \text{ miles} \times \frac{63,360 \text{ inches}}{1 \text{ mile}}} \quad (2.4)$$

The miles in the denominator cancel out, as do the remaining inches in the numerator and denominator, leaving equation (2.5).

$$\frac{1}{x} = \frac{3}{633,600} \quad (2.5)$$

Remember that the numerator of an RF is always 1. In this case, you must also reduce the numerator from 3 to 1 by dividing the numerator and denominator by 3. To solve for x, you invert the fractions to get the correct map RF of 1:211,200, as shown in equation (2.6).

$$x = \frac{633,600}{3}$$

$$x = 211,200 \quad (2.6)$$

The conversions are even easier with metric units. If the word scale is "4 centimeters to 1 kilometer," equation (2.7) would be as follows.

$$\frac{1}{x} = \frac{4 \text{ centimeters}}{1 \text{ kilometer}} \quad (2.7)$$

To convert kilometers to centimeters, first convert the denominator knowing that there are 100,000 centimeters in a kilometer, as in equation (2.8).

$$\frac{1}{x} = \frac{4 \text{ centimeters}}{1 \text{ kilometer} \times \frac{100,000 \text{ centimeters}}{1 \text{ kilometer}}} \quad (2.8)$$

In this case, the kilometers in the denominator cancel out, as do the remaining centimeters in the numerator and the denominator, leaving equation (2.9).

$$\frac{1}{x} = \frac{4}{100,000} \quad (2.9)$$

To solve for x, invert the fraction and divide 100,000 by 4 to get the correct RF of 1:25,000, as shown in equation (2.10).

$$x = \frac{100,000}{4}$$

$$x = 25,000 \quad (2.10)$$

Converting an RF to a word scale

Sometimes you may find yourself in the opposite situation. You know the RF but want to know what the map scale represents in terms of "inches to a mile" (or "miles to an inch") or "centimeters to a kilometer" (or "kilometers to a centimeter"). If it is a "miles to an inch" word scale that you want, merely divide the number of inches in a statute mile, or 63,360, by the denominator of the RF.

The conversion of an RF of 1:200,000 to an "inches to a mile" word scale is shown in equation (2.11).

$$\frac{63,360 \text{ inches/mile}}{200,000}$$

$$\frac{63,360}{200,000} \text{ inches} = 1 \text{ mile}$$

$$0.32 \text{ inches} = 1 \text{ mile} \quad (2.11)$$

If it is a "miles to an inch" word scale that you want, divide the denominator of the RF by the number of inches in a mile, or 63,360. For example, to convert an RF of 1:200,000 to a "miles to an inch" word scale, use equation (2.12).

$$\frac{200,000}{63,360 \text{ inches/mile}}$$

$$1 \text{ inch} = \frac{200,000}{63,360} \text{ miles}$$

$$1 \text{ inch} = 3.16 \text{ miles} \quad (2.12)$$

As noted earlier, this number is often rounded off so that the word scale is "3 miles to 1 inch."

To convert an RF of 1:200,000 to a word scale of "centimeters to a kilometer," use equation (2.13).

$$\frac{100,000 \text{ centimeters/kilometer}}{200,000}$$

$$\frac{100,000 \text{ centimeters}}{200,000} = 1 \text{ kilometer}$$

$$0.50 \text{ centimeters} = 1 \text{ kilometer} \quad (2.13)$$

This calculation gives a word scale of "0.5 centimeters to 1 kilometer."

If you want a "kilometers to one centimeter" word scale, divide the denominator of the RF by 100,000. To convert an RF of 1:200,000 to a word scale of "kilometers to a centimeter," use equation (2.14).

$$\frac{200,000}{100,000 \text{ centimeters/kilometer}}$$

$$\frac{200,000}{100,000} \text{ kilometers} = 1 \text{ centimeter}$$

$$2 \text{ kilometers} = 1 \text{ centimeter} \quad (2.14)$$

The word scale is then "2 kilometers to 1 centimeter."

Sometimes you may want to create a scale bar from a word scale, or an RF. Imagine that you have a map at a scale of three-quarters of a mile to an inch, or 1:47,520. A map at this scale is considered a large-scale map, so your scale bar should have one-mile (or one-kilometer) increments because it will facilitate ground distance measurements. Because an inch represents 3/4 mile, each mile covers 4/3 inches on the map. So, you mark off 4/3-inch intervals on your scale bar to indicate the mile increments. To make it more useful universally, you can also show kilometers.

Determining map scale

No matter what sort of map scale (word scale, RF, or scale bar) you encounter, you can change it to the type of scale you want. But what if you come across a map that has no scale depicted at all or an RF or word scale that seems incorrect because the map appears to have been enlarged or reduced after it was created? Lack of scales on maps or incorrect scales occur more often than you might expect. For instance, you may want to know the scale of a photocopied segment of a map in which no scale is shown. Digital maps that you copy from the internet are another good example, because they often are scans of original maps that do not include an RF or a scale bar. Even if the RF is visible, it may not be correct because the map is displayed at a different pixel density than when it was created. For example, a map created at 150 pixels per inch but displayed on your computer monitor at a default density of 72 pixels per inch is slightly more than twice as large in scale as the RF indicates.

You can figure out the map scale if you know the ground distance between any two points on the map. Simply measure the distance between these two points on the map, whether the map is printed or displayed on your computer screen. The ratio of map to ground distance, in the same units of measurement, is the map's scale. But how do you find the ground distance between the two points? One method is to use some ground feature whose length is known, as described in the next section. Keep in mind that scale may vary across the extent of the map, so for small-scale maps in particular, the scale you come up with is only accurate around the ground feature that you selected.

Determining map scale from a known ground feature

Some ground features have standard lengths. If you can identify one of them on your map, you can easily figure out the map scale. A regulation US football field, for example, is 100 yards long. If the map distance of the field is 0.5 inches, 0.5 inches on the map represents 100 yards on the ground. To determine the map scale, convert yards to inches and reduce the numerator to 1 using equation (2.15).

$$\frac{1}{x} = \frac{0.5 \text{ inches}}{100 \text{ yards} \times \frac{3 \text{ feet}}{1 \text{ yard}} \times \frac{12 \text{ inches}}{1 \text{ foot}}}$$

$$\frac{1}{x} = \frac{0.5 \text{ inches}}{3{,}600 \text{ inches}}$$

$$x = \frac{3{,}600 \text{ inches}}{0.5 \text{ inches}}$$

$$x = 7{,}200 \quad \text{(2.15)}$$

This calculation gives an RF of 1:7,200. In Canada, a football field is 110 yards, so you must rework the calculations for the different length. This method works best for large-scale maps because most features of standard lengths are short, and their lengths and widths are accurately shown only on large-scale maps.

Determining map scale from the equator

For a world map, you can use the fact that the earth's circumference is approximately 25,000 statute miles (the actual figure is 24,901.46 statute miles or 40,075.017 kilometers for an authalic sphere using the semi-major axis of the WGS84 ellipsoid). If, for example, the equator is shown in its entirety and is measured on the map or shown on your screen as eight inches long, you can find the map's RF using equation (2.16).

$$\frac{1}{x} = \frac{8 \text{ inches}}{24{,}901.46 \text{ miles} \times \frac{63{,}360 \text{ inches}}{1 \text{ mile}}}$$

$$x = 24{,}901.46 \times \frac{63{,}360}{8}$$

$$x = 197{,}219{,}563 \quad \text{(2.16)}$$

Rounding off this number gives an RF of 1:197,219,600, to seven significant digits.

The word scale for "miles to an inch" is calculated as equation (2.17).

$$1 \text{ inch} = \frac{197{,}219{,}600}{63{,}360} \text{ miles}$$

$$1 \text{ inch} = 3{,}112.68 \text{ miles} \quad \text{(2.17)}$$

This value can be rounded to "3,113 miles to 1 inch."

In metric units, eight inches is 20.32 centimeters, and the earth's authalic sphere circumference is 40,075.017 kilometers, so the calculation is expressed in equation (2.18).

$$\frac{1}{x} = \frac{20.32 \text{ centimeters}}{40{,}075.017 \text{ kilometers} \times \frac{100{,}000 \text{ centimeters}}{1 \text{ kilometer}}}$$

$$x = 40{,}075.017 \times \frac{100{,}000}{20.32}$$

$$x = 197{,}219{,}572 \quad \text{(2.18)}$$

As you can see, this calculation also gives the same RF of 1:197,219,600, to seven significant digits. To determine the word scale in metric units, use equation (2.19), or "1,972 kilometers to 1 centimeter."

$$1 \text{ centimeter} = \frac{197{,}219{,}566.9}{100{,}000} \text{ kilometers}$$

$$1 \text{ centimeter} = 1{,}972.20 \text{ kilometers} \quad \text{(2.19)}$$

Remember that this scale is only valid at the equator because the scale may be much larger or smaller at other places on a small-scale map.

Determining map scale from reference material

If you can't find a feature of standard length on your map, you can still determine scale if you turn to other reference material, such as road maps, topographic sheets, atlases, or other maps of a similar scale to yours. From these sources, you should be able to find the distance of something on your map—for instance, the distance between two prominent road intersections or two towns or two mountain peaks. Using this information, you can compute the map scale using the calculations described in this section for determining map scale from a known ground feature.

For example, on a map with a scale of 1:100,000, you might measure the map distance between two road intersections as 0.7 centimeters and as 0.5 centimeters on your map. Because the ratio of measured distances equals the ratio of the two RFs, you can use the proportion in equation (2.20).

$$\frac{\frac{1}{100,000}}{\frac{1}{x}} = \frac{0.7 \text{ centimeters}}{0.5 \text{ centimeters}} \quad (2.20)$$

This proportion reduces to equation (2.21).

$$\frac{x}{100,000} = \frac{0.7 \text{ centimeters}}{0.5 \text{ centimeters}}$$

$$x = \frac{0.7 \text{ centimeters}}{0.5 \text{ centimeters}} \times 100,000$$

$$x = 140,000 \quad (2.21)$$

This calculation gives an RF of 1:140,000.

Determining map scale from the spacing of parallels and meridians

It is not always convenient, or even possible, to find a known ground feature or to use other reference materials to determine map scale. But on many maps, especially those of small scale, parallels and meridians are shown. You can determine map scale by finding the ground distance between these lines if they are evenly spaced on the map. Finding the ground distance between parallels is simple because a degree of latitude varies only slightly from pole to equator. Small-scale world maps usually use the authalic sphere as the approximation to the earth, so parallels are equally spaced at 69.093 miles (111.195 kilometers) per degree of latitude. From appendix table A.3, you see that on the WGS84 ellipsoid, the first degree of latitude is 68.703 statute miles (110.567 kilometers) north or south of the equator, whereas the distance from the 89th parallel to the adjacent pole is 69.407 miles (111.699 kilometers). Thus, the variation in a degree of latitude from equator to pole is only 0.704 miles or 1.132 kilometers. This difference is so small that you can safely ignore it for small-scale maps.

Say, for example, that on a screen display of a small-scale world map you find two parallels separated by two degrees of latitude, and you use an app with a screen distance measurement function to find a map distance of five inches between the parallels. To find the ground distance between these parallels, multiply the number of degrees separating them by the length of a degree, in equation (2.22).

$$2° \times \frac{69.093 \text{ miles}}{1°} = 138.18 \text{ miles} \quad (2.22)$$

Then you can find the ratio between map distance and ground distance, in equation (2.23).

$$\frac{1}{x} = \frac{5 \text{ inches}}{138.18 \text{ miles}}$$

$$\frac{1}{x} = \frac{5 \text{ inches}}{138.18 \text{ miles} \times \frac{63,360 \text{ inches}}{1 \text{ mile}}}$$

$$\frac{1}{x} = \frac{5 \text{ inches}}{8,755,100 \text{ inches}}$$

$$x = \frac{8,755,100}{5}$$

$$x = 1,751,020 \quad (2.23)$$

This calculation gives an RF of 1:1,751,020.

In metric units, five inches is 12.7 centimeters, so you have equation (2.24).

$$2° \times \frac{111.195 \text{ kilometers}}{1°} = 222.40 \text{ kilometers} \quad (2.24)$$

Again, you can find the ratio between map distance and ground distance, in equation (2.25).

$$\frac{1}{x} = \frac{12.7 \text{ centimeters}}{222.40 \text{ kilometers}}$$

$$\frac{1}{x} = \frac{12.7 \text{ centimeters}}{222.40 \text{ kilometers} \times \dfrac{100{,}000 \text{ centimeters}}{1 \text{ kilometer}}}$$

$$\frac{1}{x} = \frac{12.7 \text{ centimeters}}{22{,}240{,}000 \text{ centimeters}}$$

$$x = \frac{22{,}240{,}000}{12.7}$$

$$x = 1{,}751{,}180 \quad (2.25)$$

This calculation gives an RF of 1:1,751,180, which is nearly the same value—the difference lies in rounding the numbers.

However, two parallels are not always shown on a map. If two meridians are shown, you can still compute the map scale. The complication is that, because meridians converge at the poles, the distance between a degree of longitude varies from 69.09 miles (111.20 kilometers) along the equator to zero at either pole (see appendix table A.6). The distance between meridians, called the **longitudinal distance**, depends on the mapped area's latitude. If you again assume an authalic sphere for the earth's shape in making a small-scale map, there is a simple functional relationship between latitude and longitudinal distance: longitudinal distance decreases by the cosine of the latitude. To find the longitudinal distance for a degree of longitude at a given latitude, multiply the length of a degree of latitude (69.093 miles or 111.195 kilometers) by the cosine of the latitude. This relationship is written mathematically as equation (2.26).

Longitudinal distance $= \cos(\text{latitude}) \times 69.093$ miles

or

Longitudinal distance $=$
$\cos(\text{latitude}) \times 111.195$ kilometers $\quad (2.26)$

How can you use this equation to determine map scale? The procedure is straightforward if viewed as a series of simple steps. Begin by measuring the map distance between two meridians. For example, on the aeronautical chart segment in figure 2.7, you find that at 53° N, meridians that span 0.5 degrees (30 minutes) of longitude on the map are 2.66 inches (6.75 centimeters) apart.

Next, determine the ground distance between these two meridians. To do that, use a calculator to find the cosine of 53° and multiply that by the length of a degree of latitude (69.093 miles), in equation (2.27).

$$\cos(53°) \times 69.093 \text{ miles}$$

$$= 0.6018 \times 69.093$$

$$= 41.58 \text{ miles} \quad (2.27)$$

This calculation gives a longitudinal distance of 41.58 miles. For kilometers, use equation (2.28).

$$\cos(53°) \times 111.195 \text{ kilometers}$$

$$= 0.6018 \times 111.195$$

$$= 66.92 \text{ kilometers} \quad (2.28)$$

So, the longitudinal distance is 66.92 kilometers.

You now know that one degree of longitude at 53° N covers a ground distance of 41.58 miles or 66.92 kilometers. But because the measured map distance spans one-half rather than one degree of longitude, the ground distance in this problem is half the computed value, or equation (2.29).

$$0.5 \times 41.58 = 20.79 \text{ miles or } 33.46 \text{ kilometers} \quad (2.29)$$

You now have all the information necessary to find the map scale—the distance between two meridians on the map and the corresponding ground distance. Simply compute the ratio between map and ground distance, in equation (2.30).

$$\frac{1}{x} = \frac{2.66 \text{ inches}}{20.79 \text{ miles}}$$

$$\frac{1}{x} = \frac{2.66 \text{ inches}}{20.79 \text{ miles} \times \dfrac{63{,}360 \text{ inches}}{1 \text{ mile}}}$$

$$\frac{1}{x} = \frac{2.66 \text{ inches}}{1{,}317{,}254 \text{ inches}}$$

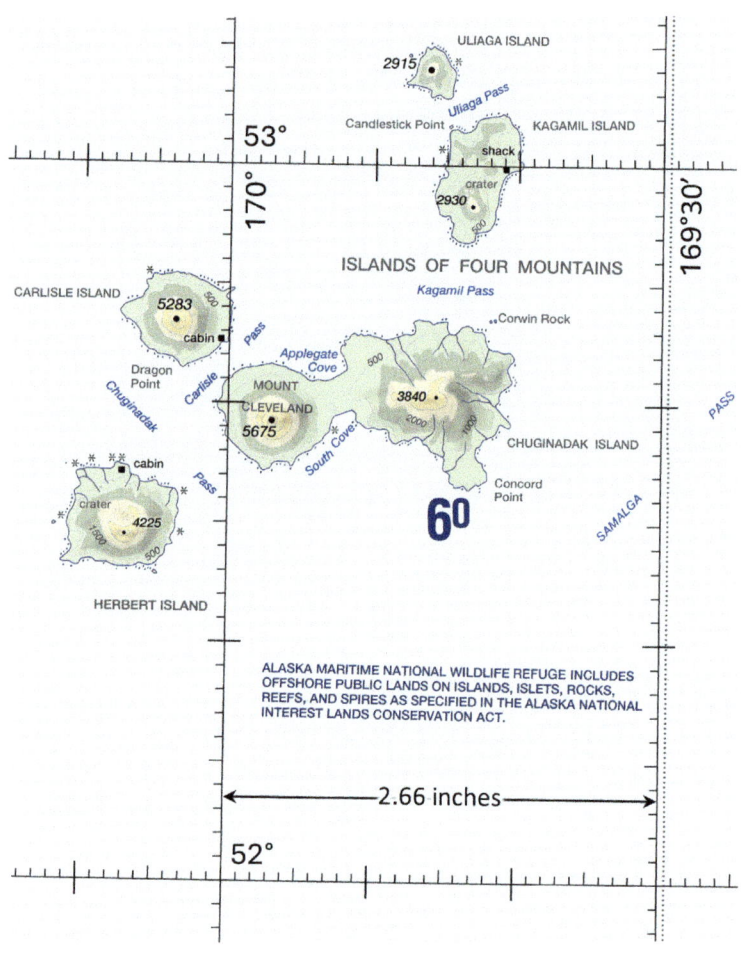

Figure 2.7. You can begin to compute the RF of this segment of the 1:500,000-scale Dutch Harbor aeronautical chart for an area in the Aleutian Islands by measuring 0.5 degrees of longitude as 2.66 inches long on the chart (latitude and longitude ticks are spaced one minute apart on the chart). Chart courtesy of the National Aeronautical Charting Office.

$$x = \frac{1,317,254}{2.66}$$

$$x = 495,208.4 \quad (2.30)$$

This calculation gives an RF of 1:495,208, which is rounded to 1:495,000. In metric units, 2.66 inches equals 6.756 centimeters, so you have equation (2.31).

$$\frac{1}{x} = \frac{6.756 \text{ centimeters}}{33.46 \text{ kilometers}}$$

$$\frac{1}{x} = \frac{6.756 \text{ centimeters}}{33.46 \text{ kilometers} \times \dfrac{100,000 \text{ centimeters}}{1 \text{ kilometer}}}$$

$$\frac{1}{x} = \frac{6.756 \text{ centimeters}}{3,346,000 \text{ centimeters}}$$

$$x = \frac{3,346,000}{6.756}$$

$$x = 495,234 \quad (2.31)$$

This calculation gives an RF of 1:495,234, which is also rounded to 1:495,000.

Determining map scale from ground resolution and pixel density

You may be faced with finding the scale of an image map that is made from imagery obtained by one of the remote sensing devices described in chapter 11. You can compute the image map scale if you know the **ground resolution** (GR) of a **pixel** in the image and the **pixel density** (PD) of the image map on your device. Say that your image map is made from satellite imagery with a 10-meter ground resolution, meaning that each pixel in the image covers a 10 meter × 10 meter square on the ground. You find that the image map is displayed on your computer display at 50 pixels per centimeter. Then use equation (2.32), as follows.

$$RF_{denominator} = GR \times PD \quad \text{(2.32)}$$

Here, GR and PD are both in the same units of measurement. In this example, a ground resolution of 10 meters per pixel is the same as 1,000 centimeters per pixel, or equation (2.33).

$$GR = \frac{1,000 \text{ centimeters}}{1 \text{ pixel}} \quad \text{(2.33)}$$

The PD is 50 pixels per centimeter, so you have equation (2.34).

$$GR \times PD = 1,000 \text{ centimeters/pixel} \times 50 \text{ pixels/centimeter} = 50,000 \quad \text{(2.34)}$$

The RF for your image map is 1:50,000.

Selected readings

Quattrochi, D. A., and M. F. Goodchild. 1997. *Scale in Remote Sensing and GIS*. Lewis.

Raposo, P. 2017. "Scale and Generalization." In *The Geographic Information Science and Technology Body of Knowledge*, edited by John P. Wilson. https://gistbok-topics.ucgis.org/CV-03-004.

US Army. 2011. "Scale and Distance." In *US Army Map Reading and Land Navigation Handbook*. Army Field Manual FM 3-25.26.

CHAPTER 3
Map projections

Much has been written about **map projections**, which allow us to geometrically transform the earth's spherical or ellipsoidal shape onto a flat map surface. Yet people still find this subject one of the most bewildering aspects of map use. Many people readily admit that they do not understand map projections. This shortcoming can have unfortunate consequences, especially in today's global society, because it hinders their ability to understand how people, plants and animals, and other features and phenomena are distributed across the earth. It also makes them easy prey for advertisers, special-interest groups, politicians, and others who, through lack of understanding or by design, use map projections in potentially deceptive ways.

There is an infinite number of map projections, but some are better suited than others for certain uses. How, then, do you go about distinguishing one projection from another and choosing among them? One way is to organize the wide variety of projections into a limited number of **map projection families** based on shared attributes. Two approaches commonly used include classifying the projections into families based on their geometric distortion properties (relating to distance, shape, direction, and area) and examining the nature of families based on the surface used to construct the projection (a plane, cone, or cylinder), which, in turn, helps you understand the pattern of spatial distortion over the map surface. The two approaches work together because the map user is concerned, first, with what spatial properties are preserved (or lost) and, second, with the pattern and extent of distortion. This chapter closely examines the distortion properties of map projections and the surfaces used in the creation of map projections.

Before we discuss map projection properties and families, we can help clarify the issue of map projections with a discussion of the logic behind them and of why projections are necessary. We begin our discussion with globes, a form of map that you have probably looked at since childhood. Your familiarity with globes makes it easy for us to compare them with flat maps.

Globes versus flat maps

Of all maps, a **globe** (a sphere on which a map of the earth is depicted) gives the most realistic picture of the earth as a whole. You may also have seen **celestial globes** that show the sky above the earth with its stars, planets, and constellations; however, in this chapter, we focus on globes of the earth.

Basic geometric properties, such as distance, direction, shape, and area, are preserved because the globe is the same scale everywhere (figure 3.1). However, globes have several

Figure 3.1. A typical world globe showing political features, such as countries and country names. Courtesy of RäthGloben, Germany.

disadvantages. They do not let you view all parts of the earth's surface simultaneously—the most you can see is a **hemisphere** (half of the earth). Neither are globes useful for seeing the kind of detail you might find on a road map that you keep in your car, a topographic map that you carry when hiking, or a navigation map that you use on your phone.

Globes also are bulky and do not lend themselves to convenient handling and storage like paper maps or the maps you use on a digital device. You would not have as many handling and storage problems if you used a baseball-size globe. But such a tiny globe is of little practical value for map reading and analysis because it would have a scale of approximately 1:125,000,000. Even a globe that is two feet (60 centimeters) in diameter is still at a scale of 1:20,000,000. It would take a globe about 40 to 50 feet (12 to 15 meters) in diameter—the height of a four-story building—to provide a map of the scale 1:1,000,000 as used for printed state highway maps. A globe that is nearly 1,800 feet (about 550 meters) in diameter—the length of six US football fields—would be required to provide a map of the same scale as the US 1:24,000-scale topographic map series (explained in chapter 2).

Another problem with globes is that the computations and techniques that are suited for measuring distance, direction, and area on spherical surfaces are relatively difficult to use. Computations on a sphere are far more complex than on a planar surface. Finally, globe construction is laborious and costly. High-speed printing presses have kept the cost of flat-map reproduction to manageable levels but have not yet been developed to work with curved media. 3D printers can now be used to make small plastic single-color globes but at relatively high cost. Therefore, globe construction is not suited to the volume of map production required for modern map use.

What about virtual globes? Virtual globes are not solid 3D objects, but rather a series of maps that look three-dimensional but are earthlike map projections on a 2D digital display. In the future, we may be able to view and analyze virtual globes projected onto a real 3D display surface. The apps that display digital globes, such as ArcGIS Earth, often provide convenient tools for measuring distances (figure 3.2).

It would be ideal if the earth's surface could be mapped undistorted on a flat medium, such as a sheet of paper or computer screen. Yet the spherical earth is not what is known as a **developable surface**, defined by mathematicians as a surface that can be flattened onto a plane without geometric distortion. The only developable surfaces in our 3D world are cylinders, cones, and planes, so all flat maps necessarily distort the earth's surface geometrically. To transform a spherical surface that curves away in every direction from every point onto a plane surface that does not curve in any direction from any point means that the earth's surface must be distorted on the flat map. Thus, map projection affects basic properties of the representation of

Figure 3.2. Apps such as ArcGIS Earth allow you to easily measure distances and perform other analysis on a virtual globe.

the earth on a flat surface, such as scale, continuity, and completeness, as well as geometric properties that relate to direction, distance, area, and shape. You want to minimize these distortions or preserve a particular geometric property at the expense of others. This conundrum is the map projection problem.

The map projection process

The concept of map projections is somewhat more involved than implied in the previous discussion. Not one but a series of geometric transformations is required. The irregular topography of the earth's surface is first defined relative to a much simpler 3D surface, and then the 3D surface is projected onto a plane. This progressive flattening of the earth's surface is illustrated in figure 3.3.

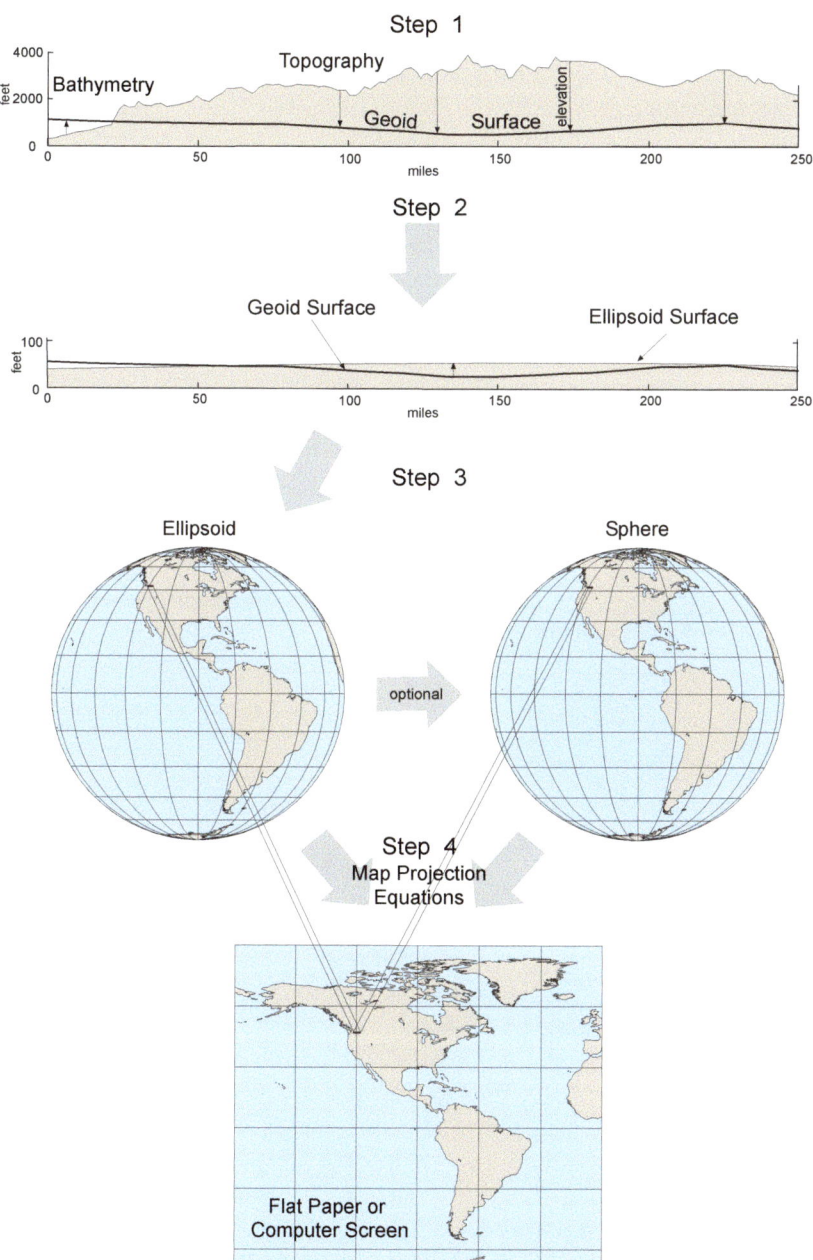

Figure 3.3. The map projection process involves a number of steps: (1) the land elevation or sea depth of every point on the earth's surface is defined relative to the geoid surface; (2) the elevations and depths on the geoid are projected onto the more regular ellipsoid surface to define the positions with geodetic latitude and longitude; (3) optionally, only for small-scale maps, geodetic latitudes and longitudes may be converted into spherical coordinates (the small difference between the ellipsoid and sphere is not apparent at this scale); (4) finally, the ellipsoidal or spherical coordinates are transformed into planar (x,y) map coordinates through map projection equations.

The first step is to define the earth's irregular surface topography as elevations of land (**topography**) and depths of water (**bathymetry**) relative to a more regular surface known as the geoid, as discussed in chapter 1. The geoid is the surface that would result if the average level of the world's oceans (mean sea level) was extended under the continents. It serves as the vertical reference datum, or starting reference surface, for elevation data on maps. (See chapter 1 for more on datums.)

The second step is to project the slight **geoidal undulation** (the difference between the geoid and the ellipsoid) on the more regular oblate ellipsoid surface. This new surface serves as the basis for the geodetic control points and as the datum for the geodetic latitude and longitude coordinates found on maps (see chapter 1 for more information about oblate ellipsoids, control points, and geodetic latitude and longitude.)

An optional step in making a small-scale flat map or globe is to mathematically transform geodetic coordinates into **geocentric coordinates** on a sphere, usually equal in surface area to the ellipsoid—that is, an authalic sphere. (See chapter 1 for more on the authalic sphere.)

The last step involves projecting the ellipsoidal or spherical surface on a plane through map projection equations that transform geographic or spherical coordinates into planar (x,y) map coordinates. The greatest distortion of the earth's surface geometry occurs in this step.

Map projection properties

With an understanding of the map projection process, you can see that it ultimately leads to distortion of the earth's geometry on the flat map. We will look at how the properties of scale, completeness, correspondence relations, and continuity, which are a result of this distortion, are affected by the map projection process.

Scale

Because of the stretching and shrinking that occurs in the process of transforming the spherical or ellipsoidal earth surface into a plane, the stated map scale (see chapter 2 for more information on scale) is true only at selected points or along particular lines called **points** and **lines of tangency**, which we talk about later in this chapter. Everywhere else, the scale of the map is actually smaller or larger than the stated scale.

To grasp the idea of scale variation in map projections, you first must realize that there are, in fact, two map scales.

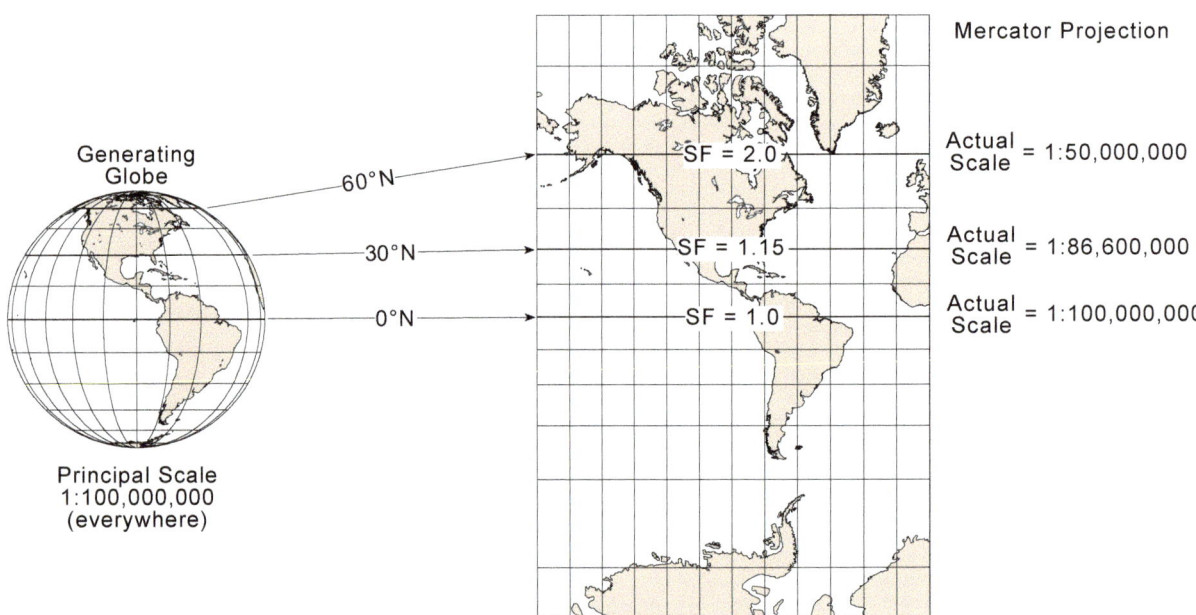

Figure 3.4. Scale factors (SFs) greater than 1.0 indicate that the actual scales are larger than the principal scale of the generating globe. On the Mercator projection map on the right, the principal scale is identical to the actual scale at the equator, where the map projection surface touches the generating globe.

One is the **actual scale**—the scale that you measure in the vicinity of any point on the map; it differs from one location to another. Variation in actual scale is a consequence of the geometric distortion that results from flattening the earth.

The second is the **principal scale** of the map. You can think of the principal scale as the constant scale of what cartographers call a **generating globe**—a globe that is reduced to the scale of the desired flat map. You can then visualize this globe as being transformed into a flat map (figure 3.4), knowing that the actual transformation is through map projection equations. The constant scale of the generating globe is the principal scale stated on the flat map, which is correct only at the points or lines at which the generating globe touches the projection surface.

To understand the relationship between the actual scale and the principal scale at different places on the map, we compute a ratio called the **scale factor (SF)**, which is defined in equation (3.1):

$$SF = \frac{\text{Actual Scale}}{\text{Principal Scale}} \quad (3.1)$$

We use the representative fractions of the actual and principal scales to compute the SF. An actual scale of 1:50,000,000 and a principal scale of 1:100,000,000 thus give an SF as follows in equation (3.2):

$$SF = \frac{\frac{1}{50,000,000}}{\frac{1}{100,000,000}}$$

$$SF = \frac{100,000,000}{50,000,000}$$

$$SF = 2.0 \quad (3.2)$$

If the actual and principal scales are identical, the SF is 1.0, leading to **unity**. But as you saw earlier, because the actual scale varies from place to place, so does the SF. An SF of 2.0 on a small-scale map means that the actual scale is twice as large as the principal scale (see figure 3.4). An SF of 1.15 on a small-scale map means that the actual scale is 15 percent larger than the principal scale. On large-scale maps, the SF should vary only slightly from 1.0, following the general rule that the smaller the area being mapped, the less the scale distortion.

Completeness

Completeness refers to the ability of map projections to show the entire earth. You will find the most obvious distortion of the globe on world maps that do not show the whole world. Such incomplete maps occur when the equations used for a map projection cannot be applied to the entire range of latitude and longitude. The Mercator projection world map (see figure 3.22) is a classic example. In this map projection, y-coordinates for the North and South Poles are positive and negative infinity, respectively, so the map usually extends to only the 84th parallel north and 80th parallel south. (See note 1 at the end of the chapter.) Omitting these high latitudes may be acceptable for maps that show political boundaries, cities, roads, and other cultural features, but the data for physical phenomena, such as average temperatures, ocean currents, and landforms, is usually global and should therefore be shown on a map of the entire earth. The gnomonic projection (see figure 3.18), discussed later in this chapter, is an even more extreme example because it is limited mathematically to covering less than a hemisphere (the SF is an infinitely large 90 degrees away from the projection center point). So, one of the considerations when choosing a map projection is the ability to show your complete area of interest.

Correspondence relations

You might expect that each point on the earth is transformed into a corresponding point in the map projection. The agreement between points on the earth and points on the projected surface is called **correspondence**. **Point-to-point correspondence** allows you to shift attention with equal facility from a point on the earth to the same point on the map, and vice versa. However, this desirable property cannot be maintained for all points in many world map projections. As figure 3.5 shows, one or more points on the earth may be transformed into straight lines or circles on the boundary of the map projection, most often at the North and South Poles. You may notice that the SF in the east–west direction must be infinitely large at the poles in this projection, yet it is 1.0 in the north–south direction. Other map projections, such as the orthographic and Lambert azimuthal equal-area projections (see figures 3.16 and 3.20), discussed later in the chapter, allow for point-to-point correspondence for one hemisphere of the earth.

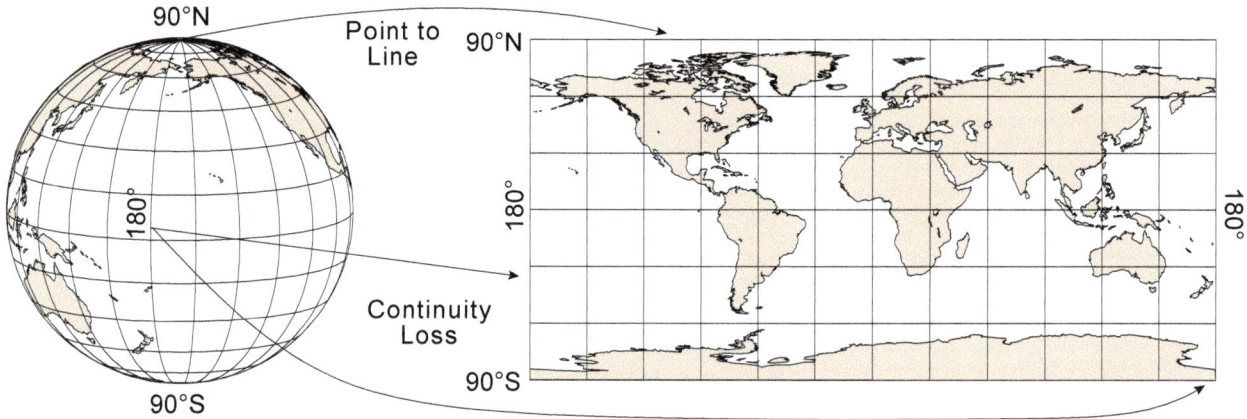

Figure 3.5. The point-to-point correspondence between the earth and the map projection may become point-to-line at some locations as along the top and bottom edges of the map on the right. There may also be a loss of continuity, in which the same line on the globe forms two edges of the map, such as the left and right edges of the map on the right, both of which are the 180° meridian.

Continuity

To represent an entire spherical surface on a plane, the continuous spherical surface must be interrupted at some point or along some line. These breaks in **continuity** (preservation of spatial proximity at all locations) form the map border in a world projection. Where the mapmaker places the discontinuity is a matter of choice. On some maps, for example, opposite edges of the map are the same meridian (figure 3.5, *right*). Using the same meridian for opposite edges of the map means that features next to each other on the ground are found on opposite sides of the map. This blatant violation of proximity relations is a source of confusion for many map users. Similarly, a map may show the North and South Poles with lines that are as long as the equator, as on the map on the right in figure 3.5. Maps of individual continents or nations almost always show these areas without breaks in continuity. To do otherwise needlessly complicates reading, analyzing, and interpreting the map.

Map projection families

As we mentioned previously, it is sometimes convenient to group map projections into families based on geometric distortion and the projection surface so that you can distinguish and choose among them. First, we will examine geometric distortion.

Map projection families based on geometric distortion

You can gain an idea of the types of **geometric distortions** that occur in map projections by comparing the graticule (latitude and longitude lines) on the projected surface with the same lines on a globe (see chapter 1 for more on the graticule). Figure 3.5 shows how you might make this comparison (keeping in mind that, in the figure, the drawing of a globe on the left is, in actuality, a map projection of a hemisphere). Ask yourself several questions: To what degree do meridians converge? Do meridians and parallels intersect at right angles? Do parallels shorten with increasing latitude? Are the areas of quadrilaterals in the projection the same as on the globe?

You can use these observations to better understand how cartographers assign the types of geometric distortion you see in the map projection into the categories of distance, shape, direction, and area. Now we look at each type of distortion, beginning with variations in distance.

Distance

The preservation of spherical great-circle distance in a map projection, called **equidistance**, is, at best, a partial achievement of distance preservation (see chapter 1 for more on great circles). For a map projection to be truly distance-preserving, the scale must be equal in all directions from every point, which is geometrically impossible to achieve on a flat map. Because the scale varies continuously from location to location in a map projection, the great-circle distance between the two locations on the globe must be distorted in the projection.

Long-distance route planning is based on knowing the great-circle distance between the starting point, usually at the center of the map, and the ending point; and equidistant maps are excellent tools for determining these distances on the earth. For example, meteorologists use equidistant maps to determine the distance between weather stations. One obvious drawback is that the map projection must be changed every time the starting point is moved.

Equidistant projections are used to show the correct distance between a selected location and any other point in the projection. Although the correct distance will be shown between the selected point and all others, distances between all other points are distorted. Consequently, no flat map can preserve both distance and area. For some equidistant projections, such as the azimuthal equidistant projection that you will learn more about later in this chapter, cartographers make the SF a constant 1.0 radially outward from a single point, such as the North Pole (see figure 3.19). Great-circle distances are correct along the lines that radiate outward from that central point only.

Shape

When angles on the globe are preserved at individual points on the map (thus preserving shape), the projection is called **conformal**, meaning "correct form or shape." Unlike the property of equidistance, conformality can be achieved at all points in conformal projections. To attain the property of **conformality**, the local map scale must be the same in all directions from a point. A circle on the globe thus appears as a circle on the map. But to achieve conformality, it is necessary to either enlarge or reduce the scale by a different amount at each location on the map (figure 3.6). This change in scale means, of course, that the map area around each location must also vary. If you could draw tiny circles on the earth, they would always appear as circles on the map, but their sizes would differ on the map. Because tiny circles on the earth are projected as circles, all directions from the center of the circles are correct in the projected circles. The circles must be tiny because the constant change in scale across the map means that they would be projected as ovals if they were hundreds of miles in diameter on the earth. You can also see the distortion in the appearance of the graticule—although the parallels and meridians intersect at right angles, the distance between parallels varies. Typically, there is a smooth increase or

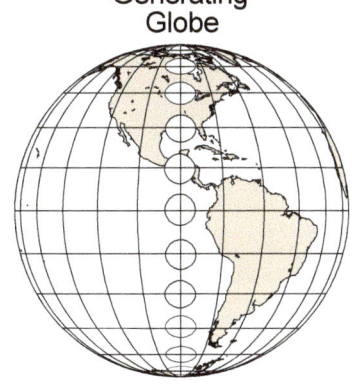
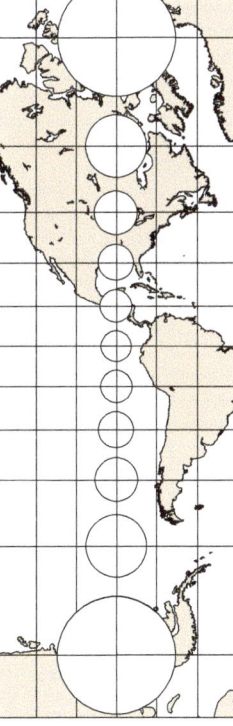

Figure 3.6. On a map that has a conformal projection, like the Mercator projection, identical tiny circles (greatly magnified here) centered at points on the generating globe are projected to the flat map as different-size circles according to the local SF. The appearance of the graticule in all the projected circles is the same as on the globe.

decrease in scale across the map. Conformality applies only to directions or angles at—or in the immediate vicinity of—points. Thus, shape is preserved only in small areas. Conformality does not apply to areas of any great extent—the shape of large regions can be greatly distorted.

Conformal maps are best suited for tasks that involve plotting, guiding, or analyzing the motion of objects over the earth's surface. Thus, conformal projections are used for aeronautical and nautical charts, topographic quadrangles, and meteorological maps. They are also used when the shape of features is a matter of concern.

Direction

It is sometimes important to preserve directions globally. Projections that preserve global directions are called **true-direction projections**. You will also sometimes see these referred to as **azimuthal projections**. As you will see in the next section, this term is also used instead of planar to describe map projections developed on a plane surface. To avoid confusion, we use the terms *true direction* and *planar* in this book, respectively, to describe map projections that preserve directions and map projections developed on a plane surface.

No projection can correctly represent all directions from all points on the earth as straight lines on a flat map. But scale across the map can be arranged so that certain types of direction lines are straight, as is the case in which all meridians radiating outward from the pole are correct directionally. All true-direction projections correctly show the **azimuth** or direction from a reference point, usually the center of the map. Hence, great-circle directions on the ground from the reference point can be measured on the map.

True-direction projections are used to create maps that show the great-circle routes from a selected point to a desired destination (and thus the shortest distance on the earth's surface). Special true-direction projections for maps, such as the gnomonic projection (see figure 3.18) discussed later in this chapter, are used for long-distance route planning in air and sea navigation.

Area

When the relative size of regions on the earth is preserved everywhere on the map, the projection is said to be **equal area** and have the property of **equivalence**. The demands of achieving equivalence are such that the SF can be the same in all directions only along one or two lines, or from, at most, two points. Because the SF and, hence, angles around all other points will differ from 1.0 in different directions (for example, 2.0 in the north–south direction and 0.5 in the east–west direction), the scale requirements for equivalence and conformality are mutually exclusive. No projection can be both conformal and equal area—only a globe can be. And, as we also stated earlier, no flat map can preserve both area and distance.

Adjusting the scale along meridians and parallels so that reduction in one direction from a point is compensated by exaggeration in another direction creates an equal-area map projection. For example, a small circle on the globe with an SF of 1.0 in all directions may be projected as an ellipse with a north–south SF of 2.0 and an east–west SF of 0.5 so that the area of the ellipse is the same as the circle. Equal-area world maps, therefore, compact, elongate, shear, or skew circles, as well as the quadrilaterals of the graticule on the generating globe (figure 3.7). The distortion of shape,

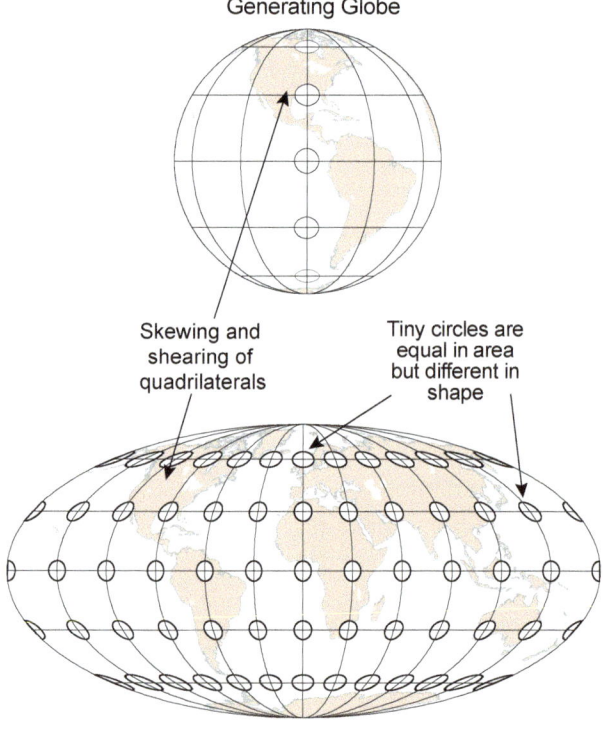

Figure 3.7. Tiny circles on the generating globe (greatly enlarged here) are projected as ellipses of the same area but a different shape on a map using an equal-area projection, such as the Mollweide projection. This distortion in shape skews the directions of the major and minor axes for each ellipse. Similarly, quadrilaterals are skewed and sheared in an equal-area projection.

distance, and direction is usually most pronounced toward the map's margins.

Despite the distortion of shapes that is inherent in **equal-area projections**, they are the best choice for tasks that call for area or density comparisons from region to region. Examples of geographic phenomena that are best shown on equal-area map projections, such as the Mollweide projection in figure 3.7, include world or continental maps of population density, per capita income, literacy, poverty, and other human-oriented statistical data.

Map projection families based on developable surfaces

So far, we have categorized map projections into families based on the types of geometric distortions that occur in them, realizing that certain geometric properties can be preserved under special circumstances. We can also categorize map projections based on the surfaces used to construct them.

As a child, you probably played the game of casting hand shadows on the wall. You can think of the surface that your shadow was projected on as the **projection surface**. You discovered that the distance and direction of the light source relative to the position of your hand influenced the size and shape of the shadow you created. But the surface on which you projected your shadow also had a great deal to do with it. A shadow cast on a corner of the room or on a curved surface, such as a lampshade, is quite different from one cast on the flat wall.

Now think of the generating globe as your hand and the map projection as the shadow on the wall. To visualize this shadow casting, imagine a transparent generating globe that has the graticule and continent outlines drawn on it in black. Then consider placing this globe at various positions relative to a light source and a surface on which you project its shadow (figure 3.8). So now you have the generating globe (your hand) being projected as a map (casting a shadow) on the projection surface (the wall).

In the case of maps, the projection surface is always flat. In cartography, this flat surface is called the **developable surface**. Three basic projection families are based on developable surfaces: planar, conic, and cylindrical projections. Depending on which type of developable surface you use and where the light source is placed, you will end up with different map projections cast by the generating globe. All projections that you can create in this light-casting way are called **true-perspective projections**.

Strictly speaking, very few projections involve casting light on planes, cylinders, or cones in a physical sense. Most projections are not purely geometric, and all use a set of mathematical equations that transform latitude and longitude coordinates on the earth into x,y coordinates on the projection surface. The x,y coordinates for a coastline, for example, are the endpoints for a sequence of short, straight lines drawn on paper or displayed on a digital device. You see the sequence of short lines as a coastline. Projections that do not use developable surfaces are more common because they can be (1) designed to serve any desired purpose; (2) made conformal, equal area, or equidistant; and (3) readily produced through computing. Yet even nondevelopable surface projections can usually be considered as variations on the use of the three basic developable surfaces.

Planar projections

Planar projections (also called **azimuthal projections**) can be thought of as being made by projecting the generating globe onto a flat plane. The plane can either touch the generating globe at a point—called the **point of tangency**—or

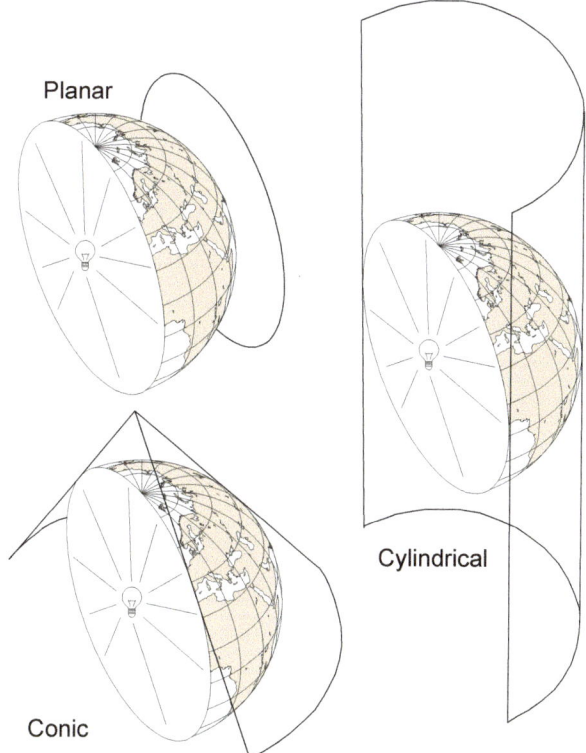

Figure 3.8. Planes, cones, and cylinders are used as developable surfaces for true-perspective map projections.

slice through the generating globe, which results in a **line of tangency** that is a small circle. (See chapter 1 for more on small circles.)

If you again consider your childhood shadow-casting game, recall that the projection surface was only one of the factors that influences the shape of the shadow cast on the wall. Another influence is the distance of the light source from the wall. Thus, there are basic differences in the **planar projection family**, depending on where the light source is located.

This family includes three commonly used **true-perspective planar projections** (figure 3.9): **orthographic** (point light source at infinity), **stereographic** (point light source on the surface of the generating globe opposite the point of tangency), and **gnomonic** (point light source at the center of the generating globe). Other projections, such as the azimuthal equidistant and Lambert azimuthal equal-area projections (described later in this chapter), are mathematical constructs that cannot be created geometrically and must be built from a set of mathematical equations; however, they can still be thought of as variations on using a plane as the developable surface.

Planar projections lend themselves to equidistant and true-direction map projections. They are commonly centered on the North or South Pole. Part of this choice may be so that lines of longitude follow true directions on the projected map. The longitude lines may also help to reinforce to the user of an equidistant map projection that distance should only be measured outward from the pole at the center of the map.

The three planar projections described in this section have a variety of appropriate uses. You will see orthographic projections in 3D-appearing depictions of the earth as if it were viewed from outer space. Ocean and air navigators have used the gnomonic projection to plot great-circle routes between seaports or airports. Geographers have used the stereographic projection for maps showing hemispheres and large continents, and earth scientists use the projection to show information about polar areas. The azimuthal equidistant and Lambert azimuthal equal-area planar projections discussed later in the chapter also have a variety of uses ranging from measuring great-circle distances between two distant locations to showing statistical data for circular-shaped continents, such as Africa and North America.

Cylindrical projections

Cylindrical projections can be thought of as being made by projecting the generating globe on a cylinder that touches or slices through the generating globe. A cylinder that touches the generating globe results in one line of tangency, usually at the equator.

As with planar projections, cylindrical projections can also be distinguished by the location of the light source relative to the projection surface. The **cylindrical projection family** includes two commonly used **true-perspective cylindrical projections** (figure 3.10): **central cylindrical** (light source at the center of the generating globe) and **cylindrical equal area** (linear light source, akin to a fluorescent light with parallel rays, along the polar axis). These projections are described in more detail later in this chapter. The whole world cannot be projected on the central cylindrical projection because the polar rays will never intersect the cylinder. For this reason, this projection is often cut off at some specified latitude north and south.

Cylindrical world projections allow most of the world where human beings reside to be projected. Lines of tangency lend themselves to follow both the equator and meridians for the transverse case to be discussed later in this chapter.

Mapmakers choose cylindrical projections for a variety of mapping problems. For centuries, the Mercator projection has been used for nautical charts ranging from

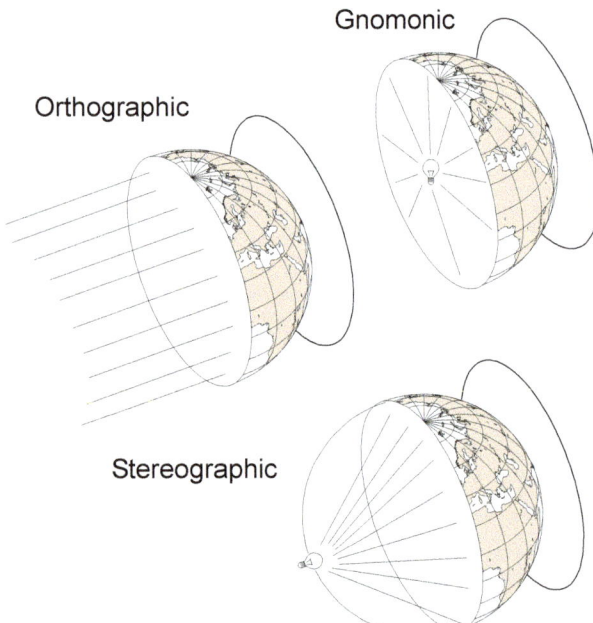

Figure 3.9. You can visualize three true-perspective planar projections (*clockwise, from left*: orthographic, gnomonic, and stereographic) as being constructed by changing the location of the light source relative to the generating globe.

large-scale harbor charts to small-scale charts of oceans. A recent form of this projection, the web Mercator, serves as the basemap for online mapping, mainly because it is square in shape. You will see in chapter 4 that the transverse aspect of the Mercator projection is used in smaller areas for grid coordinate systems requiring minimal distortion in scale across the mapped region. The rectangular shape of cylindrical projections makes them a logical choice for wall maps of the world, as well as for maps displaying digital elevation data and images collected by orbiting earth resources satellites.

Conic projections

The last family of map projections based on the surface used to construct the projection is **conic projections**, which can be thought of as projecting the generating globe onto a cone that touches or slices through the generating globe. As with a cylinder, a cone that touches the generating globe has one line of tangency. Mapmakers usually select mid-latitude parallels as the line of tangency. The **conic projection family** includes one **true-perspective conic projection**: the **central conic projection** (light source at the center of the generating globe as in figure 3.11).

Considering the size of cones typically used for map projections, the cone tends to touch or slice through the generating globe in the mid-latitudes. This allows for low-distortion map projections in the areas where much of the population of the earth lives. You can see in figure 3.11 that conic projections are ideal for mapping mid-latitude areas, such as the United States and Australia. Equal-area conic projections are commonly used for statistical maps of these two countries and similarly located nations, such as South Africa. Conformal conic projections are widely used for aeronautical charts in the mid-latitudes but also as the base for grid coordinate systems in smaller areas of greater east–west than north–south extent.

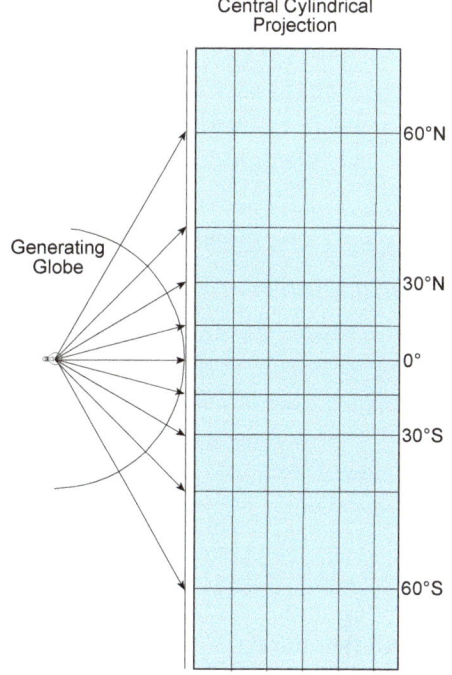

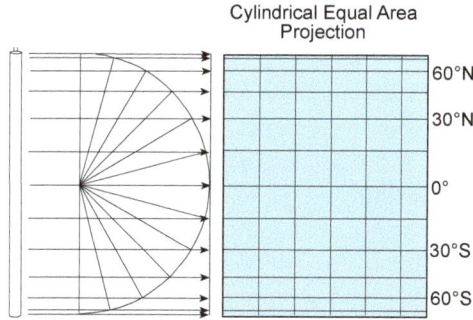

Figure 3.10. You can visualize the two true-perspective cylindrical projections (central cylindrical and cylindrical equal area) as being constructed by placing the light source at the center of or outside the generating globe, respectively.

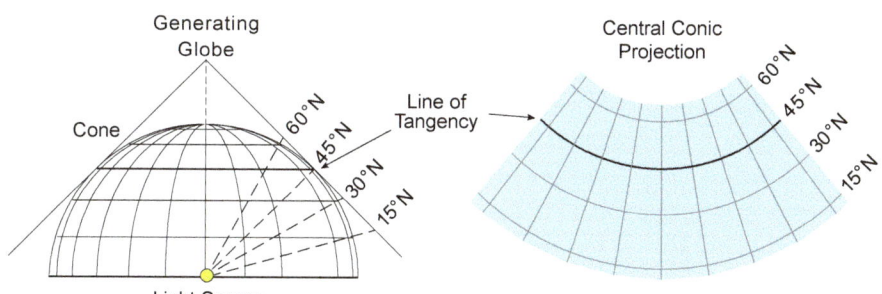

Figure 3.11. You can visualize the true-perspective central conic projection as being constructed by placing the light source at the center of the generating globe, with a cone touching the globe at the line of tangency.

We have introduced a wide variety of uses for planar, cylindrical, and conic projections. To fully understand map projection families based on the projection surface, you must also understand various projection parameters, including case and aspect, among others.

Map projection parameters

Tangent and secant case

The developable surface can either touch or slice through the generating globe so that the projection surface has either a tangent or a secant relationship with the generating globe, called the **case** of the projection. A **tangent-case** projection surface touches the generating globe at either a point (called the point of tangency) for planar projections or along a line (called a line of tangency) for conic or cylindrical projections (figure 3.12). The SF is 1.0 at the point of tangency for planar projections or along the line of tangency for cylindrical and conic projections. The SF increases radially outward from the point of tangency or perpendicularly away from the line of tangency.

A **secant-case** planar projection surface intersects the generating globe along a small-circle line of tangency (figure 3.12). Secant-case conic projections have two small-circle lines of tangency, called **standard parallels**, usually at the mid-latitudes. Secant-case cylindrical projections have two small-circle lines of tangency that are equidistant from the parallel at which the projection is centered. For example, a secant-case cylindrical projection centered at the equator might have the 10° N and 10° S latitudes as lines of tangency. Secant-case conic and cylindrical projections have an SF of 1.0 along the lines of tangency. Between the lines of tangency, the SF decreases from 1.0 to a minimum value halfway between the two lines, while outside the lines, the SF increases from 1.0 to a maximum value at the top and bottom edges of the map. So, the SF is slightly smaller than the stated scale in the middle part of the map and slightly larger at the edges. For secant-case planar projections, the SF increases outward from the circle of tangency and decreases inward to a minimum value at the center of the circle of tangency.

The advantage of the secant case is that it minimizes the overall scale distortion on the map more than in the tangent case. The SF is 1.0 along a circle instead of at a single point (for a secant-case planar projection) or two lines instead of one (for secant-case conic or cylindrical projections). As a result, the central part of the projection has a slightly smaller SF than the stated scale, and the edges don't have SFs as large as with the tangent case, thus creating a map with less overall scale distortion.

Because distortion is minimal around the point or line of tangency, the earth's curvature may often be ignored without serious consequence when using flat maps for measurement or analysis of a local area, but only if the line or lines of tangency are positioned within the area being mapped. As the distance from the point or line of tangency increases, so does the scale distortion. By the time a projection has been extended to include the entire earth, scale distortion may have greatly impacted the earth's appearance on the map away from the point or line(s) of tangency. Thus, secant-case planar, conic, and cylindrical projections are used more often than tangent-case projections, whose small area of minimal scale distortion limits their practical value.

Secant-case cylindrical projections are best suited for world maps because they have less overall scale distortion than tangent-case projections. This distortion is minimized in the mid-latitudes, which is also where most of the earth's population lives. Secant-case conic projections are best suited for maps of mid-latitude regions, especially those elongated in an east–west direction. The United States and Australia, for example, meet these qualifications and are frequently mapped using secant-case conic projections.

Aspect

Map projection **aspect** refers to the location of the point or line(s) of tangency on the projection surface (figure 3.13). A projection's point or line(s) of tangency can, in theory, touch or intersect anywhere on the developable surface. When a point of tangency is at the equator or a line of tangency is along the equator in the tangent case, or equidistant from the equator in the secant case, the resulting projection is said to be in **equatorial aspect**. When the point of tangency is at the pole, or the line or lines of tangency encircle the pole, the projection is in **polar aspect**. With cylindrical projections, the term **transverse aspect** is also used. Transverse aspect occurs when the line or lines of tangency for the equatorial aspect projection are rotated 90 degrees. In the tangent case, the line of tangency is a center meridian (figure 3.14). Any other alignment of the point or line(s) of tangency to the generating globe is in **oblique aspect**.

The aspect that has historically been used the most for the planar, conic, and cylindrical projection families is called the **normal aspect**. The normal aspect of planar projections is polar (figure 3.13, *right*)—the parallels of

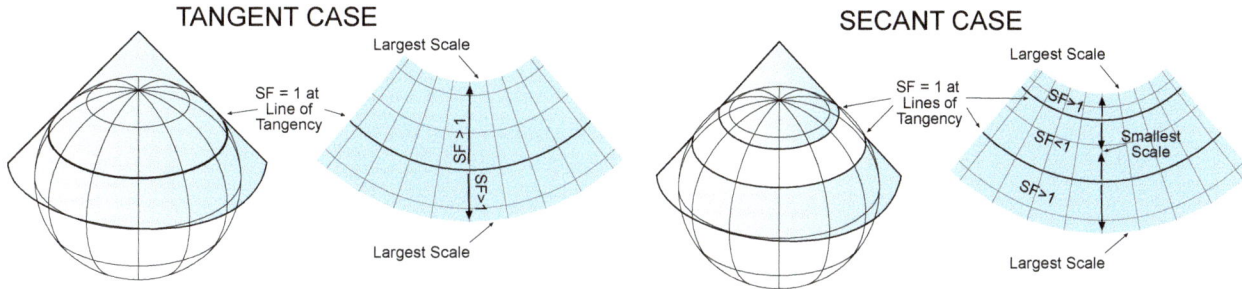

Figure 3.12. Tangent and secant cases of the three basic developable surfaces.

the graticule are concentric around the center, and the meridians radiate from the center to the edges of the map. In the normal aspect for conic projections (see figure 3.12, *bottom*), parallels are projected as concentric arcs of circles, and meridians are projected as straight lines that radiate at uniform angular intervals from the apex of the cone.

The normal aspect of cylindrical projections is equatorial (figure 3.14, *left*). You can recognize the normal aspect of cylindrical projections by horizontal parallels of equal length, vertical meridians of equal length that are also equally spaced, and right-angle intersections of meridians and parallels.

As you can see in figures 3.13 and 3.14, the choice of projection surface aspect leads to different-looking

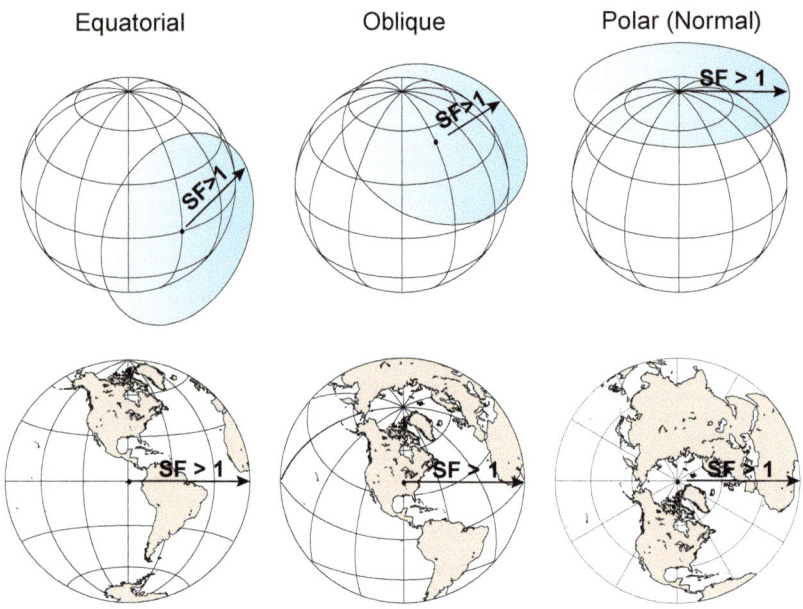

Figure 3.13. Tangent-case equatorial, oblique, and polar aspects for planar projections. Scale distortion increases radially away from the point of tangency, no matter where the point is located on the globe.

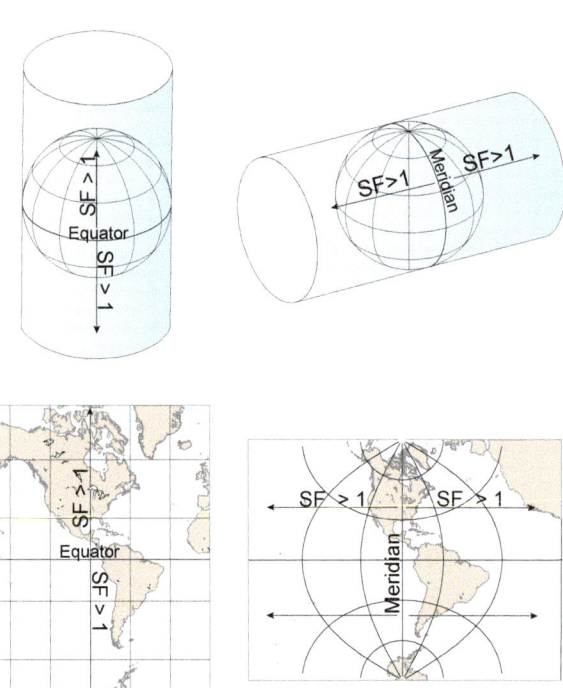

Figure 3.14. Tangent-case normal (*left*) and transverse (*right*) aspects for cylindrical projections. Scale distortion increases perpendicularly away from the line of tangency, no matter where the line is located on the globe.

appearances for the earth's land masses and the graticule. Because scale distortion always increases radially from the point or perpendicularly to the line of tangency, the distortion properties of a given projection remain unaltered when the aspect is changed from normal to transverse, polar, oblique, or equatorial. The map on the left in figure 3.14 is a good example. In the normal aspect of the tangent case, the SF for the cylindrical projection is 1.0 at the line of tangency at the equator and increases north and south perpendicular to the equator. In the transverse aspect of the tangent case (figure 3.14, *right*), the SF is 1.0 along the central meridian, that is, the line of tangency. The SF again increases perpendicularly outward from the line of tangency.

In the secant-case, transverse aspect cylindrical projection, the two lines of tangency are vertical small circles equally spaced from the central meridian (figure 3.15). The SF is 1.0 at the two lines of tangency and progressively increases outward from each line. Inward from each line of tangency, the SF progressively decreases until reaching a minimum value at the central meridian.

Other map projection parameters

Many other map projection parameters are also important because they can affect the definition, appearance, and appropriate use of the projections. The **central meridian** (also called the **longitude of origin** or, less commonly, the **longitude of center**) defines the origin of the x-coordinates. This meridian is usually placed in the center of the area being mapped. The **central parallel** (also called the **latitude of origin** or, less commonly, the **latitude of center**) defines the origin of the y-coordinates. It is commonly placed in the center of the mapped area with a false northing used to avoid negative y-coordinates. (See chapter 4 for more on false northings.) Conic projections, however, often are designed with a lower central parallel so that the origin of y-coordinates is below the mapped area, which makes all y-coordinates positive.

Commonly used map projections

As you have seen, one way that map projections are commonly grouped into families is based on the developable surface—a plane, cone, or cylinder. We can use these family groupings to structure our discussion of commonly used map projections.

Planar projections

Orthographic

The **orthographic projection** is how the earth would appear if viewed from a distant planet. Because the light source for this true-perspective projection is at an infinite distance from the generating globe, all rays are parallel (see figure 3.9, *left*). This projection appears to have first been used by astronomers in ancient Egypt, but it came into widespread use during World War II, with the advent of the global perspective provided by the age of air travel. It is even more popular in today's space age, often used to show land cover, land and ocean topography (see figure 3.16), and data obtained from remote sensing satellites. In fact, the generating globe and half-globe illustrations in this book are orthographic projections, as is the globe on the front cover of this book. The main drawback of the orthographic projection is that only a single hemisphere can be seen at a time. In the past, showing the entire earth required two maps, often of the Northern and Southern or Western and Eastern Hemispheres.

Stereographic

Projecting a light source from the antipodal point on the generating globe to the point of tangency creates the **stereographic projection** (see figure 3.9, *bottom*). This projection is conformal, so shape is preserved in small areas.

The Greek scholar Hipparchus, credited with proposing the system of parallels (see chapter 1), is also credited with inventing this projection in the second century BC. It is now most used in its polar aspect and secant case for maps of polar areas (figure 3.17). It is the projection surface for the universal polar stereographic grid coordinate system for polar areas, as you will see in chapter 4. A disadvantage of

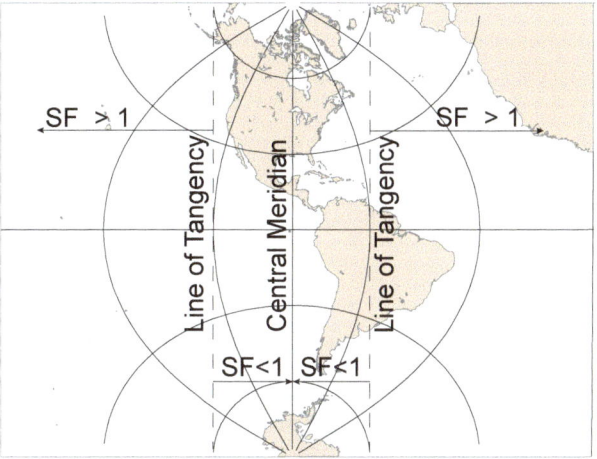

Figure 3.15. Secant-case transverse cylindrical projection. Scale factors increase perpendicularly away from the two small circle lines of tangency while decreasing perpendicularly inward to a minimum at the central meridian.

Figure 3.16. The orthographic projection is often used to show the spherical shape of the earth. Data courtesy of Tom Patterson, Equal Earth (see note 2 at the end of the chapter).

the stereographic projection is that it is generally restricted to one hemisphere. If it is not restricted to one hemisphere, the distortion near the edges increases to such a degree that the geographic features in these areas are basically unrecognizable. In past centuries, it was used for atlas maps of the Western or Eastern Hemisphere.

Gnomonic

Projecting the generating globe on a planar surface with the light source at the center of the generating globe produces the **gnomonic projection** (see figure 3.9, *right*). One of the earliest map projections, the gnomonic projection was first used by the Greek scholar **Thales of Miletus** in the sixth century BC for showing different constellations on **star charts**, which are used to plot planetary positions throughout the year.

The gnomonic projection has the distinction of being the only projection with the useful property of all great circles on the globe being shown as straight lines on the map (figure 3.18). Because a great-circle route is the shortest distance between two points on the earth's surface, the gnomonic projection is especially valuable as an aid to navigation. Its major disadvantages are extreme distortion of shape and area away from the point of tangency and the inability to project a complete hemisphere. This projection severely distorts shape compared with the polar stereographic projection in figure 3.17, even though less than a hemisphere is shown.

Azimuthal equidistant

The **azimuthal equidistant projection** in the tangent case and polar aspect has the distinctive appearance of looking like a dart board with equally spaced parallels and straight-line meridians that radiate outward from the pole (figure 3.19). This arrangement of parallels and meridians results in all straight lines drawn from the point of tangency being great-circle routes. Equally spaced parallels means that great-circle distances are correct along these straight lines because the north–south SF along meridians is 1.0.

The ancient Egyptians apparently first used this projection for star charts, but during the age of air travel it also became popular for use by pilots planning long-distance air routes. In the days before electronic navigation, the flight planning room in major airports had a wall map of the world in an oblique-aspect azimuthal equidistant projection centered on the airport. You may also find these interesting and useful maps in the public areas of some airports. All straight lines drawn from the airport show the correct distances (that is, they are correctly scaled great-circle routes). This planar projection is one of the few

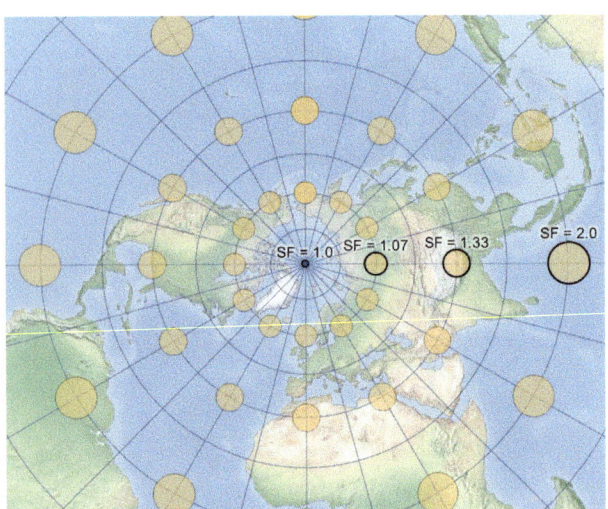

Figure 3.17. Polar stereographic projection of the Northern Hemisphere. Because this is a conformal projection, tiny circles on the generating globe are projected as circles of the same size at the point of tangency to four times as large at the equator on the edge of the map.

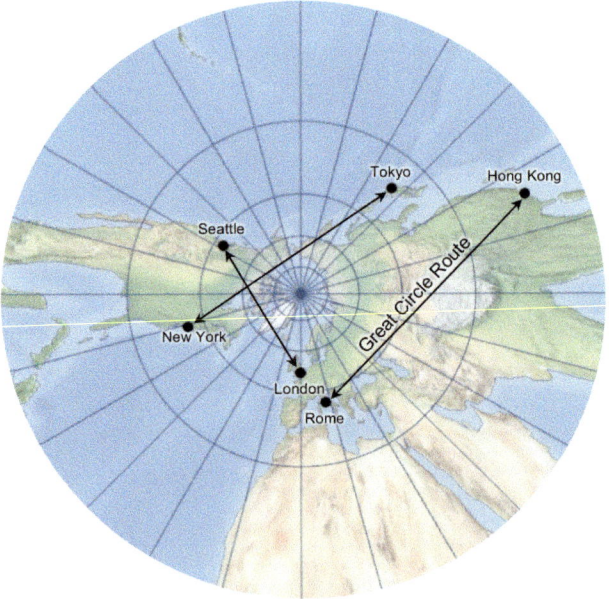

Figure 3.18. Polar gnomonic projection of the Northern Hemisphere from 15° N to the pole. All straight lines on the projection surface are great-circle routes.

that can show the entire surface of the earth, although the distortion in shape and area toward the edges of the map is extreme because the east–west SF increases to infinity at the South Pole, which is shown as a circular line.

Lambert azimuthal equal area

In 1772, the mathematician and cartographer **Johann Heinrich Lambert** published equations for the tangent-case planar **Lambert azimuthal equal-area projection**, which, along with other projections he devised, carries his name. Although this map projection can show the entire earth, there is great distortion along the edges. This equal-area projection preserves land features at their true relative sizes, so it is well suited for thematic hemisphere maps and thematic maps of polar regions. Because of the extreme distortion at the edges, it is often clipped into a circular or other shape that is smaller than the projection's extent, as shown in figure 3.20A. It has been used historically in commercial atlases. More recently, this projection is used for statistical maps of continents and countries that are basically circular in overall extent (figure 3.20B), such as Africa and North America. You will also see the oceans shown on maps that use the polar, equatorial, or oblique aspects of this projection, such as the south polar aspect map in figure 3.20B that shows the Southern, Indian, South Pacific, and South Atlantic oceans as a single large ocean.

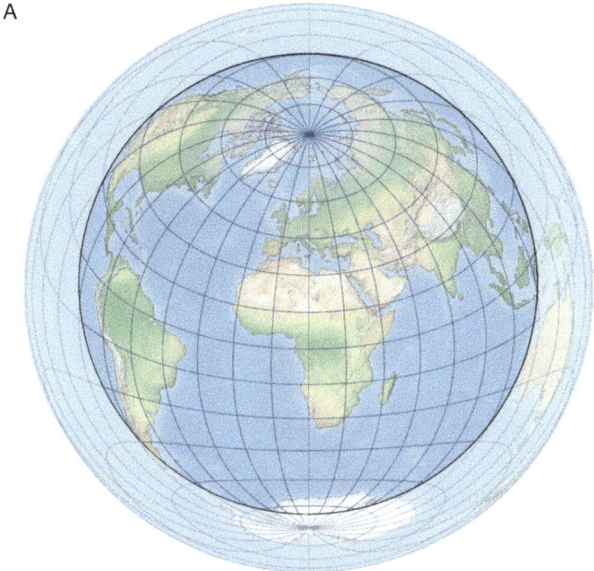

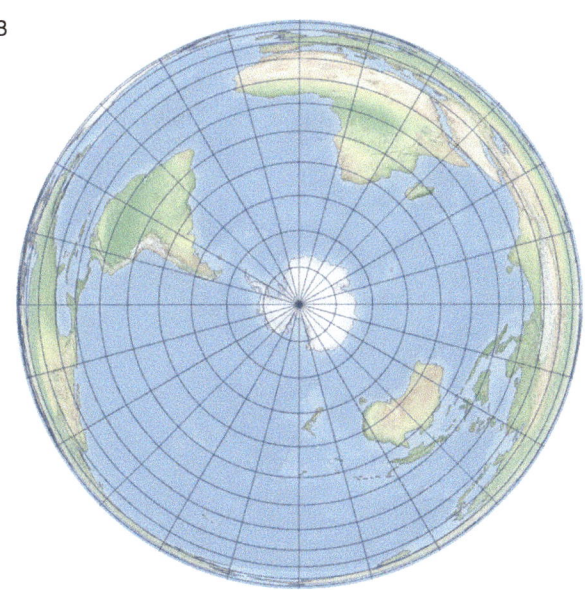

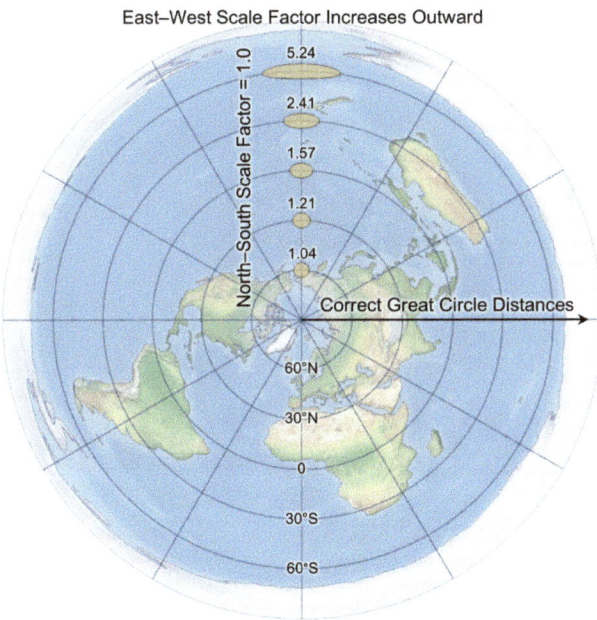

Figure 3.19. Polar-aspect azimuthal equidistant projection for a map of the world. Great-circle distances are correct along straight lines outward from the point of tangency at the pole.

Figure 3.20. The Lambert azimuthal equal-area projection is best suited for thematic maps of polar regions and thematic hemisphere maps. The first map (*A*), with the central meridian set to 15° E and the latitude of origin at 25° N, shows how the distorted edge may be clipped off. The second map (*B*) is a polar aspect projection of the Southern Hemisphere.

Cylindrical projections

Equidistant cylindrical

The **equidistant cylindrical projection** or **equirectangular projection** is also called the **plate carrée** (meaning "flat square" in French) or also sometimes the **geographic projection**. This simple map projection, nearly 2,000 years old, is attributed to **Marinus of Tyre**, a Roman geographer, cartographer, and mathematician credited with founding mathematical geography. He is thought to have constructed the projection in about AD 100. Parallels and meridians are mapped as a grid of equally spaced horizontal parallels and vertical meridians, with lines that represent the parallels twice as long as the meridians (figure 3.21). The equal spacing of parallels means that the projection is equidistant in the north–south direction, with a constant SF of 1.0. In the east–west direction, the SF increases steadily from a value of 1.0 at the equator to infinity at each pole, which is projected as a straight line (and therefore has point-to-line correspondence, as we saw earlier in this chapter).

Owing to its ease of construction, the equidistant cylindrical projection was often used in the past, and its simplicity of projecting the spherical earth on a flat map has made it the projection of choice for use in computer mapping apps—particularly geographic information systems (GIS) and web pages. You may see world maps that show elevation data or satellite imagery made in this projection.

Mercator

Invented by **Gerhardus Mercator** in 1569, the **Mercator projection** is a tangent-case cylindrical conformal projection. As with all conformal projections, shape is preserved in small areas. This projection offers a classic example of how a single projection can be used both poorly and well. Looking at the projection (figure 3.22), you can imagine Mercator starting the construction of his projection with a horizontal line to represent the equator and then adding equally spaced vertical lines to represent the meridians. Mercator knew that meridians on the globe converge toward the poles so that meridians he drew as parallel vertical lines must become progressively more widely spaced toward the poles than they would be on the generating globe. He progressively increased the spacing of parallels away from the equator so that the increase matched the increased spacing between the meridians. As a result of this extreme distortion toward the poles, he cut off his projection at 80° N and S. This spacing of parallels produced a conformal map projection. More importantly, it produced the only projection in which all lines of constant compass direction, called **rhumb lines**, are straight lines on the map.

Navigators who used a magnetic compass immediately saw the advantage of plotting courses on maps using the Mercator projection, because any straight line they drew was a line of constant compass **bearing**, or the azimuth of a track. This geometric property meant that they could plot

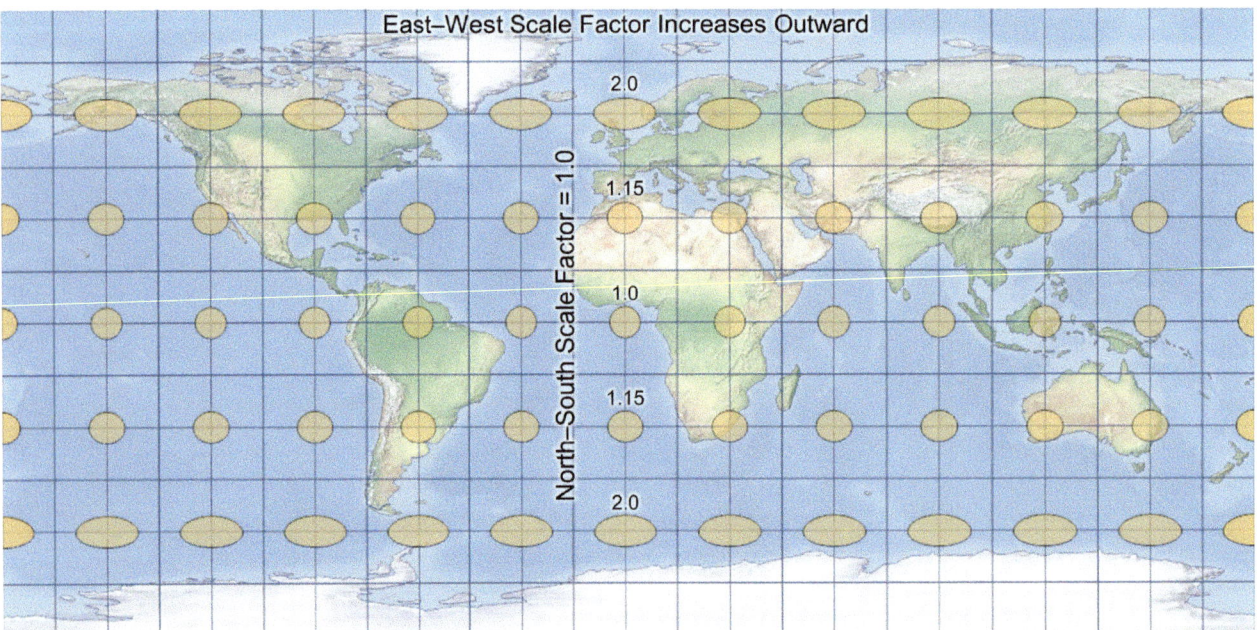

Figure 3.21. Equidistant cylindrical world map projection with the standard parallel set to the equator.

a course on the map and simply maintain the associated bearing during passage to arrive at the plotted location. It is not hard to understand why navigators preferred a map on which compass bearings appeared as straight lines. The Mercator projection has been used ever since for nautical charts, such as small-scale navigational charts of the oceans. Large-scale nautical charts used for coastal navigation can be thought of as small rectangles cut out of a world map made in the Mercator projection.

Of course, using these maps for navigation is not as simple as plotting a single straight line and then maintaining this heading. Recall that the gnomonic projection is the only projection in which all great circles on the generating globe are shown as straight lines on the flat map. Lines drawn in a Mercator projection show constant compass bearing, but they are not the same as the great-circle route, which is the shortest distance between two points on a globe. Therefore, the Mercator projection is often used in conjunction with the gnomonic projection to plot navigational routes.

The use of the Mercator projection in navigation is an example of a projection used for its best purpose. An example of a poor use of the Mercator projection is for wall maps of the world. As mentioned earlier, this projection is often cut off at 80° N and S, which creates a rectangular map with a height-to-width ratio that fits walls well. The problem, of course, is the extreme scale enlargement and consequent area distortion at higher latitudes. The area exaggeration of Canada, Greenland, northern Europe, and Russia gives many people an erroneous impression of the size of the land masses in the Northern Hemisphere.

Web Mercator

The **web Mercator projection**, a variation of the Mercator world projection, rose to prominence in computer mapping when it was adopted for use in Google Maps in 2005. It is now used by virtually all major online map providers, including Esri. Fundamentally, it is the Mercator world projection with latitude limits of 85.05° N and S. These bounding parallels make the projection a square. It is mathematically similar to the Mercator world projection on the authalic sphere but uses latitude-longitude coordinates based on the WGS84 ellipsoid for map production. These ellipsoid coordinates, coupled with slightly different mathematical formulas, make the web Mercator projection not precisely conformal.

The difference between the web and world Mercator projections is imperceptible on small-scale world maps, but on large-scale maps of local areas, the difference becomes noticeable. The discrepancy becomes progressively larger

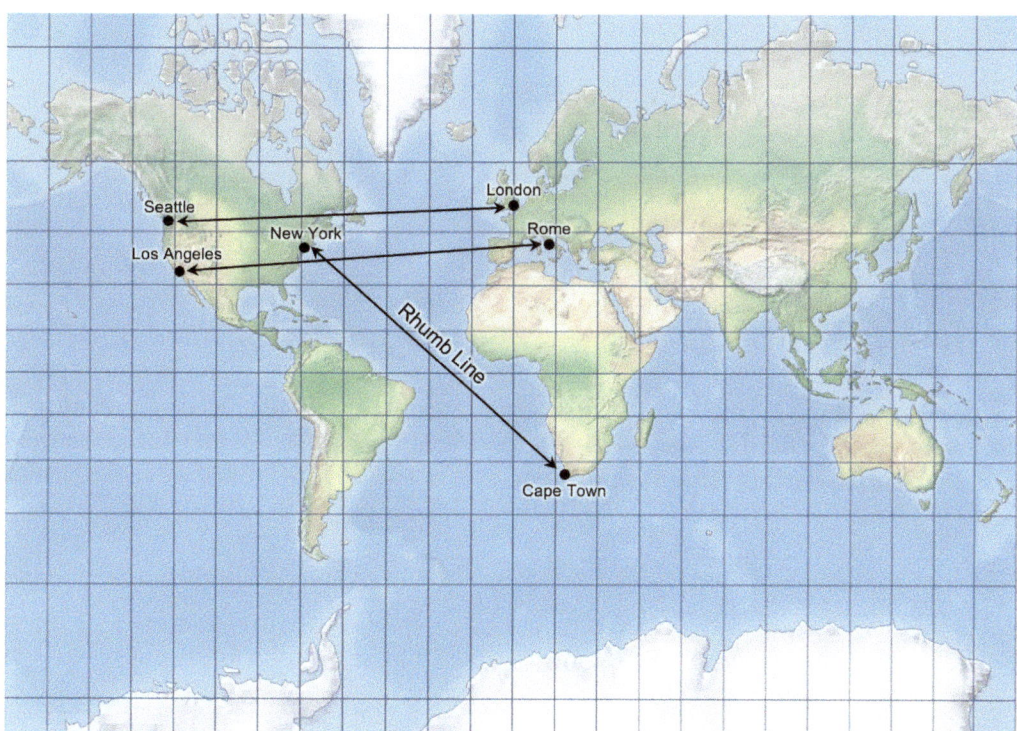

Figure 3.22. Mercator projection shown with rhumb lines between selected major world cities.

farther north or south of the equator, and it can reach as much as 35 kilometers (about 22 miles) at its northern and southern limits.

Transverse Mercator

In 1772, the same year that Lambert constructed his azimuthal equal-area projection, he also constructed the **transverse Mercator projection**, along with the Lambert conformal conic projection described in the next section. In Europe, the transverse Mercator projection is called the **Gauss-Krüger projection**, in honor of the mathematicians **Carl Gauss** and **Johann Krüger**, who later worked out formulas describing its geometric distortion and the equations for making it on the ellipsoid. Lambert's idea for the transverse Mercator projection was to rotate the Mercator projection by 90 degrees so that the line of tangency becomes a pair of meridians—that is, any selected meridian and its antipodal meridian (see figure 3.14, *top right*). The resulting projection, the transverse Mercator projection, is conformal, but rhumb lines are no longer straight lines. Along the line of tangency, the SF is 1.0, and the scale increases perpendicularly away from the line of tangency.

You are likely to see the transverse Mercator projection used to map north–south strips of the earth called **gores** (figure 3.23), which are used in the construction of globes because narrow north–south strips of the earth are projected with no local shape distortion and little distortion of area. Because printing the earth's surface directly on a spherical surface is difficult, a map of the earth is printed as flat, elongated gores that are cut out along the meridians that form their edges and then pasted on a spherical base surface.

The narrow, six-degree-wide zones of the universal transverse Mercator grid coordinate system (described in chapter 4) are based on a secant-case transverse Mercator projection. North–south trending zones of the US state plane coordinate system (also explained in chapter 4) are also based on secant cases of the projection. Most historical 1:24,000-scale USGS topographic maps use the transverse Mercator projection.

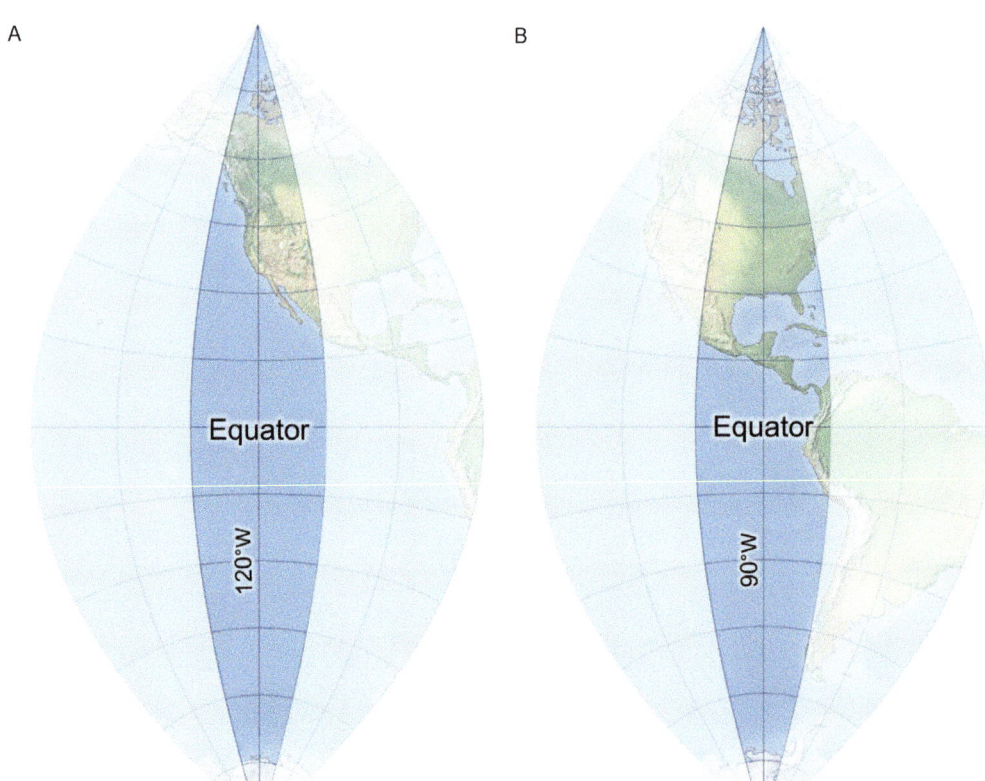

Figure 3.23. Two gores of the globe in transverse Mercator projections that are 30 degrees wide at the equator and centered at 120° W (*A*) and 90° W (*B*). To make a world globe, the gores are trimmed to their edge meridians and then pasted on the sphere used to make the globe.

Conic map projections

Lambert conformal conic

The **Lambert conformal conic projection** is another widely used map projection devised mathematically by Lambert in 1772. It is a secant-case, normal-aspect conic projection with its two standard parallels placed to minimize the map's overall scale distortion. The standard parallels for maps of the conterminous United States are placed at 33° N and 45° N to minimize scale distortion within the mapped area with a maximum of less than 3 percent at the map's edges (figure 3.24).

Although the transverse Mercator projection is used as the basis for the state plane coordinate system zones in north–south trending states in the United States, the Lambert conformal conic projection is used as the basis for east–west zones (explained in chapter 4), such as Oregon and Wisconsin.

One major use of the Lambert conformal conic projection is for aeronautical charts in mid-latitude countries—all US 1:500,000-scale sectional charts are in this projection. Air navigators prefer navigational charts that use conformal projections, which preserve local shapes and directions. Equally important is that straight lines drawn on the charts are close to being great-circle routes over the earth's surface (figure 3.24). The Mercator projection, of course, is also conformal, but the Lambert conformal conic projection has the advantage of much smaller scale distortion in the mid-latitudes (and, of course, no distortion along the two standard parallels).

Albers equal-area conic

Mathematically devised in 1805 by the German mathematician **Heinrich C. Albers**, the **Albers equal-area conic projection** was first used in 1817 for a map of Europe. Today, this projection is most often used for statistical maps of the conterminous United States and other mid-latitude east–west trending regions, such as Europe and Australia. For example, you will see this projection used for statistical maps created by the US Census Bureau and other federal agencies in the United States. The reason for the widespread use of the Albers projection is simple—people looking at maps that use this projection can assume that areas on the map are proportionately true to their areas on the earth.

The secant-case version that has been used for nearly 100 years for the conterminous United States has standard parallels placed at 29.5° N and 45.5° N (figure 3.25). This placement reduces the scale distortion to less than 1 percent at the 37th parallel in the middle of the map, and

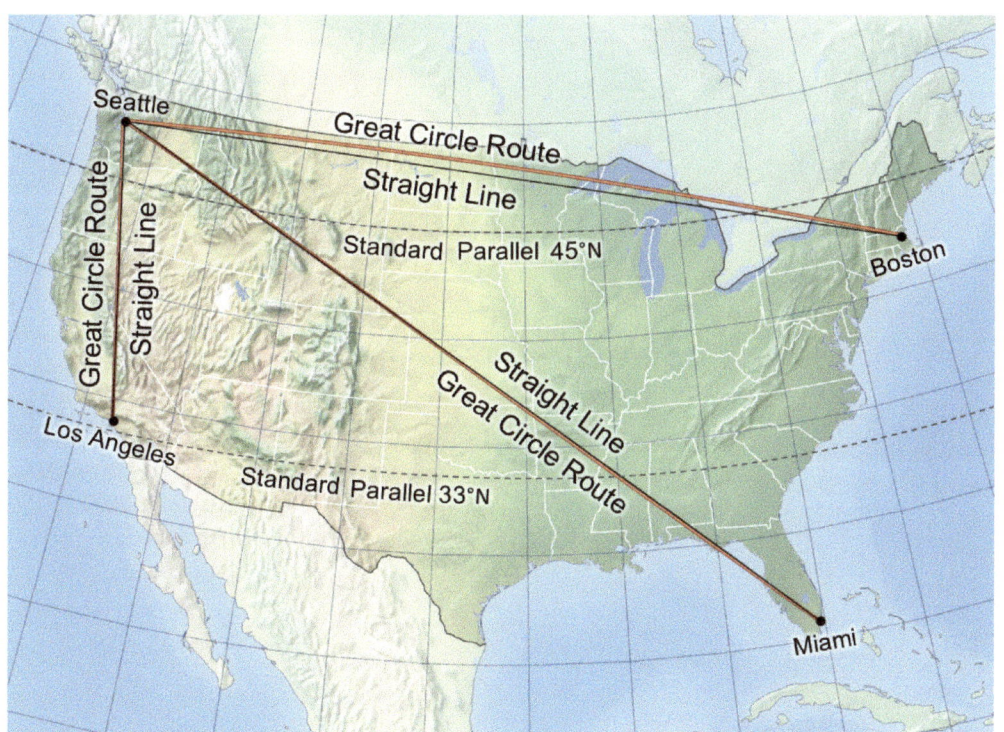

Figure 3.24. Lambert conformal conic projection used for a map of the conterminous United States. Straight lines are close to being great-circle routes, particularly on north–south paths.

to 1.25 percent at the northern and southern extents of the country. It also gives the minimum average distortion in shape for the area mapped, as measured by what cartographers call **angular deformation**. Maps of Alaska use standard parallels of 55° N and 65° N, and maps of Hawaii use standard parallels of 8° N and 18° N. Some USGS products, such as the national tectonic and geologic maps, also use the Albers projection, as do reference maps and satellite image maps of the country at a scale of around 1:3,000,000.

Pseudocylindrical and other projections

You have seen that map projections can be classed into families based on the nature of the surface used to construct the projection, resulting in planar, cylindrical, and conic projections. We also explained how map projections can be classified based on their geometric distortion properties relating to distance, shape, direction, and area. Next, we look at a last set of projections that don't strictly fit into one or both of the families we've already described. For example, **pseudocylindrical map projections** are similar to cylindrical projections in that parallels are horizontal lines and meridians are equally spaced. The difference is that all meridians except the vertical-line central meridian are curved instead of straight. We also examine a few map projections that are of special interest for a variety of reasons, including orthophanic projections.

For many purposes, a projection that looks "right" is more important than a projection that rigidly provides fidelity to area, distance, shape, or direction. A right-appearing projection is called **orthophanic**, a **compromise projection** that is neither equal area nor conformal nor equidistant but rather is an attempt to balance these geometric properties.

Mollweide

If you have seen a world map in the shape of an oval, most likely you were looking at a map made using the **Mollweide projection**. This projection was invented in 1805 by the German mathematician **Carl B. Mollweide** as an elliptically shaped equal-area projection with the equator as the standard parallel and the prime meridian as the central meridian (figure 3.26). Parallels are projected as horizontal lines, but they are not equally spaced as on the generating globe. Instead, the distance in spacing is reduced toward the poles, and the overall distortion in shape is less than in other equal-area world projections.

The elliptical shape of this projection makes it look more earthlike, making it a popular projection for showing human-oriented statistical data for the world. The Mollweide projection is used for world maps that show a range of global phenomena, from population to land cover to major diseases. Cartographers have devised other versions of this projection by adjusting the central meridian to better show the oceans or to center attention on a particular continent.

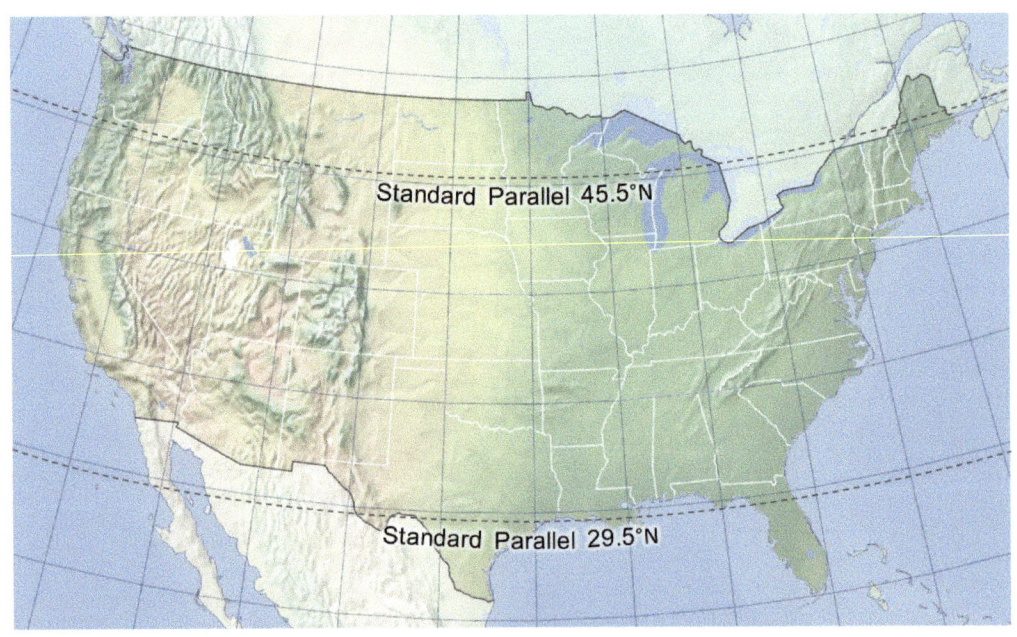

Figure 3.25. Albers equal-area conic projection of the conterminous United States with standard parallels at 29.5° N and 45.5° N, resulting in maximum scale distortion of 1.25 percent at the north and south extents of the conterminous states.

Sinusoidal

According to some sources, **Jean Cossin** of Dieppe, France, appears to be the originator of the **sinusoidal projection**, which he used to create a world map in 1570. Others suggest that Mercator devised it because it was included in later editions of his atlases. Like the Mollweide projection, it is an equal-area projection of the world with straight parallels and curved meridians, but the distinctive feature of this projection is the pointed shape of the poles (figure 3.27). It was used by **Nicholas Sanson** (circa 1650) of France for atlas maps of the world and continents and by **John Flamsteed** (1729) of England for celestial maps. Hence, you may also see it called the **Sanson-Flamsteed projection**.

In addition to correctly portraying the relative areas of continents and countries, the sinusoidal world projection has an SF of 1.0 along the central meridian, and the east–west SF is 1.0 anywhere on the map. This projection is equidistant in the east–west direction and in the north–south direction along the central meridian, but only in these directions. Severe shape distortion is also evident at the edges of the map.

Goode homolosine

World map projections can also be made by combining different projections along certain parallels or meridians. One example is the **uninterrupted Goode homolosine projection**, or more commonly called the **Goode homolosine projection** (figure 3.28, *top*). Constructed in 1923 by American geography professor **J. Paul Goode**, it is a composite of two equal-area pseudocylindrical projections. The sinusoidal projection is used for the area from 40° N to 40° S latitude, whereas Mollweide projections are used for areas from 40° N and 40° S to the respective poles. Because

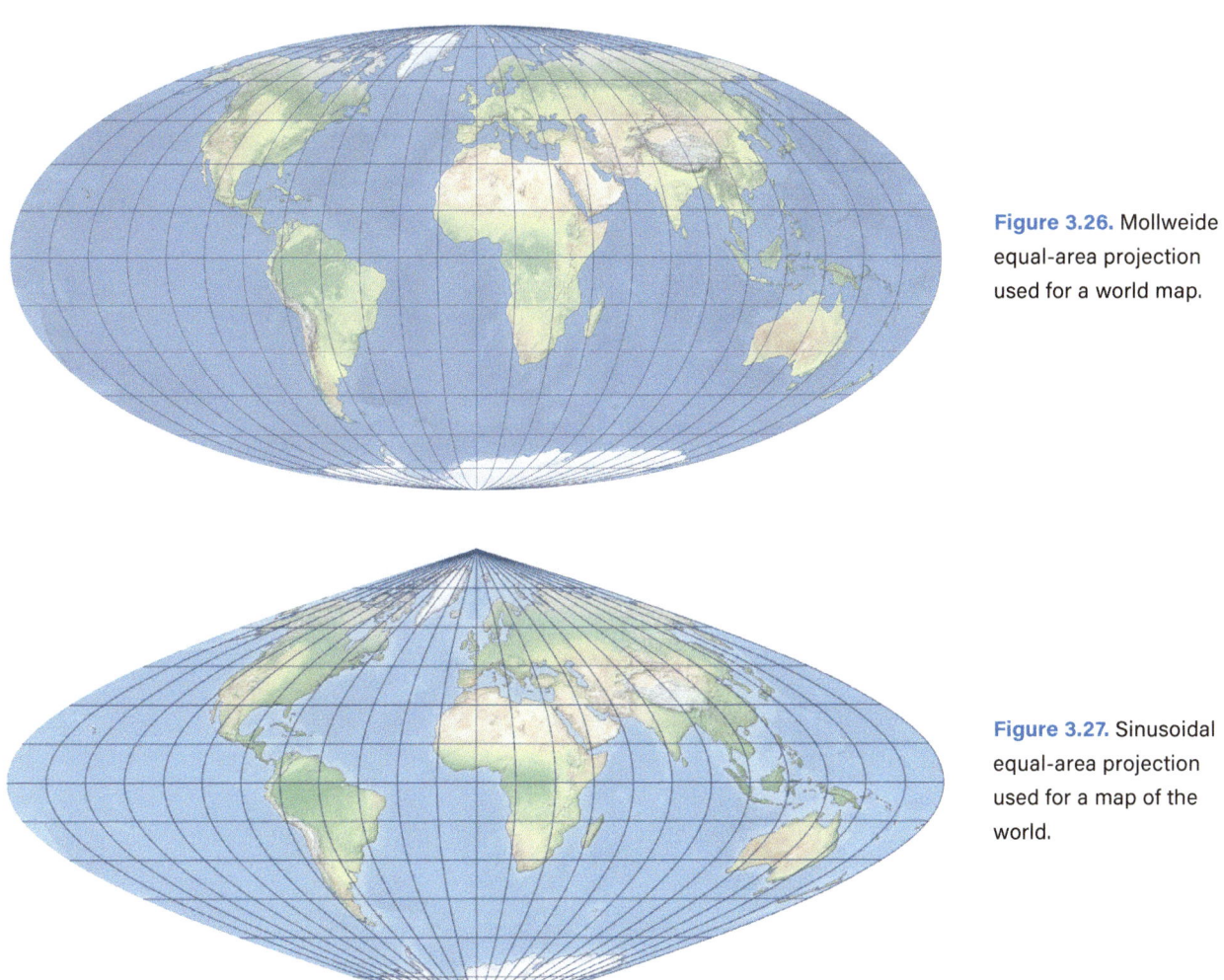

Figure 3.26. Mollweide equal-area projection used for a world map.

Figure 3.27. Sinusoidal equal-area projection used for a map of the world.

the two component projections are equal area, the homolosine projection is also equal area. Compare the maps in figures 3.26 and 3.27 with the map at the top of figure 3.28—the shapes of the continents look less distorted in the homolosine projection than in the Mollweide and sinusoidal projections, particularly in the polar areas.

An **interrupted projection** is one in which the generating globe is segmented to minimize the distortion within any **lobe** (section) of the projection. Shape distortion at the edges of the map can be lessened considerably by interrupting the composite projection into lobes that are pieced together along a central line, usually the equator. With interruption, the less geometrically distorted parts of the projection are used within each lobe.

The **interrupted Goode homolosine projection** (figure 3.28B), sometimes shortened to the **interrupted homolosine projection**, was created in 1923 by Goode from the uninterrupted homolosine projection. It is an interrupted pseudocylindrical equal-area composite map projection used for world maps. The projection is a composite of 10 Mollweide and sinusoidal projection pieces that form six interrupted lobes. The two lobes at the top and four at the bottom are Mollweide projections from 40° N or 40° S to the adjacent pole, each with a different central meridian. The four interior regions from the equator to 40° N and 40° S are sinusoidal projections, each with a different central meridian. The two northern sections are usually shown

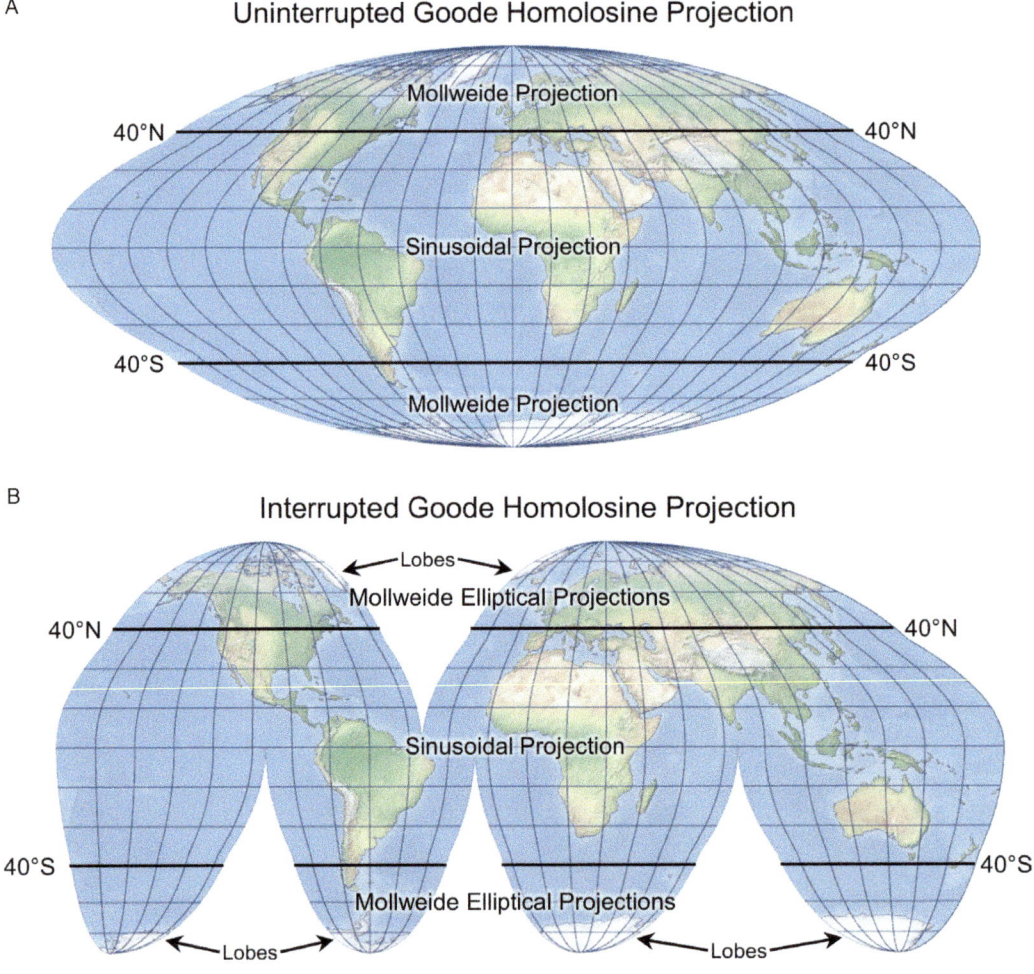

Figure 3.28. Uninterrupted Goode homolosine projection (*A*) and interrupted Goode homolosine equal-area world projection (*B*).

with some land areas repeated in both regions to show the Greenland land mass without interruption.

You will find the interrupted homolosine projection used historically for maps in commercial world atlases, most notably *Goode's World Atlas*, used as an educational resource since it was first published in 1923, to show a variety of global information. It is a popular projection for showing physical information about the entire earth, such as land elevations and ocean depths, land cover and vegetation types, and satellite image maps. In this projection, either land areas or water areas can be shown nearly in their entirety (figure 3.29), depending on the placement of the central meridians for each of the lobes. When viewing these maps, remember that the continuity of the earth's surface is lost by interrupting the projection, which may not be apparent if the graticule is left off the map.

Robinson

In 1963, the prominent American academic cartographer and professor of geography **Arthur H. Robinson** constructed a pseudocylindrical projection, the **Robinson projection**, which is neither equal area nor conformal but a compromise projection that makes the continents look right. For this orthophanic projection (figure 3.30), Robinson visually adjusted the horizontal parallels and curving meridians until they appeared suitable for use as a world map projection for atlases and wall maps. To make this adjustment, Robinson represented the poles as horizontal lines that are a little over half the length of the equator. You may have seen this projection on world maps or wall maps of the world that show the shape and area of continents with far less distortion than maps made with the Mercator world projection.

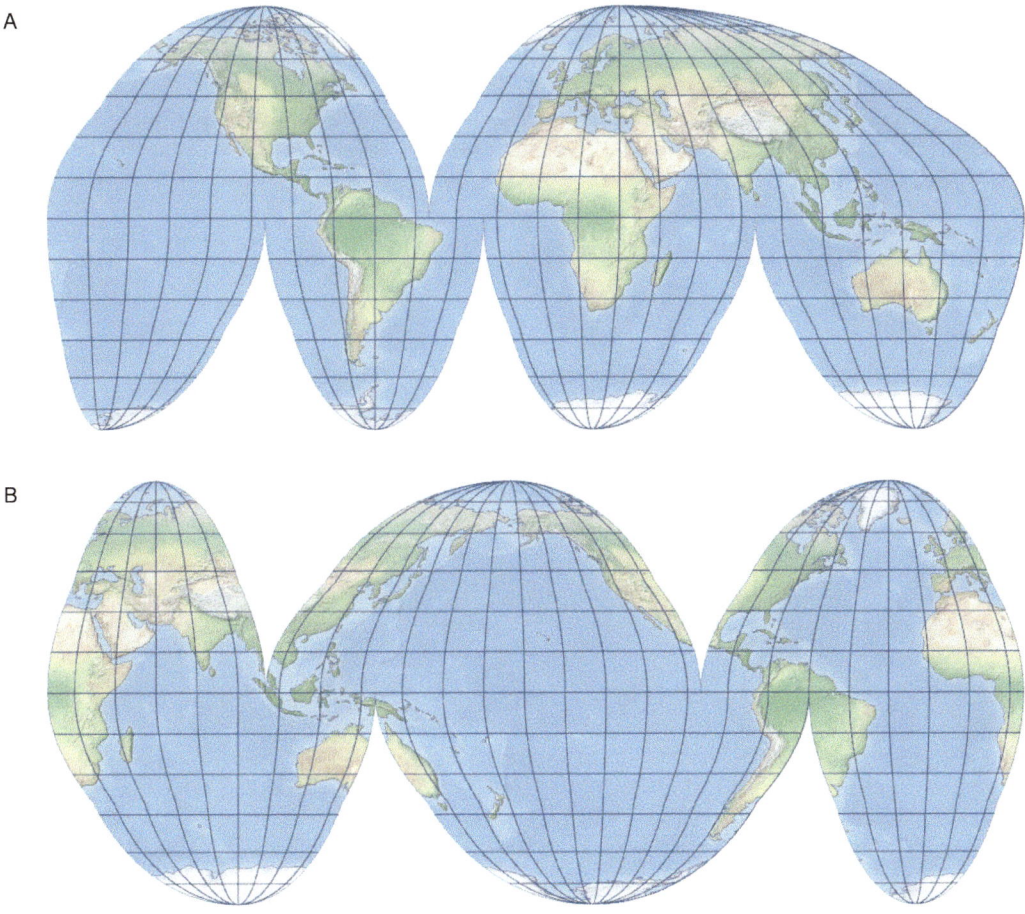

Figure 3.29. Interrupted homolosine projection showing land areas uninterrupted except for Antarctica (*A*) and interrupted Goode homolosine projection showing oceans uninterrupted except for the Arctic and Southern Oceans (*B*).

Natural Earth

The Natural Earth projection was designed by Tom Patterson in 2007 for displaying physical data on world maps (figure 3.31). It is a compromise pseudocylindrical map projection for world maps. The projection has rounded meridians that meet the horizontal pole lines, suggesting that the earth has a rounded shape. The Natural Earth projection is neither conformal nor equal area. As with the Robinson and other compromise pseudocylindrical projections, shapes, areas, distances, and directions are distorted throughout the projection. Area distortion increases with latitude and does not change with longitude. Angular distortion is moderate near the center of the map and increases toward the edges. Distortion values are symmetric north and south of the equator and east and west of the central meridian.

Winkel tripel

Map projections may also be constructed as mathematical combinations of two or more projections. Perhaps the best-known example is the **Winkel tripel projection** constructed in 1921 by the German cartographer **Oswald Winkel**. The

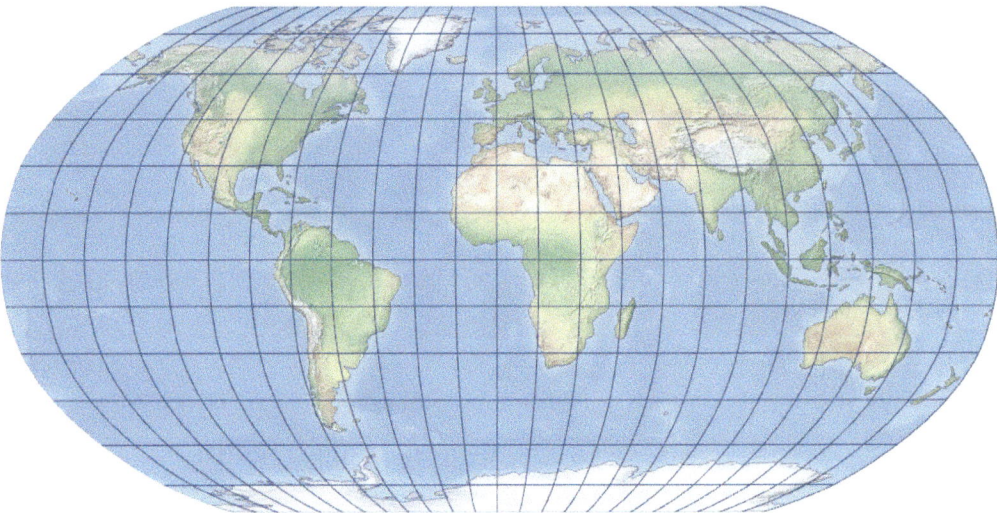

Figure 3.30. Robinson pseudocylindrical compromise projection used for a world map.

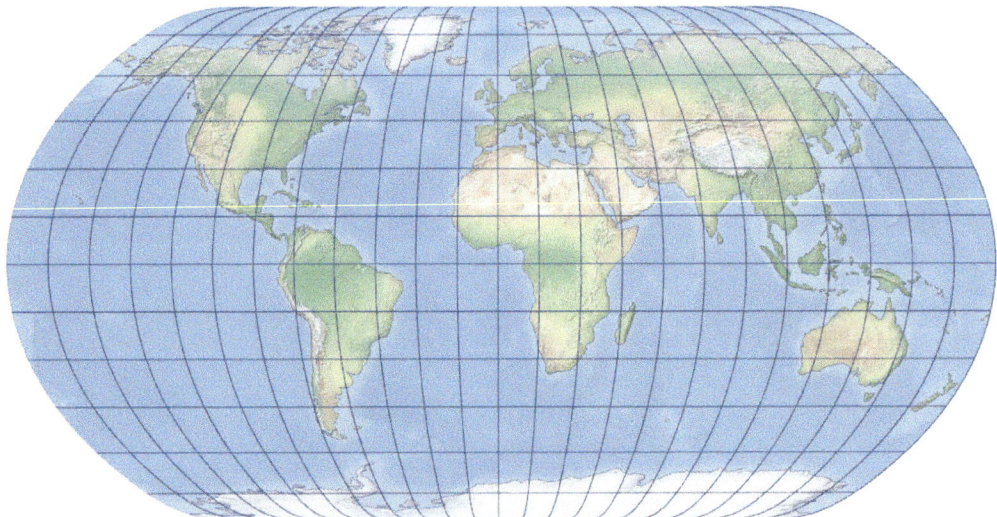

Figure 3.31. Natural Earth pseudocylindrical compromise projection used for a world map.

term **tripel** is not someone's name but rather a German word that means a combination of three elements. Winkel used the term to emphasize that he had constructed a compromise projection that was neither equal area nor conformal nor equidistant, but rather that it minimized all three forms of geometric distortion. The Winkel tripel projection looks similar to the Robinson projection, but if you look closely, you will see that parallels are not the straight, horizontal lines characteristic of pseudocylindrical projections. Rather, they are slightly curving, nonparallel lines (figure 3.32).

The Winkel tripel projection was not used widely until 1998, when the National Geographic Society announced that it was adopting the projection as its standard for maps of the entire world. As a result, use of the Winkel tripel projection has increased dramatically. One advantage of using this projection for web maps is that the **aspect ratio** (the width-to-height ratio) of 16:10 is convenient for on-screen display because most high-definition computer monitors have an aspect ratio of 16:10.

Oblique-perspective projections

True-perspective projections, such as the orthographic, stereographic, and gnomonic projections described earlier, are **vertical-perspective projections** because the center point of projection is the point of tangency of the projection surface. Any other alignment produces an **oblique-perspective projection** of the earth. In figure 3.33, the vertical-perspective developable surface is a horizontal plane that is tangent to the generating globe at the North Pole, the polar-aspect projection point of tangency. The mapped area extends outward from the point of tangency to the parallel at which rays from the light source are seen on the generating globe as being on the horizon (tangent to the globe).

An oblique-perspective, polar-aspect projection has the same light source above the North Pole, but the developable surface is tilted and secant to the generating globe. In figure 3.33, the developable surface intersects the globe at the North Pole and at about 37° N. The rays from the light source to the parallels on the generating globe intersect the vertical-perspective developable surface in a symmetrical manner to the right and left of the North Pole (the ever closer spacing of rays intersecting the developable surface is the same). Now examine how the rays from the light source intersect the oblique-perspective developable surface to the right and left of the North Pole. The rays to the right of the pole are more closely spaced on the oblique developable surface than those to the left of the pole. We use the term **foreshortening** for this compression and extension

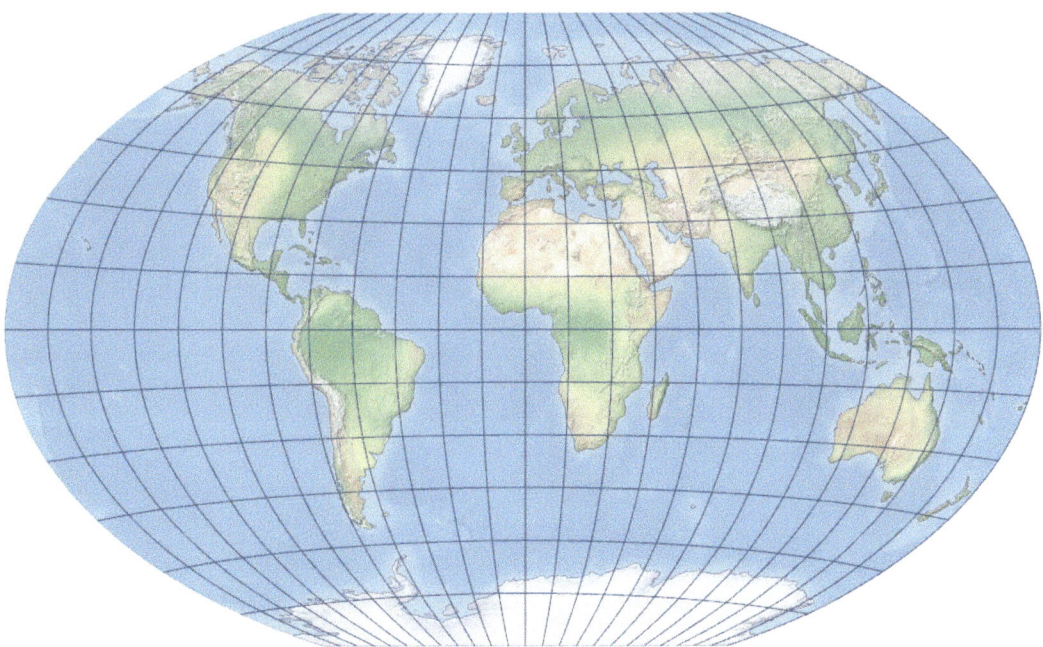

Figure 3.32. Winkel tripel projection used for a world map.

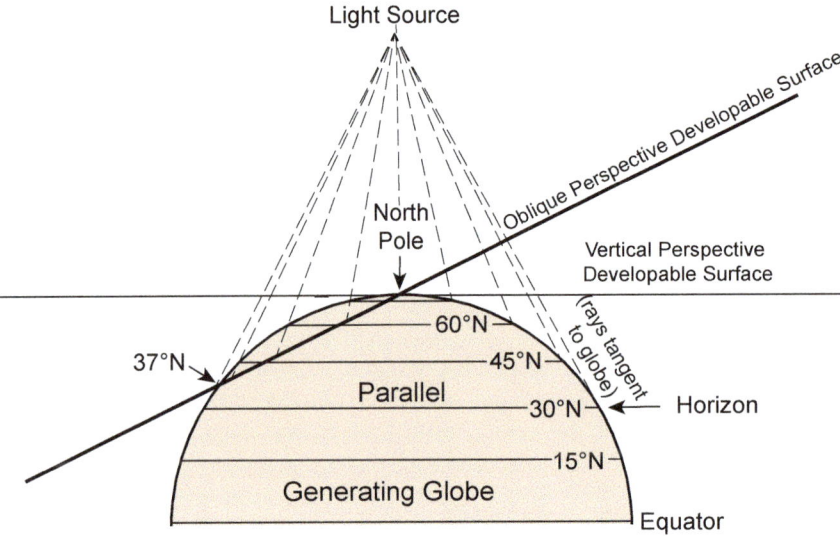

Figure 3.33. Geometry of vertical- and oblique-perspective map projections. Adapted from Snyder, 1987, *Map Projections: A Working Manual*, p. 170.

of the projected positions of features equally spaced on the generating globe. Foreshortening is what gives oblique-perspective projections the appearance of viewing part of the earth obliquely from a point above the surface of the earth. Tilting the oblique-perspective developable surface even more and lowering the light source closer to the globe creates an **oblique-perspective map** of a smaller area, such as California and Nevada as shown in figure 3.34.

Web apps such as ArcGIS Earth use oblique-perspective projections to show the globe as it appears from space. In these apps, you can interactively pan and zoom, flying over the earth from the perspective of a spacecraft or an airplane.

ArcGIS also supports scene views that can be either local or global. Global scenes are used to display data that spans the globe or if viewing the curvature of the earth is important. Global scenes show data in a global context, allowing map users to zoom in and out and see the entire earth. Global scenes support the WGS84 geographic coordinate system. For other spatial reference systems, a local scene can be used to avoid the distortion from being reprojected for global scene views. Local scenes are used to view data at a smaller extent. Because data can remain in a projected coordinate system, local scenes can be used for editing, analysis, or measurement. Local scenes support geographic, projected, and custom coordinate systems. Often local scenes are used to visualize extents such as cities or construction sites.

Map projections on the sphere and ellipsoid

You saw in chapter 1 that mapmakers have a general rule that small-scale maps can be projected from a sphere, but large-scale maps must always be projected from an ellipsoidal surface using a datum such as WGS84. Small-scale world or continental maps normally use coordinates based on a spherical approximation of the ellipsoidal earth because it is much easier to construct these maps from spherical geocentric coordinates. Equally important, the differences in the plotted positions of spherical and corresponding geodetic coordinates are negligible on small-scale maps.

Large-scale maps are projected from an ellipsoidal surface because, as you saw in chapter 1, the spacing of parallels decreases slightly but significantly from the poles to the equator. We note in chapter 1 that on the WGS84 ellipsoid, the distance between two points that are one

Figure 3.34. Oblique-perspective projection used for a map of North America.

degree apart in latitude at the equator, between 0° and 1°, is 110.567 kilometers (68.703 miles), a little more than a kilometer shorter than the 111.699-kilometer (69.407-mile) distance between two points at the North or South Poles between 89° and 90°.

Now we examine the differences in length when you project three lines that span one degree of geodetic latitude using the transverse Mercator projection, one along a meridian at the North Pole, the second at the 45th parallel, and the third at the equator (figure 3.35). At a scale of 1:1,000,000, a line the length of a degree of latitude from 89° to the adjacent pole is projected as slightly over one millimeter longer than a line that is one degree of latitude in length from the equator to 1°. The difference in these two lengths that you can see at the bottom of the lines may seem minimal, but on the ground, it represents slightly over a kilometer (0.62 miles), as you would expect from the distances listed in the previous paragraph.

At larger map scales, this slightly greater than a kilometer difference in ground length for a degree of latitude at the equator and the poles becomes more noticeable in the transverse Mercator projection—for instance, slightly

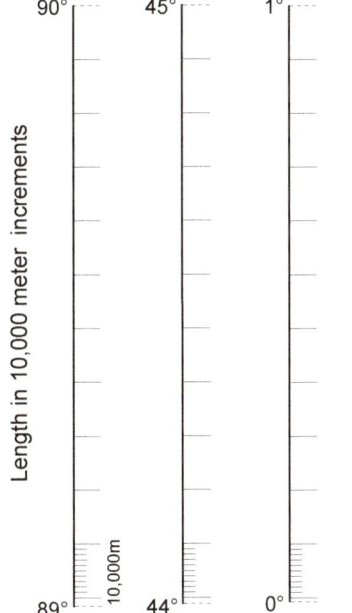

Figure 3.35. Transverse Mercator map projections for lines one degree of geodetic latitude in length on the WGS84 ellipsoid between 89° and 90°, 44° and 45°, and 0° and 1°.

over one centimeter on polar and equatorial maps at a scale of 1:100,000. Topographic and other maps at this scale and larger are projected from an ellipsoid so that accurate distance and area measurements can be made. The same holds true for large-scale equidistant projections, such as the polar-aspect azimuthal equidistant projection, because the spacing of parallels in the projection must be slightly lessened from the pole outward to reflect their actual spacing on the earth.

As discussed in chapter 1 and in appendix table A.2, the only place on the earth where 69.09 miles (111.20 kilometers) per degree spacing of parallels on the authalic sphere is the same as on the WGS84 ellipsoid is in the mid-latitudes close to the 45th parallel. Imagine examining two maps that straddle the 45th parallel made at the same scale with the transverse Mercator map projection, one based on the authalic sphere and the second based on the WGS84 ellipsoid. Lines that are one degree of latitude in length on the earth are projected as north–south lines of the same length in the two projections.

Map projection selection

We have described different map projections with differing geometric properties made for different map uses. Some GIS software, such as ArcGIS Pro, makes selecting the right projection easier by providing interactive filters. You narrow the number of appropriate projections for your map by selecting the spatial extent, projection property, and geographic coordinate system. You can also specify that you want to use a map projection that is defined by a particular geographic coordinate system, such as WGS84 (see chapter 1 for more on these systems).

Alternatively, you can use the information in table 3.1 as a guide to selecting an appropriate projection. With this table, map projection selection is a three-step process, beginning with determining the size of the region to be mapped—the entire world, a hemisphere, or a continent or smaller. You next select the projection property to be maintained: conformality, equivalence, equidistance, or another special property listed in the table. You then choose the desired projection characteristic, such as constant scale along the equator or a meridian. These three decisions lead you to a recommendation on an appropriate projection for your map. For example, the Mollweide projection is appropriate for the entire world when equal area and not being noninterrupted are important selection properties and characteristics.

Table 3.1. Map projection selection guide

Region mapped	Property	Characteristic	Appropriate projection
World	Conformal	Constant scale along the equator	Mercator
		Constant scale along a meridian	Transverse Mercator
	Equal area	Noninterrupted	Mollweide
			Sinusoidal
		Interrupted	Goode's homolosine
	Equidistant	Centered on a pole	Polar azimuthal equidistant
		Centered on a city	Oblique azimuthal equidistant
		Straight rhumb lines	Mercator
		Compromised distortion	Robinson
			Winkel tripel
Hemisphere	Conformal		Stereographic
	Equal area		Lambert azimuthal
	Equidistant		Azimuthal equidistant
	Global view		Orthographic
Continent or smaller	Conformal	East-west region along the equator	Mercator
		East-west mid-latitude region	Lambert conformal conic
		North-south region	Transverse Mercator
		Extent at an angle	Oblique Mercator
		Equal north-south, east-west extent	Oblique stereographic
	Equal area	East-west region along equator	Cylindrical equal area
		East-west mid-latitude region	Albers equal area conic
		Equal north-south, east-west extent	Lambert azimuthal
	Local view	Oblique view of a region	Oblique perspective
	Straight navigational routes	All great circles are straight lines	Gnomonic azimuthal

Modified from Snyder, 1987, *Map Projections: A Working Manual* (see selected readings for this chapter).

Selected readings

Abee, M. 2021. "The Spread of the Mercator Projection in Western European and United States Cartography." *Cartographica* 56 (2): 151–65.

Battersby, S. 2017. "Map Projections." In *The Geographic Information Science & Technology Body of Knowledge*. Edited by John P. Wilson. https://gistbok-topics.ucgis.org/CV-03-006.

Battersby, S. E., M. P. Finn, E. L. Usery, and K. H. Yamamoto. 2014. "Implications of Web Mercator and Its Use in Online Mapping." *Cartographica* 49 (2): 85–101. DOI: 10.3138/carto.49.2.2313.

Bugayevskiy, L. M., and J. P. Snyder. 1995. *Map Projections: A Reference Manual*. Taylor & Francis.

Finn, M. P., E. L. Usery, L. Woodard, and K. H. Yamamoto. 2017. "The Logic of Selecting an Appropriate Map Projection in a Decision Support System (DSS)." Chapter 10 in *Choosing a Map Projection*. Lecture Notes in Geoinformation and Cartography. Springer Nature Link. DOI: 10.1007/978-3-319-51835-0_10.

Lally, N. 2022. "Sculpting, Cutting, Expanding, and Contracting the Map." *Cartographica* 57 (1): 1–10.

Lapaine, M., and N. Frančula. 2022. "Map Projections Classification." *Geographies* 2 (2): 274–85.

Monmonier, M. 2004. *Rhumb Lines and Map Wars: A Social History of the Mercator Projection*. University of Chicago Press.

Mulcahy, K., and K. C. Clarke. 2001. "Symbolization of Map Projection Distortion: A Review." *Cartography and Geographic Information Science* 28 (3): 167–81. DOI: 10.1559/152304001782153044.

Osborne, P. 2013. "The Normal and Transverse Mercator Projections on the Sphere and the Ellipsoid with Full Derivations of All Formulae." *The Mercator Projection*. www.scribd.com/document/277668197/The-Mercator-Projection.

Šavrič, B., and B. Jenny. 2016. "Automating the Selection of Standard Parallels for Conic Map Projections." *Computers & Geosciences* Vol. 90, Part A: 202–12. https://doi.org/10.1016/j.cageo.2016.02.020.

Snyder, J. P. 1987. "Map Projections: A Working Manual." Professional Paper 1395. US Geological Survey. https://pubs.usgs.gov/publication/pp1395.

Snyder, J. P. 1993. *Flattening the Earth: A Thousand Years of Map Projections*. University of Chicago Press.

Snyder, J. P., and P. M. Voxland. 1989. "An Album of Map Projections." Professional Paper 1453. US Geological Survey. https://pubs.usgs.gov/publication/pp1453.

Notes

1. The spherical version of equation (3.3) is as follows.

$$y = R \times \ln\left(\tan\left(\frac{\pi}{4} + \frac{\varphi}{2}\right)\right) \quad (3.3)$$

Here, R is the earth's radius, and φ is the latitude in radians (there are 2π radians in a 360° circle, so that 1 radian is approximately 57.295 degrees). The value of equation (3.3) at 90° ($\pi/2$ radians) is infinity and at −90° (−$\pi/2$ radians) is minus infinity.

2. Data for map projection figures courtesy of Tom Patterson, Equal-Earth.com.

CHAPTER 4
Grid coordinate systems

You can pinpoint locations on maps in several ways. You saw in chapter 1 that the latitude-longitude graticule has been used for more than 2,000 years as the worldwide locational reference system. Geocentric latitude and longitude coordinates on the sphere or geodetic latitudes and longitudes on the oblate ellipsoid, still key to modern position finding, are not as well suited for making measurements of length, direction, and area on the earth's surface. The basic difficulty is that latitude-longitude is a coordinate system that gives positions on a curved spherical surface. It would be much simpler if we could designate location on a flat surface, using horizontal and vertical lines spaced at regular intervals to form a grid. We could then use the coordinates from the grid of intersecting straight lines to determine locations, lengths, areas, and directions.

In chapter 3, we explained that most maps are created by projecting the earth's spherical surface onto a flat surface, such as a sheet of paper or a computer screen. The advantage of the flat map projection surface is that you find locations using a simple two-axis coordinate reference system. This coordinate system is the basis for the grid of horizontal and vertical lines on the map.

A coordinate system mathematically placed on a flat-map projection surface is called a **grid coordinate system**. To devise such a system for large areas, you must deal somehow with the earth's curvature. From chapter 3, you know that transferring something spherical to something flat always introduces geometric distortion. But you also know that map projection distortion caused by the earth's ellipsoidal shape is minimal for smaller regions. If you superimpose a grid on flat maps of small areas, you can achieve positional accuracy that is good enough for many map uses. All grid coordinate systems are based on Cartesian coordinates, invented in 1637 by the French philosopher and mathematician **René Descartes**.

Cartesian coordinates

The familiar **Cartesian coordinate system** has divisions on a horizontal **x-axis** and a vertical **y-axis** in which the x- and y-axes cross at the system's **origin** (figure 4.1). The Cartesian coordinate system is divided into four **quadrants** (I to IV) based on whether the values along the x- and y-axes are positive or negative. This system allows any location on the map to be defined precisely by giving its two coordinates (x,y). Mapmakers use only quadrant I for grid coordinate systems so that all coordinates are positive numbers relative to the (0,0) grid origin (see figure 4.1). To use only the first quadrant, they position the origin of the grid on the map to the southwest of the entire mapped area, as you will see for the various systems described in this chapter.

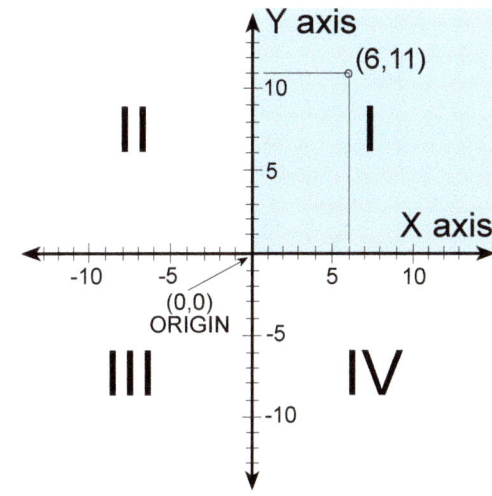

Figure 4.1. The structure of the Cartesian coordinate system. The basic notation for this system includes the use of a negative sign for the x- and y-coordinates along the edges of quadrants II, III, and IV. Cartographers use only quadrant I (shaded) on maps because it is desirable to have all coordinates as positive numbers.

Grid coordinates

Defining map positions using grid coordinates has a definite advantage over using the latitude, longitude coordinates on the ellipsoid (see chapter 1 for more on latitude, longitude coordinates). Grid coordinates based on the Cartesian coordinate system are especially handy for map analysis procedures, such as finding the distance or direction between locations or determining the length or area of a mapped feature. Next, we examine in depth one commonly used grid coordinate system—the **universal transverse Mercator (UTM)** system—that is found on many topographic maps, military maps, and navigational charts, among other types of maps. We then look in detail at the state plane coordinate system, which is found on many large-scale maps of the United States.

Universal transverse Mercator system

We have said that you can achieve positional accuracy if the grid you superimpose on a flat map is of a relatively small area. To use a grid coordinate system worldwide, you can divide the earth into zones in which the grid is repositioned. If enough zones are defined, you can ensure reasonable geometric accuracy. In this system, a **zone** is a well-defined region on the earth between defined limits, usually two meridians of longitude or parallels of latitude. The best-known grid coordinate system used internationally is the UTM system. The UTM system was developed by the French military at the end of World War I and was soon adopted worldwide. The accuracy of UTM grid coordinates determined from topographic maps sufficed for most military applications during World War II and later conflicts, but today GNSS receivers provide highly accurate geodetic latitude and longitude positions that are easily converted to UTM grid coordinates.

The UTM system covers the world from 84° N to 80° S. Sixty north–south zones are used, each six degrees of longitude in width (figure 4.2). Each zone has its own central meridian and uses a secant-case transverse Mercator projection centered on the zone's central meridian (see chapter 3 for more on this map projection). This map projection makes it possible to achieve a scale distortion (see chapter 3 for more on scale distortion) of one part in 1,000 or less, because scale factors (see chapter 3 for more on scale factor) range from 0.9996 at the central meridian to 1.001 at the zone boundary on the equator.

The UTM zones are numbered from west to east (*left to right* in figure 4.2), beginning with zone 1 from 180° W to 174° W. Zones 10 through 19 cover the conterminous United States, zones 28 through 39 cover Africa, and so on. The area of the part of each zone above the equator uses the designation *N* (for example, 10N), and the part of each zone below the equator uses *S* (such as 10S). The north

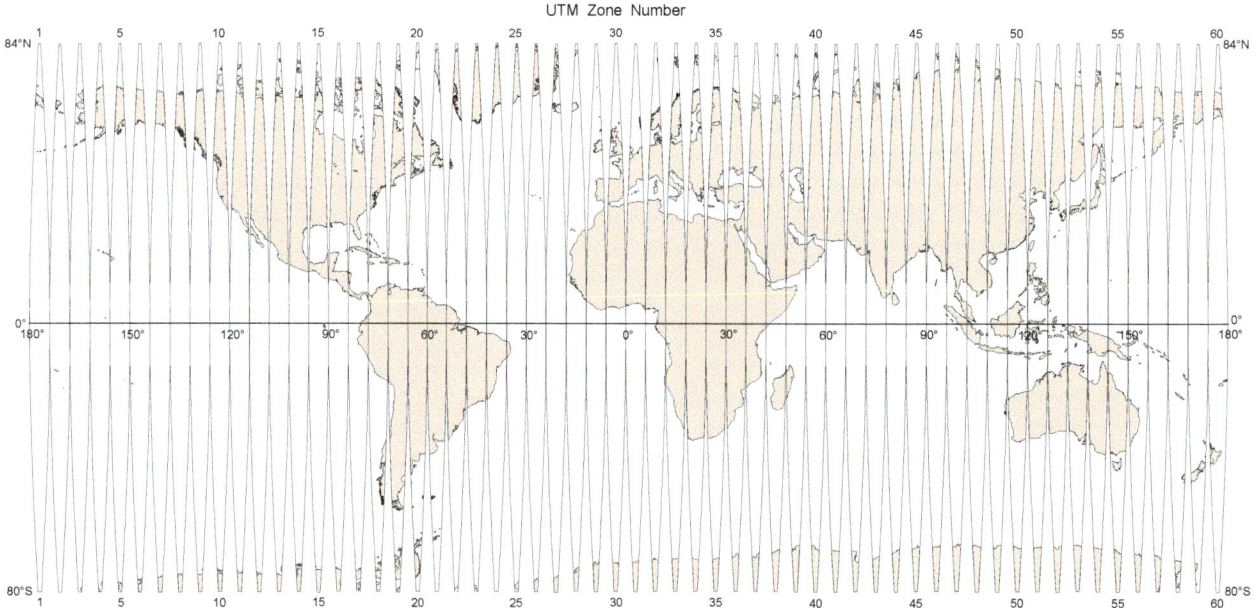

Figure 4.2. The 60 zones of the universal transverse Mercator (UTM) grid coordinate system.

and south portions of the zones have separate origins. To understand how the origins are specified, it is useful to first understand a few terms and concepts.

The x-coordinates east of the zone's origin and the y-coordinates north of the origin are called eastings and northings, respectively. An **easting** is the x-coordinate, or the distance east from the origin. In the Northern and Southern Hemispheres, an easting value of 500,000 meters (written as 500,000mE) is assigned to the central meridian of each UTM zone. This value, called the **false easting**, is added to ensure that there are no negative eastings in the zone.

Similarly, a **northing** is the y-coordinate that indicates the distance north of the origin. For areas in the Northern Hemisphere, a northing value of 0mN is assigned to the equator so that all northings are positive numbers. Because no false northing value is added, a UTM northing is simply the distance in meters north of the equator. For areas in the Southern Hemisphere, the equator is given a **false northing** of 10,000,000mN. This large false northing value places the origin of the zone very close to the South Pole so that there are no negative y-values in the southern part of the UTM zone.

Taking zone 10, which covers much of the West Coast of the United States, as an example (figure 4.3), you can examine the origin at the equator for the Northern Hemisphere (Zone 10N). The central meridian for this zone is 123° W, and the longitude range is 120° W to 126° W. The origin lies on the equator 500,000 meters west of the central meridian at 123° W. For zone 10 in the Southern Hemisphere (Zone 10S), the zone's origin lies 500,000 meters west and 10,000,000 meters south of the intersection of the equator and the central meridian at 123° W.

Because the equator has different northings for the Northern and Southern Hemispheres, every location that lies exactly on the equator has two UTM grid coordinate pairs, one for the northern zone and one for the southern zone. For example, the coordinates for the intersection of the equator central meridian of zone 10 are as follows:

500,000mE, 0mN in the Northern Hemisphere (for zone 10N)

500,000mE, 10,000,000mN in the Southern Hemisphere (for zone 10S)

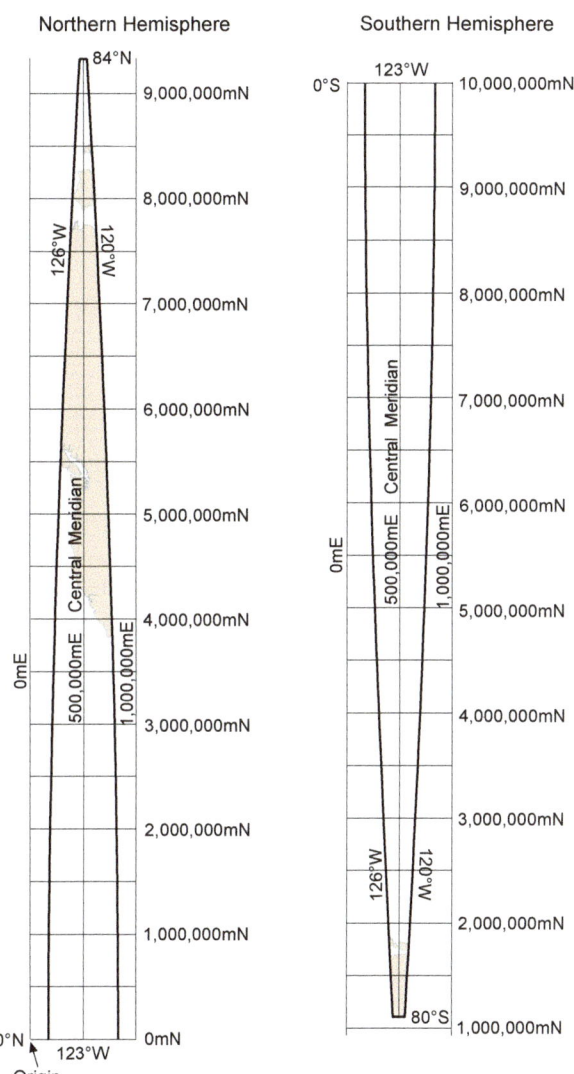

Figure 4.3. The complete universal transverse Mercator grid for zone 10.

Knowing these terms and concepts, you can now designate a location in the UTM grid coordinate system. Because north is conventionally at the top of the map, it may be helpful to remember a simple rule: Always read coordinates right and then up. You give the easting in meters, the northing in meters, the zone number, and the zone hemisphere (North or South). Thus, you designate the location of the Oregon State Capitol in Salem as follows:

497,595mE, 4,976,116mN, Zone 10N

The near-global extent (84° N to 80° S) of the UTM grid makes it a valuable worldwide referencing system, except for the polar areas. The UTM grid is found on most foreign topographic maps and on all recent USGS quadrangles in the topographic, orthophotomap, and orthophoto quad series (see chapter 11 for a description of orthophotomaps and orthophoto quads). All GNSS receiver manufacturers have UTM grid coordinates as an option with their receivers.

UTM coordinates differ when different horizontal datums are used (see chapter 1 for more on datums), so you should check the datum information on the map and in the GNSS receiver to know which datum is being used to record the coordinates.

Because meridians and not administrative boundaries delimit UTM zones, it usually takes more than one UTM zone to cover an administrative area completely. For example, the West Coast of the United States falls into zones 10 and 11 (see figure 4.2).

Universal polar stereographic system

As mentioned earlier, UTM grid zones extend from 84° N to 80° S. The UTM system was not extended to each pole because the 60 zones converge at the poles. To have complete global coverage in the remaining polar areas, a complementary grid coordinate system called the **universal polar stereographic (UPS) system** was created.

The UPS system is based on a secant-case, polar-aspect stereographic projection, which is a conformal projection that preserves shapes and correctly shows the angles of meridians radiating outward from the pole (see chapter 3 for more on this projection). The UPS coordinate system is divided into two zones: the North Polar Zone (NPZ) and the South Polar Zone (SPZ). The NPZ covers the area north of 84 degrees north latitude, and the SPZ covers the area south of 80 degrees south latitude (figure 4.4). This asymmetry is due to the larger extent of the Antarctic land mass relative to Greenland. Each zone is also extended an additional 30 minutes of latitude to provide some overlap with the UTM grid.

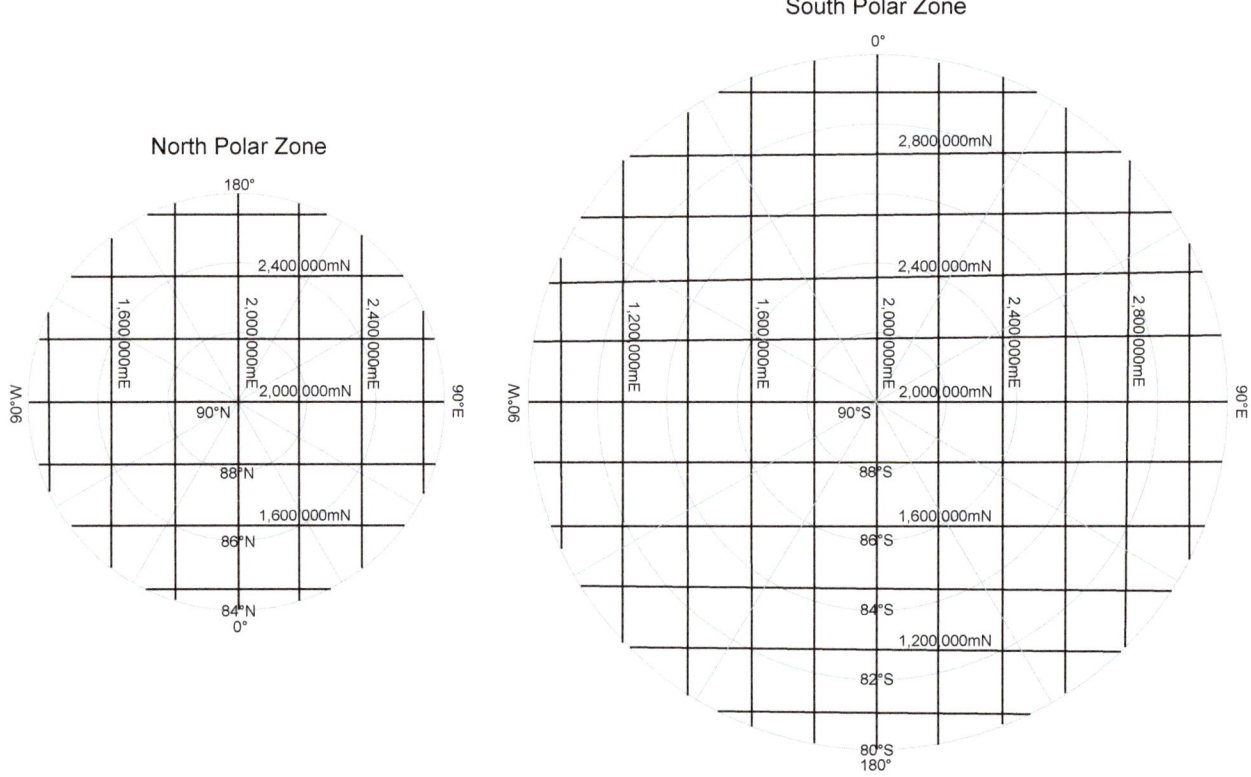

Figure 4.4. North and South Polar Zones of the UPS grid coordinate system.

UPS coordinates are also expressed as eastings and northings in meters. Each zone's origin is its respective pole, which is assigned a false easting and false northing of 2,000,000 meters. So, for example, the coordinates of the NPZ origin at both the North and South Poles are 2,000,000mN and 2,000,000mE. The reference line for the y-axis for both grids is the line from the North Pole to the South Pole along the prime meridian and 180° meridian. This makes the reference line for the x-axis the 90° E and W meridians.

The UPS projection is particularly useful for mapping and navigation in polar regions because it provides an accurate representation of angles and shapes. This is important for tasks such as measuring distances, determining directions, and creating accurate maps of the areas near the poles.

State plane coordinate system

The **state plane coordinate system (SPCS)**, sometimes shortened to **SPC**, was created in the 1930s by the land surveying profession in the United States to define property boundaries and simplify computation of land parcel perimeters and areas. The idea was to completely cover the United States and its territories with grids superimposed on map projections appropriate for each state so that the maximum scale factor does not exceed one part in 10,000. This level of accuracy could not be achieved if there was only one grid covering the whole country. The solution was to divide each state into one or more zones and make a separate grid for each zone.

SPCS 27 and 83

The United States was originally divided into 134 zones, with from one to ten zones for each state, each zone having its own map projection based on the Clarke 1866 ellipsoid and NAD 1927, or NAD27, geodetic latitudes and longitudes (see chapter 1 for details on this datum) using feet as its primary unit of measure. This original system is referred to as **SPCS 27**. Because of technological advancements in measuring the earth's shape and size, the states later converted to NAD 1983, or NAD83, and meters in the **SPCS 83** system. This conversion process resulted in the 125 zones shown in figure 4.5. For states with more than one zone, the names North, South, East, West, and Central are used to distinguish the zones. Texas has North, Central, and South zones, but also two additional zones—North Central and South Central. And the two interior zones of Wyoming's four are Central West and Central East. Alaska and Hawaii use numbers for their zones. In California, Roman numerals are used instead. The New York Long Island zone is unique, as are the Massachusetts Mainland and Island zones.

Secant-case Lambert conformal conic projections are used for zones of predominantly east–west extent, and secant-case transverse Mercator projections are used for zones of greater north–south extent. The only exception is the Alaska Panhandle, which, because of its east–west orientation, uses an oblique Mercator projection.

The logic of the SPCS zones is simple. Zone boundaries follow county boundaries because surveyors must register land surveys within a county. The map projection for each zone has its own central meridian roughly in the center of the zone. An origin is established to the west and south of each zone to ensure that all y-coordinates (northings) are positive numbers. The zone center is usually 2,000,000 feet west of the central meridian for Lambert conformal conic zones and 500,000 feet west of the central meridian for transverse Mercator zones. So, the central meridians usually have an x-coordinate (false easting) of either 500,000 feet for the transverse Mercator or 2,000,000 feet for the Lambert conformal conic zones, thus ensuring that all x-coordinates are positive numbers.

To further illustrate how zones are defined, we will look at the case of Oregon. The Lambert conformal conic is the map projection for Oregon's north and south SPCS zones (figure 4.6). The central meridian is the same for both zones, with an easting of 2,500,000 feet in the original SPCS 27 and 2,500,000 meters in the newer SPCS 83. The x-axis for the origin in each zone, called the latitude of origin, is a parallel that is just south of the counties in the zone (see chapter 3 for more on this and other projection parameters). The intersection of this parallel and the central meridian has a northing of 0 meters and an easting of 2,500,000 meters in SPCS 83.

You read and record SPCS coordinates in the same manner as UTM coordinates—right and then up. Specifically, the correct form of SPCS notation is to give the easting in feet or meters, the northing in feet or meters, the state, and the zone name. For example, the location of the Oregon State Capitol dome in Salem, Oregon, in its abbreviated form is 2,300,329mE, 144,443mN, Oregon, North Zone.

As mentioned earlier, the SPCS was modernized by switching to NAD83 and the GRS80 ellipsoid. Zones

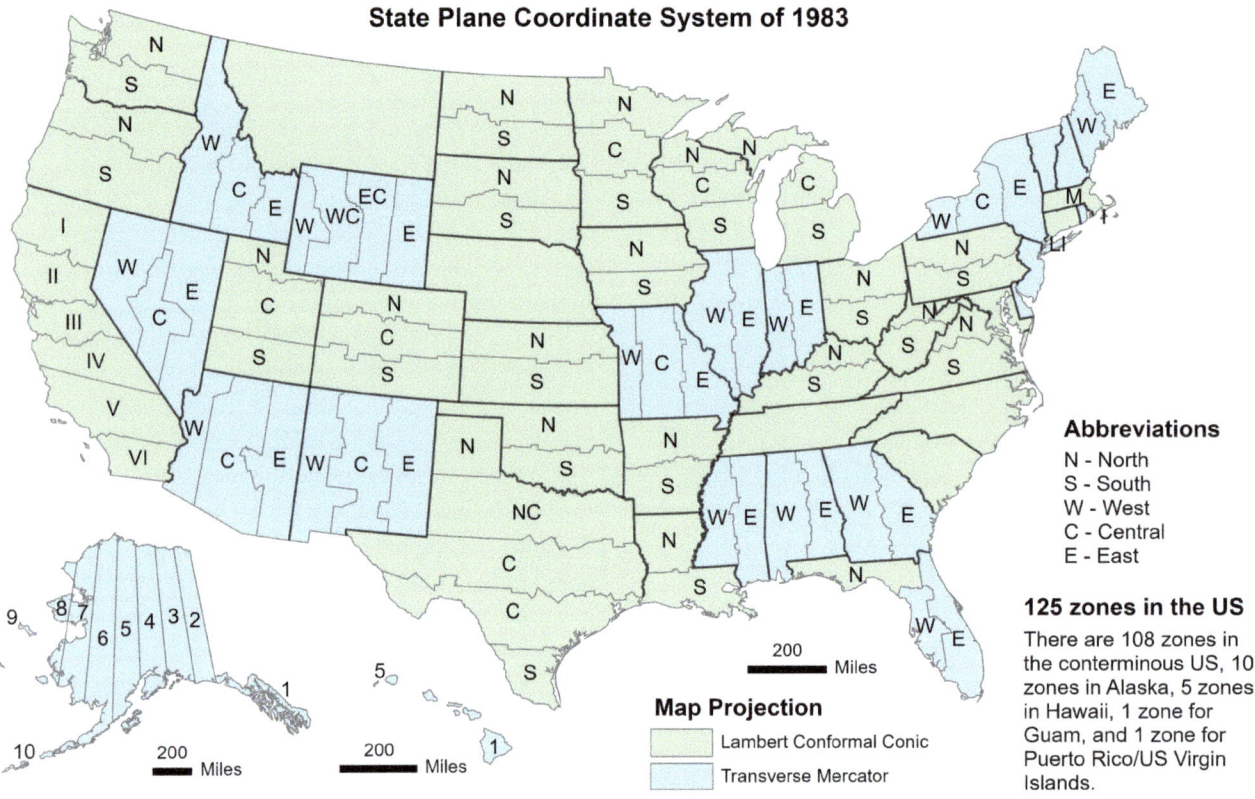

Figure 4.5. Zones of the State Plane Coordinate System of 1983. The transverse Mercator projection is used for zones that are oriented north–south, and the Lambert conformal conic projection is used for zones oriented east–west. Multiple zones within a state are identified by North, South, Central, East, and West (and their combinations in Texas and Wyoming), by numbers in Alaska and Hawaii, and by Roman numerals in California. Massachusetts has a mainland zone (M) and an island zone (I). New York also has a Long Island zone (LI). States without letters or numbers have only one zone.

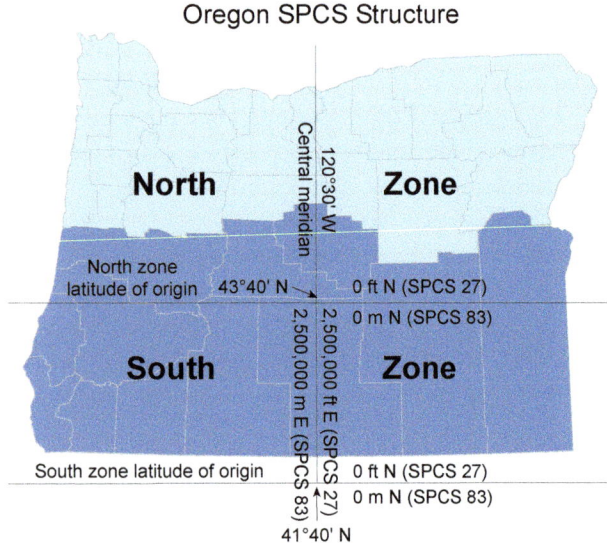

Figure 4.6. North and south zones of the Oregon SPCS 83.

were redefined in metric units, so the intersection of the Oregon central meridian and the latitude of origin now has an easting of 2,500,000 meters and a northing of 0 meters. These changes in values and the switch from Imperial to metric units make it easy to tell whether the map is based on NAD27 or NAD83. Because American surveyors prefer Imperial units, almost all states have redefined their NAD83 SPCS units in feet. Consequently, you should check the map legend to see whether old SPCS metric- or new Imperial-unit grid coordinates are used on the map.

Each SPCS zone is a separate entity with its own grid definition—a fact that frustrates and discourages use across zone boundaries. Nevertheless, the SPCS grid is useful for some map analysis applications, such as distance and direction finding. SPCS zone parameters can also be entered into GNSS receivers as a user-defined grid.

SPCS2022

When it was created, SPCS 27 served to support surveying, engineering, and mapping activities throughout the US and its territories, and SPCS 83 is still used today for surveying and engineering projects, land records and property ownership boundaries, and emergency management and public safety applications. However, the SPCS in its 1983 form is largely obsolete as far as current surveyors and other professional map users are concerned. One reason is that the accuracy of one part in 10,000 for locating points on the ground is now easily exceeded using modern GNSS-based surveying methods. The design of a new system, the **State Plane Coordinate System of 2022 (SPCS2022)**, is currently under review and will be officially adopted in 2026. Because SPCS2022 will be based on the modernized National Spatial Reference System (NSRS), SPCS2022 will be referenced to the four 2022 horizontal reference frames that will be released as part of the modernization (see chapter 1).

A key goal of SPCS2022 is to reduce map projection scale distortion within SPCS zones so that the maximum scale factor does not exceed one part in 20,000 (0.99995 to 1.00005 scale factor range) or 50 parts per million (ppm). To meet this goal, the design for the SPCS was modified in four primary ways. First, linear distortion is minimized at the **topographic surface** (the ground surface) instead of the reference ellipsoid surface used in the older systems. Second, every state and territory has a statewide zone, and most states also have several smaller zones that have less distortion than the statewide zone. Third, population density is considered to further reduce distortion in populated places.

Fourth, rather than having the National Geodetic Survey (NGS) define its zones, states could develop their own zones using map projections that make the difference between the distance on the map and the distance on the ground negligible. We will look at these modifications in more detail, starting with the number and types of zones.

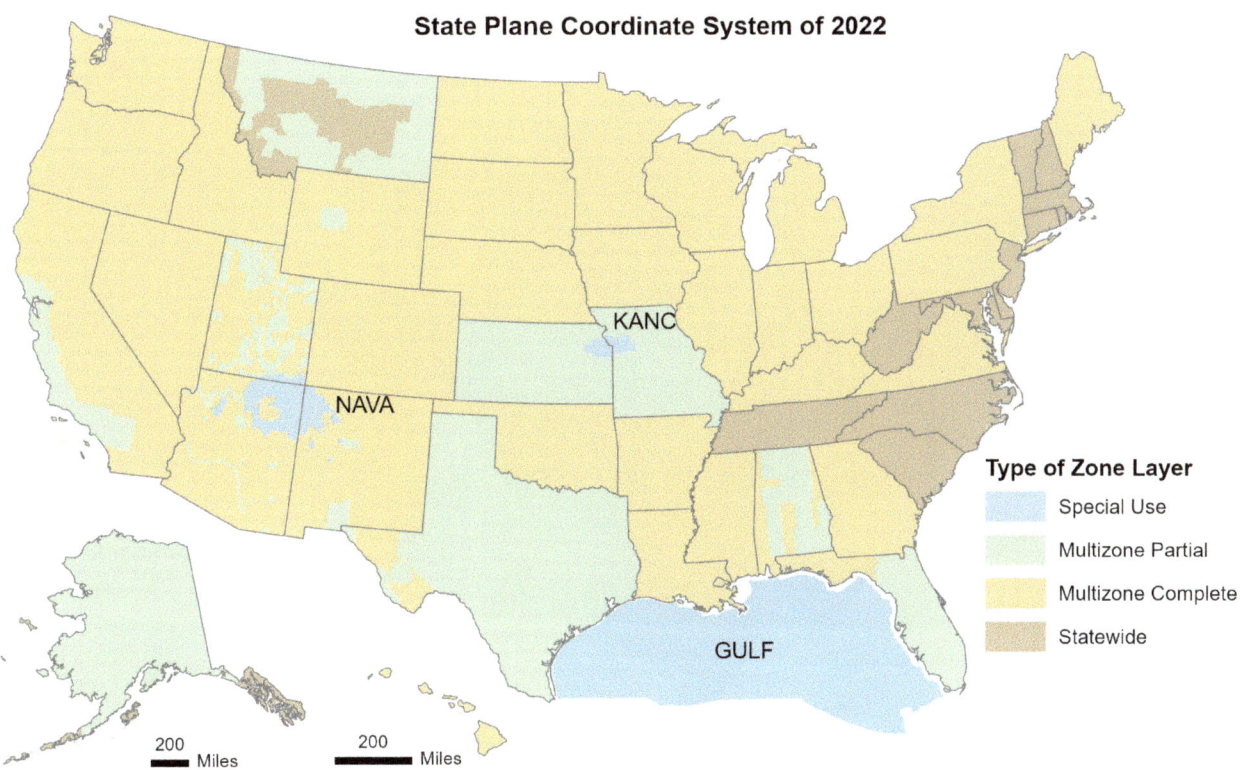

Figure 4.7. Along with the statewide zone layer, states can have a multizone layer with complete or partial coverage of the state. Three special use zone layers have also been designated. Data courtesy of the National Geodetic Survey.

Zones

A **zone** is a region on the surface of the earth within which a projected coordinate reference system is used and whose extents are usually based on meeting a specified linear distortion requirement. A **zone layer** can include a single zone or multiple zones. There are three types of zone layers: statewide, multizone, and special use zone layers.

The **statewide zone layers** for each state and territory were designed by the NGS to provide complete coverage for the entire state (figure 4.7). One zone also covers both Puerto Rico and the US Virgin Islands, and another covers both Guam and the Mariana Islands. States can also have **multizone layers** (also called **layered zones**) that contain smaller subzones covering all or part of the state. Most (28) states have two layers: their statewide zone layer and a **complete coverage multizone layer** (or **multizone complete zone layer**) in which the subzones cover the entire state. States with a zone layer that only covers part of the state—or a **partial-coverage multizone layer** (**multizone partial zone layer**)—are often in mountainous western states, so that subzones are defined for areas that are different from the rest of the state. For example, using population data from the US Census Bureau, subzones may be created in valley areas where most of the population is typically located.

SPCS2022 specifications allow for the development of special use zones for well-defined geographic regions that fall within two or more states. Because they cover more than one state, they cannot be included in any one state's zone layers. The three special use zones in SPCS2022 were designed for the Gulf of Mexico (GULF in figure 4.7), the Navajo Nation (NAVA), and the Kansas City, Kansas, and Kansas City, Missouri, metropolitan area (KANC). The Gulf zone covers the area of offshore waters for five states. The NAVA zone covers the entire Navajo Nation, which spans parts of Arizona, New Mexico, and Utah.

SPCS2022 has more zones (967) than either SPCS 27 (131) or SPCS 83 (125), but the number of SPCS2022 zones varies greatly between states (figure 4.8). Small or narrow states in the eastern US with single zones in the

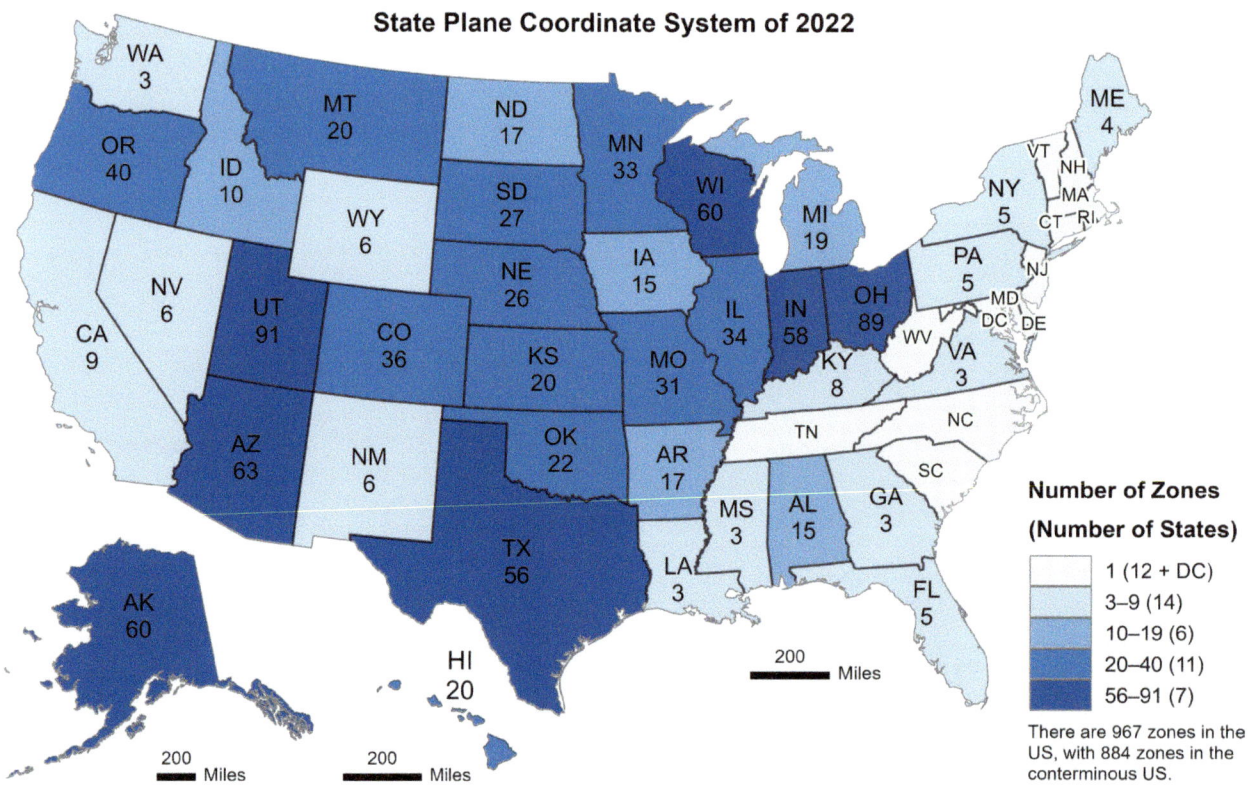

Figure 4.8. The number of zones in each of the 50 states in the State Plane Coordinate System of 2022 (SPCS2022). Data courtesy of the National Geodetic Survey.

original system have only their statewide zone in the new system. States such as Virginia and Washington have their two original zones, plus the new statewide zone. At the other extreme, Utah's three original zones have been replaced by the statewide zone, as well as a new multizone partial and a multizone complete layer, which together include 90 subzones.

Minimizing distortion

The most important difference between the older SPCS systems and SPCS2022 is that the new system is designed so the distance on the map is closely equal to the distance on the earth. To achieve this, linear distortion is minimized at the topographic surface rather than the ellipsoid surface, as with the older systems. **Linear distortion** is the amount by which a distance or length in a projected coordinate reference system (the **grid distance**) differs from the actual horizontal distance on or near the topographic surface of the earth (the **ground distance**). The **topographic surface** is the earth's ground surface. Thus, the goal of SPCS2022 is to minimize the difference between grid and ground distances.

To better understand the differences between the systems, we will first look at how linear distortion is evaluated in SPCS 27 and 83. In these systems, linear distortion is determined by comparing the grid distance computed from grid coordinates with the **ellipsoid distance** (the distance on the curved ellipsoid surface—see chapter 1 for more on ellipsoids computed from geodetic latitude and longitude). Linear distortion can be positive, which means that the distance on the projected map is greater (longer) than the distance on the ellipsoid, or negative—the grid map distance is less than the ellipsoid distance. This approach of evaluating linear distortion relative to the ellipsoid was rejected for SPCS2022 because, although using the ellipsoid makes the calculations easier, as the NGS geodesists noted, people do not live or work on the ellipsoid surface. Linear distortion is also evaluated in relation to the topography. Taking the topographic surface into account by adding the ellipsoid height, the horizontal ground distance is compared with the grid distance. As shown in figure 4.9, linear distortion due to the height of the topographic surface can be greater than that due to curvature of the ellipsoid surface.

In SPCS2022, the basic approach is to define the map projection so that the orientation (by choice of map projection), scale (using the scale factor), and location (by specifying the central meridian) of the projection axis results in the least amount of linear distortion for the zone. As with the original SPCS, each zone uses a secant-case conformal projection—either Lambert conformal conic or

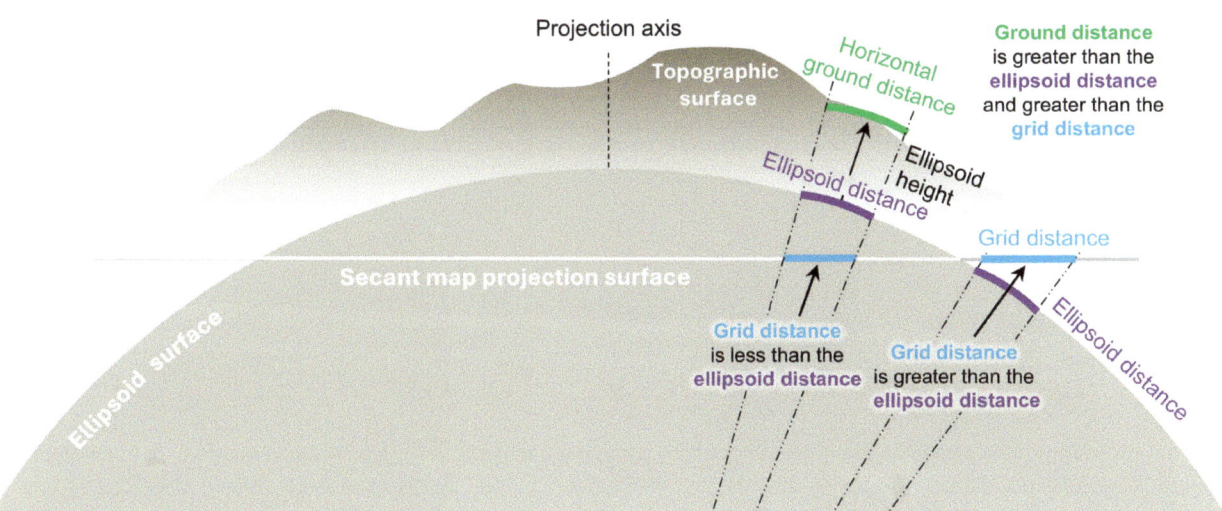

Figure 4.9. The design approach for SPCS 27 and 83 minimizes distortion with respect to the ellipsoid. Adapted from a figure by Dan Martin (National Geodetic Survey).

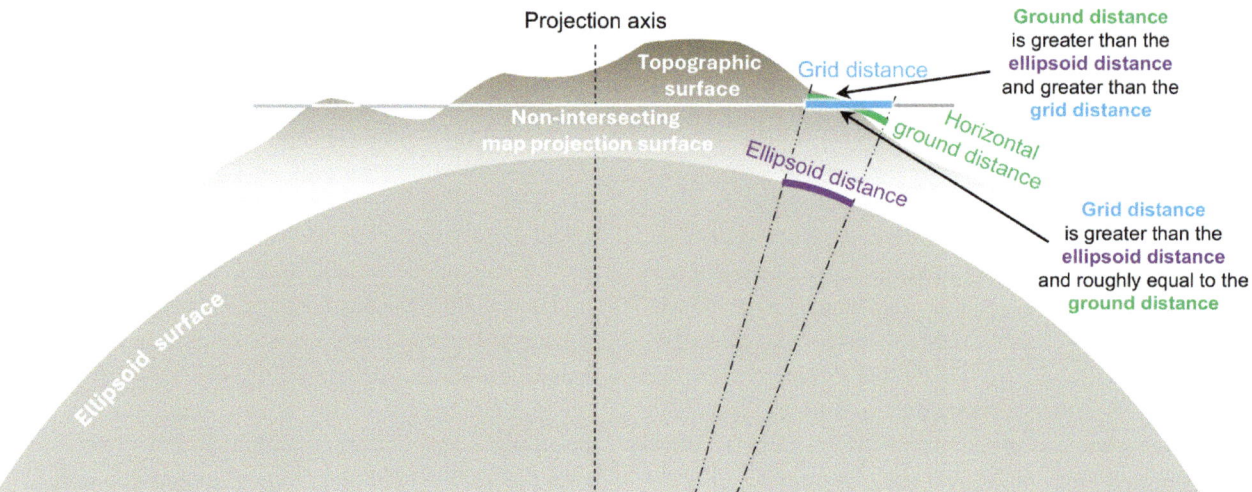

Figure 4.10. The design approach for SPCS2022 minimizes distortion with respect to topography by using a map projection surface that is above the ellipsoid and closer to the topographic surface, resulting in a nonintersecting projection surface. Adapted from a figure by Dan Martin (National Geodetic Survey).

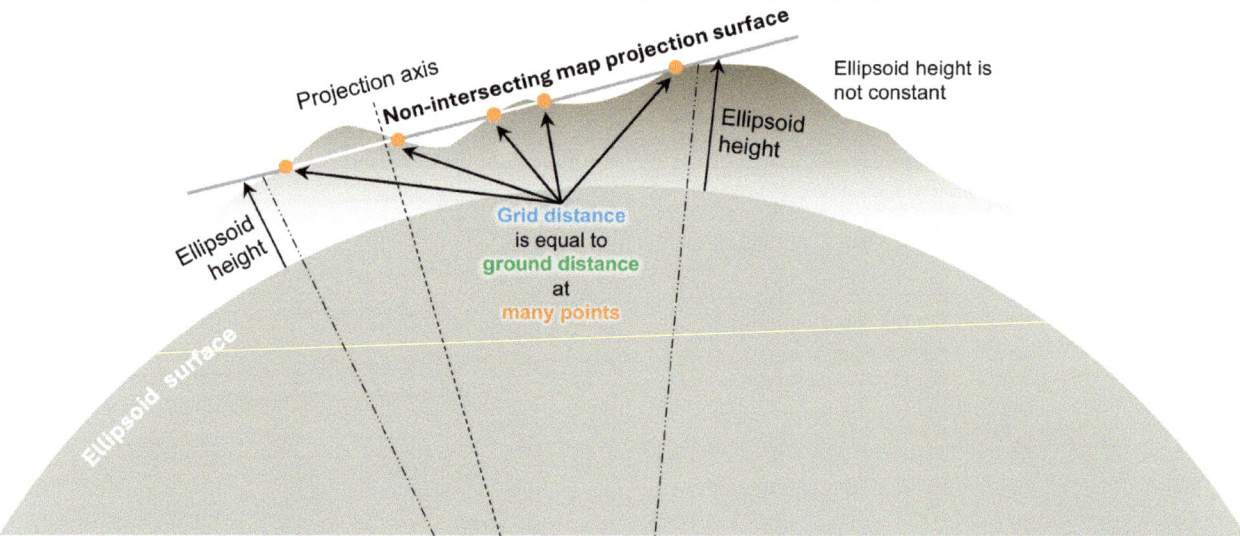

Figure 4.11. Modifying the location of the projection axis by changing the central meridian (latitude of origin) of the projection reduces variation in the distortion within a zone. Adapted from a figure by Dan Martin (National Geodetic Survey).

transverse Mercator, depending on the zone's orientation (a Hotine oblique Mercator projection is used for zones that are not orthogonal). The **projection axis** is the latitude of origin (or central parallel) for the Lambert conformal conic projection, and for the transverse Mercator projection, it is the central meridian. For the oblique Mercator projection, it is the central line (or skew axis). In the new system, the scale factor of the map projection is modified to fit the projection surface to the topographic surface (see chapter 3 for more on the scale factor). For almost 80 percent of the zones, this involves using a scale factor of greater than one to raise the projection surface above the ellipsoid, as illustrated in figure 4.10. This results in a **nonintersecting map projection surface**. Scale factors of less than one result in a projection surface that intersects the ellipsoid.

The result is that grid distance is equal to ground distance at points where the projection surface and topographic surface intersect. This only works for limited distances, and grid distance will be smaller for areas with greater variation in the topographic surface (that is, elevations) and larger for areas with little elevation change. For very large areas, there is no single constant height to scale the projection surface. The only way to reduce the variation in distortion is to change the location of the projection axis by modifying the central meridian (latitude of origin) of the projection. In figure 4.11, the projection axis is modified to fit the topographic surface so that grid distance is equal to ground distance at many points and the overall variation in linear distortion is decreased. As a result, the distortion range and mean for the zone are minimized. Although the distortion is minimized, the ellipsoid heights are not constant, which is true for the projection surface in figure 4.10 as well. This shows that a single ellipsoidal height should not be used to scale the projection for large areas. Thus, the need for multiple zones.

To assure the least distortion within a zone, the NGS set width limits for the zones—the larger the zone width, the larger the maximum allowable distortion. The NGS also set limits on the range of topographic height within a zone. Typically, the area of a zone decreases as the range in topographic height increases. Table 4.1 gives zone width and height ranges corresponding to various linear distortion ranges, expressed in terms of parts per million (ppm) or feet per mile. You can see in row 5 of table 4.1 that for a zone to meet the linear distortion criterion of no more than ±50 ppm, the zone can be no wider than 112 mi. (180 km) and have a topographic height range of no more than 2,090 ft. or (637 m). Because many states are larger than that, the state must be divided into several smaller areas, each meeting the maximum allowable linear distortion criterion.

Low-distortion projections

We now turn to Oregon as an example of a state that defined its own zones using **low-distortion projections (LDPs)**. LDPs are designed so that the distortion is low enough that the difference between grid distance and ground distance is negligible. LDPs are used to reduce linear distortion at the topographic surface because mapping activities are undertaken on the ground, not on the ellipsoid, and engineering and surveying applications often require actual ground distances. Additionally, as we have seen, the topographic surface can be far above the ellipsoid, resulting in greater linear distortion.

Oregon's state mapping officials worked with the NGS to create several high-accuracy SPCS2022 zones to replace its North and South zones in the original SPC system. Notice in figure 4.12 that the Oregon SPCS2022 consists of 40 zones. The statewide zone (not shown) is essentially the same as the Oregon Lambert system in figure 4.16. It and the 39 smaller zones together form the Oregon coordinate reference system (OCRS). Each of the 39 smaller

Table 4.1. Maximum linear distortion for different zone widths and topographic heights

Scale factor range	Maximum linear distortion in parts per million (and as a ratio)	Maximum linear distortion in feet per mile	Maximum width of a zone	Topographic height range
0.99999 to 1.00001	±10 ppm (1:1,000,000)	±0.05	50 mi. (81 km)	418 ft. (127 m)
0.99998 to 1.00002	±20 (1:50,000)	±0.11	71 mi. (114 km)	836 ft. (255 m)
0.99997 to 1.00003	±30 (1:33,333)	±0.16	87 mi. (140 km)	1,254 ft. (382 m)
0.99996 to 1.00004	±40 (1:25,000)	±0.21	100 mi. (161 km)	1,672 ft. (510 m)
0.99995 to 1.00005	±50 (1:20,000)	±0.26	112 mi. (180 km)	2,090 ft. (637 m)

zones is named after a city or geographic feature it includes, and all zones overlap adjacent zones. The zones are rectangular in shape, and all are quadrilaterals bounded by parallels and meridians except for the two slanted zones—one along the Pacific coast and one along the Columbia River from the Pacific Ocean to east of Portland.

The 39 smaller SPCS2022 zones for Oregon were created largely in response to land surveyors and GIS professionals in local jurisdictions wanting a zone for their local area of higher population density. These zones were designed to meet the NGS guideline for SPCS2022 that zones should have a linear distortion less than ±50 ppm (±0.26 feet per mile), so grid distances on the map very closely match the same distances measured on the ground.

The mathematics behind creating an LDP are beyond the scope of this book, but we show the geometric basis

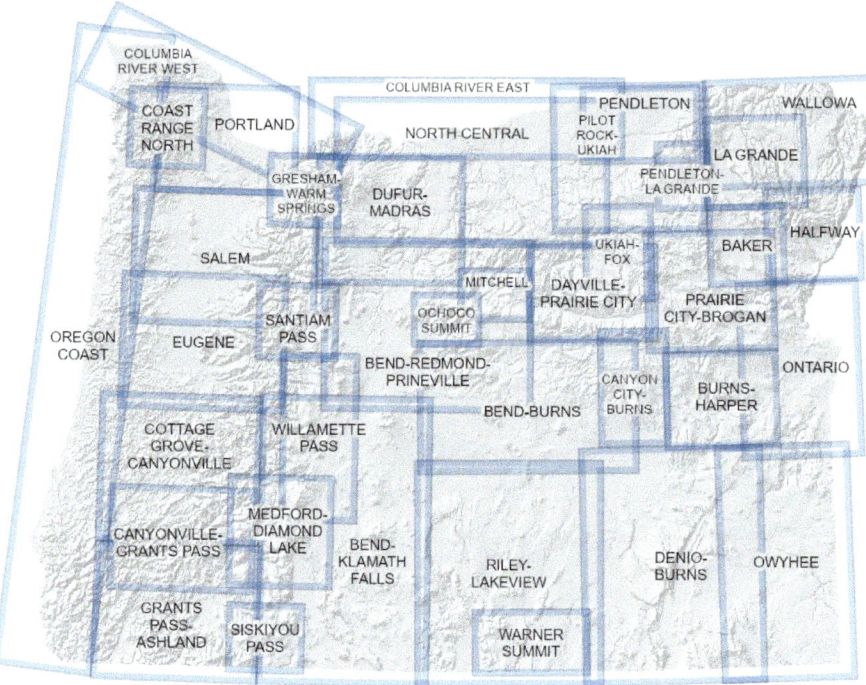

Figure 4.12. The Oregon coordinate reference system includes the 39 zones on this map and one statewide zone. Each zone is named for a geographic feature within the zone, and each overlaps adjacent zones. Data courtesy of Michael Dennis (National Geodetic Survey) and the Oregon Department of Transportation, Geometronics Unit.

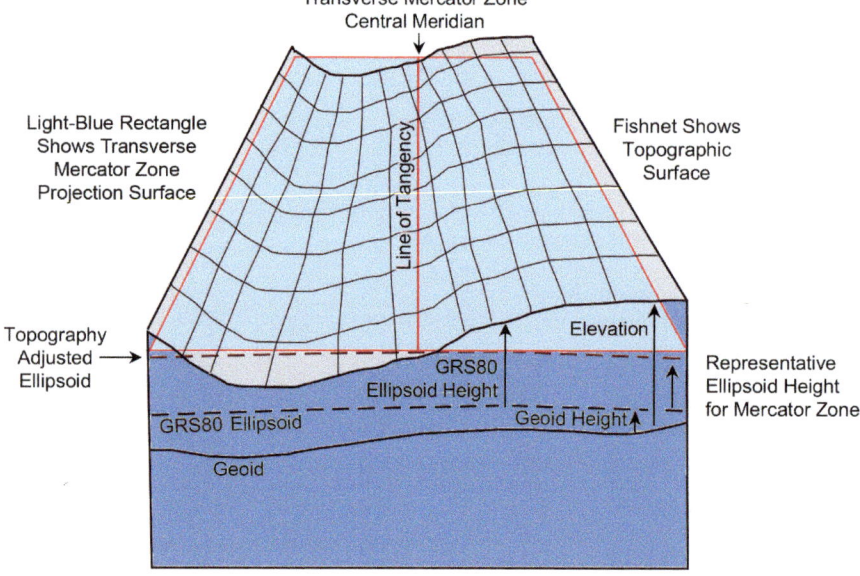

Figure 4.13. A low-distortion projection adjusts the ellipsoid surface in the vicinity of the SPCS2022 zone upward to a representative ellipsoid height for the zone.

for this type of projection in figure 4.13. The idea is to raise the tiny segment of the GRS80 ellipsoid surface surrounding the center of the zone upward by a representative ellipsoid height for the area covered by the zone, thereby defining a topography-adjusted ellipsoid for the zone. The representative height could be based on the average elevation within the zone, the average elevation within the most populated part of the zone, or some other criterion.

Ellipsoid heights at points in the zone are simply the difference between the elevation and geoid height at each point. The digital data required to do these calculations throughout the zone is readily available from government sources, including the USGS 1/3 arc-second digital elevation model (DEM) dataset for all 50 United States (see chapter 10 for more on DEMs). The ground spacing of data points in this dataset is approximately 10 meters north–south by 7 meters east–west over Oregon's latitude range. You can compute the GEOID18 height (discussed in chapter 1) for any geodetic latitude and longitude in the conterminous United States using a freely available online calculator provided by the NGS. The geoid height is essentially constant throughout any OCRS zone, so only one calculation is needed.

We now examine the Salem zone of the OCRS in detail.

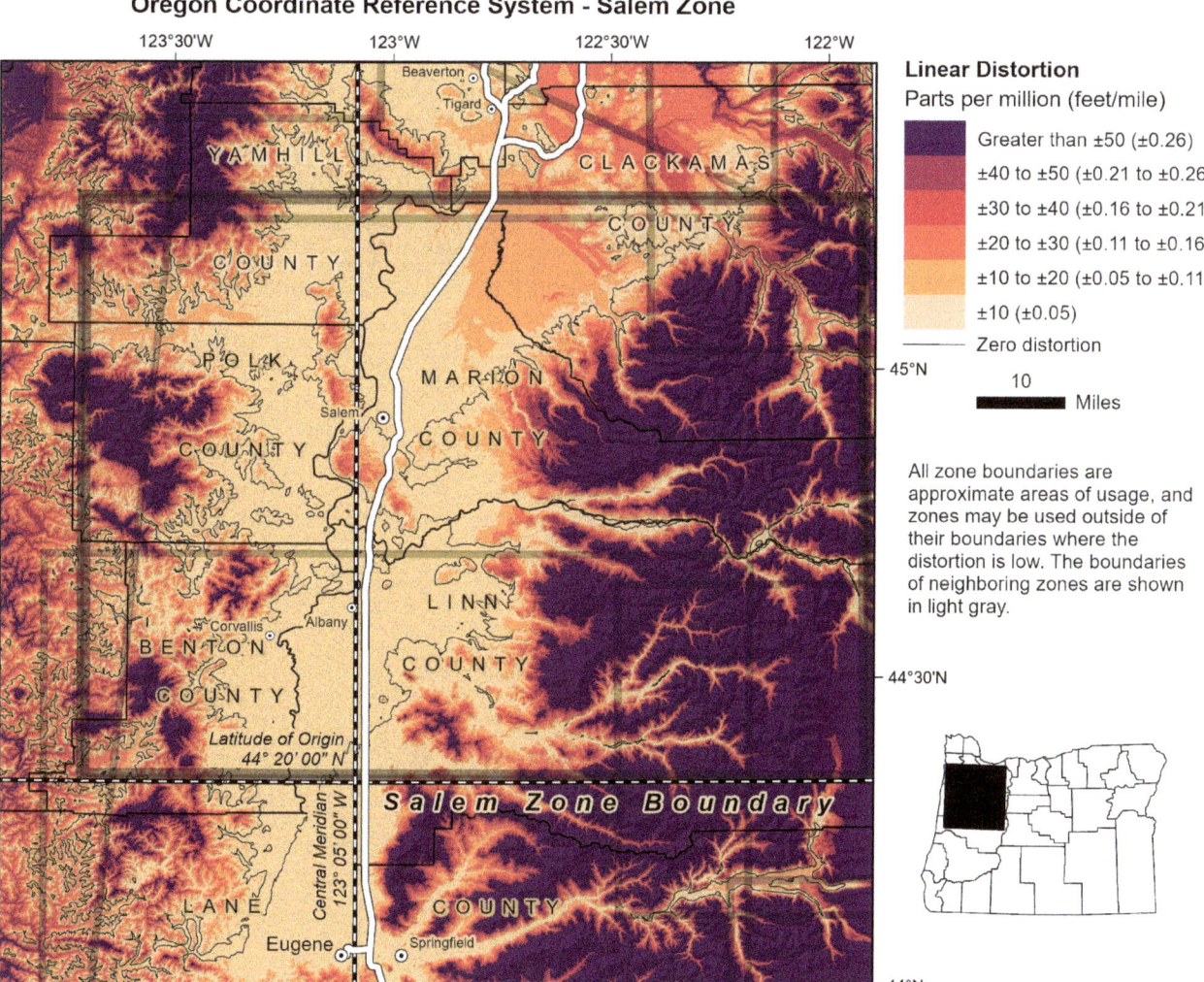

Figure 4.14. The Salem zone of the OCRS. The most distorted areas are shown in dark red and purple, both with distortion exceeding 50 ppm. The progression of orange through dark purple colors shows linear distortion increasing from ±10 PPM to ±50 PPM. Data courtesy of Michael Dennis (National Geodetic Survey) and the Oregon Department of Transportation, Geometronics Unit.

The Salem zone (figure 4.14) covers approximately 1 degree of latitude by 1.5 degrees of longitude (70 × 70 miles), considerably less than the 112-mile maximum zone width for a projection with linear distortion of less than 50 ppm. The projection axis for the transverse Mercator projection for the zone is the central meridian at 123°05′ W, which puts it in the middle of the Willamette Valley, where most people within the zone live.

This is a low-distortion projection because the GRS80 ellipsoid surface within the zone has been raised to a representative ellipsoid height of 209 feet (63.8 m) and the geoid height throughout the zone is computed as 76 feet (23.2 m). Adding these values gives a topography-adjusted ellipsoid at an elevation of 285 feet (87 m), which is close to the average elevation in this area of the Willamette Valley. The thin black lines on the map are 285-foot contour lines (see chapter 10 for more on contour lines), which also are lines of zero linear distortion because at this elevation the map projection surface coincides with the topography-adjusted ellipsoid. The light-orange areas on the map have elevations close to 285 feet, and the map legend shows the linear distortion to be in the lowest category of ±10 ppm (±0.05 feet per mile).

The average elevation within the entire zone is much greater than 285 feet because it extends from the Coast Range on the west to the Cascade Mountains on the east. The greatest linear distortion (more than ±50 ppm) is found in the dark purple areas because they are farther away from the zone's projection axis and are a thousand feet or more above the topography-adjusted ellipsoid for the zone.

Modern GNSS-based surveying methods make it possible to measure the distance between two locations on the ground as 1,000 feet apart (approximately 1/5 of a mile) with an extremely high 10 ppm (0.05 feet per mile) measurement accuracy. At this measurement accuracy, the actual ground distance is somewhere between 999.99 and 1,000.01 feet. A typical map scale for showing large-scale subdivision plats, engineering plans, and cadastral maps (see chapter 5 for more on these maps) is 1:1,200 (100 feet per inch). At this map scale, two locations 1,000 feet apart on the ground will be exactly 10 inches apart on the map if the local area is in a zero linear distortion region of the zone. If the locations are in an area of the zone with a linear distortion of exactly ±10 ppm, they would be 10.0001 inches apart on the map. If the actual ground distance between the locations was 1,000.01 feet, they would be 10.00010001 inches apart on the map. From these calculations, you may conclude that the plotted positions of locations on the SPCS2022 zone map will be exact to within our ability to shown them graphically, because positions in error by 1/10,000th of an inch on the map (and certainly 0.00010001 of an inch) are far closer than the narrowest line size we could draw on the map. The same holds true for all locations within the zone, as calculations for a ±50 ppm linear distortion result in the same conclusion.

US state grids

The widespread use of GIS in state and local government has facilitated the development and use of grid coordinate systems tailored to each state's needs. States that fall into two UTM zones often create a special **state grid** by shifting the central meridian of a UTM zone to the center of the state and the origin to a point southwest of the state. For example, Wisconsin, which falls about equally within UTM zones 15 and 16, routinely records and reports data in the **Wisconsin transverse Mercator (WTM) system** (figure 4.15). The UTM and WTM grid coordinate systems have the same geometric accuracy, but the WTM grid has a different central meridian in the center of the state (90° W instead of 93° W), and its coordinates are based on a false easting of 520,000 meters and a false northing of −4,480,000 meters.

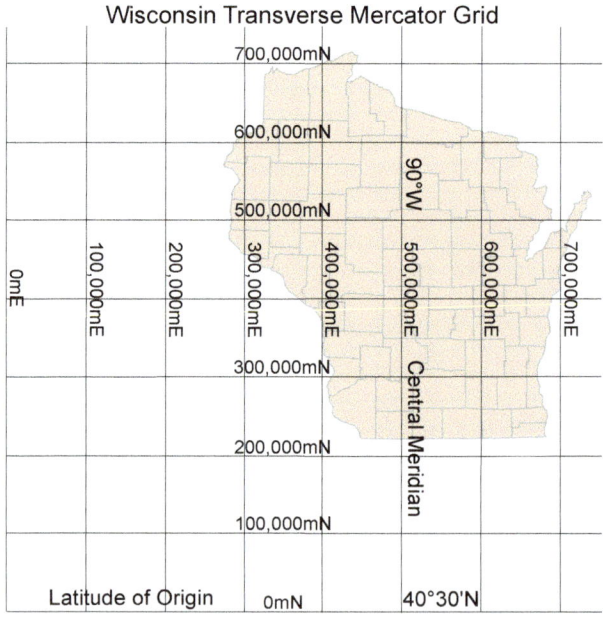

Figure 4.15. The Wisconsin transverse Mercator system.

The same approach can be taken when a state has two SPCS zones, as do 20 states seen in figure 4.5 (not including Massachusetts with its Mainland and Island zones). The **Oregon Lambert system** is a good example (figure 4.16). In the new system, a single grid has replaced Oregon's two SPCS zones based on a Lambert conformal conic projection. The new grid coordinate system uses the central meridian common to both the north and the south zones, but the origin is southwest of the state. So, for example, the easting value at the 120°30′ W central meridian (400,000mE) for Oregon is entirely different from the SPCS value (2,500,000mE).

Figure 4.17 shows how state plane and UTM grids appear on two sections of the 2016 Madison West 1:24,000-scale topographic quadrangle. Black ticks along the outer margin indicate SPCS 10,000-foot increments. The northings or eastings of these ticks are given by the value at one tick on each edge of the map (for example, a 370,000-foot northing at the tick on the lower-left edge of the Madison West quadrangle in figure 4.17). Other ticks are spaced every 10,000 feet on the 1:24,000-scale topographic maps. USGS topographic maps produced before 2017 are annotated with NAD83 State Plane Coordinate values, but the ticks and annotations were dropped on

Grid coordinate determination on maps

Grid coordinate system appearance on maps

You have seen that UTM grid coordinates appear on US and foreign topographic maps and that state plane coordinates are printed on quadrangles of the USGS topographic series. The way the grid looks varies among map series, so we will focus first on US large-scale topographic maps. Topographic maps created for use by the military have UTM grid lines superimposed at one-kilometer intervals, whereas civilian maps show both UTM and SPCS grids by grid ticks along the edges of each map.

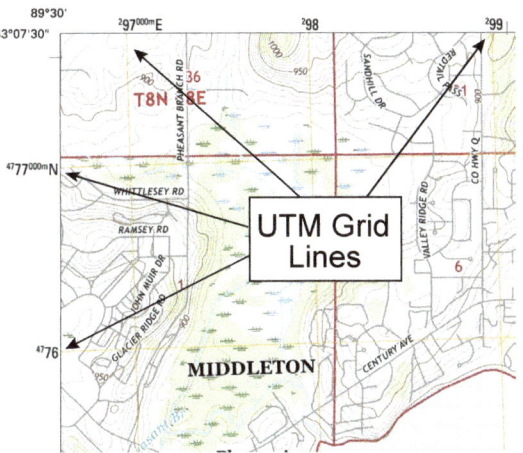

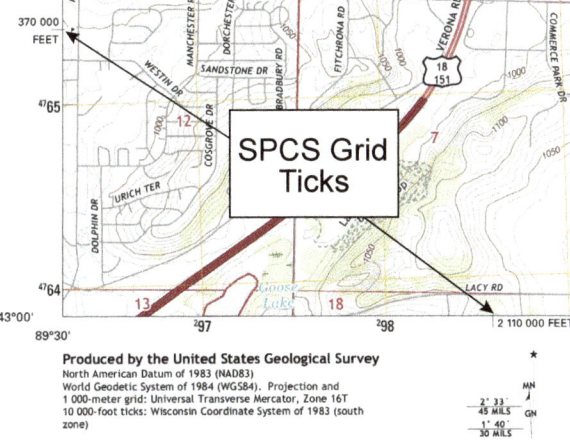

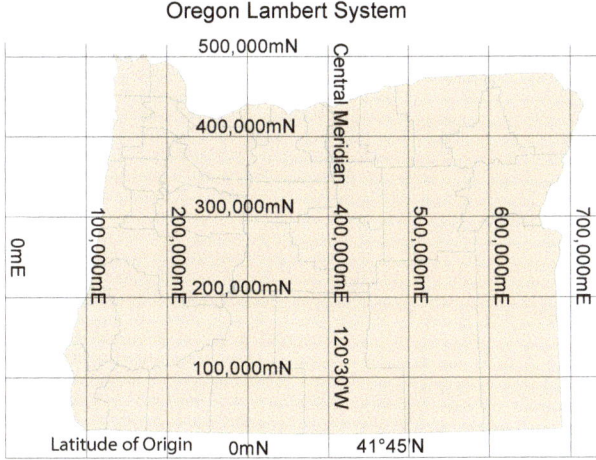

Figure 4.16. The Oregon Lambert system combines two SPCS grids into a single Lambert conformal conic grid coordinate system that covers the entire state.

Figure 4.17. Appearance of UTM 1,000-meter lines and SPCS 10,000-foot grid ticks in the top left (*top*) and bottom left (*bottom*) on the historic Madison West, Wisconsin, USGS 1:24,000-scale topographic quadrangle. Maps courtesy of the US Geological Survey.

more recent maps, although the current maps also show the US National Grid, discussed later in this chapter.

UTM grid lines are spaced at 1,000- or 10,000-meter intervals, depending on the map scale. On USGS quadrangles of the 1:24,000-scale topographic series, 1,000-meter grid lines are shown using thin orange lines. These lines are labeled, with their easting and northing values in black, along the map margin. The **principal digits**—those that show changes in the eastings and northings in kilometers across the map area—are printed in larger type, with the three trailing digits (000) dropped from all but one label along each edge of the map that gives the full coordinate value.

Grid orientation

On topographic and other maps with quadrilateral extents, grid lines are rarely oriented parallel to the edges of the page. You can see why by looking at UTM zone 10 in figure 4.3. Horizontal UTM grid lines intersect the zone boundaries perpendicularly only at the equator. Moving toward either pole, the converging zone boundary meridians intersect the horizontal grid lines at increasingly acute angles to a maximum of slightly less than three degrees from perpendicular at the top and bottom of the zone. Now look at the southwest and southeast corners of USGS 1:100,000-scale topographic maps that are at the same 43° N latitude but at the center and east edge of UTM zone 10 (figure 4.18). On the map on the left, at the 123° W central meridian for zone 10, notice that the UTM grid lines shown in light gray are vertical and horizontal (the vertical 500,000mE line at the central meridian is not shown on the map because it is the same line as the meridian). There is a slight clockwise rotation (two degrees at this latitude) of the east UTM edge meridian to vertical on the map on the right. This slight rotation of grid lines from horizontal at zone boundaries is known as **grid convergence**.

Grid convergence at zone boundaries is, therefore, zero at the equator and maximum at the top and bottom of the UTM zone. You may wonder how to deal with grid

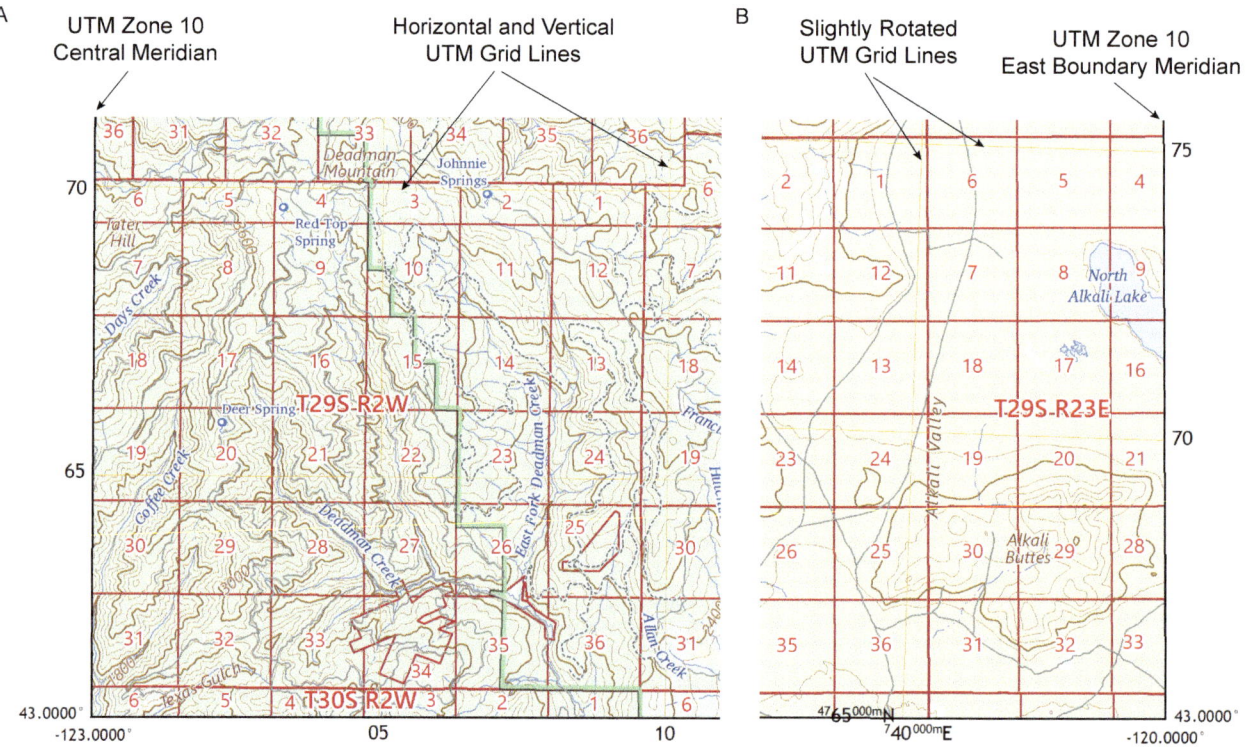

Figure 4.18. UTM grid lines at the zone center (*A*) are vertical and horizontal but are slightly rotated clockwise to the east boundary meridian of the zone (*B*) when they are drawn as orange lines on the map. These segments of USGS 1:100,000-scale topographic maps in UTM zone 10 are of Diamond Lake (*A*) and Christmas Valley, Oregon (*B*). Maps courtesy of the US Geological Survey.

coordinates on a map such as the one in figure 4.19, which spans UTM grid zones 15 and 16. Measuring half your coordinates in each zone is confusing and difficult to work with if you want to calculate lengths, directions, and areas from the coordinates. The solution is to extend the zones outward to cover the entire map, which allows you to choose one of the two zones for your map work. UTM zones, for example, can be overlapped with acceptable distortion of up to 30 minutes of longitude within their neighboring zones. For the same reason, SPCS zones also extend above and below or to each edge of the US counties they cover.

Grid coordinate measurement

Although the structure of grid coordinate systems is relatively easy to understand, it may take practice to gain skill using the coordinates on maps. Sometimes, you will want to determine a feature's SPCS or UTM coordinates. At other times, you will want to find the position on the map of a feature whose coordinates are given.

For instance, assume that you want to determine the UTM coordinates of Lacy Rd and Central Park Pl on the 1:24,000-scale topographic map segment in figure 4.20. If UTM grid lines are not printed on the map, you can use the marginal grid ticks and a straightedge to construct the grid lines that lie immediately to the south, north, east, and west of the road intersection. First, note the coordinate values of the grid lines to the west and south of the intersection (305,000mE and 4,764,000mN, respectively).

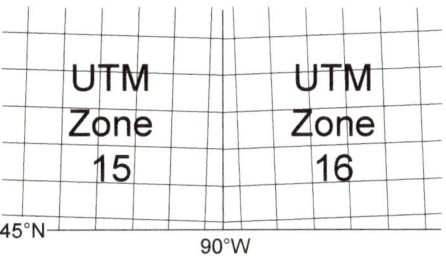

Figure 4.19. UTM grid lines usually are not parallel with graticule lines, so they intersect at a slight angle. Maps that span two grid zones always have grid convergence, a problem solved by extending one or both grids across the entire map.

Next, measure the map distance from these lines to the intersection (1.39 and 2.43 centimeters, respectively) and form ratios between these values and the grid interval distance in map units (4.17 centimeters per kilometer for a 1:24,000-scale map). Multiply these proportions (0.333 and 0.583) by the grid interval distance in ground units (1,000 meters) and add the results to the west and south grid line values. Thus, the UTM coordinates of the road intersection are found in equations (4.1) and (4.2):

$$\text{easting} = 305,000\text{m} + 333\text{m}$$
$$= 305,333\text{mE} \quad (4.1)$$

$$\text{northing} = 4,764,000\text{m} + 583\text{m}$$
$$= 4,764,583\text{mN} \quad (4.2)$$

Figure 4.20. To determine the UTM coordinates of the intersection of Lacy Rd and Central Park Pl or to plot the intersection's location from UTM coordinates, follow the steps outlined in the text. Map courtesy of the US Geological Survey.

Now imagine that you want to plot the location of a feature for which you know the grid coordinates. The problem is essentially the reverse of determining the UTM coordinates of a mapped feature, such as the road intersection in figure 4.20, which you now know is at 305,333mE, 4,764,583mN.

First, determine the UTM northing and easting for the grid lines that fall immediately below and to the left of the coordinate, respectively (4,764,000mN, 305,000mE). If these grid lines are not drawn on the map, use the grid ticks and a straightedge to draw them. Next, subtract the grid line value immediately west of the easting from the easting (305,333 m – 305,000 m = 333 m), and subtract the grid line value immediately south of the northing from the northing (4,764,583 m – 4,764,000 m = 583 m). Form proportions between these differences and the 1,000-meter grid interval distance (0.333 and 0.583, respectively). Next, multiply these proportions by the grid interval in map units (4.17 centimeters per kilometer) to obtain the easterly and northerly differences in map units (1.39 cm and 2.43 cm, respectively). Finally, plot these distances from the grid lines to the west and south. The two plotted lines will intersect at the road intersection.

If you are working with a single coordinate system over and over on maps of a certain scale, it may pay to construct a simple measurement aid called a **roamer**. Commercially available roamers are clear plastic devices that have calibrated rulers etched into their surface (figure 4.21). The rulers match standard topographic map scales. Check your local map or outdoor recreation store to get one of these handy products. You can also construct your own roamer by marking off the grid interval distance along each edge of

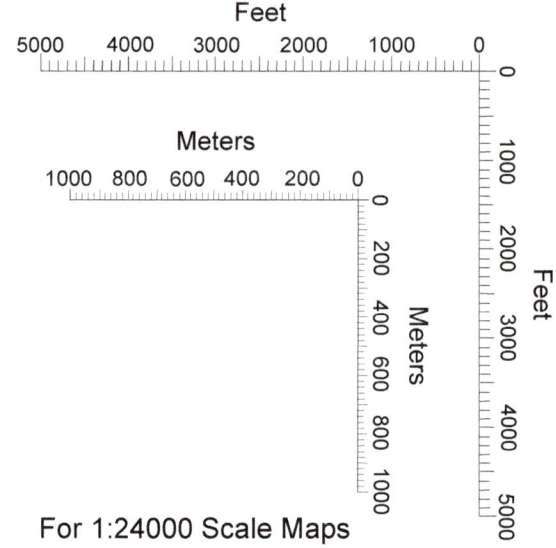

Figure 4.21. Roamers for determining SPCS and UTM coordinates on 1:24,000-scale topographic maps.

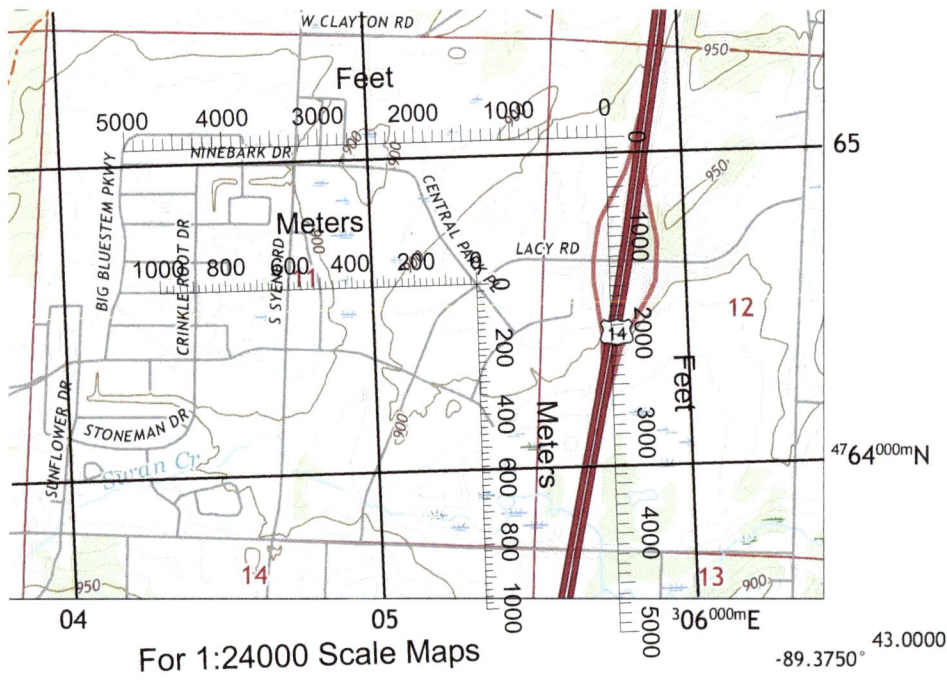

Figure 4.22. You can use a roamer to determine rectangular coordinates for a feature or plot a feature's position from known coordinates, such as the road intersection on this topographic map. Map courtesy of the US Geological Survey.

a piece of transparent plastic, starting at the corner. Then divide these distances into units that are fine enough for the precision of your measurements. Millimeters or tenths of inches normally suffice. You can also copy the roamer in figure 4.21 to a transparency and use it in your map work, but make sure it is first correctly scaled to your map.

By aligning the roamer with the north–south and east–west grid lines, you can determine the coordinates of map features and plot coordinate locations quickly and accurately (figure 4.22). Of course, you will need a separate roamer for each map scale and grid system. Therefore, you may want to put the UTM and SPCS roamers for standard map scales in opposite corners of the same transparency.

Grid cell location systems

So far, the grid coordinate systems that we have described all use x,y coordinates. Other reference systems can be found on maps, including **grid cell location systems**, which are based on the location of a feature inside a cell within the grid. These location systems consist of columns and rows that are identified by alphanumeric codes. In this section, we explore several grid cell location systems, starting with the **Military Grid Reference System (MGRS)**.

Military Grid Reference System

The US MGRS, developed by the US Army to cover the entire world, is a grid cell location system used by militaries in North American Treaty Organization (NATO) nations. The US devised this system to minimize the confusion of using the long positional coordinates (up to 17 digits and letters) and in the UTM and SPCS grid coordinate systems. The MGRS is based on cells rather than point locations. It extends the UTM and UPS systems by substituting single

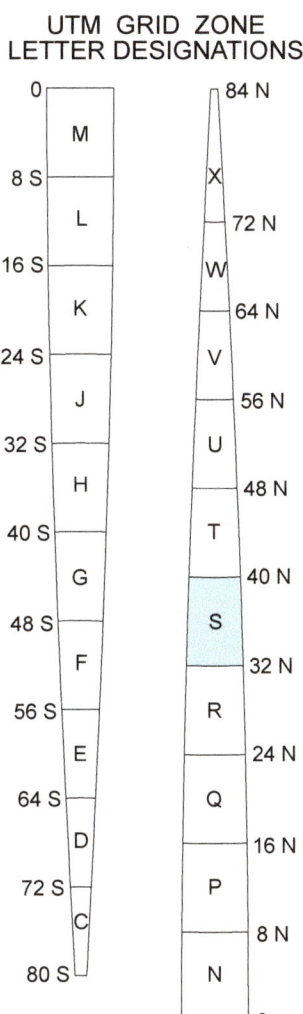

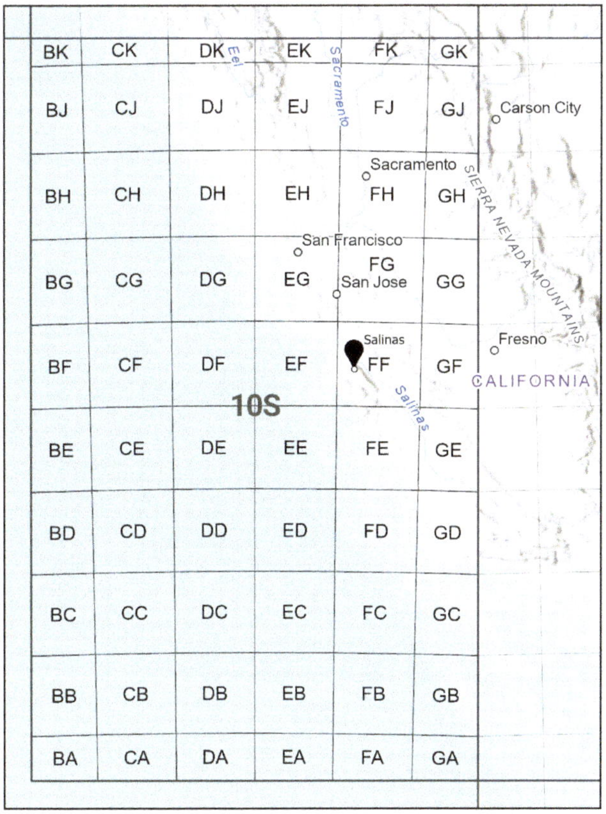

Figure 4.23. In the MGRS, each of the 60 UTM zones is first divided into eight-degree quadrilaterals of latitude and lettered from south to north.

Figure 4.24. Each of the quadrilaterals in figure 4.22 is divided into 100,000-meter cells that are given a two-letter code. Shown here is the 10S quadrilateral. The cell that contains Salinas, California, is FF.

letters for several numerals. Here we discuss only the UTM version of the MGRS.

In the MGRS, each of the 60 UTM zones (figure 4.23) is divided into 19 quadrilaterals covering eight degrees of latitude and two quadrilaterals covering the northernmost 12 degrees and southernmost 8 degrees of latitude. Quadrilaterals are assigned the letters *C* through *X* consecutively, beginning at 80° S latitude. The letters *A and B* are used for quadrilaterals on the south UPS grid, and *Y* and *Z* are used on the north grids. The letters *I* and *O* are omitted to avoid possible confusion with the numerals 1 and 0. Each quadrilateral is identified by an alphanumeric code that refers first to one of the 60 UTM zone numbers and then to the quadrilateral letter (figure 4.24). This is called the **grid zone designation**. Salinas, California, which we will use throughout this section as an example, has a grid zone designation of 10S (UTM zone 10, quadrilateral S).

Each UTM zone quadrilateral is divided into 100,000-by-100,000-meter cells, known as **100,000-meter-square cells**, and each cell is identified by a two-letter code called the **100,000-meter square identifier**. The first letter is the column designation, and the second letter is the row designation. For example, the 100,000-meter square identifier for the cell that contains Salinas, California, is FF. This cell locates our point of interest to within 100 square kilometers.

The 100,000-meter-square cells are further divided into 10,000-meter-square cells, and those are divided into 1,000-meter-square cells (figure 4.25).

To identify more precise locations for features, the

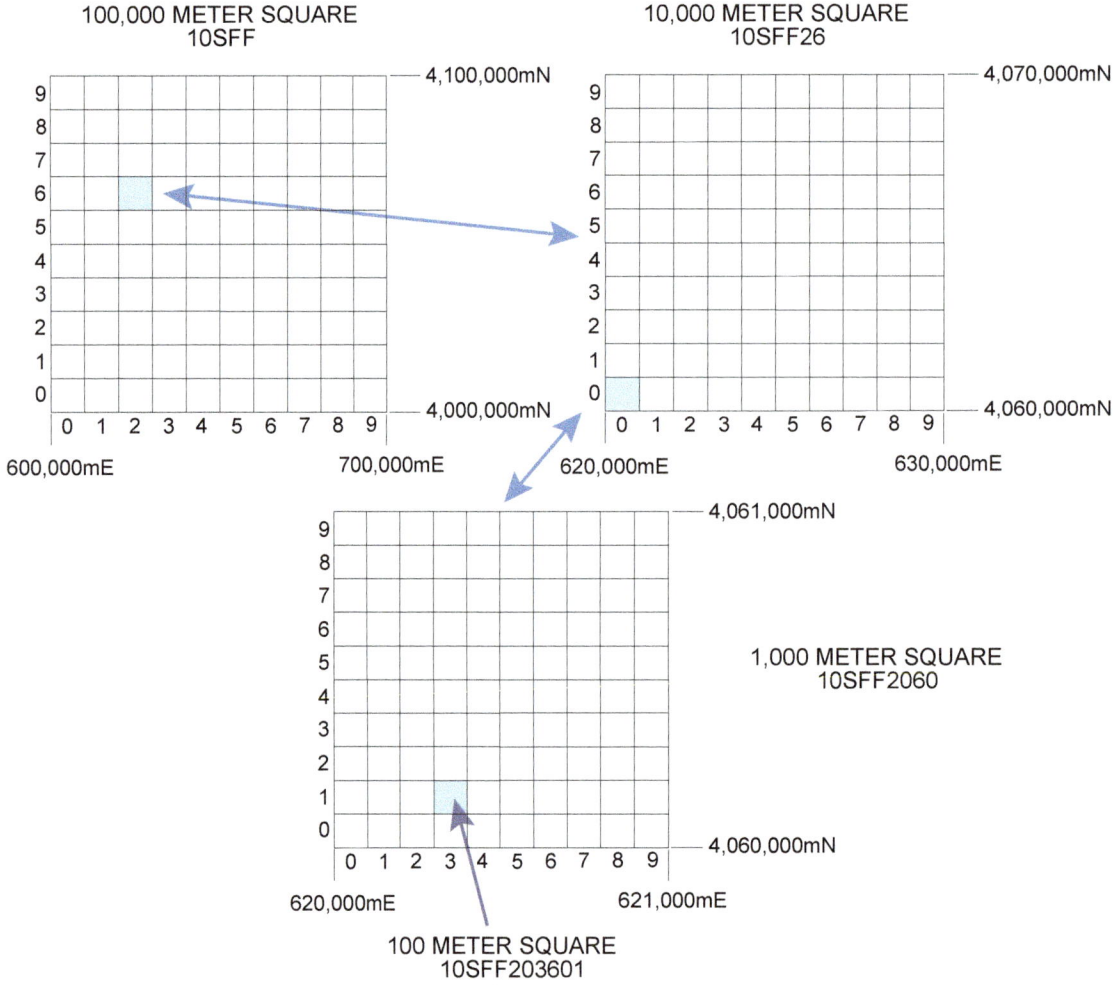

Figure 4.25. MGRS designations within 100,000-meter-square cells involve progressive subdivisions by 10 and the use of standard UTM numerals. The highlighted cells indicate the location of downtown Salinas, California.

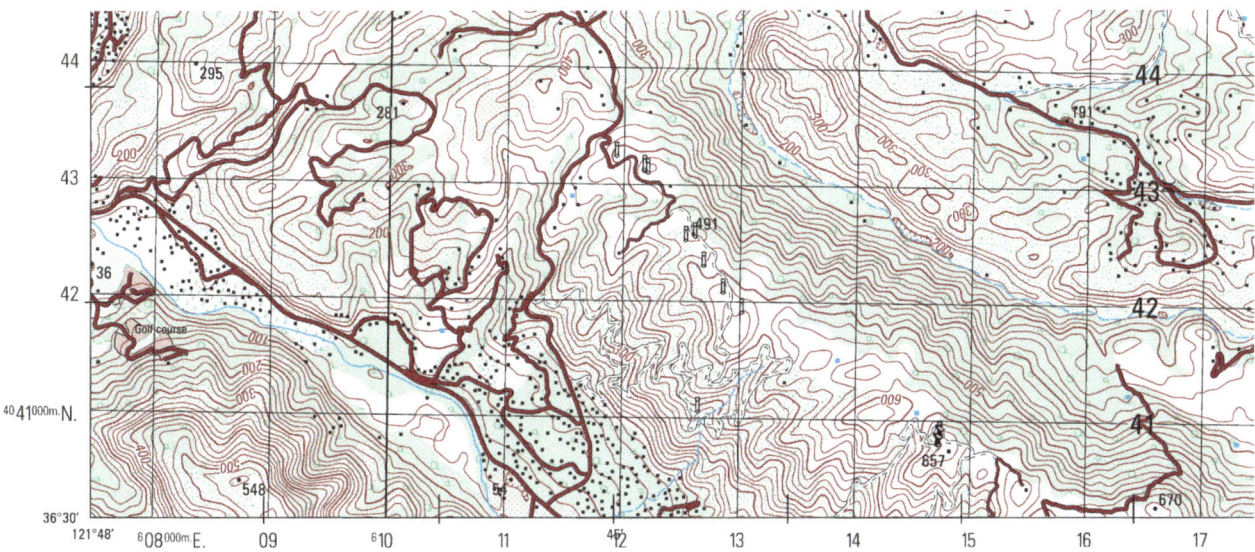

Figure 4.26. This corner of a 1:50,000-scale map shows the UTM easting of 608,000mE and the northing of 4,041,000mN. The MGRS grid cell labels reflect the 10,000-meter designation in the UTM labels (08 east and 41 north) and increase by 1,000-meter increments toward the right and top of the map sheet. These labels are also printed in large type along the grid lines on the map. On this map segment, they are shown along the horizontal grid lines on the right.

MGRS uses the standard UTM numerals. You can see in figure 4.26 that the MGRS grid on the map (shown with gray lines and labels) coincides with the UTM grid (shown with its traditional labels at the bottom left). The numbers for the MGRS cells on the map itself and the tics and labels along the edges correspond to the UTM easting and northing shown near the bottom left. Thus, from the easting of 608,000mE, which equates to 08 in the MGRS system, the 1,000-meter MGRS grid labels along the bottom read 09, 10, and so on toward the right edge of the map. The same is done for the labels along the left edge of the map, reading from the bottom to the top. You will also find these labels printed in large type along grid lines within the map, as shown along the horizontal grid lines on the right in figure 4.26.

For further exact positioning, you can imagine dividing a single cell into 10 rows and 10 columns (figure 4.27). On this 50,000-scale map with MGRS grid divisions at 1,000 meters and your subdivision of a 1,000 meter-by-1,000-meter cell into 100 smaller cells, you can identify the location of a feature to within 100 square meters (328 square feet). Consider how this works for the 548-meter spot elevation in figure 4.27. Start with the 1,000-meter MGRS grid number for the vertical grid line west of the feature (08 in the figure), and then count the number of imaginary cells to the feature (7 in the figure). Do the same

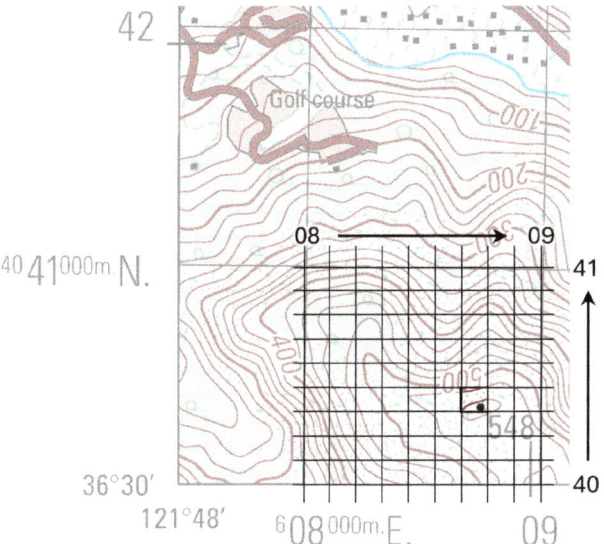

Figure 4.27. To identify a feature's position to within 100 square meters, read the vertical grid line numbers left of your point of interest and estimate tenths (100 meters) from the grid line to the point. For the 548-meter spot elevation on this map segment, that yields 08 7. Do the same for the vertical grid lines numbers, starting below the point. In this example, the result is 40 4. The 100-meter reference for the point is thus 087404.

for the horizontal grid line south of the feature (40 and 4). This gives you the **100-square-meter reference** for the feature as 087404. The **100,000-square-meter reference** is FF087404, and the full MGRS reference is 10SFF087404.

US National Grid

In 2005, the **US Department of Homeland Security (DHS)**, a department of the federal government charged with protecting the United States from terrorist attacks and responding to natural disasters, proposed that the **US National Grid (USNG)** be used for all US government geographic referencing and mapping. This proposal was put forward in response to the ever-increasing use of portable GNSS-enabled devices, GNSS-enhanced cell phones, and sensors for **automated vehicle location (AVL)** technology. Users of these devices require a nationally uniform grid reference system for accurate and consistent identification, communication, and mapping of ground coordinates. Identifying ground locations to 10- or 1-meter resolution is important in high-accuracy navigation, as well as in scientific studies and land management activities, especially in urban areas. Consequently, GNSS receivers—from recreational to survey-grade instruments—calculate and display geographic positions to this level of precision in USNG format. A single national grid reference system is easier to use because it is not necessary to convert between different grid systems, such as UTM and state plane coordinates.

The USNG (officially known as the **US National Grid for Spatial Addressing**) is a grid cell location system that is based on the UTM coordinate system and the alphanumeric grid square referencing system used in the MGRS. As mentioned earlier in the chapter, 10 UTM zones, numbered 10 through 19, span the conterminous United States, from 126° to 66° west longitude. The MGRS divides each UTM zone into quadrilaterals that are eight degrees in latitude, and the letters *R*, *S*, *T*, and *U* identify the four quadrilaterals from 24° to 56° north latitude. The land area of the conterminous United States falls within 32 of the quadrilaterals shown in figure 4.28.

To better understand the USNG, we will look at a particular quadrilateral in a UTM grid zone, such as quadrilateral 17R, which covers the Florida peninsula. As with the MGRS, this quadrilateral is divided into 100,000-by-100,000-meter cells, and each cell is identified by a two-letter code. The first letter is the column designation, and the second is the row designation. For example, Saint Lucie Inlet, approximately 30 miles (50 kilometers) north of West Palm Beach, Florida, on the southeast coast lies in 100,000-meter cell NL, or 17RNL in USNG notation (figure 4.29).

You can subdivide cell NL into 10,000-by-10,000-meter and then 1,000-by-1,000-meter cells in the manner shown for the MGRS in figure 4.25. Here is how to give the MGRS and USNG coordinates for finer subdivisions

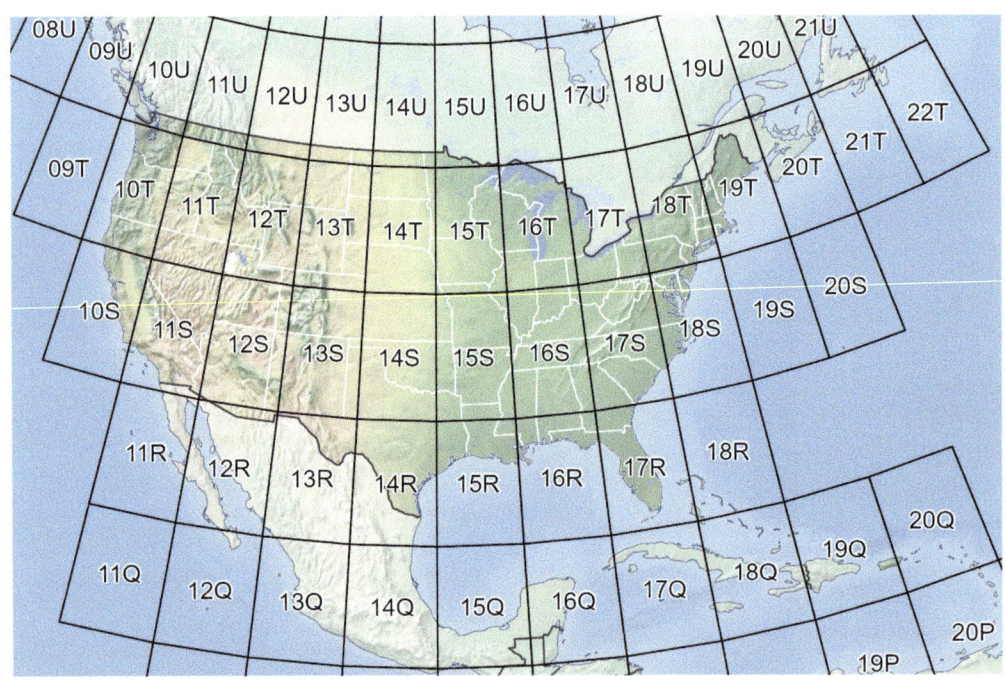

Figure 4.28. Codes for the 40 6° × 8° quadrilaterals from the MGRS that form the US National Grid.

of 1,000-by-1,000-meter cell 17RNL7513 in the northwest corner of the Saint Lucie Inlet, Florida, topographic map in figure 4.29. Reading right and then up, the six-digit coordinate for the 100-by-100-meter cell in the top left of the cell is 17RNL750139, the eight-digit coordinate for the 10-by-10-meter cell in the bottom left is 17RNL75001300, and the 10-digit code for the 1-by-1-meter cell in the top right is 17RNL7599913999.

Reference grids

Sometimes, you will see a different kind of grid on a map—one that does not have geographic coordinates, but rather one that divides the map extent into an array of cells identified by letters along the x-axis and numbers along the y-axis (or vice versa). These **reference grids** are independent of the map's coordinate system. Because each column is labeled, for example, with a letter, and each row is labeled with a number, every square in the array has a unique identifier. The features on the map are usually keyed to a place

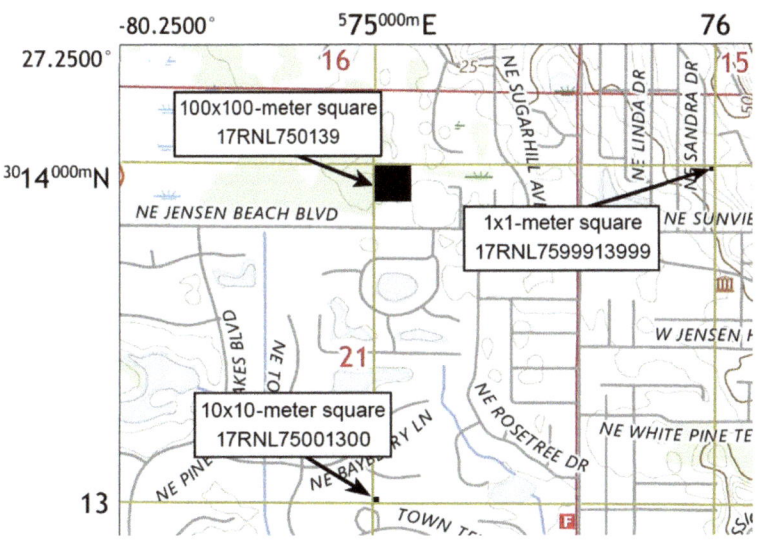

Figure 4.29. Northwest corner of the Saint Lucie Inlet, Florida, 1:24,000-scale USGS topographic map, annotated with coordinates for 100-by-100-meter, 10-by-10-meter, and 1-by-1-meter cells in the USNG. Map courtesy of the US Geological Survey.

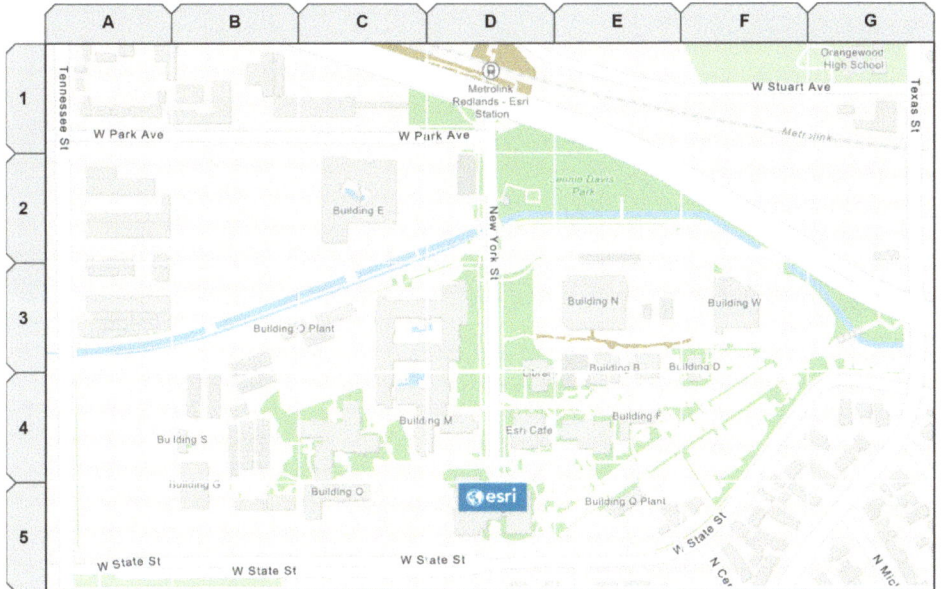

Figure 4.30. The reference grid for the Esri campus identifies the location of features by column letter and row number.

name index that identifies the location of the feature by its grid cell's letter-number combination. If you are looking for a particular feature, look it up in the index or gazetteer to find its grid cell identification. In figure 4.30, the main building on the Esri campus is in cell D-5.

You have probably seen reference grids superimposed on city street, state highway, recreational, or atlas maps. Reference grids can be applied to rectangular map frames only because any other shape would result in nonequal-area divisions.

Finding a particular feature within a grid cell can be difficult if the density of names is great or the cell is large in relation to the size of the feature you are trying to find. It can also be difficult to find a feature that spans many cells, such as an interstate highway, river, or mountain range. In these cases, the location is sometimes identified by the cell locations at which the feature starts and ends on the map. The problem of searching for a difficult-to-find feature is even greater when only margin ticks indicate cell boundaries. Remember, too, that each grid cell is specific to the map for which it is drawn. If people ask you where a street is and you tell them D-5, the information won't be useful unless they have the same map you do. Keep this tip in mind when you use maps that include insets. Because each inset may have its own grid cell locator system, the same place can be represented in several grid cells.

Selected readings

Armstrong, M. L., J. Thomas, K. Bays, and M. L. Dennis. 2017. *Oregon Coordinate Reference System Handbook and Map Set*, version 3.01. Oregon Department of Transportation, Geometronics Unit. www.oregon.gov/odot/ETA/OCRS/OCRS-Handbook-User-Guide.pdf.

Chrisman, N. R. 2017. "Calculating on a Round Planet." *International Journal of Geographical Information Science* 31, no. 4: 637–57.

DeMers, M. N. 2016. "Cartesian Coordinate Systems." In *International Encyclopedia of Geography: People, the Earth, Environment and Technology: People, the Earth, Environment and Technology*, 1–12. Wiley & Sons.

Dennis, M. 2018. *The State Plane Coordinate System: History, Policy, and Future Directions*. NOAA Special Publication NOS NGS 13, National Oceanic and Atmospheric Administration, National Geodetic Survey. https://geodesy.noaa.gov/library/pdfs/SP_NOS_NGS_13.pdf.

Dennis, M. 2023. "The Future Is Here: Introducing the State Plane Coordinate System of 2022." In *Proceedings of FIG Working Week 2023, Protecting Our World, Conquering New Frontiers*. www.fig.net/resources/proceedings/fig_proceedings/fig2023/papers/cinema03/CINEMA03_dennis_12044.pdf.

Helm, C., and Klenke, P. 2019. "On the Measurement of the Earth: New Approaches to Our Planet's Distinctive Shape." *Cartographica* 54 (4): 145–92.

Maher, M. M. 2018. *Lining Up Data in ArcGIS: A Guide to Map Projections*, third edition. Esri Press.

Martin, D. 2019. *Future of State Plane Coordinates*. Microsoft PowerPoint presentation for the National Geodetic Survey and the New Jersey Society of Professional Land Surveyors. www.ngs.noaa.gov/web/science_edu/presentations_library/files/spcs2022_njspls_2019.pdf.

National Oceanic and Atmospheric Administration. 2023. *Procedures for Design and Modification of the State Plane Coordinate System of 2022*, NGS 2023-1214-03-A3. https://geodesy.noaa.gov/INFO/Policy/files/SPCS2022-Policy.pdf.

Rollins, C. M., and T. H. Meyer. 2019. "Four Methods for Low-Distortion Projections." *Journal of Surveying Engineering* 145 (4).

US Army. 2011. "Grids." In *US Army Map Reading and Land Navigation Handbook*. Army Field Manual FM 3-25.26.

CHAPTER 5
Land partitioning

In chapter 4, we examined grid coordinate systems, which allow you to use alphanumeric values to identify the location of a point or cell on the surface of the earth. Many people confuse grid coordinate systems with land partitioning systems, and because of this confusion they often experience less than satisfactory results. In this chapter, we clarify the differences between these systems by exploring land partitioning systems, focusing primarily on systems used in the United States.

Borrowing from others in Egypt, Etruria, and Greece, the Romans were among the first to develop land partitioning systems, primarily to subdivide the land for taxation purposes. **Land partitioning** is the division of property into what we now call **parcels**, or **tracts**, which are areas that have some implication for landownership or land use. Property ownership patterns, which can be seen in the landscape today, are the result of land partitioning systems.

One of the first steps in partitioning the land is to divide an area into parcels. The parcels are then recorded on **plats**—maps that show the boundaries of a piece of land and how it is divided. A **subdivision** is a piece of land divided from a larger parcel into smaller lots. A **platted subdivision** is therefore a subdivision that is mapped to show the lots. Landownership, zoning, taxation, and resource management are just a few uses of such a system, and several land partitioning systems have been developed over time. Whatever system is used, it must be conceptually simple enough to be generally understood and technically simple enough to be readily implemented in the field at the time of land settlement or development.

Land partitioning systems have been used throughout the world, and many of them were brought to the United States by the people who settled here. For example, the metes-and-bounds system was used in Great Britain since the late 17th century, and then it was applied in other land jurisdictions based on English common law, such as India, South Africa, and Zimbabwe, as well as the original 13 US colonies. Settlers to the United States who were not British brought with them other land partitioning methods. As a result, a variety of partitioning systems were soon in use throughout the land. A geometrically irregular scheme, called the metes-and-bounds system, was used in the colonial states, Texas, and portions of Louisiana. Other irregular systems, such as French long lots, Spanish and Mexican land grants, and donation land claims, were used in other areas. Regular systems of partitioning the land were also developed, such as the Public Land Survey System (PLSS), described later in this chapter, which was later the model for Canada's system. In this chapter, we look at these land partitioning systems in detail, starting with irregular systems and then moving on to the more regular systems.

Irregular systems
Metes and bounds

Much of the land in the United States that was colonized before 1800 was characterized by widely scattered settlements. This settlement pattern was particularly evident in areas where the climate was moderate, and agriculture was widely practiced. These conditions encouraged people to settle far apart rather than cluster together, as they did in the urban areas. But the way that the areas were settled differed. English settlements in the relatively humid Northeast and Southeast were located mainly with respect to soil and timber resources. Consequently, the shape of parcels was decided primarily by the geography of the local setting—people naturally picked the best land they could, regardless of its shape. The convenience of a land parcel description was only an afterthought. An irregular parcel, such as the one shown in figure 5.1, is harder to describe than a simple geometric shape, such as a square.

The problem was to define these irregular parcels well

enough to make clear what their boundaries were. The solution in the early days of land settlement was to start from an established point, called the **point of beginning (POB)**, and then describe the property lines, typically in a clockwise fashion, back to the POB. Boundaries of the parcel were described relative to natural landmarks, such as trees or rivers, or artificial structures, such as roads or stakes. The parcel in figure 5.1A, for example, can be given the following description based on landmarks and structures:

> "Parcel beginning at the intersection of Jamaica and Brambury Roads, thence northeasterly along Brambury Road to the point where Brambury Road crosses Raccoon Creek, thence southeasterly along Raccoon Creek to the big tree, thence southwesterly to the big rock pile, thence southeasterly to the stake in the ground, thence southwesterly along the line of the property now or formerly owned by John Smith to the south corner of the fence, thence northwesterly along Jamaica Road back to the point of beginning."

With this method, it was relatively straightforward to describe a property's boundary. This legal property description was tied to geographic features and remained useful as long as neighbors agreed with the place names, and as long as the landmarks still existed in the landscape. Some natural landmarks, such as trees, may be missing, and others, such as streams, may have moved by the time the parcel description was later retraced. Artificial landmarks may have been moved or removed.

Over time, methods were developed for more precise **surveying** to measure angles and distances on the ground so that property boundaries could be accurately plotted on a map. These measurements were added to the descriptions, leading to the current form of land description called **metes and bounds**, in which each property line or **course** is described by its "metes" and its "bounds." The **metes** is the math that describes the property, and the **bounds** state what each line is bounded by or what natural or artificial landmarks are fixed at its corners or angles. The math, or the metes, contains three parts: a direction call; an angle described in degrees, minutes, and seconds; and a distance measured using some well-known unit, such as feet or chains. Earlier descriptions cited distance in **chains**—a chain is 66 feet (22 yards), and its subdivisions are called **links**. Today, distances are recorded in feet and tenths or hundredths of a foot. The property in figure 5.1B described by metes only would be as follows:

> "Parcel beginning at the intersection of Jamaica and Brambury Roads, thence North 43 degrees, 57 minutes, 30 seconds East, a distance of 2,039.06 feet, thence South 59

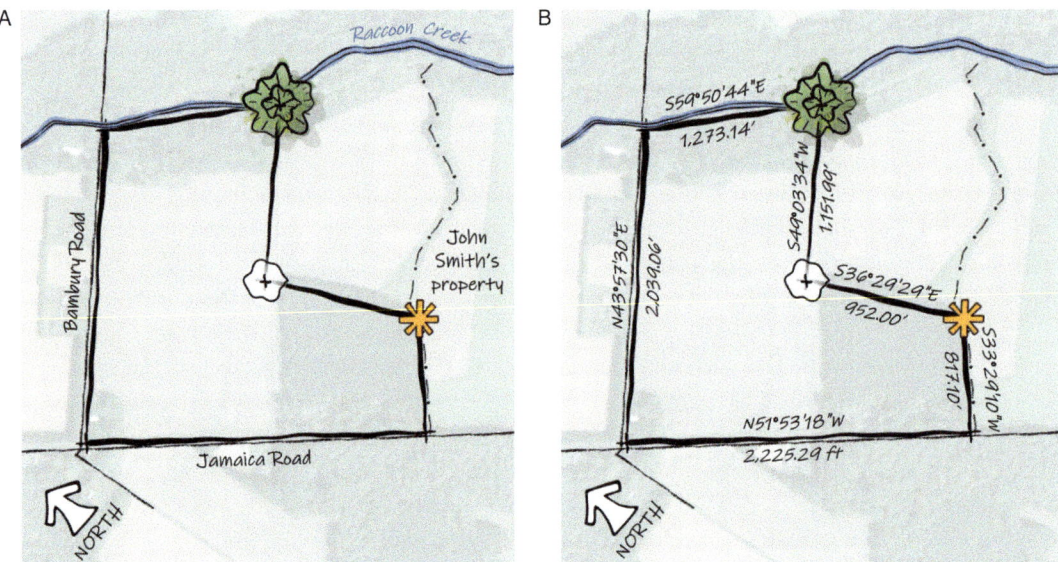

Figure 5.1. Illustration of a parcel that could be described by its bounds, or the features that bound the parcel, such as the roads, rivers, or fences, or by the landmarks at the corners of or angles in the property lines, such as trees, rock piles, and stakes (*A*). Illustration of the same parcel's metes, or the math describing the direction, angle, and length of each property line (*B*).

degrees, 50 minutes, 44 seconds East, a distance of 1,273.14 feet along Raccoon Creek, thence ... back to the point of beginning."

For the first course in this example, the direction call is North/East, the angle is 43 degrees, 57 minutes, 30 seconds (or 43°57'30"), and the distance is 2,039.06' (or feet). The description would also likely include the parcel's size in common areal units used in the measurement of land, such as **acres** (1/640th of a square mile, or 43,560 square feet) or **hectares** (10,000 square meters or 2.471 acres).

The system of metes and bounds used today as a legal description of a parcel includes both the metes and the bounds. A metes-and-bounds description for the property in figure 5.2 would read as follows:

> "Parcel beginning at the intersection of Jamaica and Brambury Roads, thence North 43 degrees, 57 minutes, 30 seconds East, a distance of 2,039.06 feet, along Brambury Road to the point where Brambury Road crosses Raccoon Creek, thence South 59 degrees, 50 minutes, 44 seconds East, a distance of 1,273.14 feet along Raccoon Creek to the big tree, thence ... back to the point of beginning."

It is possible to define a property by citing the bounds or identifying the metes, but these property definitions do not consider the differing opinions of surveyors. When differences in their math come up, the bounds become an essential part of the property description. Regardless of differences in math, if the same recognized bounds are cited, it is clear that the lines are the same. Thus, metes and bounds are considered the most accurate description of a parcel in some jurisdictions, and the system is still used today for the states of Maine, New Hampshire, Vermont, Massachusetts, Rhode Island, Connecticut, New York, New Jersey. Maryland, Virginia, North Carolina, South Carolina, Georgia, Tennessee, Kentucky, Texas, and parts of Ohio.

Areas surveyed by metes and bounds are often easily identified on topographic maps and aerial images (figure 5.3). The characteristic pattern of parcels that are irregular in shape and size and features intersecting at oblique angles tells you that this is most likely an area that uses metes and bounds to partition the land.

French long lots

French settlements in Quebec and Louisiana from the 1630s until around 1760 had a unique land partitioning method that resulted from the French feudal system of giving land grants, called **seigneuries**, to military officers, Catholic clergy, and other elite citizens. Initially, the French settlers were more interested in the fur trade than in farming, so land settlement usually took place along rivers, bayous, and other water features that were the main routes to the interior and the chief source of transportation, commerce, and communication for the French. Boundaries ran back from the waterfront as roughly parallel lines, creating narrow ribbons or **long lots**, also called **arpent sections** or **French arpent land grants**. (An **arpent** is a French measurement that equals approximately 192 feet or 58 meters in length or 0.84 acres or 0.34 hectares in area). Each seigneury, or long lot, had a narrow frontage on the river (about 350 to 500 feet wide), but it extended to up to 10 times as deep off the back slopes of levees along the river. This land partitioning system allowed each settler to have a dock on the river, a home on the natural levee formed by the river, and a narrow strip of farmland that often ended at the far end of the lot at the edge of a marsh or swamp. Over time, the settlers turned to farming and found that the long-lot system of land partitioning had the advantage of giving each farm equal access to the types of soil found on the floodplain.

People given a seigneury often subdivided the lot into smaller parcels, but to continue the tradition of providing access to the river, the parcels often became so narrow that they were no longer practical to farm. Nonetheless, their

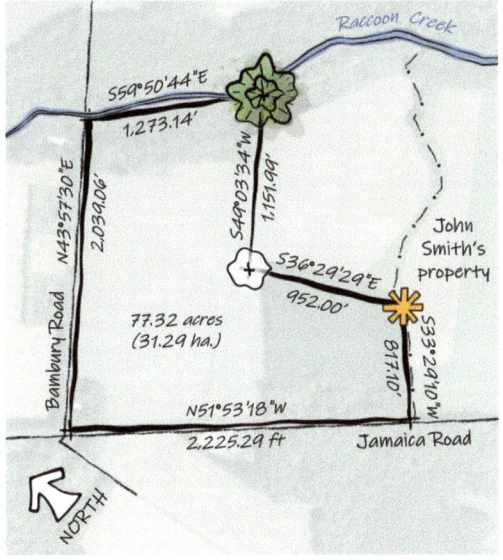

Figure 5.2. Illustration of a parcel that could be described by both its metes and its bounds.

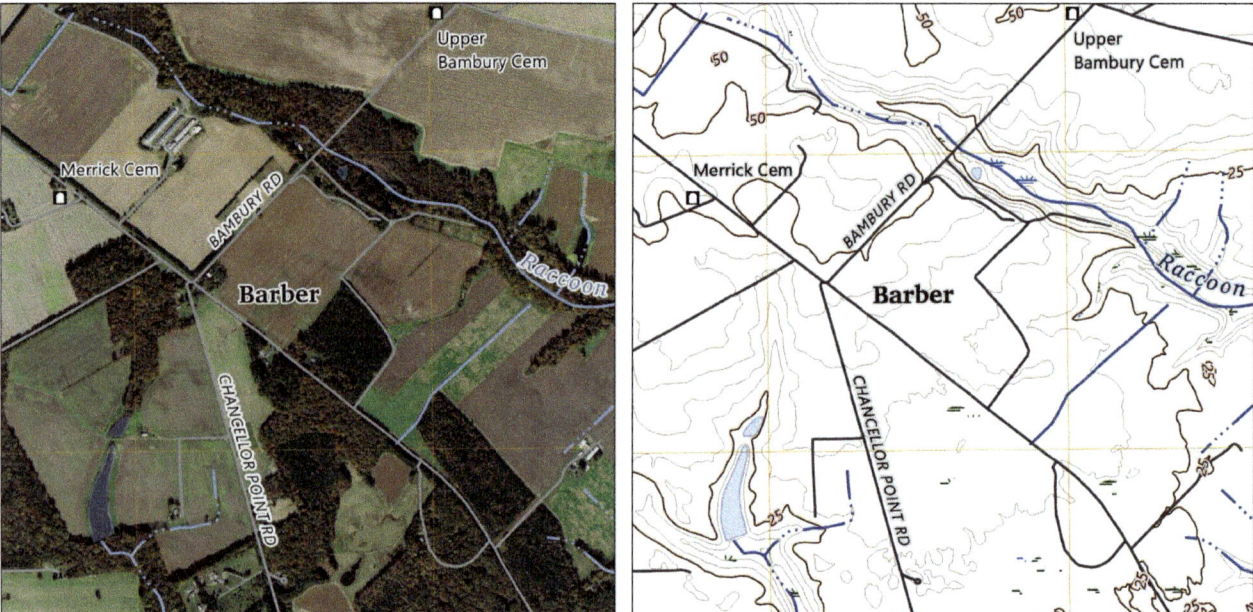

Figure 5.3. This segment of the 1:24,000-scale topographic map and orthophotomap for Trappe, Maryland, shows evidence of metes-and-bounds land partitioning. The irregular sizes, shapes, and angles of landscape features reflect the original settlement pattern and metes-and-bounds land survey in this area. Courtesy of the US Geological Survey.

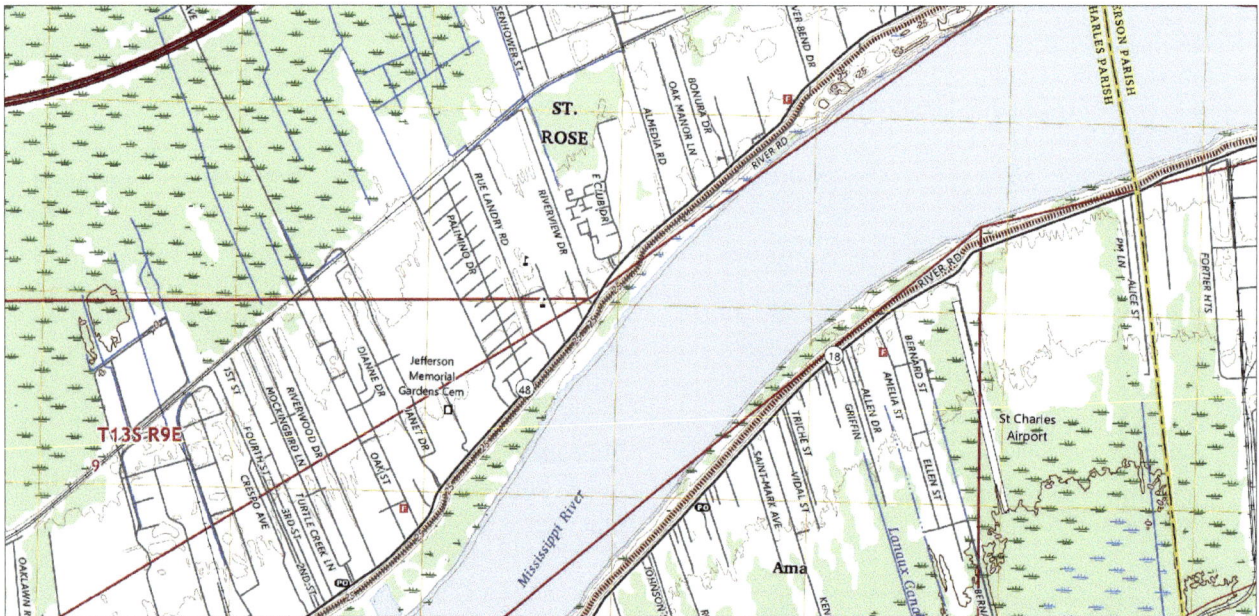

Figure 5.4. This segment of the 1:24,000-scale topographic map for Luling, Louisiana, shows evidence of French long lots. Short roads perpendicular to the river's edge access the many narrow properties along the river. Courtesy of the US Geological Survey.

boundaries still exist legally and are recorded on the maps of the areas of early French settlement, such as Detroit, Michigan; Green Bay, Wisconsin; Vincennes, Indiana; and along the Mississippi River in Louisiana (figure 5.4).

Spanish and Mexican land grants

A **land grant** is an area of land whose title was given to its owner before the territory was part of the United States. A **title** is the legal right to use and modify a property or transfer all or a portion of it to others. After the territory was acquired by the US government, the title of the land previously granted to the owner was confirmed officially. From the late 1600s to around 1850, when much of the southwestern United States was part of the Spanish Empire and, later, the Mexican Republic, three types of land grants were issued by the Spanish and Mexican governments. **Pueblo grants** issued to communities of Native Americans were among the earliest lands granted, and today Native American land areas in the United States administered as federal reservations are often based on these land grants in New Mexico and other states. **Private grants** of property were issued by the government to individuals as a reward for service to the government. **Community grants** were made to groups of settlers such that individuals in the group were given small tracts of land within the community grant to settle and cultivate, but most of the land was held in common for grazing, timber, and other purposes. To receive a land grant, you had to (1) physically step on the land, (2) run your fingers through the soil, and (3) make a commitment to live on the land, use it to feed your family or graze your livestock, and defend it with your life, if necessary.

Land grant boundaries on maps are easy to recognize when they align with physical features, such as hilltops, rivers, and **arroyos** (dry creeks or streambeds that fill after sufficient rains). Because these land grants were in predominantly arid areas where settlers were concerned with water resources, many land grants straddled rivers and lakes. The boundaries of large grants are shown with a thick red line on current 1:24,000-scale USGS topographic maps (figure 5.5).

Donation land claims

Enacted by the US Congress to promote settlement in the Oregon Territory (Oregon, Washington, Idaho, and parts of Wyoming and Montana), the **Donation Land Claim Act of 1850** affected settlement patterns across the territory. The act legitimized any existing claims, called **donation land claims (DLCs)**, and set the terms for claims to White settlers and Native Americans of mixed blood who arrived between December 1, 1850, and December 1, 1853.

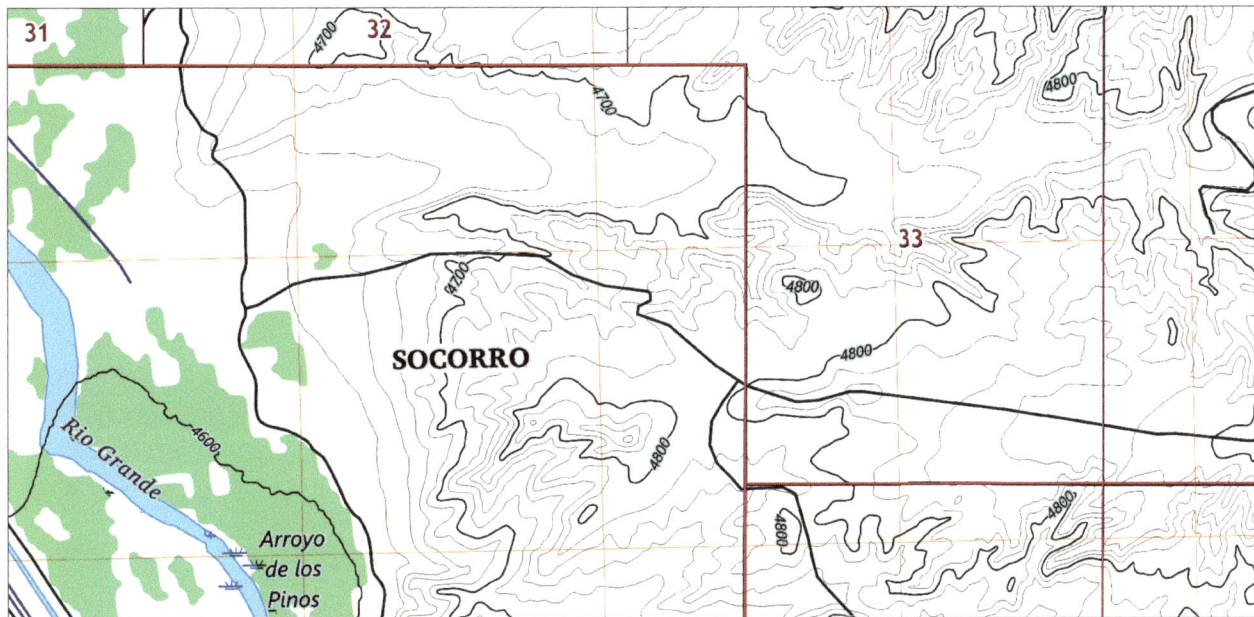

Figure 5.5. This segment of the 1:24,000-scale topographic map for Loma de las Cañas, New Mexico, shows the Socorro community land grant with a thick red line. The regular grid of the PLSS, explained later in this chapter, ends abruptly at the boundary of the Socorro land grant where 70 families originally settled in 1815. Courtesy of the US Geological Survey.

Claims of 320 acres (almost 130 hectares) were granted to individual settlers, and claims of 640 acres were granted to married couples. Settlers were required to live on the land and cultivate it for four years. Under the later **Homestead Act of 1862**, citizens and persons who filed their intentions to become citizens were given 160 acres of public land if they fulfilled certain conditions, including building a home on the land, residing there for five years, and cultivating the land. By law, the land parcels had to be surveyed with north–south and east–west boundaries to conform to the PLSS (described later in this chapter) if the PLSS survey was completed in the area.

DLCs were shown on maps made by the **General Land Office (GLO)**, which was created in 1812 as an independent agency of the US government that was responsible for administering the nation's public-domain lands. In 1946, it became the **US Bureau of Land Management (BLM)**. DLCs on the GLO maps were overlaid on the PLSS and assigned numbers greater than 36 because 1 to 36 were reserved for section numbers in the PLSS. The irregular PLSS boundaries shown in red on the historical USGS map at the top of figure 5.6 are not shown on the current map at the bottom. Instead, only PLSS sections are shown in red.

The land partitioning systems we have discussed so far are irregular partitioning schemes in which the land was allotted before regular surveys were made. These irregular systems encouraged inefficient partitioning of land—at least from the government's point of view. The first people into a region had a virtual monopoly over the choicest lands. The only land left available to settle later often

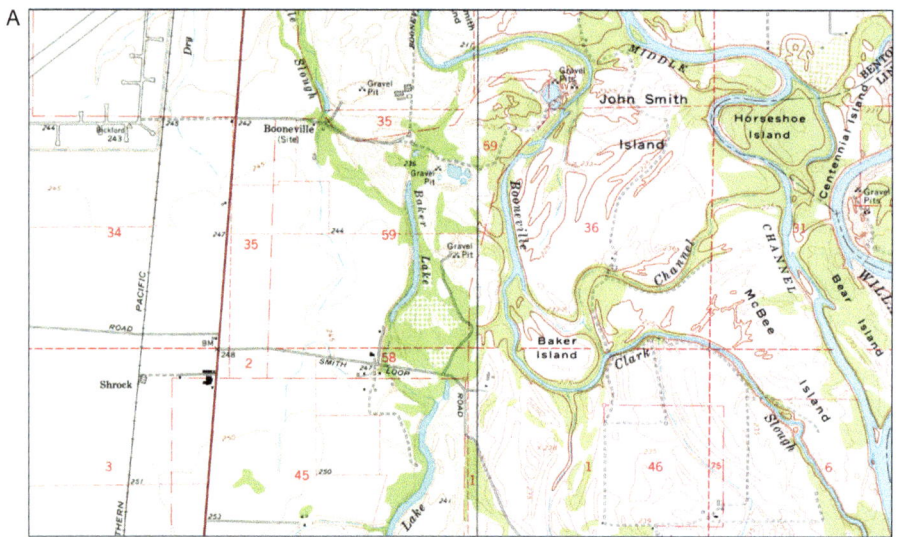

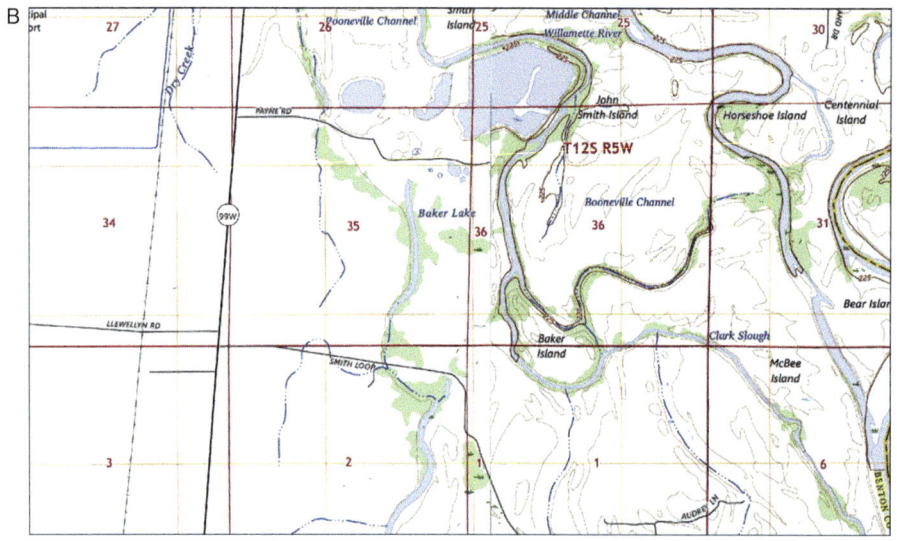

Figure 5.6. The segment of a historical 1:24,000-scale USGS topographic map for Greenberry, Oregon, shows the irregular boundaries of donation land claims (*A*). The second segment of the map at the bottom shows the same area on a current USGS map (*B*). The DLCs have been eliminated and the regular sections of the PLSS are shown instead. Courtesy of the US Geological Survey.

had steep slopes, poor soil, no access to water, or swamps, depending on the location. In many areas, fragments of poor land remained unowned and unwanted long after a region was fully settled.

Furthermore, land claims were often later found to overlap, and boundary descriptions were often in conflict or in error. In homogeneous environments, it was hard to establish accurate borders in the first place owing to the inability to distinguish geographic features. In any setting, boundary mistakes naturally occurred as the land passed to subsequent owners and the environment changed. Good fences, as Robert Frost pointed out in his "Mending Wall" poem (1914), did make good neighbors, because they clarified the location of the boundaries. It used to be the custom for neighboring landowners to walk together around their parcels each spring. The excursion was far more than a social outing—it ensured that all owners agreed on the borders between their lands. Problems arose, however, when later generations relinquished the boundary-walking tradition. Boundary landmarks moved or disappeared over time—streams shifted, trees died or were cut down, rock piles were removed, and stakes disappeared. The result is a long history of legal battles over property boundaries of irregularly settled land.

Regular systems

The alternative to partitioning the land into irregular parcels was to **survey** and divide the land in a regular pattern before settlement. Surveying land into geometrically similar parcels in a systematic or regular manner is an ancient practice. The pattern of agricultural land subdivision in both ancient Egypt and China was a grid, and cities in these—and succeeding—cultures often were planned with a rectangular grid of streets and lots. Another example is from the Roman Empire, when agricultural land in northern Italy and elsewhere throughout the empire was subdivided into a grid of square parcels using a rudimentary form of land surveying.

In the United States, rectangular grids of streets and lots were used for planning towns in many of the colonies, occasionally in the South but mostly in the North. Town sites were laid out and surveyed, and plat maps of the parcels were prepared and recorded, all before settlement. These plat maps were part of the foundation of the US **Public Land Survey System (PLSS)**, the dominant public land partitioning system in the United States. We will look at this system in detail.

US Public Land Survey System

In 1783, the US Congress of the brand-new confederation of 13 states was faced with an urgent need for a national land policy. It had to devise a way to manage the vast lands east of the Mississippi River that were ceded by Great Britain to the United States after the Revolutionary War. Quick action was important for several reasons—land was promised to Revolutionary War soldiers, a source of income was necessary to run the new country, and future states had to be carved out of the wilderness.

By the time the lands north and west of the Ohio River (known as the **Northwest Territory**, covering all of modern-day Ohio, Indiana, Illinois, Michigan, and Wisconsin and the northeastern part of Minnesota) were opened to settlement in the late 1700s, the newly formed US government had come up with what seemed to be an orderly way to transfer land to settlers. The solution was called the **Land Ordinance of 1785**, which established the PLSS. Otherwise known as the **Township and Range System**, this plan called for the regular, systematic partitioning of land into easily understood parcels before settlement. The legal description for each parcel was carefully recorded, and settlers were able to select surveyed parcels to their liking.

The first step in establishing the land partitioning system in these areas was to arbitrarily select an **initial point**. Each of the 37 initial points in the PLSS were established

Figure 5.7. Initial points were surveyed and defined physically with survey markers. The current marker at the intersection of the Willamette principal meridian and Willamette baseline in Portland, Oregon, is shown here. The original marker, called "The Willamette Stone," was a small stone obelisk placed in this spot on June 4, 1851. Courtesy of The Center for Land Use Interpretation.

by surveyors sent into the field, usually soon after treaties and other agreements were signed to tie the new territory to the PLSS. The locations of initial points were marked with permanent identifiers, such as the survey marker shown in figure 5.7 that marks the initial point of the Oregon survey near Portland. Government land surveyors then determined the parallel and meridian that intersected at the initial point. The parallel was called the **baseline** or **geographer's line** (surveyors were called geographers in those days), and the meridian was called the **principal meridian**. Each principal meridian was given a number or name, although the baselines were not.

Thirty-two principal meridians and baselines were established within the conterminous United States (figure 5.8), and five were established in Alaska. The testing ground for extending the PLSS from the initial point was Ohio, in which seven methods were experimented with before the final structure of the system was developed using the first principal meridian and baseline in the northwest corner of the state. The resulting system was then used to survey the rest of Ohio using additional baselines and then the rest of the nation to the west and south of Ohio.

At first, smaller areas east of the Mississippi River were surveyed from their initial points. Later, surveys in the West covered larger areas, and often an area surveyed from an initial point covered one or more states (for example, see Oregon and Washington in figure 5.8). Texas has its own system, which is based on early Spanish land grants, discussed earlier in this chapter. Some of the small surveys in the West were for federal Native American reservations at the time.

In figure 5.8, the entire area covered by a principal meridian and baseline seems to be surveyed completely, but this complete coverage is not the case. In fact, some land has never been surveyed under the PLSS, mainly because it was deemed unsuitable for settlement, or it was reserved

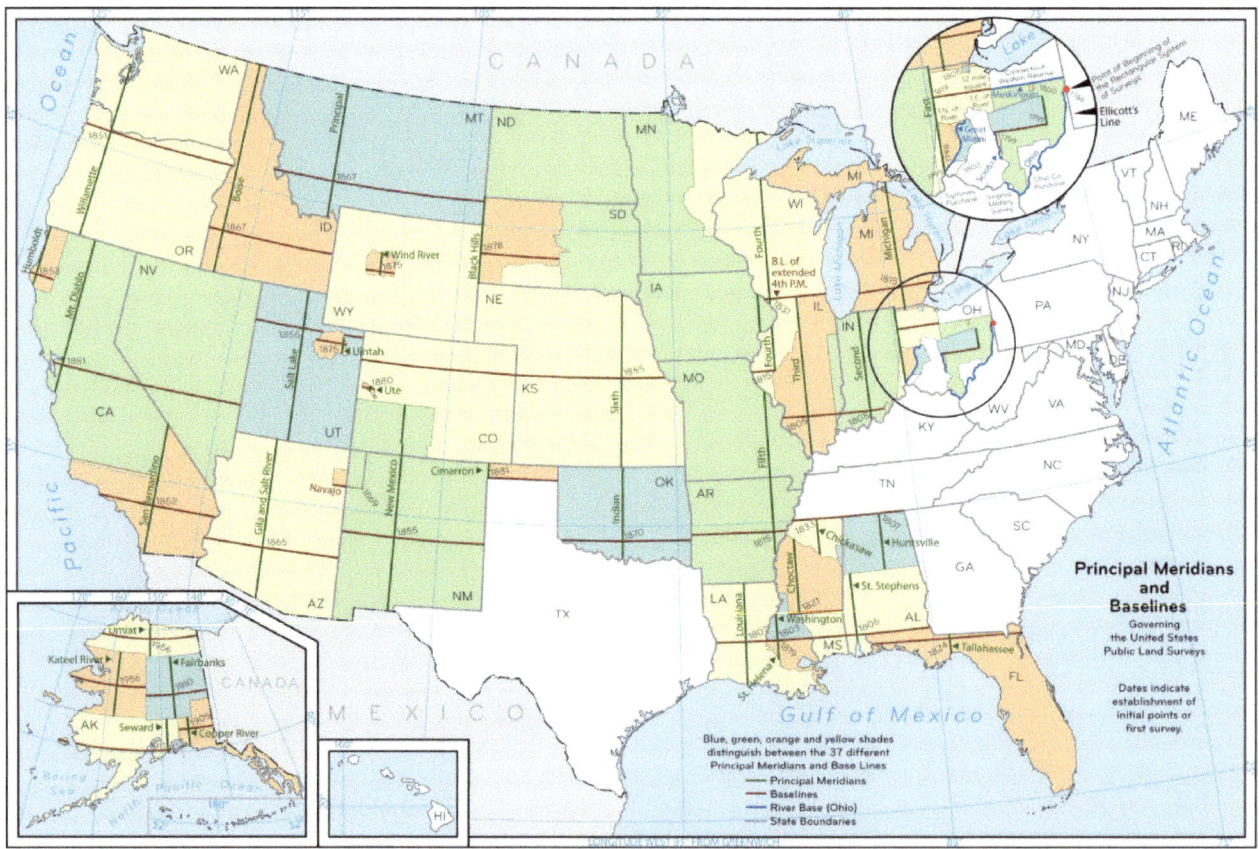

Figure 5.8. Initial points mark the intersection of the principal meridians (shown with black lines) and baselines (shown with red lines) of the PLSS for public land surveys within the conterminous United States. Areas surveyed from each initial point are shown with different colors. Areas that are not colored were never surveyed under the PLSS. Courtesy of the Bureau of Land Management.

for national forests, Native American reservations, or other government use.

Townships and sections

Once an initial point, principal meridian, and baseline were established, **range lines** were surveyed along meridians at six-mile intervals east and west of the principal meridian. **Township lines** were surveyed along parallels at six-mile intervals north and south of the baseline (figure 5.9). The six-by-six-mile quadrilaterals bounded by these intersecting township and range lines were called **survey townships** (sometimes called **congressional townships** and often shortened to **townships**). Each township is identified with a township and range designation. Township

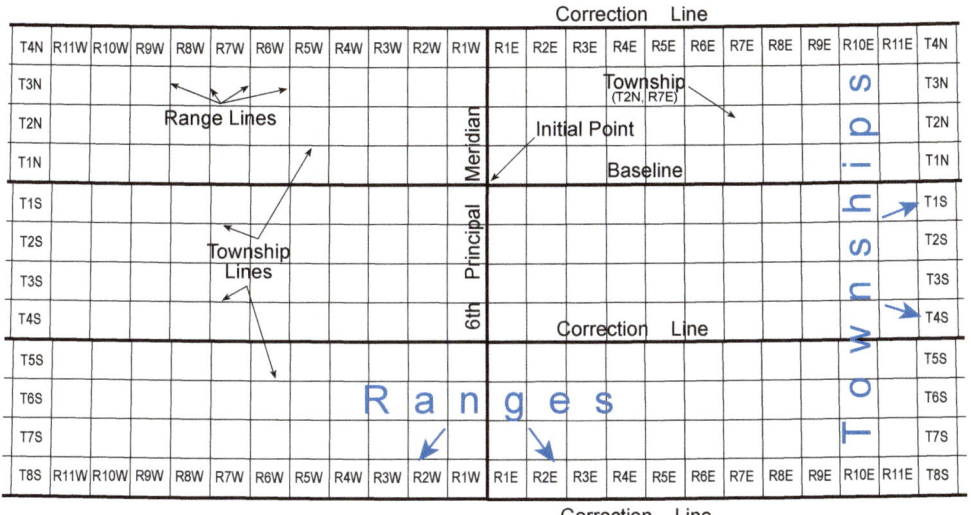

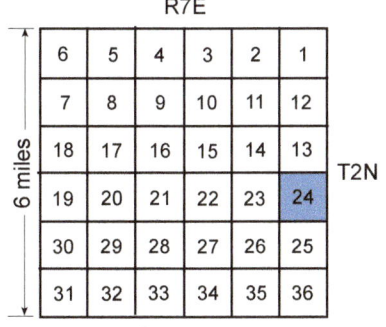

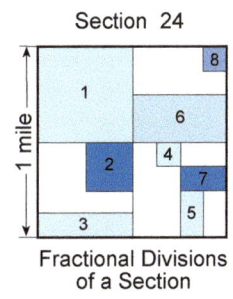

Correction Line
Range Lines
Township (T2N, R7E)
Initial Point
Baseline
6th Principal Meridian
Township Lines
Correction Line
Ranges
Townships
Correction Line
R7E
T2N
6 miles
Township Section Numbering
Section 24
1 mile
Fractional Divisions of a Section
TYPICAL FRACTIONAL LAND DIVISIONS
(1) NW¼, Sec. 24, T2N, R7E, 6th. P.M.
(2) NE¼, SW¼, Sec. 24, T2N, R7E, 6th. P.M.
(3) S½, S½, SW¼, Sec. 24, T2N, R7E, 6th. P.M.
(4) NE¼, NW¼, SE¼, Sec. 24, T2N, R7E, 6th. P.M.
(5) W½, SE¼, SE¼, Sec. 24, T2N, R7E, 6th. P.M.
(6) S½, NE¼, Sec. 24, T2N, R7E, 6th. P.M.
(7) S½, NE¼, SE¼, Sec. 24, T2N, R7E, 6th. P.M.
(8) NE¼, NE¼, NE¼, Sec. 24, T2N, R7E, 6th. P.M.

Figure 5.9. The structure and notation of the PLSS provides a regular system of identifying land parcels as small as 10 acres (four hectares).

designations indicate rows that are six miles wide north or south of the baseline, and range designations indicate columns that are six miles wide east or west of the principal meridian. The township notation includes the township or range, then the row or column number, and finally a suffix to indicate the direction away from the principal meridian or baseline. So, T3N, R2W identifies the township in the third row north of a baseline and in the second column west of a principal meridian. The word *township* is used in three ways: to identify surveyed lines, as the shortened name for six-by-six-mile surveyed areas, and as the term for the row north or south of a baseline (in units of six miles).

The surveyors knew they would encounter a problem. Because meridians converge toward the North Pole, the east–west distance between the meridians decreases (see chapter 1 for more on converging meridians). Thus, the regular township grid could not be extended indefinitely north or south from the initial survey point without townships becoming distorted in shape and area. To reduce the problem of unequal township dimensions, **correction lines** were established at every fourth township (24 miles or 38.6 kilometers from the baseline or another correction line), along which new range lines were surveyed to the east and west of the principal meridian at six-mile intervals. This pattern of baselines and correction lines is seen on the map at the top of figure 5.9. The length of township lines north of the baseline decreases progressively until the correction line is reached, at which point they are resurveyed and will appear offset on larger-scale maps. This systematic decrease in length means that township boundaries are not truly six miles on a side; nor are they exactly 36 square miles in area, because their areas progressively decrease northward from the baseline and correction lines.

Each township is further subdivided into 36 parcels called **sections**, which are approximately one square mile (640 acres or 2.59 square kilometers) in area. Sections are numbered 1 to 36, depending on their position in the township (figure 5.9). A row-by-row zigzag method of numbering the sections was adopted, beginning with section 1 in the top right and ending with section 36 in the bottom right. This method of numbering is used so that every section shares a common side with the section with the number before and after it. For example. in the Township Section Numbering example in figure 5.9, section 24 shares common sides with sections 23 and 25.

Fractional divisions

A section can be divided successively into halves or into smaller units, such as **half sections** (one-half of one section) and **quarter sections** (one-half of one-half of one section). Half or quarter sections can be further subdivided to produce even smaller units. For example, a **quarter-quarter section** is 1/16th of a section (one-quarter of one-quarter of one section). An easy way to correctly identify divisions of sections is to think of a compass placed at the center of the section. The compass direction from the center of the section to the center of each quarter section is northeast, southeast, southwest, or northwest. Similarly, half sections are identified as north, south, east, or west of the section center. With this system, each land parcel's legal description is unique and unambiguous. In figure 5.9, for example, parcel 4 (the 10-acre or 0.04-square-kilometer piece of land in section 24) is described in abbreviated form as follows:

"NE¼, NW¼, SE¼, Sec. 24, T.2N, R.7E, 6th P.M."

In expanded form, the description reads as follows:

"The northeast quarter of the northwest quarter of the southeast quarter of section 24 of township 2 north, range 7 east, 6th principal meridian."

To locate a parcel from its legal description, the trick is to read backward, beginning with the principal meridian and working through the township and range to the section and then to the fractional division. It is useful to understand the structure and notation of the system because the PLSS is the basis for most **property abstracts** (legal documents that chronicle transactions associated with a particular parcel), deeds, and other landownership documents in much of the United States.

The PLSS is still used today by a wide range of stakeholders involved in land administration, management, and development in the United States. Landowners and surveyors use the system to identify land parcels and boundaries. Real estate agents may use the PLSS to market and sell properties. Government agencies rely on the PLSS to manage land resources, enforce zoning regulations, and plan for infrastructure development. And legal professionals may use the PLSS in property transactions, land disputes, and other legal matters.

Government lots and platted lots

There are some exceptions to the use of fractional land divisions in areas covered by the PLSS. One exception occurs when most of a small parcel (less than 10 acres or four hectares) borders a body of water so that the parcel is not a complete quadrilateral. In this case, a subdivision of the section called a **government lot** is assigned instead. These lots may be regular or irregular in shape, and their acreage may vary from that of regular fractional divisions of a section. A government lot is designated by its lot number, such as Lot 3 in figure 5.10. Parcels in a platted subdivision within a section, called **platted lots**, are also specified by platted lot numbers. Other exceptions are irregularly shaped parcels that are often, but not necessarily, small, which are given descriptions in metes and bounds rather than in PLSS notation.

Survey irregularities

Irregularities in the original survey occurred because of the relatively crude instrumentation of the period, rugged terrain or, in some cases, sloppy work by surveyors who were paid based on total miles surveyed (figure 5.11). These irregularities have persisted, primarily because historical boundaries as originally surveyed hold legal precedence over newer surveys of these boundaries.

Variations in the shape and size of sections resulting from survey irregularities were not the only obstacles to partitioning the land into square-mile sections. As we noted earlier, convergence of meridians toward the poles on the ellipsoidal earth made the surveyor's job of laying out a regular grid of exactly one-square-mile sections within a township an impossible task. To systematize the distribution of shape and area variations from the ideal, surveyors determined section corners (see figure 5.14) in a standardized order so that variations accumulated along the western and northern tiers of sections within each township. A **section corner** is the point of intersection of two survey lines defining one corner of a section land parcel in the PLSS.

Another source of irregularity in the PLSS grids is the point at which surveys that start from different principal meridians meet, particularly when the boundary is an irregular feature such as a river. For example, the boundary between areas surveyed from the Michigan principal meridian and the fourth principal meridian in Wisconsin follows the state line between Wisconsin and the Upper Peninsula of Michigan, so you will find misalignment and partial townships and sections all along the Montreal River boundary between the two states (figure 5.12).

An additional source of irregularity in the pattern of the PLSS grid is apparent in areas where townships were not surveyed systematically outward from the initial point. A good example is the surveying of townships in Oregon along the baseline for the Willamette principal meridian (see figure 5.8). The earliest townships were surveyed outward from the initial point in Portland, but independent

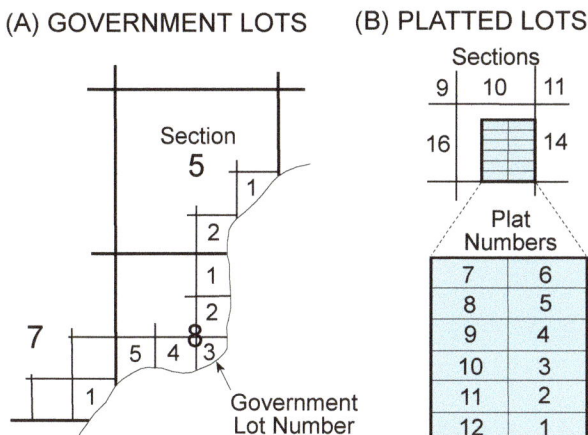

Figure 5.10. Land divisions for small parcels, such as government lots (*A*) and plotted lots (*B*), are designated with lot numbers rather than PLSS fractional division identifiers.

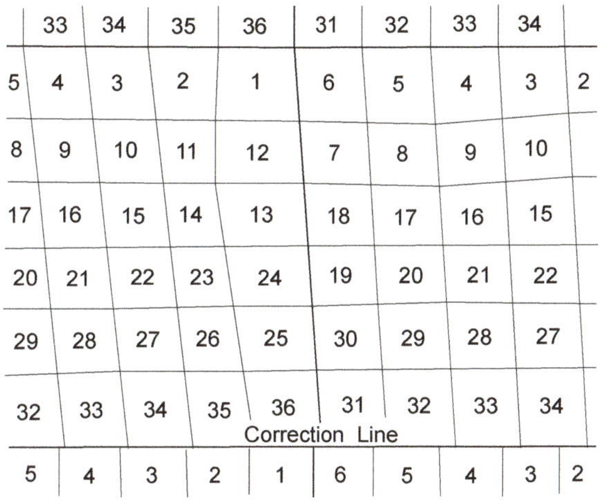

Figure 5.11. Survey inaccuracies are evident in this segment of the PLSS section grid.

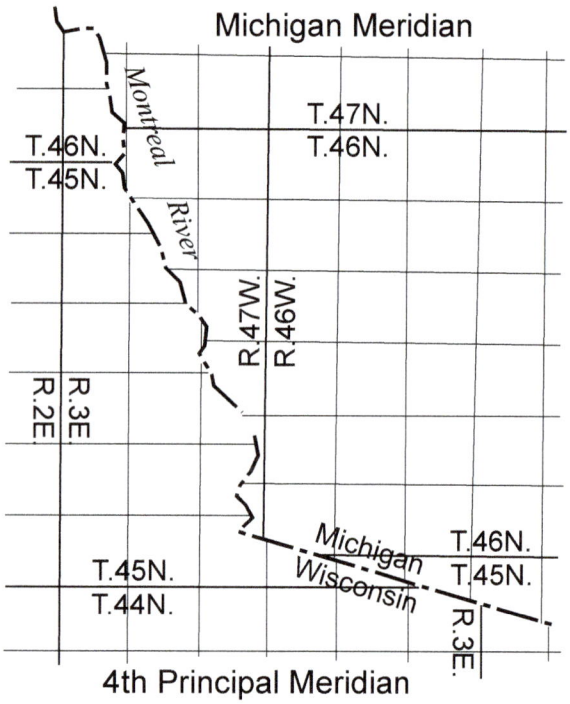

Figure 5.12. Irregularities in PLSS township and section boundaries may be due to the convergence of surveys from different principal meridians.

surveys also were completed at the coast and from three separate starting points along the baseline in eastern Oregon. These other starting points were calculated so that in theory they would mesh perfectly with the surveys from the initial point. But inherent inaccuracies in surveys of these early periods ended up making this perfect meshing impossible to achieve. The many independently surveyed parts of PLSS grids often result in a mismatch at the points or lines where they come together. Along the boundary where these independent surveys converge, some confusing land partition descriptions can occur. This confusion is especially true for two land parcels that straddle the boundary, and it is complicated by the fact that two property deeds are required.

A similar problem may occur when the PLSS grid converges with earlier land partitioning systems—this type of survey boundary mismatch can happen in regions that were not surveyed at the time of settlement or in which boundaries were defined in an older system, such as DLCs or Spanish land grants. Figure 5.13 shows the many areas where the DLCs and PLSS surveys meet in Oregon.

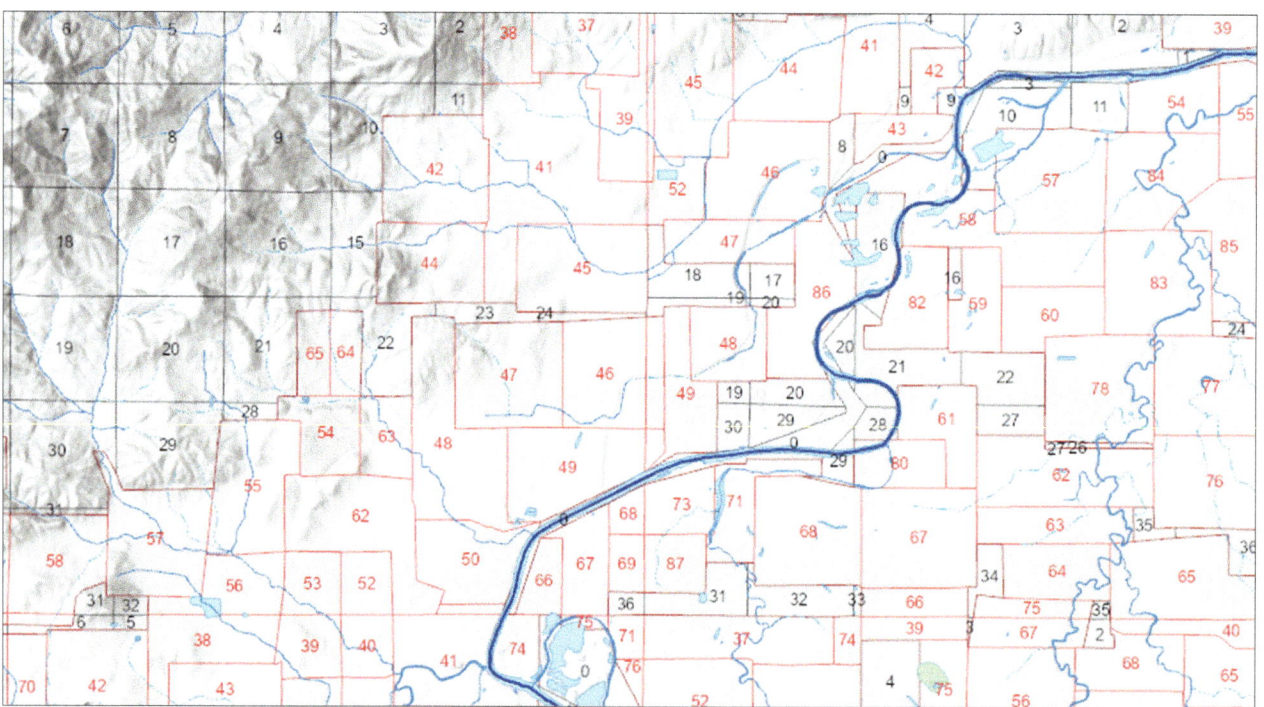

Figure 5.13. The often-irregular arrangement of the DLCs (in red) contrasts with the regular grid of the PLSS (in gray), as is evident in this area of the Willamette Valley in Oregon, between Albany to the north and Corvallis to the south.

PLSS appearance on maps

The PLSS grid is found on many types of maps. On maps made by government agencies, for example, townships and sections are often included because they are so closely associated with the boundaries of civil townships, counties, and other political units. The PLSS grid perhaps finds its fullest expression on large-scale 1:24,000-scale and 1:100,000-scale USGS topographic maps. In figure 5.14, the red PLSS information shown on historical maps created before 2006 is shown in black on the later maps. Also, on these map segments, the marginal township notation is not shown because the Pacific Ocean is along the western edge of the map, and the PLSS is used for land areas only.

Township and section lines are also found on many maps produced by government agencies that deal with land resources, such as the BLM, described earlier, and the **US Forest Service (USFS)**, which manages public lands in the form of national forests and grasslands. Privately produced road and recreation atlases often include the PLSS grid. Surveys of city and rural lots in PLSS areas also are tied to section or fractional-division corners. These corners are commonly shown on land subdivision plat maps and tax

Figure 5.14. The PLSS section, township, and range information is shown on both the historical (*top*) and current (*bottom*) 1:100,000 USGS topographic map for Port Orford, Oregon. Maps courtesy of the US Geological Survey.

assessor cadastral maps (described in more detail later in this chapter), both of which are used as land records.

Canada's **Dominion Land Survey**, not covered here, is the public land survey system used to partition most of Canada's prairie and western provinces into townships and sections for agricultural settlement and other purposes. Based on the PLSS, it differs in the way that townships and sections are surveyed and identified numerically.

Land records

A **land record** is any legal document recorded on paper or in electronic format that affects the title to real property. A **land record system** is a publicly owned and managed system, defined by statewide or provincial standards, that records real estate ownership, transfers, taxation, and development. Traditionally, the basic spatial unit for this recordkeeping is the land parcel. As mentioned earlier, a parcel results from the division of property into areas with some implication for landownership or land use. The land parcel is the smallest unit of ownership for uniform use, for example, an agricultural field. Physical features, resource reserves, market value, ownership, improvements, accessibility, and restrictions on land use are but a few of the attributes recorded for land parcels. In this section, we look at some of the maps that feature parcels, including subdivision plats and maps associated with the cadastre (explained later in this section).

Subdivision plats

Land surveyors determine the boundaries of lots and parcels, most often using **plane surveying** methods, in which the earth's surface is assumed to be a flat plane, similar to the metes survey illustrated in figure 5.1B. Property boundary corners are defined by the distance and bearing from the previous corner, with the first corner, or POB, defined by the distance and direction from a previously surveyed control point, such as a PLSS section corner. (See chapter 1 for more on control points.) The ordered corner-to-corner survey of the property lines for the entire parcel is known as a **traverse**.

A traverse is classified as either open or closed (figure 5.15). An **open traverse**, or **unclosed traverse**, starts at one point and ends at another point. An open traverse is suitable for linear features, such as a road or coastline, but open traverses for polygon features indicate a surveying or recording error. A **closed traverse** forms a closed loop, with the first point surveyed, or POB, also being the last.

These types of traverses provide checks against errors in the distances and bearings because they must ultimately return to the original starting point. The requirement of ending at the starting point allows surveyors to check the accuracy of their work and adjust all the points along the traverse to force an exact fit. This accuracy check makes the extra effort of closing the traverse worthwhile.

Land subdivision surveys are recorded on a **subdivision plat**, which is a map that shows subdivided lots. Plats must be legally recorded in the city or county surveyor's office, or an equivalent land management bureau. A typical subdivision plat map (figure 5.16) shows each property line, along with its distance and its bearing from the previous survey point. In addition to these property line measurements, lot numbers are often shown, along with the names and dimensions of proposed and existing streets. **Building setback lines** (the minimum distance a building must be from a property line) and **easements** (a right to use another person's property for a specific purpose without owning it) are also shown, as with the light-gray areas in figure 5.16, making the subdivision plat the best source of detailed information about any land parcel you are interested in purchasing. Subdivision plats are often the basis for compiling the cadastre in the United States and other countries.

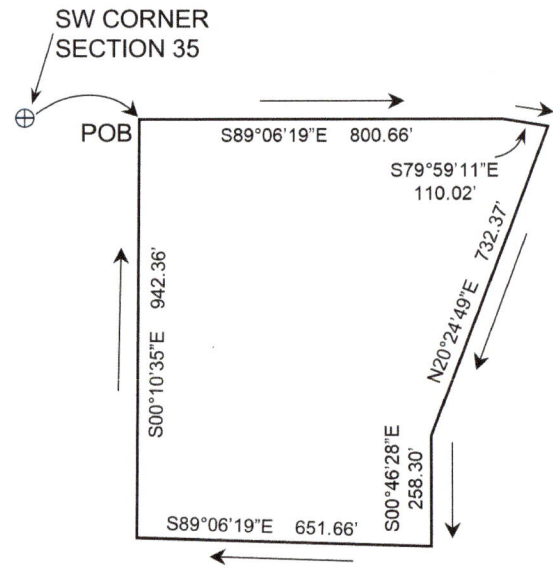

Figure 5.15. A typical closed traverse. The starting point, or POB, is determined relative to a known control point, in this case the southwest corner of PLSS section 35, and the ending point is the same as the starting point. The bearings and distances for each property line are shown.

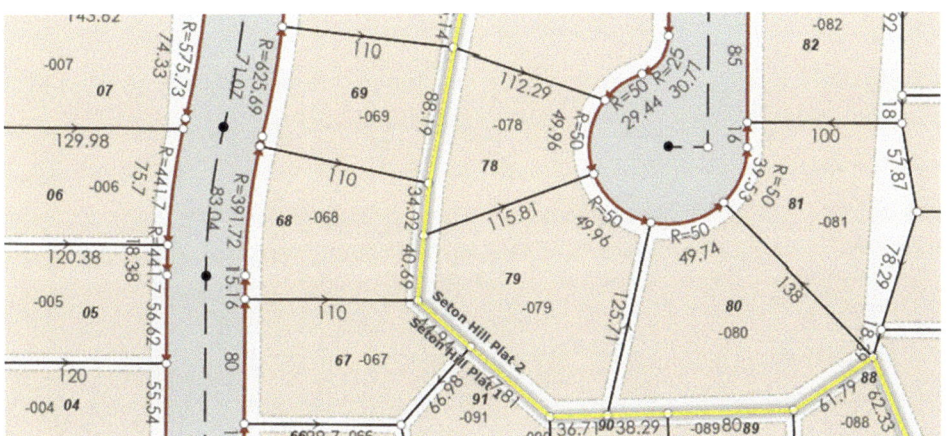

Figure 5.16. Subdivision plats show survey markers or monuments (solid black dots) and detailed land survey information for individual lots and roads in a subdivision. The yellow line on this map marks the boundary between two subdivisions, and white areas represent easements. Courtesy of City of Baltimore, Maryland.

The cadastre

The **cadastre** is a public inventory of data about properties in a county or district based on a survey of their boundaries. As the official record of the ownership, extent, and value of land parcels, it gives information about where and how much. In some places, the cadastre is the responsibility of the county government and is housed in the county treasurer's office in the county seat. In others, primarily cities, it is the responsibility of the associated municipal government office. **Cadastral maps**, with detailed boundaries, property lines, and ownership information, are essential for legal land descriptions and land rights transfer for **real property**—a parcel of land and the structures on it—plus the rights of ownership (figure 5.17). These maps serve as the foundation for urban planning and infrastructure development.

Mapping the boundaries of land parcels presents many challenges. Often in land parcel mapping, a new or updated description of the land parcel boundary adjoins a boundary that was recorded many years or decades before. The instruments and techniques used to measure land boundaries have improved over time and present challenges when trying to reconcile differences between two or more legal surveys performed at different times with differing technologies. As newer land transactions are entered, older parcel boundary descriptions can be adjusted to the new measurements, thus improving the spatial accuracy of a land information system.

Land information systems

A **land information system** is a special type of GIS used by local governments to collect, maintain, and update information about land rights, restrictions, and responsibilities. Land rights include ownership, leases, and special uses, such as mining or grazing. Examples of land restrictions include easements and restrictive covenants, such as "all lots and living units on the property shall be used for single family dwellings only." An example of a land responsibility is maintaining a parcel's green space (grass, trees, shrubs) to prevent fire hazards, for example, from power lines.

A land information system incorporates powerful analytic and graphic tools to help analyze data, generate and manage land records, produce maps and reports, plot legal descriptions, and make better land management decisions. Land information systems are an important part of local government activities because they provide current, reliable land information that is necessary for many public programs and activities, such as land-use planning, infrastructure development and maintenance, environmental protection and resource management, emergency services, social service programs, and so on.

Increasingly, you will find city and county departments providing websites for their staff and the public that allow access to the land information system data. For example, the Public San Bernardino County (California) Parcel Viewer (figure 5.18) allows users to search based on the assessor's parcel number (APN) or address. It also includes information about land-use districts and PLSS data. Users can choose basemaps and overlay additional data on the map.

Extensions of the cadastre

A **multipurpose cadastre** is an integrated land information system that comprises database layers containing legal (ownership and taxation), physical (land slope and aspect, soil characteristics, drainage, vegetation cover), and cultural (land use, number of residents, building construction, access road width and surface material, utility service,

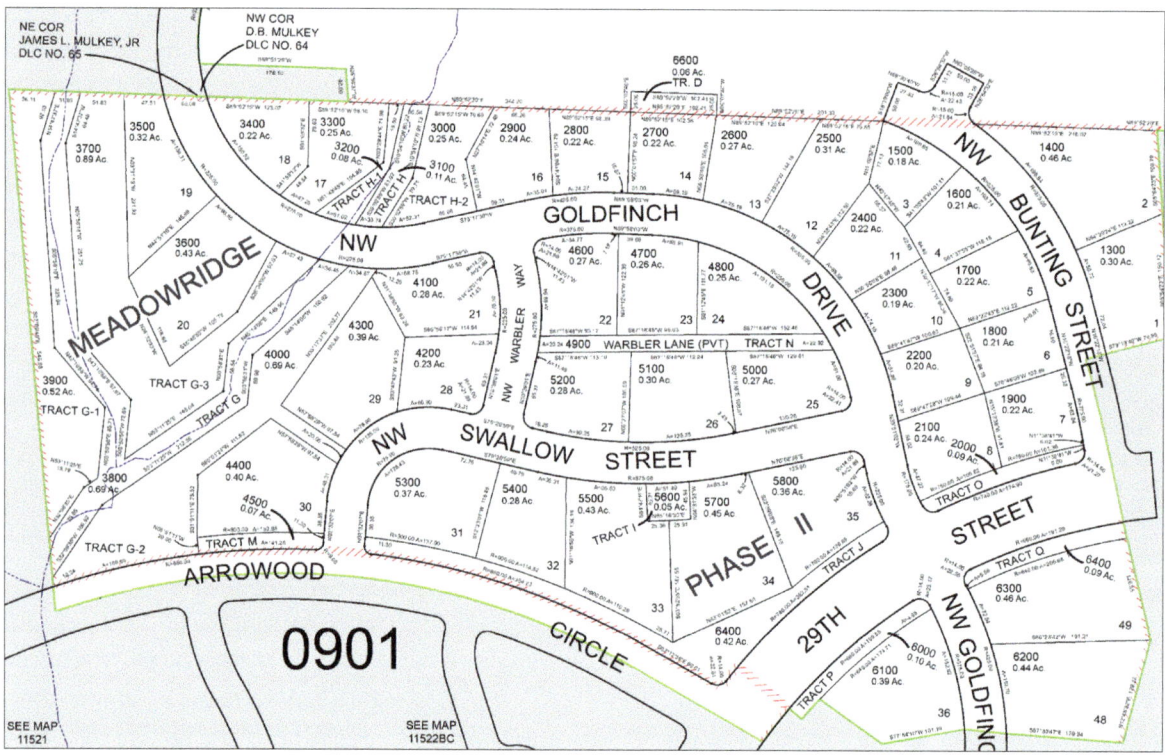

Figure 5.17. A cadastral map shows information about real property, including property boundaries and tax lot numbers that are tied to the fiscal and legal cadastre and, increasingly, the multipurpose cadastre. On this map, subdivision boundaries are shown with a red hatched symbol, and map index lines are shown in green. Courtesy of ORMAP, Benton County, Oregon.

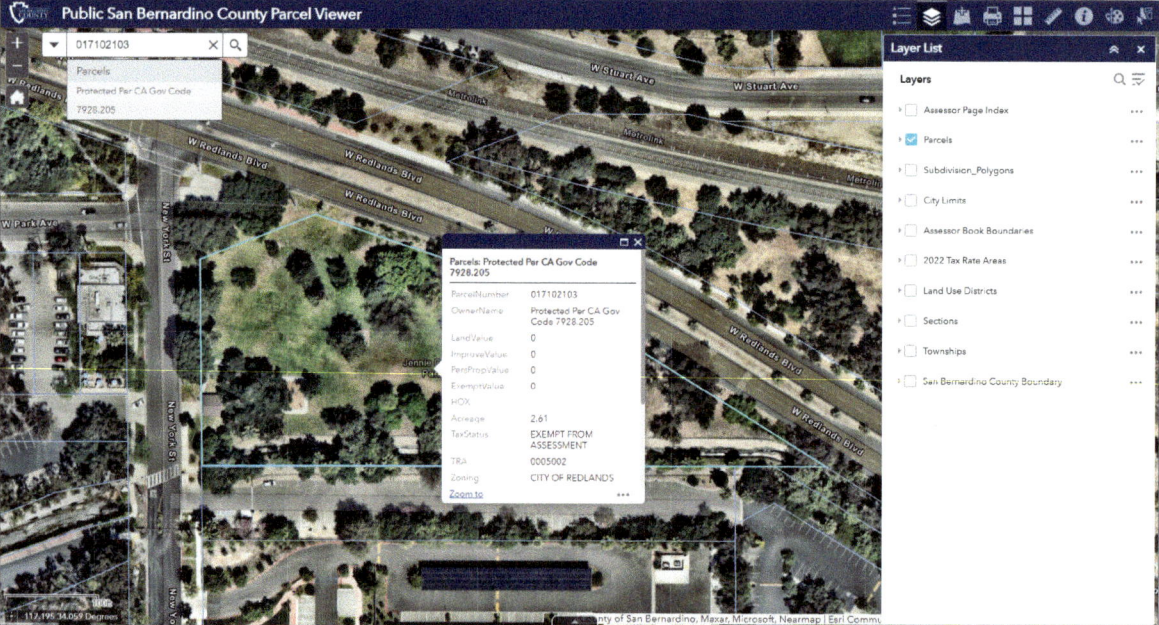

Figure 5.18. The Public San Bernardino County Parcel Viewer zoomed to an area around Jenny Davis Park in Redlands, California. The pop-up shows the assessor's parcel number (APN), acreage, tax status, and other parcel information. Some details for government-owned lots, such as this one, are not shown. Courtesy of County of San Bernardino, California.

zoning restrictions, and so on) information about land parcels. With such extensive information, a multipurpose cadastral system can answer questions about more than just landownership and taxation. For example, you could ask, "How much will that parcel increase in value if a residence is built and water and electricity services are provided?

Local governments are now beginning to extend the multipurpose cadastre to a **3D cadastre**, recognizing property rights and restrictions in 3D space. Using three spatial dimensions, the cadastre can include maps of the 3D configuration of properties to support better planning, design, construction, and management. For example, a 3D cadastre would enable more sophisticated management of the property records and legal information (such as the certificate of ownership) for each individual unit (flat or apartment) in a building (figure 5.19).

A 3D cadastre would also enable the management of above- and below-surface spaces. For example, **air rights**, the rights relating to the vertical space above a property, are a logical addition to a 3D cadastre, as are **subsurface rights**—the legal rights and privileges associated with the ownership, exploration, extraction, and utilization of resources found beneath the surface of a property.

Ownership of a surface parcel does not automatically grant air or subsurface rights. For example, a property owner may be restricted from developing a gas station over an aquifer recharge zone.

The basis for most 3D cadastres today lies in combining land information systems with **building information modeling**, or **BIM**. In this acronym, *building* refers to the creation or construction of any buildings and other assets of the built environment, such as roadways, railways, bridges, and other transportation features. Modeling refers to creating, managing, and using a 3D digital representation of the features and their attributes. Information relates to structural, architectural, operational, and other data. BIM covers the life cycle of assets, from design through construction to operation.

You may see references to further extensions of the cadastre from 3D to 4D to integrate time allowing the cadastre to record temporal activities, such as sequences in construction and planned or completed renovations. These kinds of extensions to the cadastre would help stakeholders better understand complex planning issues, intermediate development steps, and long-term possibilities.

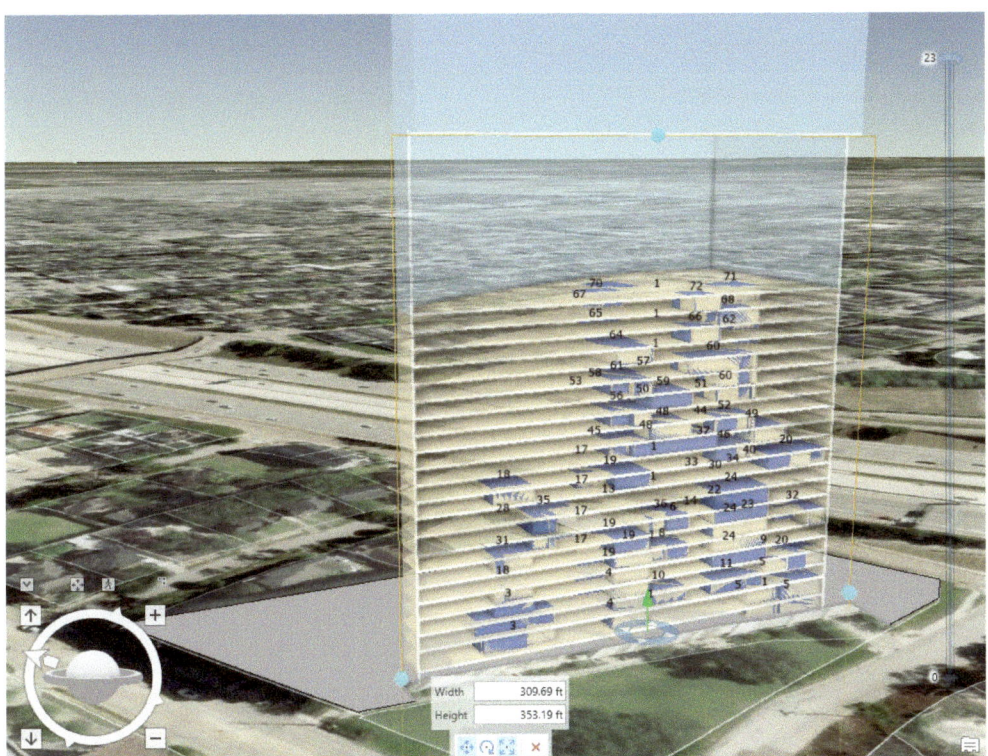

Figure 5.19. In a 3D cadastre, the interior spaces of buildings are subdivided, mapped, and given unique identifiers, as with the condominium units and their legal lot numbers shown here, so that they can be managed as unique entities.

Selected readings

Alema, A. 2022. "What Is a Metes and Bounds Property Description?" *NV5*, March 7, 2022. Accessed online at www.nv5.com/news/what-is-a-metes-and-bounds-property-description.

Crossfield, J. K. 1984. "Evolution of the United States Public Land Survey System." *Surveying and Mapping* 44 (3): 259–65.

Ebright, M. 1997. *Land Grants in a Nutshell: The Center for Land Grant Studies*. Accessed online at www.southwestbooks.org/nutshell.htm.

Hart, J. F. 1975. "Land Division in America." In *The Look of the Land*, 45–66. Prentice-Hall.

Ibragimov, L. T., F. M. Tojidinova, and B. A. Raximov. 2022. "Introduction to GIS Application in the Land Cadastre." *International Journal on Human Computing Studies* 4 (12): 5–9.

Institution of Structural Engineers. 2021. *An Introduction to Building Information Modeling (BIM)*. Accessed online at www.istructe.org/IStructE/media/Public/An_introduction_to-Building_Information_Modelling_BIM.pdf.

Johnson, H. B. 1976. *Order upon the Land: The US Rectangular Land Survey and the Upper Mississippi Country*. Oxford University Press.

Johnson, J. M., and N. R. Chrisman. 2023. "Tracing the Development of the General Land Office's 'National Map': From Lands Surveyed to Territorial Acquisitions." *Cartographic Perspectives* No. 102.

Quintero, J. R. 2004. "Land Information System." In *International Encyclopedia of the Social & Behavioral Sciences*, 1–6. Pergamon. https://doi.org/10.1016/B0-08-043076-7/99107-1.

Robillard, W. G., D. A. Wilson, C. M. Brown, and W. Eldridge. 2006. *Evidence and Procedures for Boundary Location*, 5th ed. Wiley.

US National Atlas. 2008. *The Public Land Survey System (PLSS)*. Last modified April 29, 2008. www.atlas.usgs.gov/articles/boundaries/a_plss.html.

Wilson, D. A. 2006. *Interpreting Land Records*. Wiley.

CHAPTER 6
Map design

Today, nearly anyone can make professional-looking maps. Mapmaking software and apps are widely available at low or even no cost. Sophisticated mapping techniques built into the software and apps guide users through the mapmaking process. Data for mapping is often cheap or free, and its access is made easier through online repositories and hubs. Today's mapmakers can create maps to print, use in a publication or presentation, view in a browser, access on a mobile device, embed in a website, include in a story map, and share in email or social media. The result is a paradigm shift called the **democratization of cartography**, which involves making maps available to wider audiences and empowering more people with the tools and knowledge to create and use maps effectively.

Mapmaking has broken out of the dusty corner office and out of the hands of the highly trained cartographer, who was one of the few people with access to the tools and data to make authoritative, high-quality maps. But if nearly anyone can make good-looking maps, how do we know whether the maps are useful, accurate, trustworthy, and reliable? This is the conundrum that comes with the democratization of mapmaking. In this chapter, we explore some of the most important design aspects of the intricate art and science of making maps with the intent of not only helping you succeed in making your own maps but also helping you to recognize when someone else has produced something that should only be used with caution or not at all. This requires that you have a basic understanding of **map design**—the process of arranging and presenting geographic information on a map to communicate spatial information clearly and efficiently to map readers.

Purpose, audience, and other considerations

Before designing a map, the cartographer needs to determine who will use the map, what they will use it for, and how they will use it. The **map audience** is the individual or group of people who will use or interact with the map. The **map purpose** is the primary function of the map, which, as we saw in the introduction, might be to act as a geographic refence, convey a theme, aid navigation, and more. The conditions of use are the circumstances under which the map will be used. These conditions can relate to the map's medium (for example, paper or digital screen), viewing conditions (including distance and lighting), and resolution (amount of detail shown on the display). Technical constraints, such as size of the map, ability to carry it with you, and requirements for black and white or color display, also have implications for the map's design. Designing for accessibility challenges faced by the map reader, such as visual, auditory, motor, or cognitive impairments, may also be a requirement.

A tour map, for example, is designed very differently from a utility map, although both may show similar features, such as roads and buildings. The 3D tour map of Philadelphia, Pennsylvania, in figure 6.1A is intended to entice visitors to explore an area that may be new to them. Landmarks are prominent features on the map, helping readers to navigate a potentially unfamiliar landscape. Other features on the map include historical sites (tan) and green spaces (green); tourist facilities, such as visitor centers; restaurants (wine glass), restrooms (Ⓡ), post offices (✉), and other points of interest. Transportation features include streets of historic or architectural interest (★), parking facilities (Ⓟ), traffic directions (↘), and bike rental locations (🚲). Dashed bus and colored metro stations and lines help people travel to and through the area. The map is visually captivating, drawing readers in to explore the area even if they are not there physically.

The field crews who will use the water utility map in figure 6.1B are trained professionals who are already familiar with the map's design and symbology, as well as the geographic area. The map is intended to display the complex utility network, its features, and their characteristics in a simple and straightforward manner. The map shows three utility systems simultaneously: water (blue), wastewater (green), and stormwater (orange). Aerial imagery (see chapter 11) serves as a detailed basemap over which the brightly colored line and point symbols can be easily seen. This map is designed to provide a thorough picture of specialized information to a small but expert audience so that they can perform their map use tasks efficiently and

Figure 6.1. Although both of these maps show many of the same features, they are designed for different purposes, audiences, and conditions of use. As a result, the tour map (*A*) is pictorial and eye-catching, whereas the water utility map (*B*) is minimalist and utilitarian. Courtesy of Bill Marsh, Marsh Maps (*A*), and Jeff Sprock, City of Billings, Montana (*B*).

effectively. The tour map might be designed for print so that people can carry it with them, and the utility map might better be shown digitally on a computer monitor in the office or on a mobile device in the field.

Data for map compilation

Many maps today are compiled from existing GIS datasets, but they also may be produced from original surveys, field operations, photogrammetric compilations, and **primary data** that comes from direct measurement or data collection. Data that is captured by converting maps and other geographic data sources to a format that can be used for mapping is called **secondary data**.

Maps may also be compiled from data used to make other maps, usually of a larger scale. Many national mapping agencies have developed workflows to create their smaller-scale map series from the data they compiled for their larger-scale maps. These resulting products are sometimes referred to as **derived maps**, and they may include information from secondary sources, in addition to the maps from which they are principally drawn. Thus, maps may include information from a variety of sources. For example, road maps are compiled from road surveys, topographic maps, and aerial photography, whereas city maps are often compiled from surveyors' maps originally created to support engineering plans and land development. This presents a challenge for both the mapmaker (in collating the data) and the map reader (in assessing the value of each data source as well as the mapmaker's ability to manage the data).

Map scale

Map scale, described in chapter 2, is critical to map design because it affects how much detail can be shown on the map. Map scale guides decisions about selection, generalization, and classification, as well as symbols, color, and type, all of which we explore in more detail in this chapter. Map scale is proportional to the amount of detail that can be shown. For example, as map scale decreases from 1:24,000 (figure 6.2A) to 1:100,000 (figure 6.2B), the number and types of features decreases, as does the level of detail.

Recall from chapter 2 that map scale is the ratio between distances on the map and their corresponding distances on the ground. When ground features are displayed on a map, distances between them must be reduced simply because the size of the symbols on the map is greater than the actual size of the features on the ground. If we were to map physical features in the landscape at their ground sizes, many would be so small that you could not see them on the map, depending on the map's scale. If we try to shrink a larger-scale map to a smaller scale, the symbols may become too small to see or too difficult to differentiate from other symbols (figure 6.2C). Cartographers must therefore carefully balance the variety, complexity, and density of features with the map scale. This becomes an even greater challenge for many web maps because these decisions must be revisited for each zoom level, discussed in chapter 2.

Map projection

You saw in chapter 3 that map projections are designed for particular purposes and have certain properties, so choosing the right projection is a critical mapmaking decision. Map projection selection is often one of the first decisions made in map compilation—a wrong choice can result in a bad map, no matter what other decisions the cartographer makes. And savvy map users who spot a map with the wrong projection will be dubious about the map as a whole.

A general rule of thumb relates to equal-area versus conformal projections (see chapter 3). In general, equal-area projections should be used for maps of statistical distributions, such as population density, so that the relative sizes of areas can be properly compared. Conformal projections should be used to map locations, routes, landforms, and other geographic data for which shape and angles are important. Sometimes the map projection has been decided by mapping standards, client requirements, or other considerations that are out of the hands of the mapmaker. When a choice is possible, the map selection guidelines in chapter 3 and the summary of projection uses in table 3.1 can assist you.

Default map projection parameters sometimes need to be modified for appropriate use of the projection at different scales or geographic extents. A good example is redefining the central meridian, or the origin of the x-coordinates in the map projection (see chapter 3), to the center of the mapped area. Take as an example the Lambert conformal conic map projection for the conterminous United States map in figure 6.3A. With the central meridian at 96° W, the longitude line aligns to north in the center of the map. Zooming in to Oregon, the map appears rotated about 15° clockwise (figure 6.3B). Redefining the central

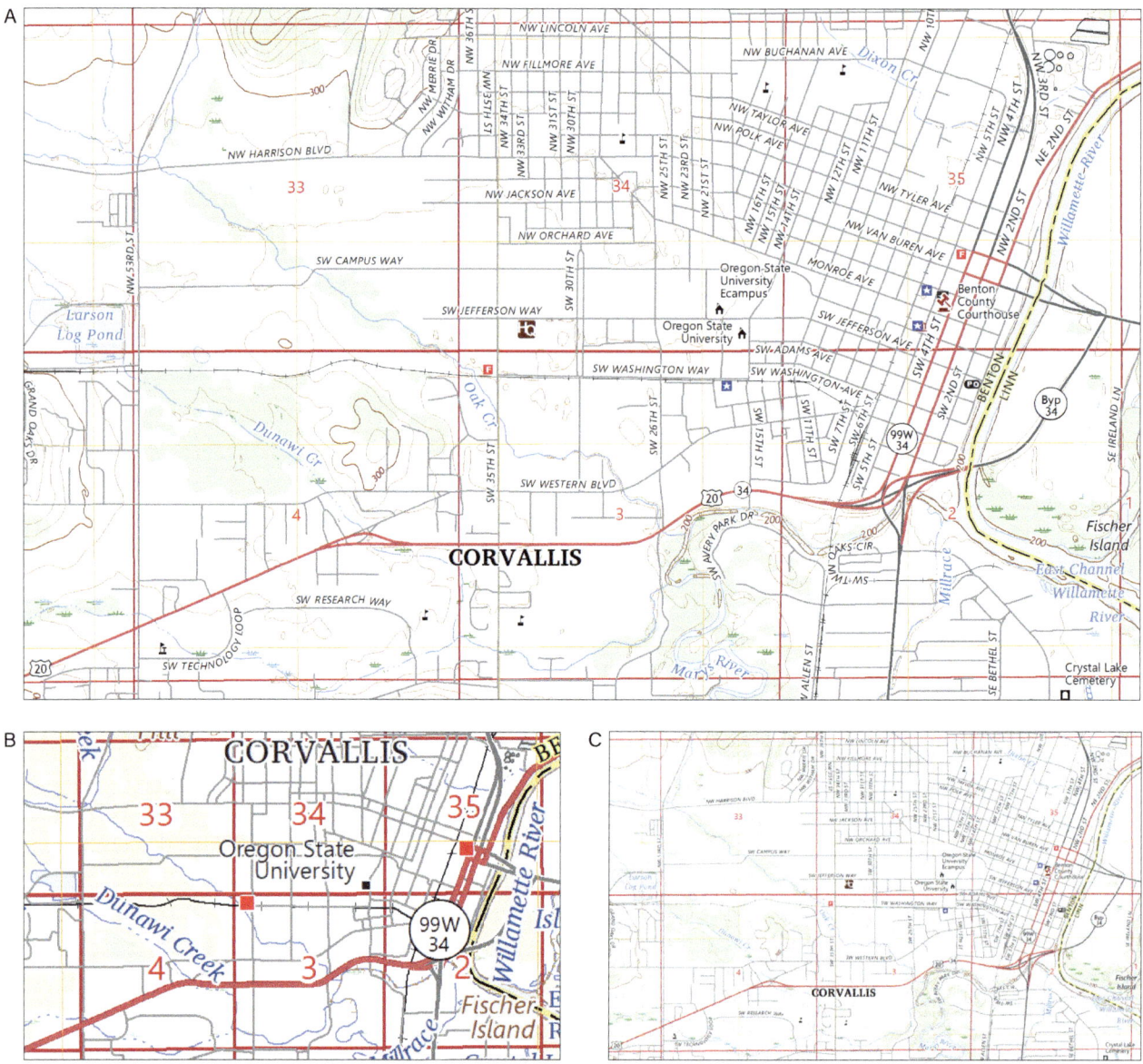

Figure 6.2. These USGS topographic maps show, at their relative map scales, the same six PLSS sections in the Corvallis, Oregon, vicinity. The detail on the 1:24,000-scale map (*A*) must be reduced by showing fewer numbers and types of features with less complexity than on the 1:100,000-scale map (*B*). Simply shrinking the map (*C*) from a larger scale to a smaller one results in a product that is illegible and useless. Courtesy of the US Geological Survey.

meridian at, or near, the center of the east–west extent of the state adjusts the map so that north is up at the center of the map extent (figure 6.3C). An eyeball estimate can be used to guess that 120° W might be a good choice for Oregon, but this can also be calculated by subtracting the easternmost longitude from the westernmost longitude, dividing by two, and adding that value to the easternmost longitude. For Oregon, the precise result is 120.513° W.

The difference between 120° W and 120.513° W will be too small to see, so eyeballing the central meridian would be perfectly acceptable in this case.

A further modification of the projection reduces distortion in the north–south direction. The **1-6 rule** is a cartographer's rule of thumb to determine the two standard parallels of a secant-case conic or cylindrical projection (see chapter 3 for secant-case projections). The rule is to divide

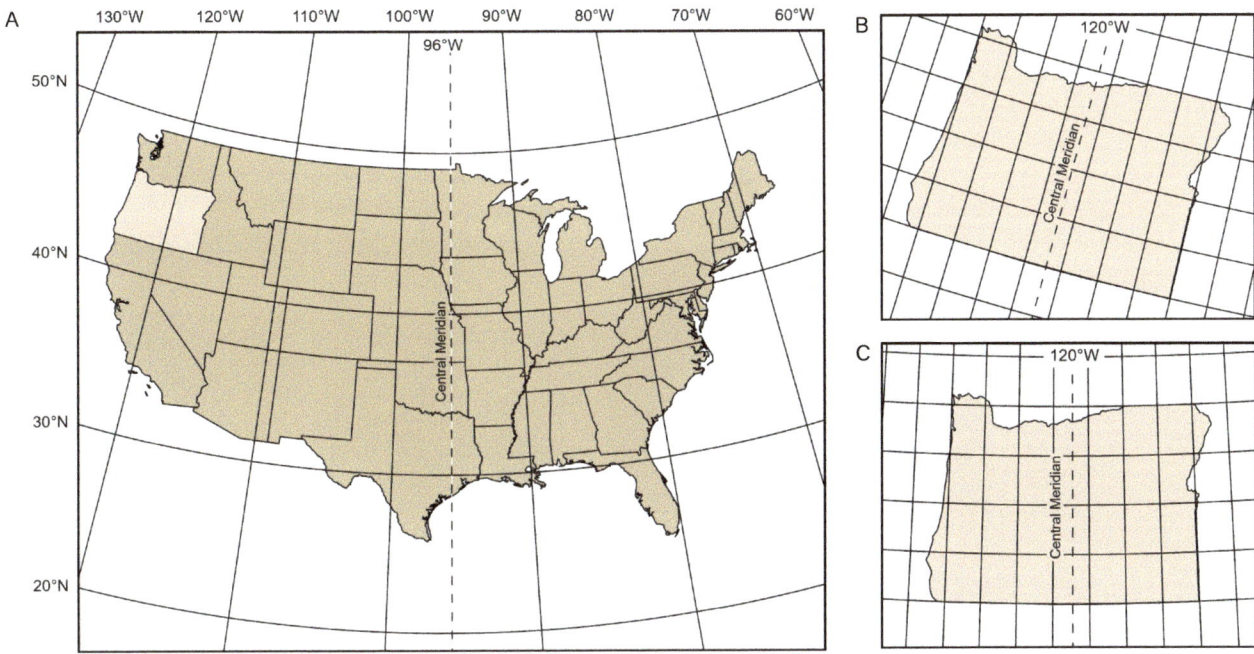

Figure 6.3. When using the Lambert conformal conic map projection for the conterminous United States centered at 96° W (A), a map of Oregon appears rotated clockwise about 15° (B). Changing the central meridian of the map projection to 120.5° W (near the east–west center of the state) adjusts the map so that north is up in the middle of the map (C).

the north–south extent of the mapped area into sixths. The first standard parallel is 1/6 from the bottom of the area, and the second is 1/6 from the top. This modification minimizes distortion between and outside the standard parallels along which, like the central meridian, the scale factor is 1. The result for Oregon is 42.72° and 45.28° N, which can be rounded to 43° N and 45° N.

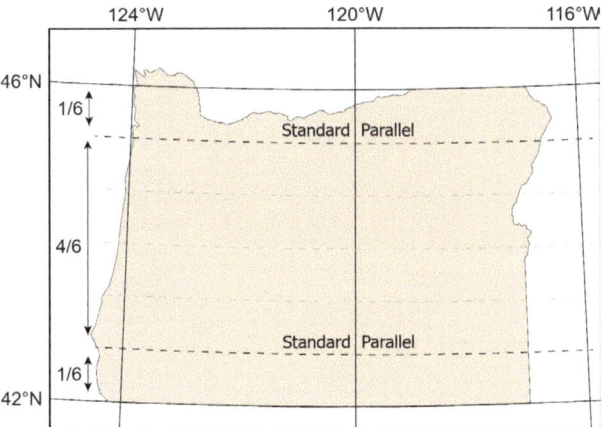

Figure 6.4. The 1–6 rule can be used to define the standard parallels for a secant-case conic or cylindrical map projection to minimize distortion in the mapped area.

Cartographic abstraction

Much of the power of maps lies in **cartographic abstraction**, in which data that has been collected about the environment is converted into a graphic representation that retains the salient aspects of the geographic features and their relationships. To do this, cartographers rely on selection, generalization, classification, and symbolization. They decide what type of and how much information to portray (selection). They eliminate or de-emphasize unwanted or unneeded detail (generalization) and summarize other details into categories (classification). They also make choices about how the information is shown graphically (symbolization). Cartographic selection, classification, generalization, and symbolization are interrelated, so mapmaking is more of an iterative process than a prescriptive one as cartographers continually refine the results they obtain through the cartographic abstraction process.

Selection

Selection in cartography is the process of deciding which features to show on the map and which to leave off. Selection decisions relate to the themes of information on the map, as well as the categories and subcategories of features within the themes. For example, on one map, you might decide to leave off all roads, but on another, you may decide to leave off only the minor roads and other less important road features.

Cartographic selection for general reference maps is much different from that for thematic maps. For reference maps, the challenge for the mapmaker is to decide which features are of greatest interest to a variety of map readers for a range of purposes. Reference maps, such as the road map of Oregon in figure 6.5A must carry a lot of information, although only some of it may be relevant for a particular map use task. Expert decisions about how to categorize and order features and labels on the map allow readers to distinguish the many types of features and their relative size or importance. This map could be used as effectively

Figure 6.5. The number and variety of features selected to be shown on reference maps, such as the recreation map (*A*), is greater than for thematic maps, such as the protected areas map (*B*). Only the reference information that provides geographic context is included on thematic maps. Courtesy of Benchmark Maps, an imprint of East View Information Services (*A*) and the US Geological Survey (*B*).

for road navigation as for planning a recreational trip or analyzing the distribution of wilderness areas.

With thematic maps, the challenge for the mapmaker is to decide which features to include without detracting from the theme. Only the features that are relevant to the theme or necessary to provide geographic context are included because superfluous information may distract, confuse, or mislead the reader. The map of protected areas in figure 6.5B provides a good example. The only things added are the features and labels for states, large cities, interstate highways, and major hydrographic features.

For any map, it is the responsibility of the mapmaker to select the features wisely. However, it is the map reader's responsibility to understand that all possible features cannot be shown.

Generalization

Generalization in cartography refers to reducing the amount of detail on a map through a change in the geometric representation of features. When the world is reduced to the scale of a map, the challenge is to remove unnecessary detail while preserving the basic geometric form of the simplified features and the salient spatial relationships among them.

The dimensionality (one dimension, two dimensions, and so on) of a feature can be used as an indicator of the appropriate level of generalization. For example, point features should appear as zero-dimensional points on a map, and linear features should appear as one-dimensional lines. When there is too much detail on the map, nearby points may coalesce and appear as blobs, and lines may collapse on themselves and appear as polygons (figure 6.6A). Feature curvature is another indicator. For example, lines that are curvilinear in nature but have sharp angles on the map (figure 6.6B) are clues that the lines are too generalized. Using data that is appropriate for the map scale is ideal (figure 6.6C), but the data might not be available. When that is the case, symbolizing a line with a thinner symbol may eliminate blobs (figure 6.6D), and showing an angular feature with a thicker line can sometimes mask its jagged edges.

Generalization in mapping is a complex topic, and entire books have been written on the subject. Because there is so much information available on generalization and so many ways to approach the subject, we severely limit our discussion here and instead provide a broad overview of generalization methods that can be used for features symbolized with points, lines, and polygons.

Generalization is used to simplify and refine features for display at smaller scales while preserving their basic geographic properties and relationships. For example, when map scale is reduced, two linear features that are next to each other in reality, such as a river and a road, may begin to coalesce on the map. Because the map should graphically communicate the real-world geographic setting, the message to map readers should be that the features are next to each other, not on top of each other. To preserve the integrity of the spatial relationship, the cartographer may offset one or both of the features.

Although myriad generalization operations have been described and developed, there is no set of operations that all cartographers agree on. The operations described here are often used in workflows to derive smaller-scale maps from larger-scale cartographic data, and all are available as geoprocessing tools in table 6.1. Some create new data that is simplified, but many modify the symbology without changing the data. This allows the original data to be preserved and managed without redundancy. Keep in mind that this is not an exhaustive set of methods to generalize features on maps. For example, simple symbology modifications can be used to display less detail, as we saw in figure 6.6D, and selection tools can be used to exclude features by category or size.

Traditionally, generalization operations affected the geometry of each class of features without regard to symbology or relationships with other features. Many of today's generalization operations perform **contextual generalization**, in which multiple features from multiple classes are considered at the same time to maintain the salient geographic pattern, density, and character of the features. This contextual approach also considers conflicts that arise between the symbols of features, which is particularly important when generalizing data for a different map scale. Tools to perform graphic conflict operations (table 6.2) can be used to detect problem areas (detect graphic conflict), move roads so their symbols do not overlap (resolve road conflicts), and move, resize, or hide buildings relative to roads (resolve building conflicts). A workflow might start with reducing the number and complexity of the features. Then conflicts between the symbolized features would be resolved. These steps might be repeated until the desired result is achieved.

To see how these generalization and other types of operations, such as selection, can be used together, we can look at a map that was reduced in scale from the **source scale** (the scale at which the data was originally captured from its

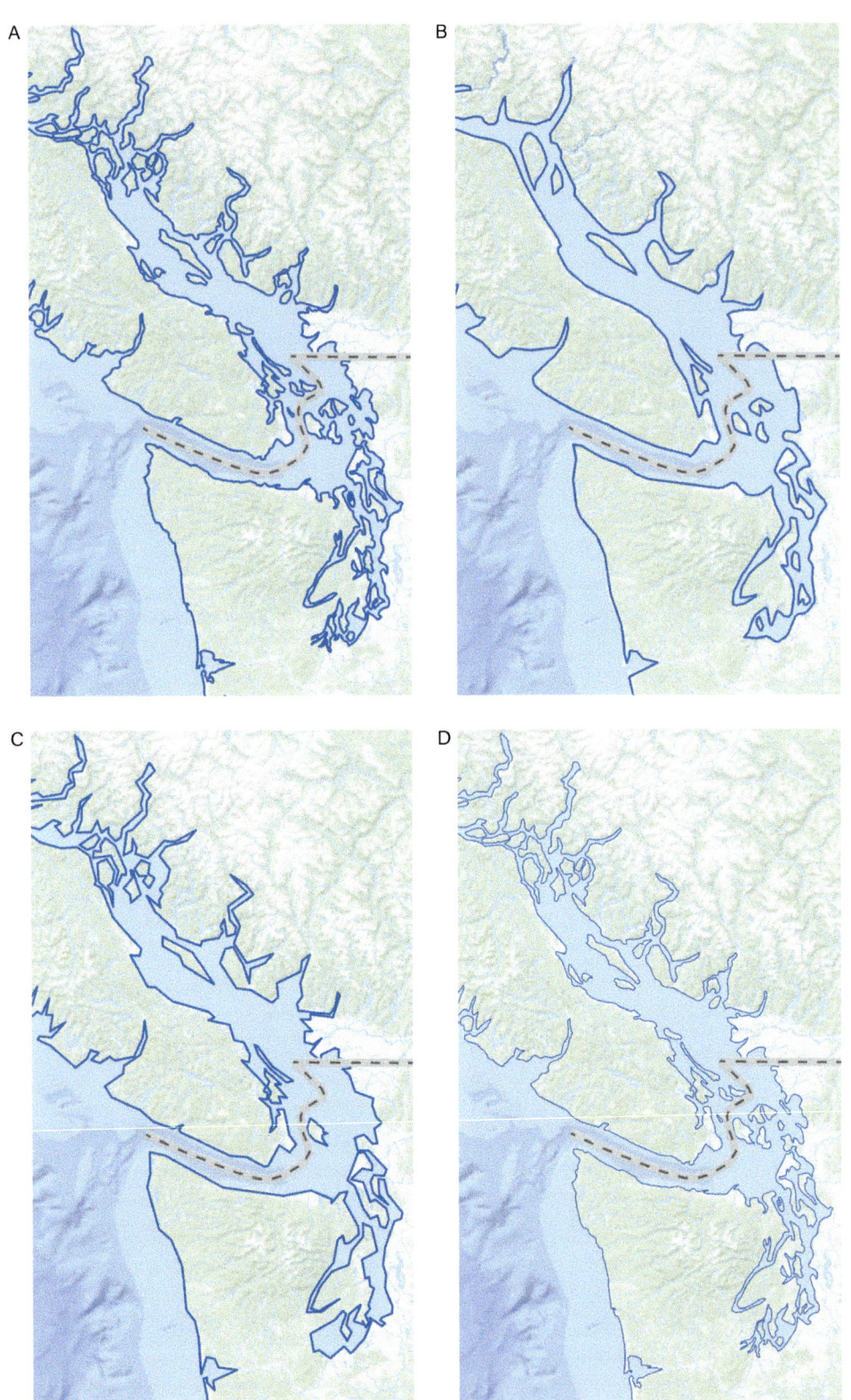

Figure 6.6. The lines in maps *A*, *B*, and *C*, each from a different source, are drawn with a one-point line, and map *D* is drawn with a 0.5-point line. At the same scale, the lines in *A* coalesce in some places, giving the appearance of additional polygons—an indication that the lines are too detailed. The lines in *B* appear too rectilinear, indicating that the data is overgeneralized. The lines in *C* are at an appropriate level of generalization. The more detailed lines in *A* can be visually improved with a thinner line symbol, as shown in *D*.

sources) of 1:25,000 to 1:100,000. The map in figure 6.7A shows the source features that were used to make the map at a smaller scale (figure 6.7B). The result is a map with much-reduced detail that still shows the proper geographic relationships among features.

Classification

When we look at the world around us, we naturally try to differentiate, group, and organize what we see to make better sense of it. We associate things that are similar and distinguish things that are different. This type of mental classification is fundamental to comprehension and learning. We can apply the same process to geographic data when we try to represent the world on a map. **Classification** is the process of grouping, ordering, or scaling data into categories or classes based on similarities in attribute values or spatial relationships. The goal is to maximize both within-group similarities and between-group differences. For cartographers, classifying data meaningfully can be a challenge, and mapmakers make better judgments about classification if they understand inherent differences in data types, data distributions, statistical measures, and classification methods. Here, we discuss the most common methods for classifying data for mapping purposes.

Table 6.1. Generalization operations

Generalization operation		
Aggregate Points / Polygons Create polygons around clustered points / polygons while retaining their essential character.	In the Aggregate examples, dark gray indicates original features, and light gray indicates generalized features. Points / Orthogonal polygons / Nonorthogonal polygons	
	Original map	**Generalized map**
Delineate Built-Up Areas Create polygons around densely clustered buildings.		
Simplify Shared Edges Reduce the number of inflections in line and polygon features, and in shared edges while retaining connectivity.		

Table 6.1 (continued)

Generalization operation	Original map	Generalized map
Smooth Shared Edges Smooth sharp angles in line and polygon features, and in shared edges while retaining connectivity.		
Collapse Hydro Polygon Replace hydrographic polygons with centerlines.		
Collapse Road Detail Replace open spaces in a road network with lines.		
Thin Road Network Eliminate smaller roads in a network while retaining connectivity and its essential character.		
Merge Divided Roads Replace double line roads with single line features.		

Source: Esri.

Chapter 6: Map design   143

Table 6.2. Graphic conflict operations

Graphic conflict operation	Original map	Generalized map
Detect Graphic Conflict Create polygons where two or more symbolized features graphically conflict.		
Resolve Road Conflicts Offset overlapping road symbols so both lanes are visible.		
Resolve Building Conflicts Move, resize, realign, or hide buildings so they align with roads.		

Source: Esri.

Qualitative data classification

Qualitative data classification refers to the process of organizing, grouping, differentiating, and categorizing qualitative data based on attributes, spatial configuration, space-time properties, or combinations of these. **Qualitative data** is nonnumerical data that describes characteristics that cannot be counted or measured. One common method for assigning geographic data to different groups is **attribute-based classification**, which determines similarity based on attributes about different types of things. As you can see in figure 6.8, parcels can be classed based on zoning, soil units can be classed into orders based on a variety of soil characteristics, land areas can be classed based on what geologic units underly the area, and elevation data can be classed into viewsheds by visibility. Attribute-based classification helps organize data, compare features, and communicate information.

Proximity-based classification assigns geographic data to groups based on their closeness to each other. For example, grouping crime incidents can help identify where to deploy more police officers, and grouping animal sightings can help expose critical habitats. The map in figure 6.9 shows groups of traffic accidents, each with a different color, to identify where to install better signage or traffic controls, such as stop signs. Proximity-based classification is helpful in revealing spatial groups and identifying outliers.

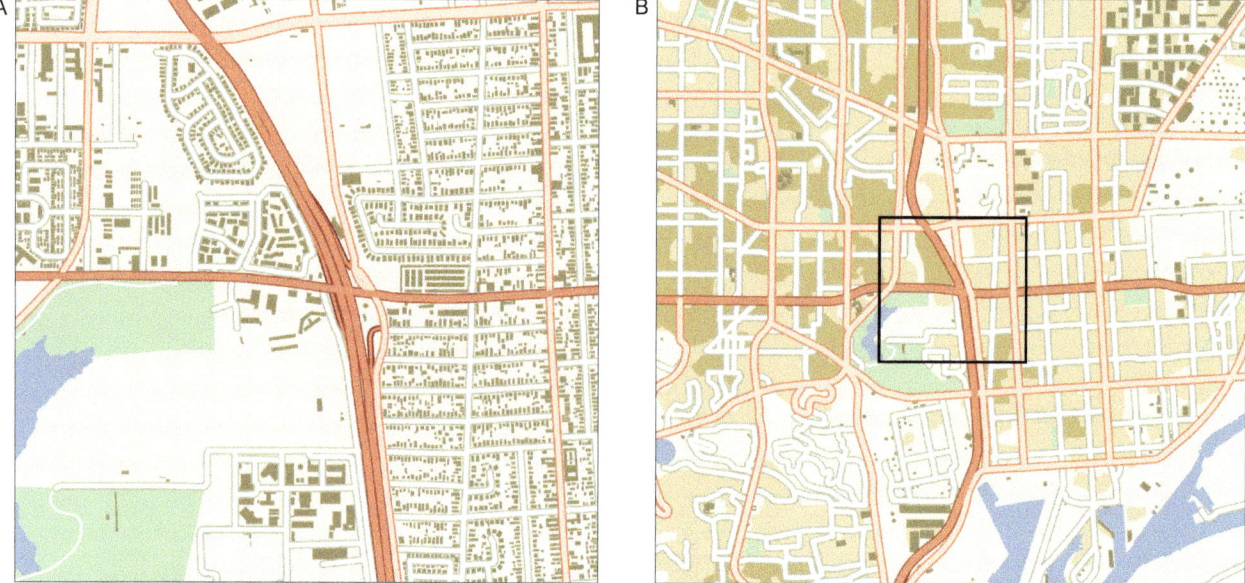

Figure 6.7. The extent of the 1:25,000-scale map (*A*) is shown with a black box on the 1:100,000-scale map (*B*). The amount of detail in the larger-scale map can be reduced using selection methods and generalization operations, such as those illustrated in tables 6.1 and 6.2. The amount of detail in the data used to create a 1:25,000-scale map (*A*) can be reduced using selection and generalization operations to create maps at smaller scales, such as the 1:100,000-scale map (*B*).

Quantitative data classification

Quantitative data classification is the process of grouping **quantitative data** (counts, amounts, magnitudes, or intensities) into classes with numerical ranges. Several standard **classing methods** (processes of grouping geographic data in categories or classes) are used to assign **class intervals** (the ranges for each numeric class). To explain and illustrate these methods, we use a set of **choropleth maps** in which areas with higher values are shown with darker colors. The maps in this section show the percentage of the working-age population (ages 18 to 64) from the 2020 census for the Great Lakes states of Wisconsin, Illinois, Indiana, Michigan, and Ohio (figure 6.10).

The number of classes and the classing method both have a significant impact on the appearance of the map, so the choices made by the mapmaker should be based on the nature of the data, not on an arbitrary design decision. This is especially true if the data is not normally distributed or there are extreme outliers. To check this, you can plot the data on a histogram, a graph that shows how many **observations** (features in the dataset) fall within each data interval (figure 6.11). A histogram with a **normal distribution** resembles a bell-shaped curve (shown in red in the figure), with most values clustering around a central region and the **mean** (average) at the center of the curve. The curve tapers off as values are farther from mean, and **outliers**, which are often thought of as values more than a certain number of standard deviations from the mean, are in the tails of the distribution.

For the data used to make the maps in this section, you can examine the histogram in figure 6.11 to see that the distribution is nearly normal, and there are some outliers in the higher percentage of population values. You can also look at the statistics, shown on the right, to learn more about the data distribution. For example, the data range is 20.8 years, with a minimum value of 49.1 and a maximum of 69.9, and the mean is 58.1.

Generally, a sound approach to data classification is to start with a common classing method and adjust the class intervals or modify certain **class breaks** (the values that separate the classes) to fit the map's purpose. (Adjusting class breaks is discussed below.) Because the number of classes will affect the appearance of the map for any classing method, we start by looking at this aspect of classing data.

Number of classes

The simplest quantitative thematic map has only two classes, for example above and below the mean, as in the two-class map in figure 6.12. Maps with more classes show more information, and the map with the most information

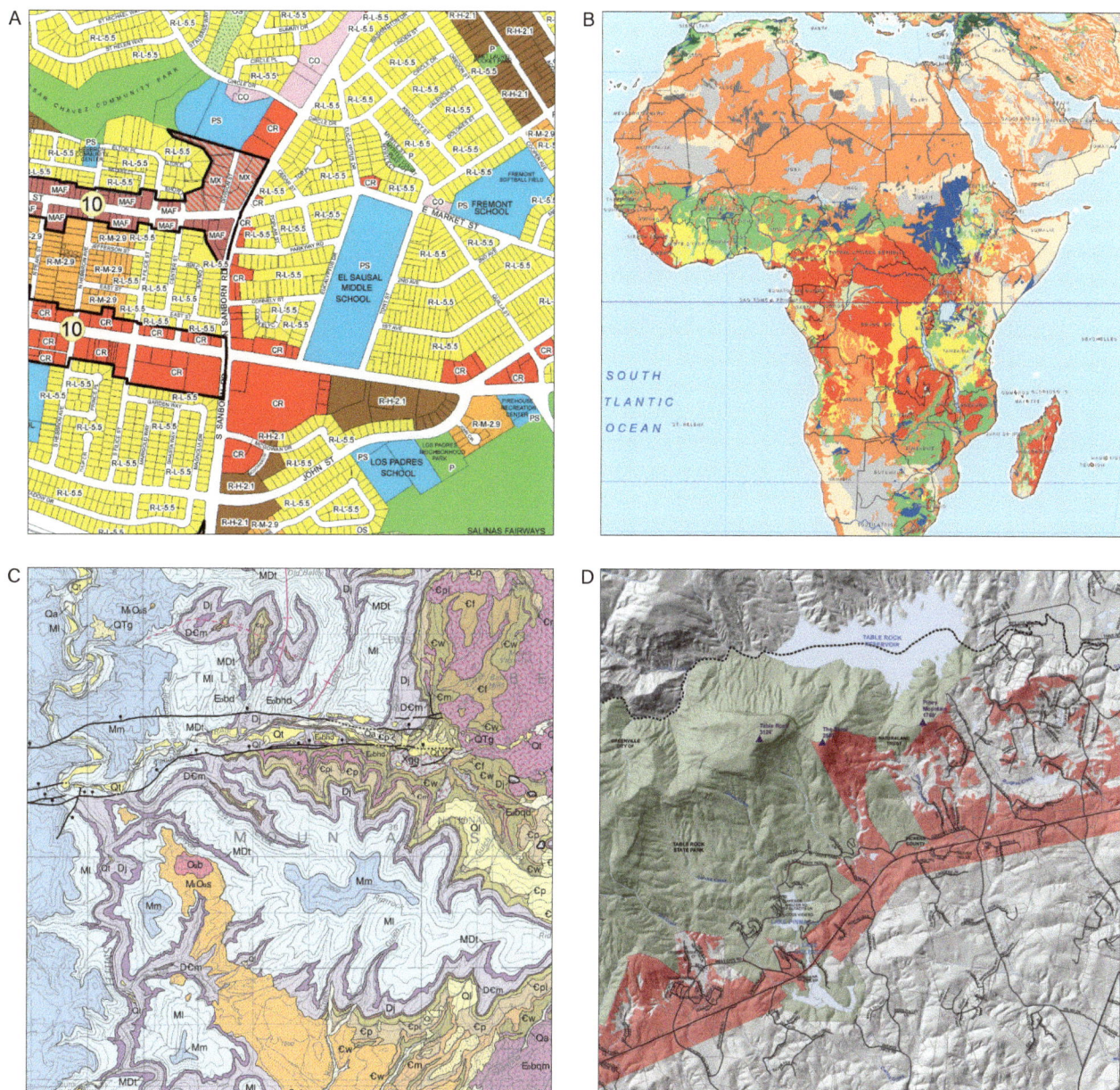

Figure 6.8. Parcel zoning (*A*), soil orders (*B*), geology (*C*), and viewsheds (*D*) are examples of classified qualitative data. Courtesy of City of Salinas, California (*A*), US Department of Agriculture (*B*), US Geological Survey (*C*), and County of Pickens, South Carolina (*D*).

has a class for each feature, as with the unclassed map in figure 6.12. You may notice that the **unclassed map** has the minimal number of counties symbolized in the lightest and darkest colors, a result of few counties having values at either end of the histogram in figure 6.11. You may wonder how meaningful the map classes you see really are, particularly when the range of data values is divided into a small number of classes. Look again at the two-class map in figure 6.12, which tells you all you need to know about which counties have values above and below the mean. But the information content of the map is minimal, and an important piece of the story is missing—with only a few classes, there is likely to be significant within-class variation that you cannot see.

Mapmakers increase the information content by using more classes. Progressive subdivision of the data

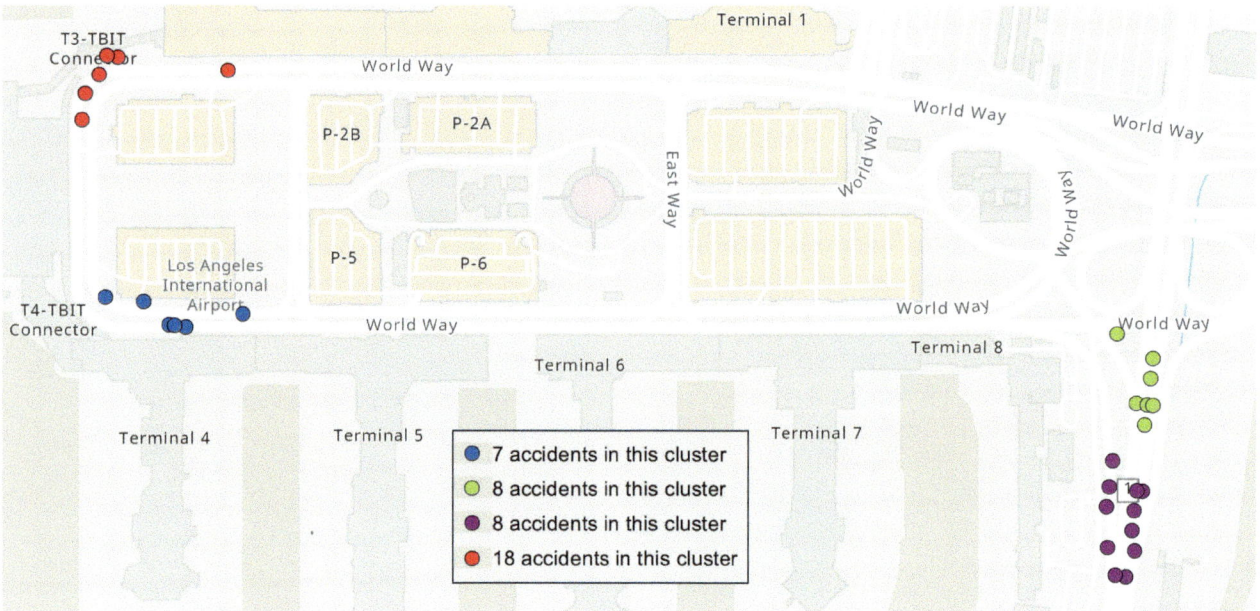

Figure 6.9. Proximity-based clustering of traffic accidents (dots of the same color) helps identify groups of points that indicate where improved traffic signage or control can be added around Los Angeles International Airport.

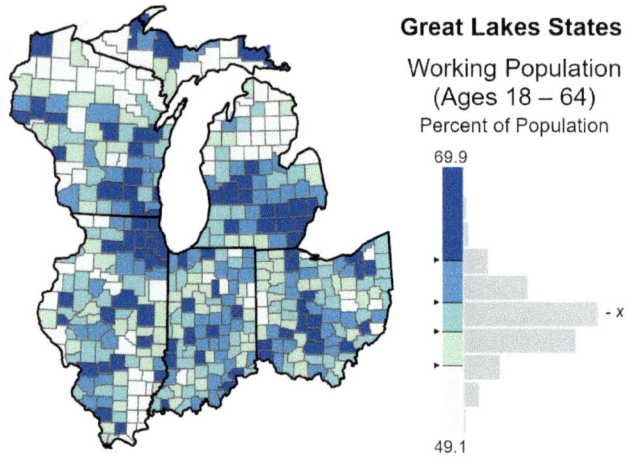

Figure 6.10. This map shows the percentage of the 2020 population that is working age (18–64) for the Great Lakes states. (Note that this map is the same as the natural breaks five-classes map in figure 6.13.)

Figure 6.11. For the data used to make the maps in this section, the distribution is nearly normal.

range into classes simply involves reducing the numeric intervals between class breaks. But this solution creates its own problems. As each additional class is added, the map pattern will change less. When many classes are used, the within-class variability is reduced, but the map is harder to read because it becomes increasingly difficult to assign class colors that readers can tell apart. Notice that for the map with eight classes in figure 6.12, it is difficult to distinguish some of the classes on the map. Thus, a cartographer's rule of thumb is for quantitative thematic maps to have three to seven classes, and for choropleth maps, like the ones shown here, to have five to seven classes.

Classing methods

Most mapmaking software and apps offer standard methods for classing data that include quantiles, equal intervals, and natural breaks. Other choices may also be offered. There is no one correct way to class quantitative data, so it is best to understand how these common methods work and what kinds of map patterns are produced when you use them. Deciding which method to use is also easier if you know something about the advantages and limitations of each method and the data distributions that each works best for. We cover these topics in this section.

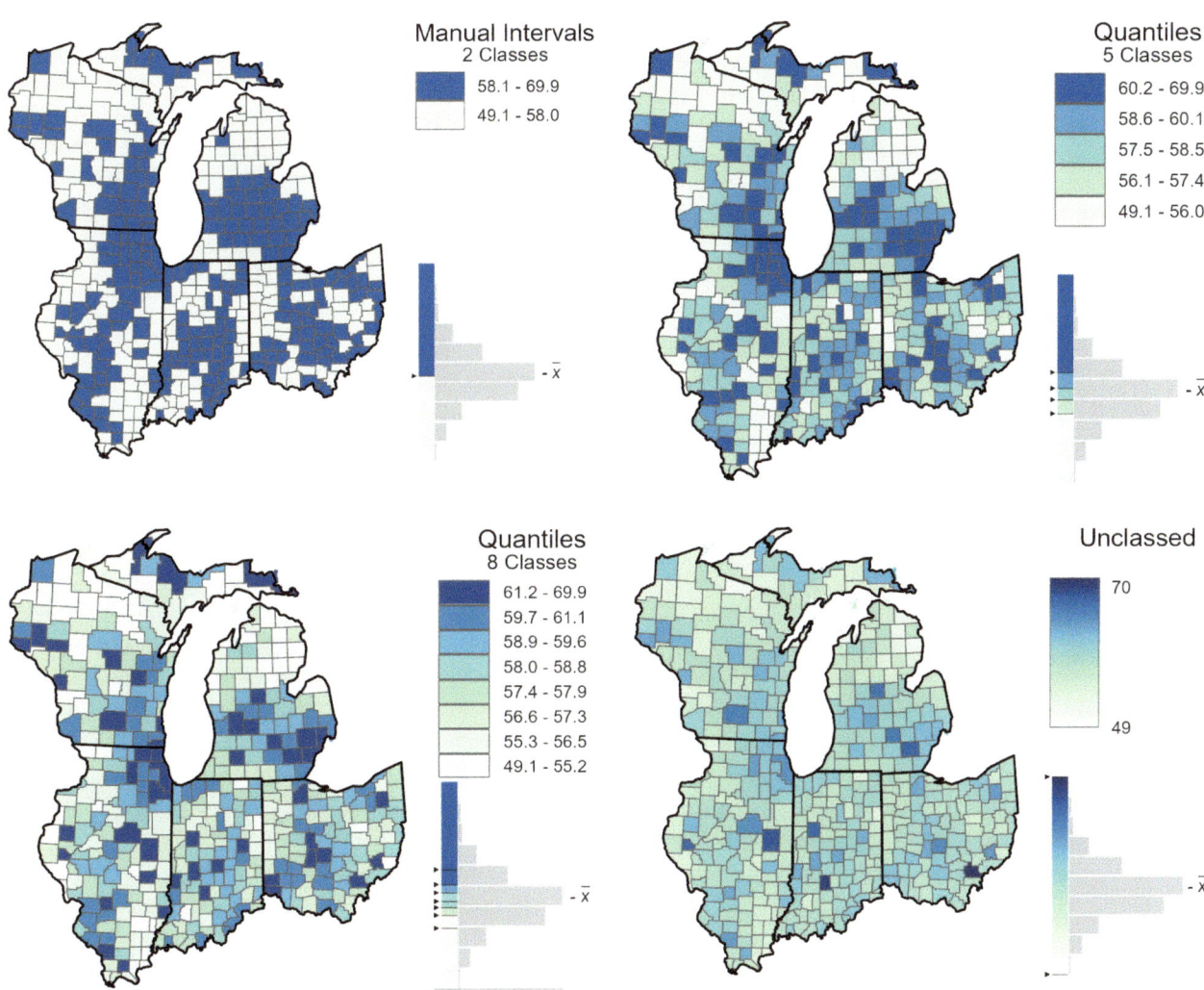

Figure 6.12. Varying the number of classes greatly alters the look of the map, which in turn affects interpretation of the data. To explore the differences, compare each of the maps by looking at the class intervals in the histograms and the colors on the maps to see relatively how many counties are in each interval.

Quantiles

With **quantile** (or **equal frequency**) **intervals**, the range of values is divided by a specified number so that each class has the same number of observations. If the number of classes is four, they are called **quartiles**; five classes are called **quintiles**; and so on. Look at the five- and eight-class quantile maps in figure 6.12. Are these maps a realistic portrayal of the distribution of data? To answer this question, compare the histograms for these maps with that of the two-class map in the same figure. Notice that the break at the mean of 58.1 for the two-class map falls near the breaks for the second and third classes with the highest values in the five- and eight-class maps, respectively. This gives you an indication that the quantile maps show about half of all counties in the classes with the highest two or three intervals. So, these maps give the impression that many counties have high values. Now look at the unclassed map. When each county is assigned its own unique color on the map, there are very few counties that have dark colors (high values).

The quantile classing method gives the most faithful portrayal of the data when the range of values for each class is approximately the same, and for choropleth maps, the areas are similar in size. Use quantile intervals to focus attention on relative rankings—the highest values (or lowest), the next highest (or lowest) values, and so on. Using quantiles is also a good way to compare maps that have different data ranges and values, because you will always be comparing high values to high values and low values to low values. Additionally, the concept of a quantile (for example, the top quarter of counties in terms of average family income) is easy to understand for most map readers. But because features are grouped in equal numbers into the classes, the resulting map can often be misleading. Features with similar values can be placed in adjacent classes or features with widely different values (due, for example, to a gap in the data distribution) can be assigned to the same class. You can minimize this within-class variation by increasing the number of classes. Also, with quantiles, outliers (extreme high or low values) are not as obvious because they are grouped with other high or low values.

Equal intervals

Another way to class quantitative data is by **equal intervals**, sometimes called **equal-range intervals**. With this method, the range of data values is divided by a specified number of classes to obtain class intervals with equal ranges. To keep the class intervals the same, the number of features in each class varies. Equal intervals are intuitively meaningful and easy to understand. Numerically constant intervals appeal to the same basic human data-handling mechanism that makes percentage figures so attractive. Our minds are comfortable with the idea of segmenting numbers into equal fractional parts. If each class contains an approximately equal number of observations, equal intervals produce the most meaningful map.

Do equal intervals give a realistic portrayal of the distribution of data? To answer this, compare the class intervals for the equal-intervals and quantiles maps in figure 6.13. On the equal-intervals map, each class has a range of about 4.16 (the actual value has more significant digits and the values in the legend have been rounded), whereas on the quantiles map, the first class of highest values has a range of 9.7, the second has a range of 1.5, and the rest of the classes also have different ranges. For equal intervals, data values should be equally distributed within their range with approximately the same number of low, medium, and high values. When the data values are equally distributed and the sizes of the areas are nearly equal, choropleth maps made with quantiles and equal intervals should look practically identical. However, you will find that geographic data that meets both of these criteria is rare.

For many themes, the data values are unevenly distributed, so it is useful to understand what equal-intervals classification does to the display of these types of data distributions. Figure 6.14 shows the counts of features in each class for the five-class, equal-intervals map in figure 6.13. Only 7 of the 437 counties are shown with the darkest blue, and only 22 are shown with the lightest green. The remaining counties are shown with intermediate colors. The result is a map that shows little variation for most of the counties but gives a good impression of the outliers.

Advantages of the quantile classing method are that class intervals can be easily calculated, and outliers can be easily seen. On choropleth maps, if the areas are about the same size, each class will have equal representation on the map, The equal-intervals method is best applied to familiar data ranges, such as percentages and temperature. This method emphasizes the amount of an attribute value relative to other values, for example to show that a shop is part of the group of shops that make up the top one-third of all sales. The primary disadvantage of the equal-intervals method is that the class intervals are assigned without regard to how the data is distributed, so some classes may have few or even no data values in them, and others may have many.

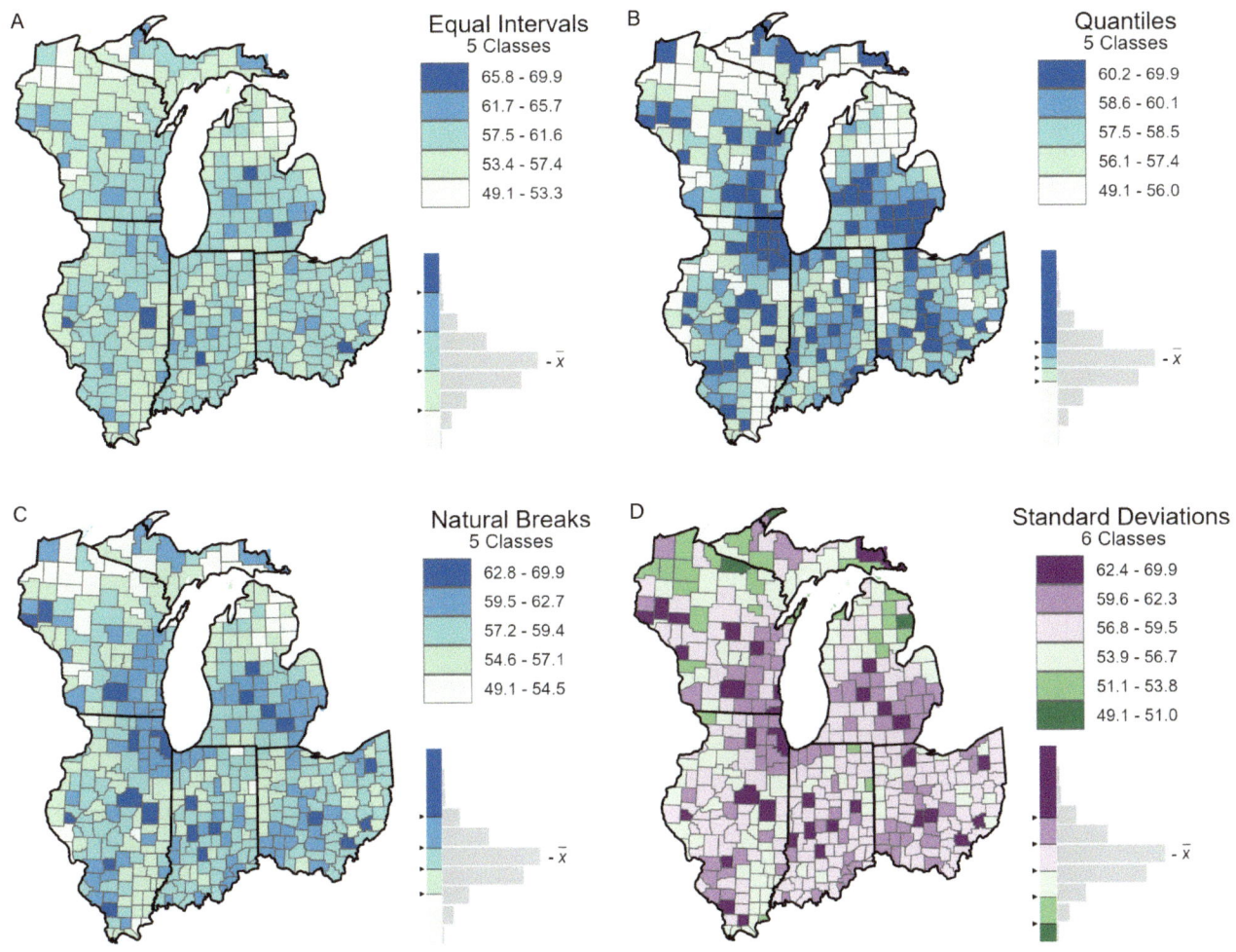

Figure 6.13. These choropleth maps were created using the equal-intervals (A), quantiles (B), natural breaks (C), and standard deviations (D) classing methods. As with the number of classes, varying the classing method greatly alters the look of the map, which affects the interpretation of the data. To explore the differences, compare the class intervals in the histograms and the colors on the maps to see relatively how many counties are in each interval using the four methods.

Natural breaks

Another classing method is to divide the data into classes using **natural breaks** in the distribution of data values. Using natural groupings in the data to set the breaks minimizes the variation within classes and maximizes the variation between classes. This classing method, described by the American cartographer and geography professor **George Jenks**, is often called the **Jenks optimization** or **Jenks natural breaks** classing method. In mapping software and apps, some form of this approach is usually referred to as natural breaks.

Does this method give a realistic portrayal of the distribution of data? Compare the natural breaks map in figure

Symbol	Upper value	Label	Count
	≤ 69.9	65.8 - 69.9	7
	≤ 65.74	61.7 - 65.7	34
	≤ 61.58	57.5 - 61.6	218
	≤ 57.42	53.4 - 57.4	156
	≤ 53.26	49.1 - 53.3	22

Figure 6.14. The number of features in each interval for the five-class, equal-intervals map in figure 6.13 are shown on the right in this table.

6.13 with those using the equal-intervals and quantiles methods. On the map with natural breaks, areas with the same color are statistically more similar to each other than to areas in other color classes. A result of using this method is to distribute the error uniformly across all areas on the map. Thus, the natural breaks map shows the nature of the data better than the other maps in the figure.

Class intervals for natural breaks are data-specific and may be difficult to explain to map readers. With the natural breaks method, each dataset results in a unique classification solution, so multiple maps are hard to compare. For maps in a series, for example population density by decade, or maps with related themes, such as diabetes and obesity, quantiles or equal intervals may be a better choice. When few classes are used, natural breaks can also mask outliers by lumping extremely high or low values into the highest and lowest classes. When many classes are used, minor gaps can lead to class breaks, which may be poorly defined or misleading.

Other classing methods also exist, such as standard deviations (figure 6.13D), which involves classing data based on the standard deviation of the data values. This method aims to create classes that reflect the variability or spread of the data distribution. Cartographers can also devise their own classing methods.

Adjusting class breaks

Cartographers generally start with one of the standard classing methods described here and then, to make the intervals less cryptic to a general audience, they adjust the class breaks to round the values and sometimes include a **critical value** in a class or a **critical break** between classes. A critical value has special relevance to the map's theme. It may be a physical aspect of the theme, such as the temperature below which a crop freezes. It may be a politically defined dividing point, such as the income level at which counties fall below the poverty line and are thus eligible for government assistance. It may also be a statistic computed from the data, such as the median or mean, as with the two-class map in figure 6.12 which shows counties with values above and below the mean.

Other adjustments can be made by first grouping extreme outliers into their own class and then classing the rest of the data using a standard method. Similarly, when a dataset has many zero or small values (or alternatively, high values), a good strategy is to group these values into their own class and then class the remainder of the dataset.

Symbols

Having selected, generalized, and classified the data for a map, cartographers turn to cartographic **symbolization**, which is the process of changing the appearance of graphic marks that represent features in the environment. In the process, the graphic marks become **symbols** that cartographers can manipulate and display in various ways. Cartographic symbolization is based on three basic building blocks: graphic marks, levels of measurement, and visual variables. Next, we look at each of these in detail.

Graphic marks

Graphic marks (or **graphic primitives**) include the points, lines, and areas (or polygons) that represent the geographic features for vector data. **Pixels** (or cells) are the graphic marks used to show raster data on a map. **Voxels** (volume pixels) are the 3D equivalent of pixels used to represent 3D data (see figure 9.24). A voxel can be visualized as a cube that represents a specific cell in 3D space.

The graphic mark on a map does not need to have the same dimensionality as the feature it represents. For example, a building, which is a three-dimensional feature on the ground, can be mapped as a zero-dimensional point or two-dimensional polygon, and a road, which has width and length, can be mapped as a one-dimensional line. The choice of which graphic mark to use is related to map scale, as marks with fewer dimensions take up less space on the map. So, at a larger scale, a building might be mapped as a polygon, but at a smaller scale, it might be shown with a point. When you read or make a map, you should understand how the graphic mark relates to the feature it represents. For cartographers, the choice of graphic mark in large part dictates how they symbolize the feature on the map, as we explain later in this chapter.

Levels of measurement

It is generally possible to determine the measurement scale for the attributes of geographic data, although we are now seeing such things as images and text blocks as feature attributes. But for most attributes, the measurement scale, or **level of measurement**, falls into one of four categories: nominal, ordinal, interval, and ratio. The nominal level of measurement is associated with qualitative information, whereas quantitative data can be at the ordinal, interval, or ratio measurement levels.

Nominal data tells you simply what type of thing a feature is. The types are often organized into groups or categories, as we saw earlier in our discussion of qualitative data classification. The term **categorical data** is also used for nominal-level data that is categorized to distinguish types of features, whereas the term **classed data** is often used when numerical data is grouped into classes with ranges of values. It is important to differentiate nominal data from data at the other levels of measurement because the symbolization methods for nominal data are different. This differentiation becomes especially important when the categories for different types of things are defined by numerals, such as the chimpanzee protected area categories in figure 6.15. The Roman numerals for each class do not represent a quantitative measure, but rather nominal-level information about the features. Can you imagine how inappropriate it would be to work with these numbers as if they were quantities?

In contrast to qualitative data, the data for quantitative thematic maps is at the ordinal, interval, or ratio levels of measurement. From an analytic perspective, these measurement distinctions are important. **Ordinal data** is ordered or ranked according to a low-to-high or less-than-to-greater-than system. How much more or less one class is than another is not specified because there are no numeric values. The map of chimpanzee priority areas in figure 6.16 is an example of a map showing ordinal data.

Interval data consists of numeric values on a magnitude scale that has an arbitrary zero value or zero point. Land elevations (figure 6.17) are an excellent example because the zero datum is arbitrarily defined as mean sea level (see chapter 1). To see how arbitrary mean sea level is, you have only to think about how sea level rises and falls—the datum shifts over time and the zero value does not denote the absence of elevation. With interval data on a map, only the numeric intervals of the classes are valid mathematically. The elevation classes in figure 6.17 are in 500-foot intervals. You can correctly conclude that the difference in

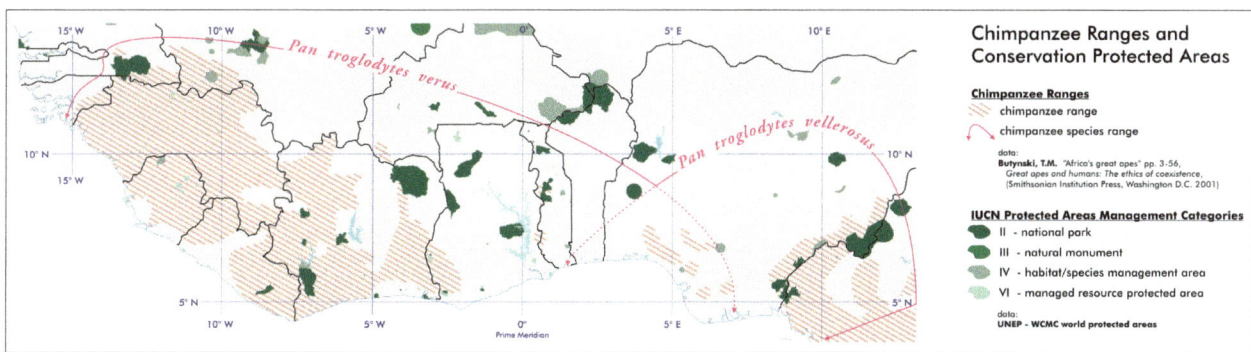

Figure 6.15. Conservation protected areas are designated by Roman numerals that represent qualitative rather than quantitative differences between the categories. Courtesy of Mark Denil, Conservation International.

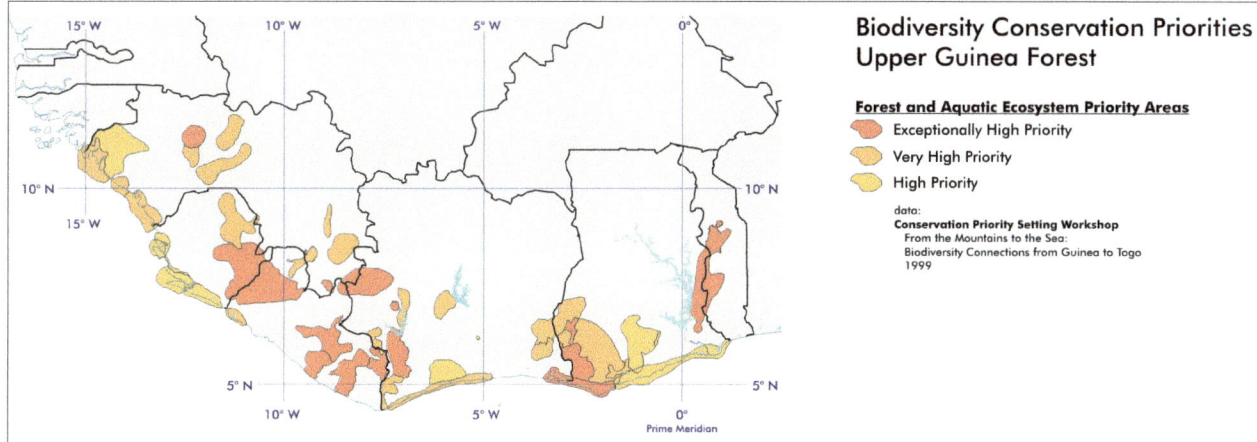

Figure 6.16. Ordinal data is shown on this map of ecosystem priority areas. Courtesy of Mark Denil, Conservation International.

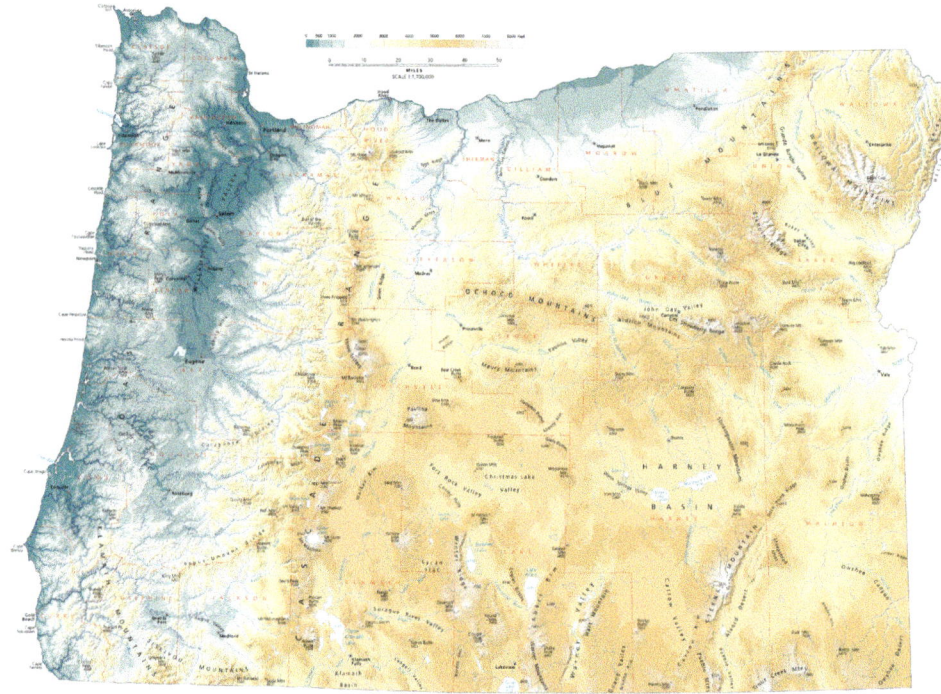

Figure 6.17. This map of landforms in Oregon demonstrates the use of interval data to show elevation. From *Atlas of Oregon*, second edition, © 2001 University of Oregon. Image courtesy of InfoGraphics Lab.

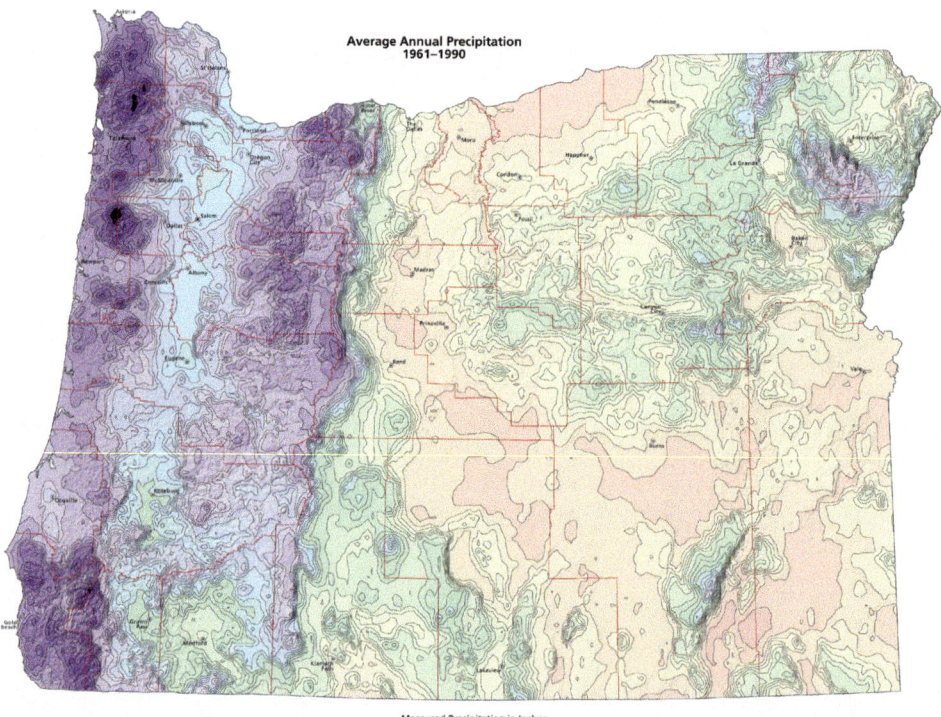

Figure 6.18. This map of precipitation for Oregon demonstrates the use of ratio data. From *Atlas of Oregon*, second edition, © 2001 University of Oregon. Image courtesy of InfoGraphics Lab.

elevation for each class is the same because the elevation interval is the same. However, it is incorrect to say that an elevation of 2,000 feet is twice as high as an elevation of 1,000 feet. To further understand this concept, imagine sea level dropping 999 feet so that the 1,000-foot elevation is now 1 foot, and the 2,000-foot elevation is 1,001 feet. The ratio of the elevation difference is now 1,001:1, but this ratio does not mean that the higher elevation is 1,001 times as high as the lower elevation.

Ratio data also consists of numeric values on a magnitude scale, but, in contrast to interval data, the zero point is not arbitrary. Instead, the zero point denotes the absence of the phenomenon. Precipitation (figure 6.18) is an excellent example, because zero means no precipitation. For ratio data, both the numeric intervals and the ratios between values are mathematically correct. So, mathematical operations, such as addition, subtraction, multiplication, and division, are valid. When you look at the map in figure 6.18, with classes such as 7–8, 8–10, and so on, for the average annual precipitation in inches, you can conclude that a value of 16 is twice that of 8, and 24 is 3 times as much.

Visual variables

Visual variables are the properties of symbols that can be altered to change their appearance and meaning on the map. The French cartographer **Jaques Bertin** first introduced the concept of visual variables for graphics in the late 1960s. Since then, cartographers have used them to guide their decision-making for symbol design. Symbols on maps are easiest to read if the mapmaker has correctly assigned the visual variables to graphic marks based on their level of measurement. To simplify our discussion, we can consider data at the nominal level to be qualitative and data at the ordinal, interval and ratio levels to be quantitative.

Qualitative visual variables

Qualitative data is descriptive data that characterizes qualities, characteristics, properties, and attributes that cannot be measured numerically. The **qualitative visual variables** give an impression of qualitative differences through the use of shape, hue, orientation, and arrangement. Each of these is illustrated in table 6.3 for point, line, and area graphic marks.

The **shape**, or form, of point symbols can encompass simple geometric shapes (circles, squares, or triangles), representations that mimic the feature (known as **mimetic symbols**), and **pictographic symbols**, such as a pictorial sketch of a building. Shape for line symbols relates to the elements that make up the line. For example, different combinations of solid or dashed lines and point symbols along the line can be used to show different kinds of linear features. As with lines, shape for area symbols relates to the elements that make up the symbol's fill. Shapes repeated along lines or within areas create **line patterns** and **area patterns**. The elements in line and area symbols also range from simple geometric to more detailed mimetic symbols.

Hue refers to the dominant wavelength of visible light that forms our color association. What we commonly recognize as the colors red, blue, and yellow are more accurately referred to as hues. Strictly speaking, using different hues should not impart a message of magnitude because red is not "more than" blue, which is not "more than" yellow. However, some colors are perceived as lighter (for example, yellow) or darker (blue), so there is some internal ordering of hues by the human visual system that may give a magnitude message. Also, some hues have inherited magnitude meanings through common usage. For example, on weather maps, red is often associated with warmer (higher) temperatures and blue with cooler (lower) temperatures. These considerations should be kept in mind when choosing hues for qualitative map symbols or when reading maps with colors.

Orientation relates to the angle or direction of the symbol or a portion of the symbol. For example, the tail on a spring symbol is oriented in the downhill direction of the spring's flow. Although orientation can be measured in angles from 0 to 360 degrees, this numeric value does not impart a magnitude message. For example, when orientation in a symbol is used to denote compass direction, north is not "more than" east, which is not "more than" south. For area symbols, orientation relates to the angle of the elements within the symbol's fill. You may find one such fill on a map but rarely more because distinguishing these types of symbols is somewhat difficult and other visual variables are usually a better choice. Orientation—and arrangement—can be used to create patterns in areas, which is known as **pattern orientation** and **pattern arrangement**.

Arrangement relates to the organization or distribution of elements within a symbol. For points, each element in a complex symbol can show a different attribute—you can find these complex symbols on nautical and aeronautical charts. Elements can also be arranged along lines or in area symbols. For example, a random arrangement of green circles in an area may be used to show scrub, and a regular array of the same symbol may show vineyards.

Quantitative visual variables

Quantitative data is numerical data that can be measured and quantified and expressed in numerical form. **Quantitative visual variables** are used to depict differences in magnitude through the use of size, value, saturation, and texture, as illustrated in table 6.4. A well-designed set of symbols using variations in one or more of these visual variables will appear to you as though they show a progression of magnitudes, from small to large or low to high.

The **size** (relative extent) of a symbol can be varied, which both the point and line examples in table 6.4 illustrate. The visual impression of varying symbol size is one of a measurable or ordered difference in the magnitude of the attribute being symbolized. In common practice, size is not used for area symbols; the size of an area feature is already dictated by its geography.

Value refers to the lightness or darkness of a hue. Lighter colors are sometimes referred to as tints—the mixture of a color with white; darker colors are called shades—the mixture of a color with black. For black and white symbols, value relates to shades of gray. For maps with a light background, the human visual system will naturally interpret a darker color as "more," so darker colors are used to represent features with greater quantities or higher values than features shown with lighter symbols. Looking at the point symbols in table 6.4, you can see that the value of the point symbols for each of the race and ethnicity categories (distinguished by hue) represents the percentage of population.

Saturation, also referred to as chroma, is the purity of a hue. Saturation is varied by mixing a pure hue with gray—the addition of gray results in lower saturation. The terms *brilliant* or *vivid* can be used to describe fully saturated colors, and *muddy* or *dull* describe low-saturation colors. In the area symbol in table 6.4, saturation is used to show percent slope for each of the aspect categories, which are differentiated by hue. In practice, saturation and value are often used together to create sets of colors ranging from light to dark to symbolize quantitative data.

As with the shape and arrangement of qualitative visual variables for area symbols, **texture**, sometimes referred to as **pattern texture**, relates to the elements within the symbol. The denser the elements, the darker the symbol because the interstitial space is reduced. So, texture works in the same way as value, with denser symbols appearing darker to show higher values.

The range of possibilities when designing cartographic symbols is virtually limitless. Although this can seem daunting to the budding mapmaker, seasoned cartographers rely on the conceptual foundation for symbol design based on graphic marks, levels of measurement, and visual variables. Tables 6.3 and 6.4 illustrate how these three concepts come together. The levels of measurement are reflected in the two figures with qualitative visual variables in table 6.3 and quantitative visual variables in table 6.4. Also, the graphic marks are represented in the columns (zero-dimensional, or 0D points, one-dimensional, or 1D lines, and 2D areas), and the visual variables are represented in the rows. For the most part, when the three concepts are considered carefully and in concert, cartographic symbol design is controlled, confined, and often conventional, although there is still room for creativity.

Bivariate and multivariate symbols

Mapmakers sometimes combine visual variables in their symbols. Mapping more than two attributes in a single symbol requires a **multivariate symbol**. Figure 6.19A shows an example for point symbols—arrows represent ocean current directions (orientation), magnitudes (size), and temperatures (hue). To emphasize a single attribute, they may use **redundant symbols**. For example, they may use a primary visual variable, such as size, augmented by a secondary visual variable, such as value, in which darkness serves to exaggerate the message that is conveyed by a larger size. This is illustrated in figure 6.19B, which shows the hours to maximum flow for river segments in the state of Arkansas. For this map, "less" is "more," because less time to maximum flow is more important because it is more dangerous. So the line sizes and values are inversely proportional to the number of hours. **Bivariate symbols** are used to map two variables (attributes). A common example is a **bivariate choropleth map**, which shows the relationship between two attributes using hue to differentiate the two attributes and value to symbolize the magnitudes within the classes (figure 6.19D).

Of course, packing more than one visual variable into a symbol makes it more complex for both the mapmaker and the map reader. In figure 6.19C, pie charts show the proportion of motor vehicle crash incidents by type using hue, and the size of the symbol shows the total number of incidents. The more complex the symbol, the harder it is for the map user to interpret, the longer it takes to

Table 6.3. Qualitative visual variables

	Point	Line	Area
Shape	[1] Single building, Tank, Mine, Tower, Dish aerial [2] Pictorial sketches	[1] Railroad, Abandoned railroad, Fence, Power transmission line [1] Bird sanctuary, Seal sanctuary	[2] Wash, Tailings, Intricate surface area, Gravel beach [1] Breakwater (loose boulders), Breakwater (masonry)
Hue	[1] Lighted red beacon, Lighted yellow beacon, Lighted green beacon [3] Phone: Public, Emergency, Roadside Assistance	[3] Motorway, Dual carriageway, Main road, Secondary road, Road generally > than 4 m wide [4] 100 m index contour: Earth 1800, Scree 1800, Glacier 1800	**Group with highest percent of population**: Asian, White, non-Hispanic, Hispanic, African American [2] Swamp, Submerged swamp, Wooded swamp, Submerged wooded swamp
Orientation	[2] Spring (orientation shows streamflow direction) [5] Wind barb (orientation shows wind direction)	In practice, rarely used	Sparse Data In practice, rarely used for multiple symbols
Arrangement	[1] Navigable water lies: N, E, S, W [1] Beacon with topmark, color, radar reflector, and designation (No2 R) Buoy with topmark, color, radar reflector, and designation (No3 G)	[5] **Front**: Warm, Cold, Occluded, Stationary [1] **Lateral mark lights**: Flashing, Long flashing, Group flashing	[2] Scrub, Orchard, Vineyard [4] Golf course, Allotment, Cemetery

Note: Examples of the qualitative visual variables are shown here for symbols on (1) NOAA nautical charts, (2) USGS 1:24,000-scale topographic maps, (3) Ordnance Survey 1:25,000-scale Explorer maps, (4) Switzerland Federal Office of Topography (swisstopo) national maps, and (5) NOAA weather maps. Sources: National Oceanic and Atmospheric Administration, US Geological Survey, Ordnance Survey, and swisstopo.

Table 6.4. Quantitative visual variables

	Point	Line	Area
Size	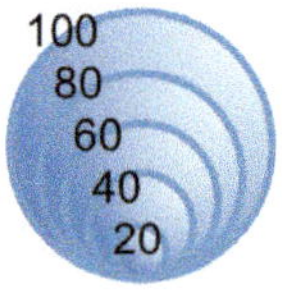 [1] Number of people by state (Urban population percent of total: 100, 80, 60, 40, 20); 34,000,000; 10,000,000; 5,000,000; 1,000,000; 500,000	[2] 10 m road, 8 m road, 6 m road, 4 m road, 3 m road, 2 m track; [3] Channels, less than 100' wide; Channels, 100' to 400' wide; Channels, over 400' wide	In practice, rarely used
Value	Race and ethnicity (percent of population): Asian, White, Hispanic, Black (50, 75, 90); Earthquake magnitude: Less than 2.5, 2.5 to 5.5, Greater than 5.5	Fire burn extent (hours): 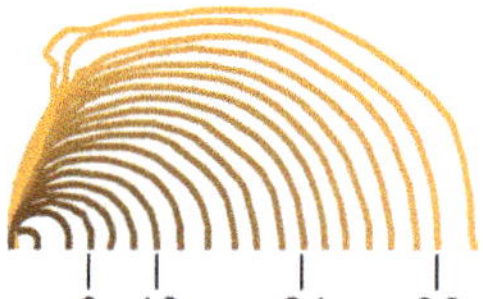 6, 12, 24, 36	[1] Change in number of people: More than 1 million; 500,001 to 1 million; 100,001 to 500,000; 0 to 100,000; Less than 0
Saturation	Excellent health, Good health, Poor health; Level of certainty: High, Medium, Low	Bathymetric depth (m): 20, 40, 60, 80, 100, 200	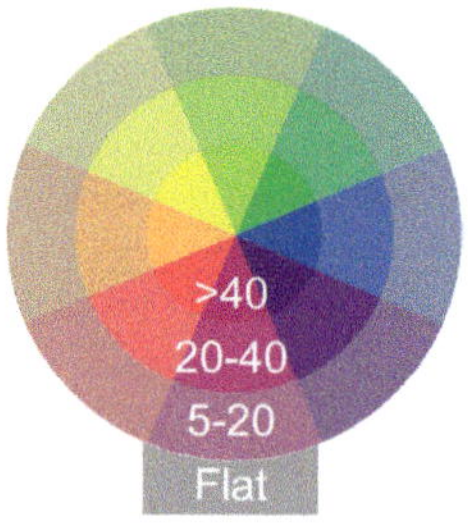 Percent slope within each aspect category: >40, 20–40, 5–20, Flat
Texture	[4] Wind speed (knots): 1–2, 3–7, 8–12, 13–17, 18–22, 23–27, 28–32, 33–37, 38–42, 43–47	[3] Depth contours: 1000, 2000, 3000, 4000; Mountain bike trail difficulty: 1, 2, 3, 4, 5	[5] 1 dot = 10,000 acres; 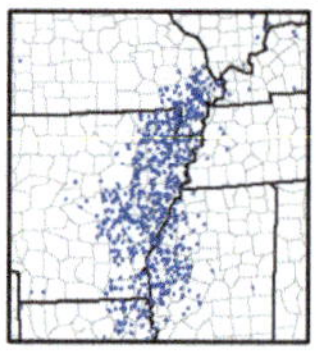 [3] Incompletely surveyed area, Unsurveyed area

Note: Examples of the quantitative visual variables are shown here for symbols on (1) US Census Bureau atlas maps, (2) swisstopo national maps, (3) NOAA nautical charts, (4) NOAA weather maps, and (5) US Department of Agriculture agricultural atlas maps. Sources: US Census Bureau, swisstopo, National Oceanic and Atmospheric Administration, and US Department of Agriculture.

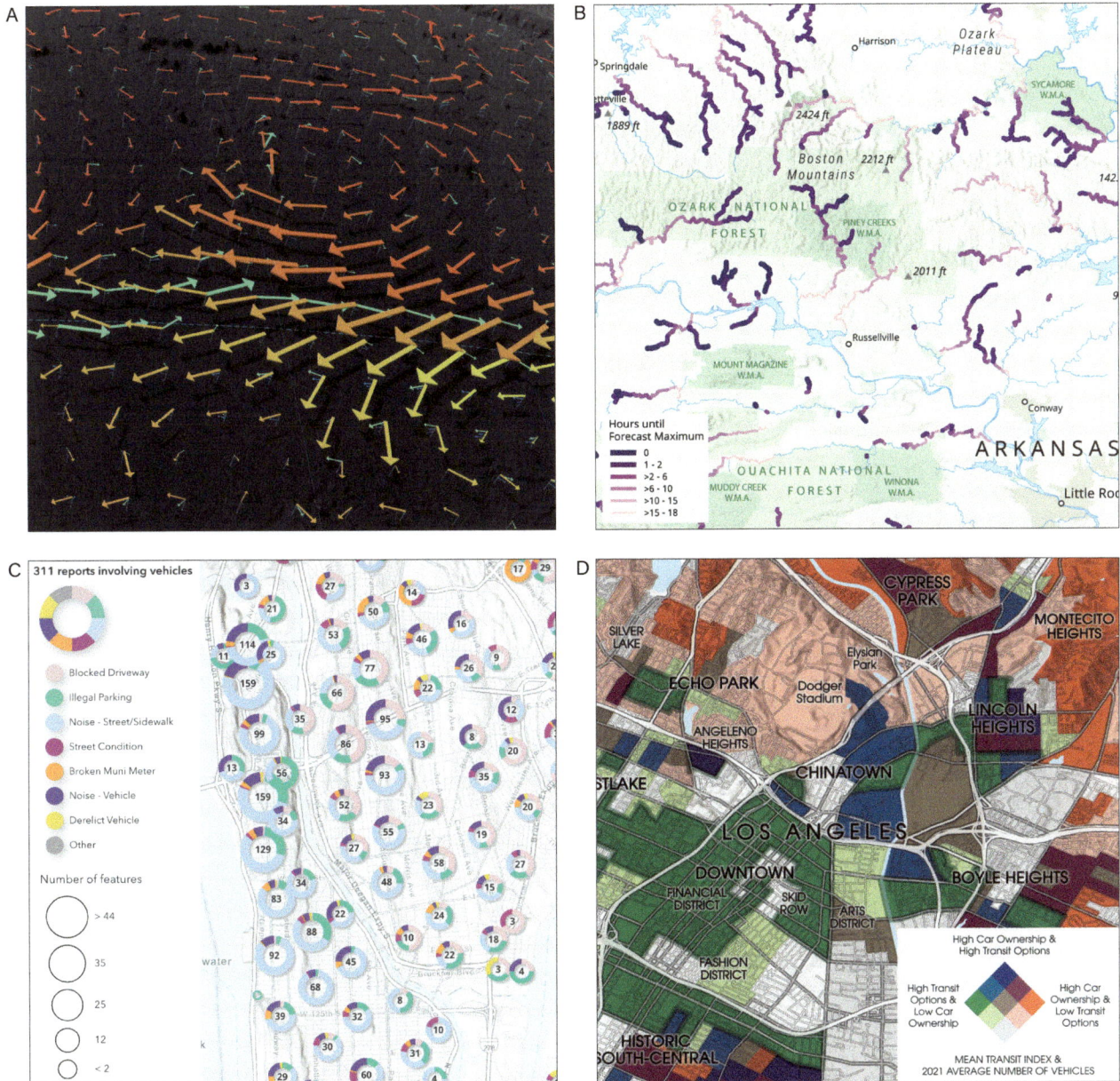

Figure 6.19. Arrows show the direction (orientation), speed (size), and temperature (hue) of ocean currents sampled at points (*A*). Redundant symbols are used on a map of national water model maximum hourly flow in Arkansas (*B*). Multivariate point symbols show a number of attributes related to New York City 311 incidents involving reported problems related to motor vehicles (*C*). A bivariate choropleth map is used to compare car ownership and transit access (*D*).

understand, and the greater the chance it will be misinterpreted. Therefore, the need for bivariate and especially multivariate symbols should be well justified.

Color

Color not only makes maps more interesting to look at, but it can also be a great aid to mapmakers because it allows more types of features to be distinguished on the map and better order to be imposed within the map. One of the major challenges for mapmakers is to use color effectively so that it aids communication rather than complicating and confusing the message. Today's computer systems with 24-bit color depth can display 16.7 million colors, making the challenge of color selection even more difficult. Mapmakers, therefore, benefit from basic knowledge about the properties of color, familiarity with common color models, and an understanding of color schemes used for different types of mapped data.

Properties of color

Although one person's conception of blue may be slightly different from another person's, most people with normal color vision do not differ substantially in their color perception. Therefore, maps can usually be safely designed based on the assumption that the map user has normal color perception. There are exceptions when designing maps for specific conditions of use (for example, to be read in low or monochromatic light) or by specific audiences (such as readers with color vision deficiency).

Several systems based on observers with normal color vision have been developed to describe color properties and define colors. These systems have to be more detailed than simply assigning a specific color to the label *blue*, for example, because the range of possible blues renders the description meaningless. Consequently, color descriptions must be more precise. This is possible because we can describe color based on the physical stimulation of color and the perception associated with viewing colors. Being able to define color with precision is important to control color throughout the mapping process, from data input to graphic display to map output. For print maps, for example, the process involves selecting a color (often from a color chart) and ensuring that the color's settings are used throughout the entire design process. Because a particular color has been specified, the mapmaker can be confident it will look correct when printed on paper.

Color specifications are often based on the three basic

Figure 6.20. The three primary elements of color are hue (*A*), value (*B*), and saturation (*C*).

elements of color: hue, value, and saturation (figure 6.20). These three elements of color must be considered relative to the medium used to display the map and the conditions of map use. For example, value can be varied to create lighter and darker versions of a hue (*A* and *B*). Saturation can change the colorfulness of a hue (*C*). However, lightness, darkness, and colorfulness are relative concepts. If a map printed on paper is seen under a bright light, the amount of reflected light is increased, and the lightness of all hues on the map will also be perceived to increase (although the relative differences between values of the hues will remain the same). This principle does not apply to a map displayed on a computer screen. In this case, the increased light of the hue is added to the light emitted by the monitor, resulting in a decrease in the relative differences in brightness in the screen image. This is one reason why problems arise when designing color for a print product on a computer and why it is best to work with color specifications rather than apparent color (an observer's perception of color, which varies depending on the light in the viewing environment).

Color systems

For cartography and other modes of graphic communication, creating different colors is often considered relative to two primary color systems. A **color system** specifies colors numerically according to their individual components. Two systems used widely in cartography are differentiated by whether the hues are mixed as ink pigments (for print map colors) or as light sources (for digital displays).

Print map colors are created through **subtractive color mixing** of three inks—cyan, magenta, and yellow (CMY)—which subtract (absorb) different wavelengths of light. These colors are referred to as the **subtractive primaries**,

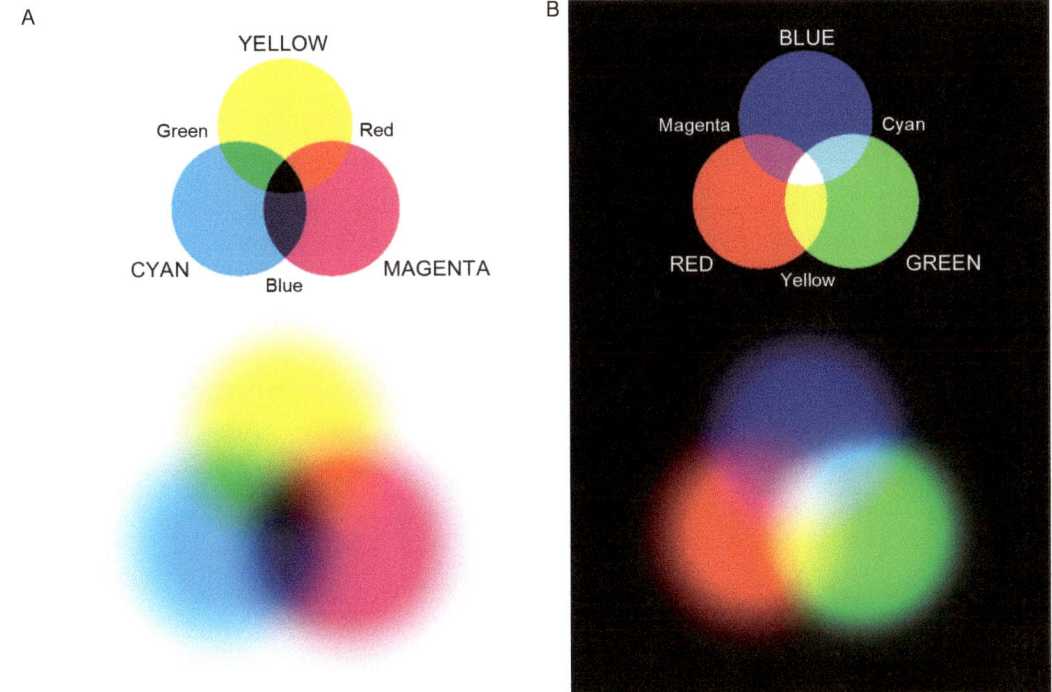

Figure 6.21. The subtractive CMY (*A*) and additive RGB (*B*) color systems. The fuzzy overlapping circles at the bottom show the wide range of intermediate colors created from intermediate percentages of the three primary colors.

or **process colors**. Overprinting cyan and magenta ink pigments on the page produces blue, cyan and yellow produce green, and yellow and magenta produce red (figure 6.21). Subtractive color mixing is used for digital printing with toners, for example laser printers, or liquid ink (for larger printers) and for **offset printing**, in which the image is transferred, or offset, from a plate to a rubber blanket and the image on the blanket is rolled onto a sheet of paper. Because ink tones are not completely pure (in practice, they do not absorb or transmit 100 percent of the theoretical wavelengths of light), mixing the three subtractive primaries does not produce pure black. So black ink is usually used as a fourth printing color. Because black is normally printed first and the other colors are keyed, or registered, to it, it is assigned the letter *K*. Thus, you will also see this system referred to as **CMYK color mixing**.

For digital displays, the **RGB system** is used, so named because the primary hues are red, green, and blue (RGB). The process of mixing colors using the RGB system is called **additive color mixing**, and red, green, and blue are the **additive primaries**. Additive color mixing is used to display colors on computer screens and other digital displays in which extremely small phosphor dots arranged on a liquid crystal display (LCD) monitor emit red, green, and blue light in varying intensities. From a distance, the varying intensities reveal a sum of light and, consequentially, a color. The background color on a monitor is black, which represents no color emission. Where all three colors overlap, the color is perceived as white. Where only two colors overlap, blue and green mixed together form cyan, blue and red produce magenta, and red and green make yellow. By altering the intensity of each of the additive primaries, a wide range of intermediate colors can be created, as shown at the bottom of figure 6.21B.

Common color models

Colors can be defined using color models, which are systems for specifying colors numerically according to their primary color components. A color model is visualized as a three-dimensional space that contains combinations of the model's primary colors to produce all possible colors. A variety of color models have been developed, including the RGB, CMY, HSV, HSL, and Lab models used in GIS. Here, we look at the first three.

The **RGB color model** specifies color based on the relative intensities of the red, green, and blue additive primaries on a computer monitor. The numerical values range from 0 to 255 for each of the three colors yielding

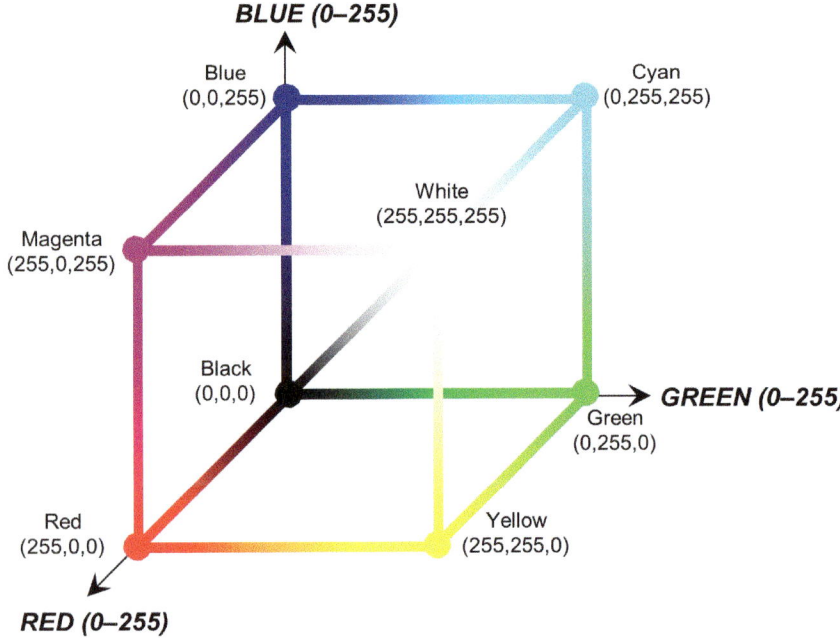

Figure 6.22. The RGB color model.

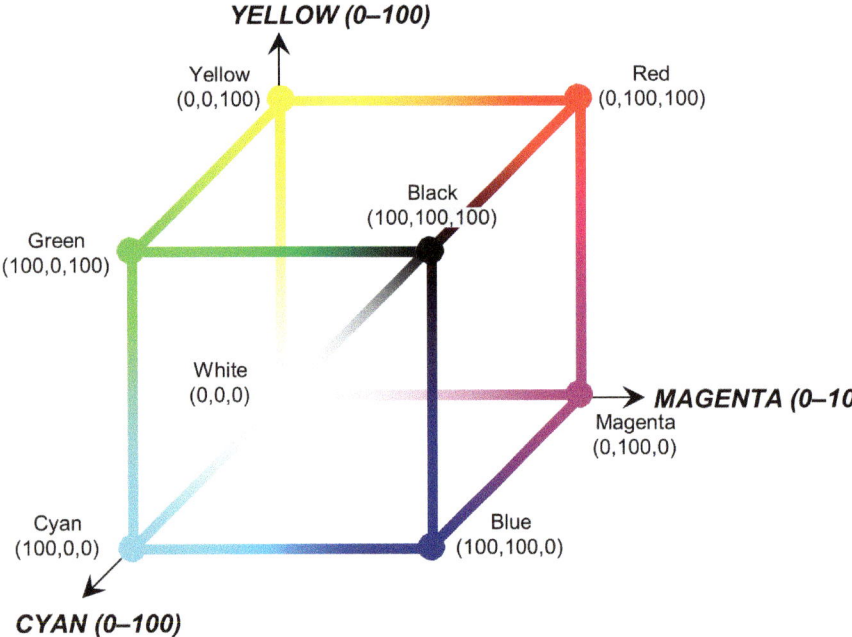

Figure 6.23. The CMY color model.

256^3 (16,777,216) possible colors. The primary colors can be represented on three axes of a cube, so that the cube becomes the color space in which any point comprising the volume of the cube can be given a numeric value for red, green, and blue that precisely specifies the color (figure 6.22). In this color model, black is defined as (0,0,0) and white as (255,255,255). Gray is scaled along the diagonal line between the two opposing corners between black and white. The secondary colors (cyan, magenta, and yellow) are situated at the remaining corners of the cube.

In the **CMY color model**, the CMY primary colors are analogous to the RGB primaries. The CMY color space can also be visualized as a cube with its primary colors at three corners, and its secondary colors (red, green, and blue) at the opposite corners (figure 6.23). Black and white are in opposite positions from their locations in the

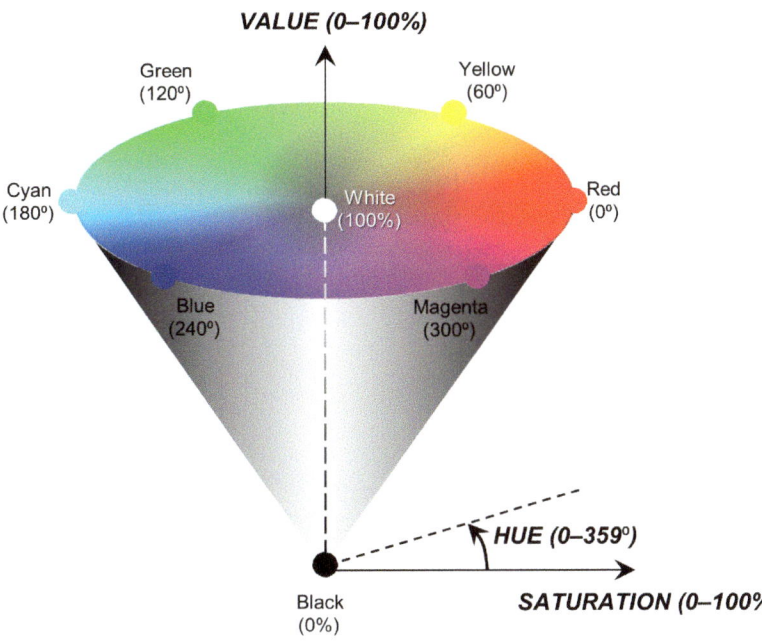

Figure 6.24. The HSV color model.

RGB color model, and gray is scaled between these two corners. Numeric values for the primary colors in this color model are expressed in percentages and range from 0 to 100 (although you will sometimes see the values ranging from 0 to 1 instead).

The **HSV color model** uses hue, saturation, and value to define colors. The HSV color model can be visualized as a cone in a cylindrical coordinate system (figure 6.24). In this color model, the hues are arranged in the circle that represents the base of the cone inverted in figure 6.28 (for visualization purposes) so that the hue element of color can be expressed as an angle varying counterclockwise from red at zero. Saturation ranges from 0 percent (gray) in the middle to 100 percent at the face, or the curved surface of the cone's exterior, where the colors are fully saturated. Value ranges from 0 percent (black) at the apex of the cone to 100 percent at the base. White is located at the center of the base as the point with 100 percent value, 0 percent saturation, and no hue. Gray tones fall along the vertical axis from this point to the apex of the cone, where black is defined as 0 percent value, 0 percent saturation, and no hue.

The HSV color model is a rearrangement of the geometry of the RGB and CMY color models. It is more intuitive to many mapmakers because it allows them to work directly with the hue, saturation, and value (lightness) visual variables, as opposed to proportions or percentages of color determined by the technology used for map production. Numeric values for the primary colors in this color model are expressed in percentages and range from 0 to 100 (although you will sometimes see the values ranging from 0 to 1 instead).

Color schemes

Mapmakers often use arrangements or combinations of preselected colors called **color schemes**, or sometimes **color progressions** or **color ramps**, to symbolize data. The colors in a scheme can be discrete, with distinct although sometimes subtle visual differences between colors, as with the seven color schemes on the left in figure 6.25. **Continuous color schemes** have a gradient that blends seamlessly between variations in a single color or between multiple colors (figure 6.25, *far right*). As more colors are added to the scheme, the visual distinctions become more difficult to detect. Therefore, color schemes generally have between three and nine distinct colors. These schemes have been developed for both qualitative and quantitative data, and for sequential and diverging data distributions, as you will see.

Qualitative color schemes (or **categorical color schemes**) have colors that vary by hue while largely holding value and saturation constant (figure 6.26A). These color schemes impart a message about a difference in type rather than ranges in magnitude. A **sequential color scheme** is used for quantitative data that is ordered from low to high. The colors in these schemes have light to

Figure 6.25. Color schemes with preselected colors can be discrete, with visual breaks between adjacent colors, as shown in the seven examples on the left, or continuous, with colors that vary gradually in a gradient, as shown in the example to the right.

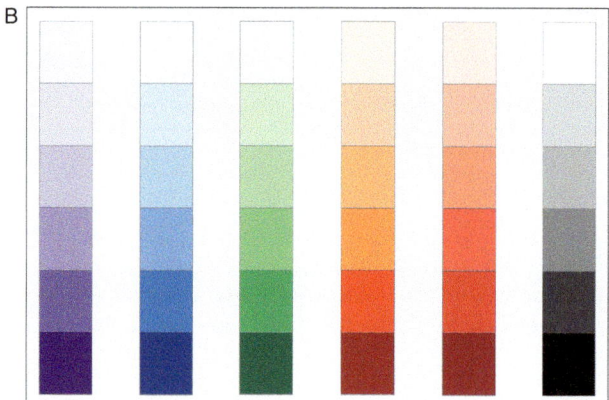

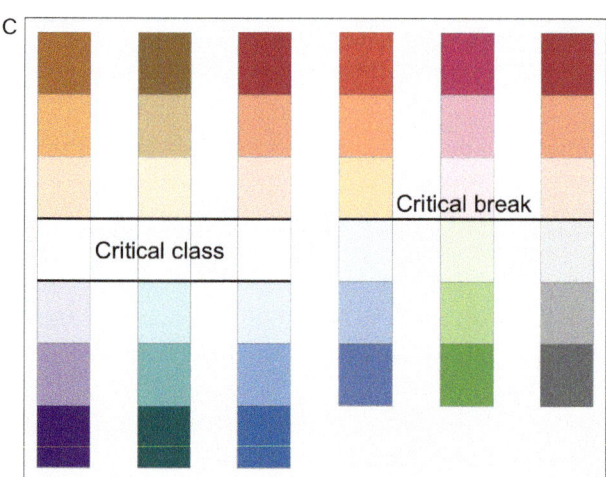

Figure 6.26. Examples of the colors in qualitative (*A*), sequential (*B*), and diverging (*C*) color schemes.

dark tones of gray or a single hue (figure 6.26B), although some schemes also have more hues (as in figure 6.25). For all schemes, the light colors for low data values progress to darker colors for high data values. A **diverging color scheme** is used for data that has a critical value or break in the midrange of the data distribution. The critical value or break is shown with a light color, and values above and below are shown with progressions of two different hues varying in value or saturation to darker colors at the extremes (figure 6.26C). The critical value or break can either be a class or the break between the two classes.

Choosing an appropriate color scheme is a challenge for many mapmakers, even seasoned professionals. A great resource is ColorBrewer 2.0, an interactive web-based app that presents color scheme options based on the mapmaker's requirements, such as the nature of data (qualitative, sequential, or diverging), number of classes, color and complexity of the background, usability for readers with color vision impairment, print friendliness, and ability to be photocopied. Like the examples in figure 6.26, Color-Brewer shows colors for symbolizing areal features; however, the same colors can also be applied to the design of point and line symbols.

Type

The great majority of maps use type (alphanumeric lettering) to convey important or additional information about mapped features. Typical uses of type on a map are illustrated in figure 6.27, including labeling places with their names ("Fort Point Rock") or numbers ("101" in the highway shield), labeling features with their values (the elevation values shown on the contours), describing the environment ("Presidio Hill"), locating a feature without using symbology ("Baker Beach"), or indicating features of

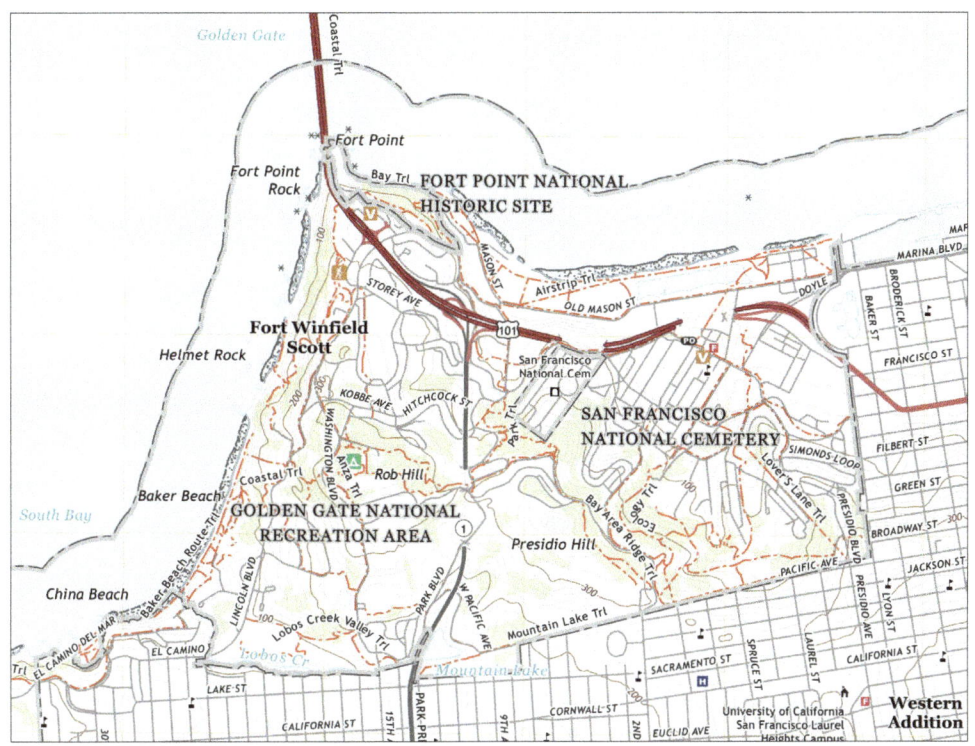

Figure 6.27. This segment of the San Francisco North, California, 1:24,000-scale USGS topographic map shows many of the uses of type on a map. Courtesy of the US Geological Survey.

Table 6.5. Typefaces and their properties

Typeface (shown in 12 pts)	Type style	Font	Size
Arial	Roman (Upright)	Arial Roman	Arial Roman 8 pts
Lucida	*Italic*	*Arial Italic*	Arial Roman 9 pts
Verdana	**Bold**	**Arial Bold**	Arial Roman 10 pts
Georgia	Condensed (Narrow)	Arial Condensed	Arial Roman 12 pts
Times	Expanded	Arial Expanded	Arial Roman 14 pts

vague extent ("South Bay"). Type is also used in titles, legends, text blocks, and other elements that accompany the map. As with color, mapmakers have developed guidelines on how to design and place type on maps.

Properties of type

As in any profession, type designers have a specialized vocabulary to describe type and its properties (table 6.5). Familiarization with this terminology makes it easier to communicate about typefaces and their properties and to recognize the underlying structure of various designs and the differences among them. A **typeface** is a group of characters—letters, numbers, punctuation marks, mathematical symbols, and other characters—that share a common design or style. Myriad, Palatino, and Tahoma are examples of typefaces, as seen in table 6.5. The characters in a typeface have the same basic design qualities, although their other properties (described below) can vary.

A **font** is a set of characters within a typeface. A font is distinguished by various styles, including form (either upright, also known as Roman, or italic, which is a uniquely designed version of a typeface that slants from left to right), weight (such as ultralight, light, demibold, and bold or black), character width (for example, narrow, which is also sometimes called condensed, and expanded), ornamentation, and designer or foundry. Examples of styles for the Verdana Pro font are shown in table 6.6. Although the fonts within a family will vary based on style

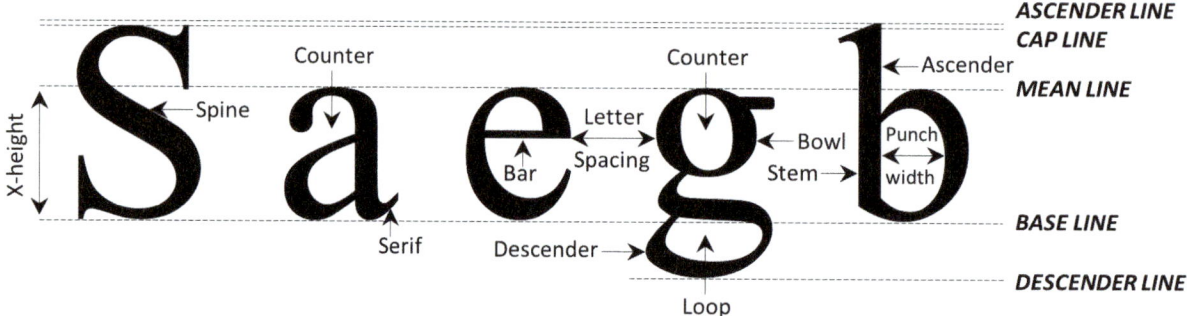

Figure 6.28. The anatomy of a font.

variations, their essentially similar design is what ties them together.

Size is another property of type. This is generally expressed as the number of points (there are 72 points in an inch) or count of pixels (which is a digital expression) representing the height of the characters. A font can also include uppercase, mixed case (also known as title case), or lowercase characters. Some fonts are designed so that all characters are in one of these cases, but for most fonts, the user can set the capitalization of the characters. The characters can also be italic or underlined. All these properties of type can be varied and give a font its unique appearance.

For mapmakers, an important consideration is whether the typeface has serifs. A serif is a short line or stroke appended to or extending from the open ends of a character (see the letter *a* in figure 6.8). Figure 6.28 shows a serif font. **Sans serif** (which means "without line") fonts do not have these appendages. Whether to use a serif or sans serif font is one of the first decisions a cartographer will often make when working with type. Serif fonts are considered decorative and appear somewhat old-fashioned, but they have also been credited with improving readability, especially for longer blocks of text, because they help the eye travel more quickly and easily across a line of type. On maps, labels are generally short, making a good case for the use of a sans serif font, which often has a more modern appearance. When considering the size for a label, consider that some sans serif fonts are more legible than serif fonts at any size, especially if the sans serif fonts have open counters, a large x-height, and wide letter spacing. These types of fonts are especially good for web maps, which are often displayed at a lower resolution on-screen than maps that are printed (especially professionally) on paper.

Table 6.6. Verdana Pro typeface font styles

Verdana Pro
Verdana Pro Black
Verdana Pro Condensed
Verdana Pro Condensed Black
Verdana Pro Condensed Light
Verdana Pro Condensed Semibold
Verdana Pro Light
Verdana Pro Semibold

Selected readings

Bertin, J. 2011. *Semiology of Graphics*. Esri Press.

Brewer, C. 2006. "Basic Mapping Principles for Visualizing Cancer Data Using Geographic Information Systems (GIS)." *American Journal of Preventive Medicine* 30 (2): S25–S36. DOI: https://doi.org/10.1016/j.amepre.2005.09.007.

Brewer, C. A. 1994. "Color Use Guidelines for Mapping and Visualization." Chapter 7 in *Visualization in Modern Cartography*, edited by A. M. MacEachren and D. R. F. Taylor. Elsevier Science.

Brewer, C. A. 2024. *Designing Better Maps: A Guide for GIS Users*, 3rd ed. Esri Press.

Brewer, C. A., and L. Pickle. 2002. "Evaluation of Methods for Classifying Epidemiological Data on Choropleth Maps in Series." *Annals of the Association of American Geographers* 92 (4): 662–81.

Brewer, C. A., and M. Harrower. 2013. ColorBrewer 2.0: http://colorbrewer2.org.

Imhof, E. 1975. "Positioning Names on Maps." *The American Cartographer* 2 (2): 128–44.

Ipatow, N., and F. Harvey. 2019. "Bertin's Graphic Variables and Online Mapmakers: An Empirical Study of Maps Produced by Prosumers and Cartographers." *Cartographica* 54 (4): 233–44.

Jenks, G. F. 1967. "The Data Model Concept in Statistical Mapping." In *International Yearbook of Cartography* 7: 186–90.

Kumar, N. 2000. "Automation and Democratization of Cartography: An Example of a Mapping System at CEM, University of Durham." *The Cartographic Journal* 37 (1): 65–77.

Slocum, T. A., R. B. McMaster, F. C. Kessler, and H. H. Howard. 2024. *Thematic Cartography and Geovisualization*, 4th ed. CRC Press.

CHAPTER 7
Map organization and layout

Map organization refers to the logical structuring and arrangement of spatial data on a map. It focuses on how geographic features, data layers, labels, and other map content are organized and presented to convey meaningful relationships and patterns. **Map composition** refers to the arrangement and design of various map elements, like titles and legends, which form a whole image, called a **layout**. Although the goal of map organization is to present map content in a clear, logical, and intuitive manner, the goal of map composition is to create a visually appealing and informative presentation. For both of these, cartographers rely on a number of basic design concepts.

Basic design concepts

In addition to determining what features and labels will appear on a map, cartographers must consider how the features and labels will be organized. This leads us to looking at a number of basic design concepts that cartographers have adopted for mapmaking, including figure-ground organization, visual hierarchy, and visual contrast.

Figure-ground organization

One of the first design decisions a cartographer often makes relates to visually separating the area of focus or main mapped area (the **figure**) from the background (the **ground**). **Figure-ground organization** is a perceptual phenomenon in which our mind and eye work together to spontaneously organize what we are viewing into two contrasting impressions—the figure, on which our eye settles, and the amorphous background behind it. Cartographers use this design concept to help map readers easily find the geographic area to focus on. Without figure-ground organization, readers can find it hard to distinguish the foreground from the background (figure 7.1A). Examples of cartographic methods to promote figure-ground organization are illustrated in figures 7.1B through 7.1E.

If the area of focus is a **closed form** (a polygon that represents the entire figure), such as a country, an island, or a park, showing only the mapped area clearly differentiates it from the background as long as the background is not the same color as the map (*B*). When mapping an area that is not a closed form or to show geographic context outside the area of focus, cartographers can use a **drop shadow** (a shadow added behind an object that creates the illusion of depth), **screening** to darken the background, **illumination** to lighten the background, or a **vignette** (a buffer around the perimeter of the figure that fades into the background). Other methods also exist, and many of the methods can be combined.

Visual hierarchy

After organizing the map's figure and ground so that the area of focus is the most prominent thing readers see, cartographers turn their attention to visually ordering the features and labels on the map. **Visual hierarchy** is the internal graphic organization or structure of the content on a map. Visual hierarchy helps mapmakers communicate the relative importance of mapped features and their labels through a visual impression that prioritizes or orders the categories of features. This visual layering of information is fundamental to a reader's ability to understand a map. Correctly applied, visual hierarchy reflects an appropriate intellectual order by graphically emphasizing the most important map features and de-emphasizing those that are less important.

The categories and order of layers in the visual hierarchy are dictated by the map's use. For reference maps and charts, categories of map features and their labels should be organized according to a hierarchy that relates to how

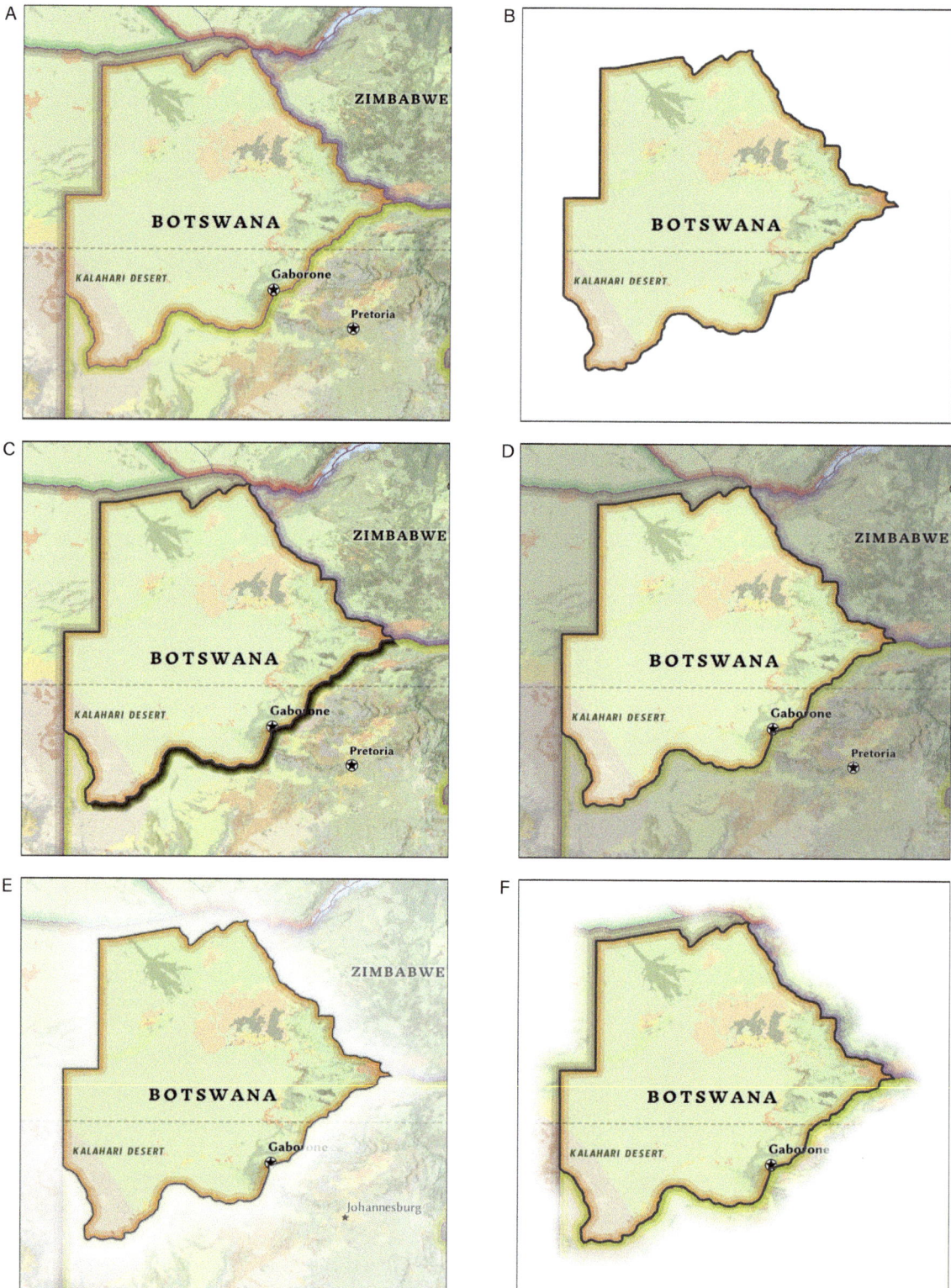

Figure 7.1. On a map with no figure-ground organization, it is difficult to know which geographic area to focus on (*A*). It is easier to know where to focus when a map uses a closed form (*B*), drop shadow (*C*), screening (*D*), illumination (*E*), or vignette (*F*).

the features exist in the physical landscape. For example, referring to the legend in figure 7.2, it is apparent on the 1:50,000-scale topographic map that terrain, shown with contours, is on the lowest visual plane, overlaid by land cover (vegetation and built-up areas); hydrography (rivers, streams, and water bodies); transportation (roads and railroads); power lines; and, on the highest visual plane, place names and other labels.

For thematic maps, reference layers should recede to the background, allowing the thematic layer to ascend to a higher visual plane. On the map in figure 7.3, the reference layers (terrain, transportation, and hydrography) are on lower visual planes than the thematic layer. The labels for the place names remain on the highest visual plane to assure their legibility.

Visual contrast

A key concept in map design is **visual contrast**—the visual distinction between two objects or between an object and its background. It is used to emphasize differences, highlight important features, and improve readability and clarity in maps. When the map has little visual contrast, it appears flat and uninteresting, and readers have a difficult time distinguishing the features on the map from each other and the background (figure 7.4A). A map with good visual contrast is seen as crisp, clean, and sharp, and the reader can easily differentiate and order the map's features and labels. Visual contrast can be achieved using many of the visual variables discussed in chapter 6, as illustrated in figure 7.4B.

The methods used to promote visual contrast relate to the conditions of map use. For example, if the map is to be used in low-light conditions, the design will necessarily change. This is apparent on mobile maps (such as those displayed on car navigation screens and smart devices) that automatically change from dark objects on a light background in daylight conditions to light objects on a dark background in low-light environments. Visual contrast is also affected by color contrast and legibility, discussed next.

Color contrast

Color contrast is the difference in the perceived color (hue, value and saturation, as discussed in chapter 6) between two objects or between an object and its background. A

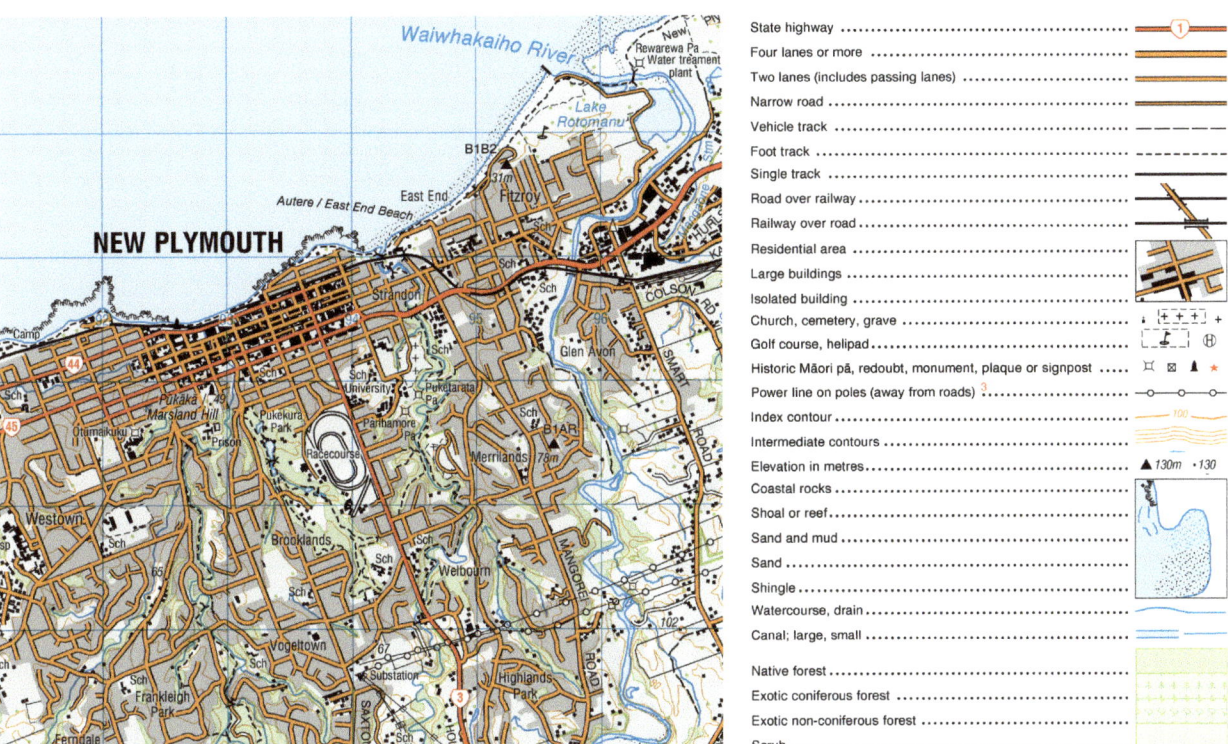

Figure 7.2. The good visual hierarchy on this segment of a 1:50,0000-scale topographic map of New Plymouth, New Zealand, is also evident in the legend. Courtesy of Land Information New Zealand.

170 Map Use: Map Reading and Design

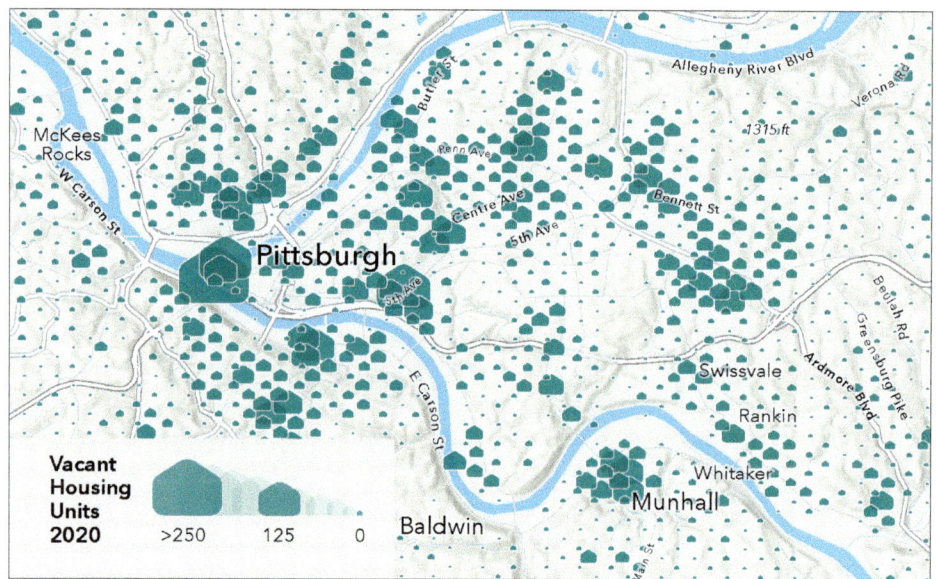

Figure 7.3. This map of vacant housing units in Pittsburgh, Pennsylvania, illustrates good visual hierarchy for the few layers of different feature types on a thematic map.

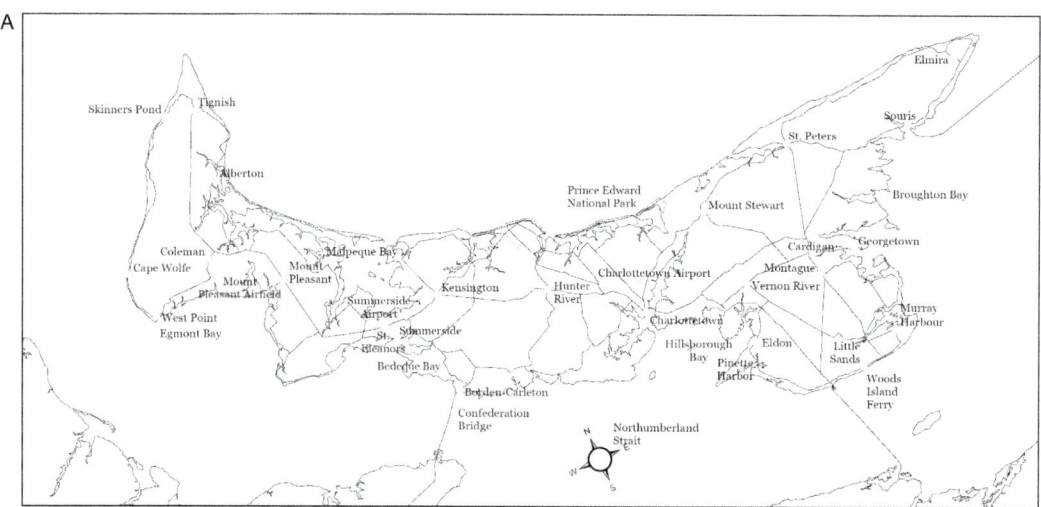

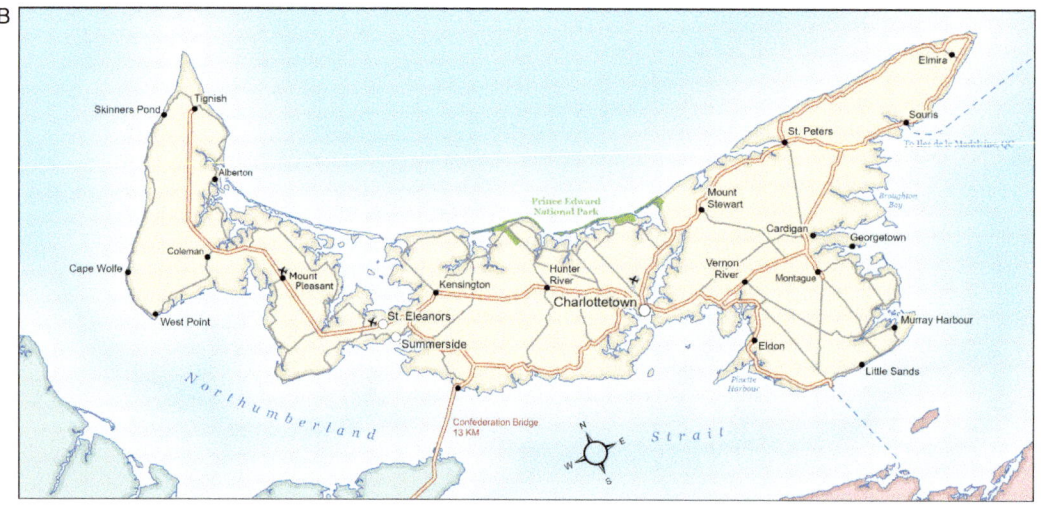

Figure 7.4. The map with no visual contrast (A) is harder to read than the map with good visual contrast (B).

color contrast metric called the **brightness difference** is a good guide for color selection that promotes visual contrast. A standard measure of color brightness, the **brightness equation**, can be computed to determine the brightness difference between map symbols and their background color. Based on the RGB color system (see chapter 6), the brightness equation from the World Wide Web Consortium (W3C) is given in equation (7.1):

$$\frac{((\text{Red} \times 299) + (\text{Green} \times 587) + (\text{Blue} \times 114))}{1000} \quad (7.1)$$

The diagram in figure 7.5 shows brightness differences for different combinations of foreground and background colors. The diagram illustrates the brightness difference between the background and the type and thin lines. Combinations that have greater brightness difference values are more legible. Using a professionally agreed-on brightness difference of 125 as the threshold for acceptable color contrast, most color combinations in the diagram are below the brightness difference threshold and hence should not be used.

You can see in figure 7.5 that black symbols on a white background have the greatest brightness difference (255). The reverse is also true because white type on a black background has the same brightness difference. Combinations of many of the colors on a white or yellow background also have acceptable brightness differences. It is not surprising that many road and other signs use these color combinations.

Legibility

Another graphic design concept that cartographers consider to assure visual contrast is **legibility**, which is the ability to be seen and recognized. The legibility of a symbol in large part relates to its size. Symbol and type size guidelines relative to viewing distance are offered in tables 7.1 and 7.2. Because type characters tend to be more complex than simpler map symbols, the sizes for the type are slightly larger than for symbols at any viewing distance. For complex map symbols, it is safer to use the guidelines for type rather than symbols to ensure legibility. For type sizes between 5 and 15 points, it takes at least a 2-point difference to distinguish between sizes. For type sizes over 15 points,

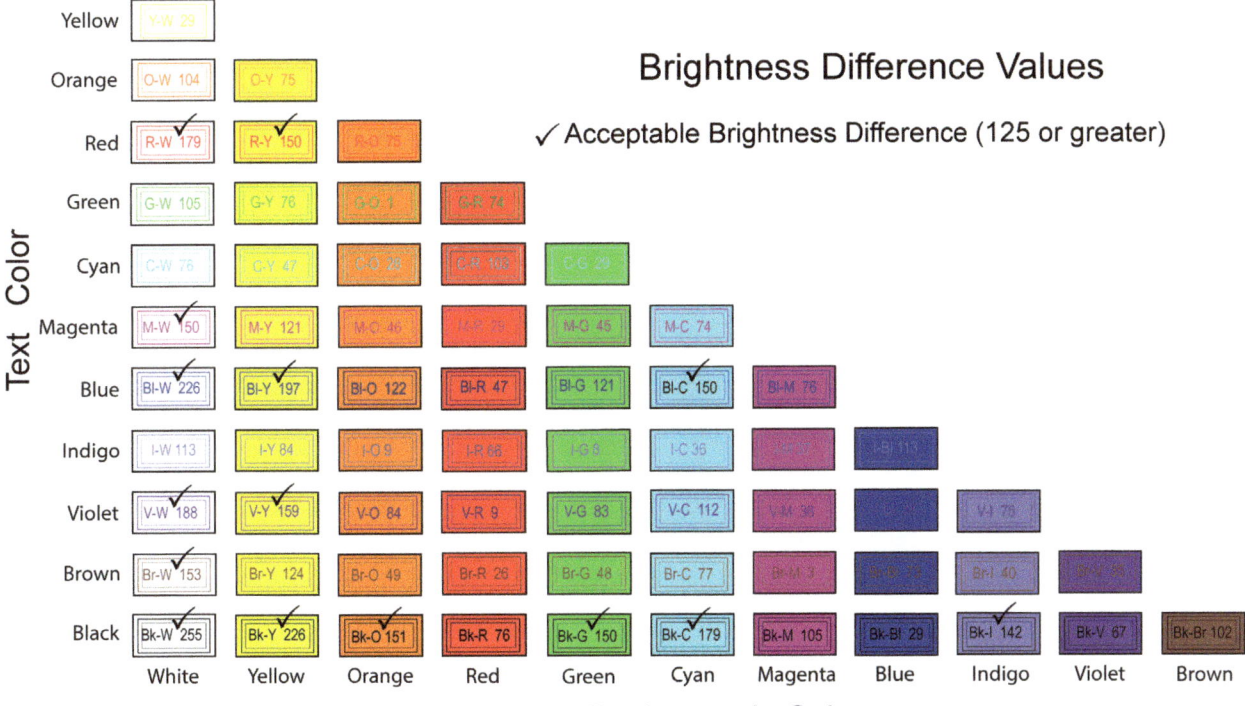

Figure 7.5. Brightness difference values for different combinations of type and lines and their background colors. A standard brightness difference of 125 or greater is the professionally agreed-on minimum acceptable value for color contrast.

Table 7.1. Recommended minimum symbol sizes

Viewing distance (feet)	Computer screen (points)	Print maps (points)	Computer screen (inches)	Print maps (inches)	Computer screen (millimeters)	Print maps (millimeters)
1.5	6	4	0.08	0.06	2.0	1.5
2	8	6	0.10	0.08	2.7	2.0
3	11	8	0.16	0.12	4.0	2.9
4	15	11	0.21	0.15	5.3	3.9
5	19	14	0.26	0.19	6.6	4.9
10	38	28	0.52	0.38	13.3	9.8
20	75	55	1.06	0.77	26.6	19.5
30	113	83	1.57	1.15	39.9	29.3
50	188	138	2.62	1.92	66.5	48.8
100	377	276	5.24	3.84	133.0	97.5

Table 7.2. Recommended minimum type sizes

Viewing distance (feet)	Computer screen (points)	Print maps (points)	Computer screen (inches)	Print maps (inches)	Computer screen (millimeters)	Print maps (millimeters)
1.5	8	6	0.11	0.08	2.8	2.1
2	11	8	0.15	0.11	3.8	2.8
3	16	12	0.22	0.17	5.6	4.3
4	21	16	0.30	0.22	7.5	5.7
5	27	20	0.37	0.28	9.4	7.1
10	53	40	0.74	0.56	18.8	14.2
20	107	80	1.48	1.12	37.6	28.4
30	160	121	2.22	1.68	56.4	42.6
50	266	201	3.70	2.79	94.0	70.9
100	533	402	7.40	5.59	188.0	141.9

differences should be at least 15 percent, but 25 percent is more desirable. Often, the default choices in mapping software reflect these basic size-difference guidelines.

Whether a map symbol can be seen and recognized from a distance depends on the visual angle. The **visual angle** is the angle between the light rays from the two ends of the viewed object as they hit the eye, as shown in figure 7.6. Vision research shows that under perfect viewing conditions, a person with 20/20 vision can distinguish objects at a visual angle of one minute (1/60th of a degree). Because this guideline assumes perfect viewing conditions and visual acuity, cartographers often assume a higher minimum visual angle—two minutes is probably more realistic.

The values in table 7.1 are minimum readable map

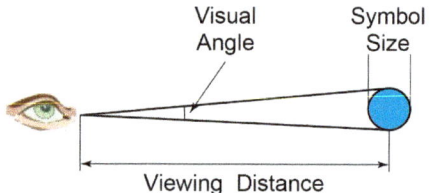

Figure 7.6. Geometric relationships among visual angle, viewing distance, and map symbol size.

symbol sizes for different map viewing distances, based on a two-minute minimum visual angle. If you are making a wall map or a map that is projected onto a screen in a large room, remember that the distance between the audience and the screen is not the same as the distance between you and your computer screen. For example, you must consider whether the person in the back of a 30-foot-long (9-meter-long) room can see the symbols projected on a screen at the front of the room. The corresponding entry in table 7.1 tells you that the projected symbol must be at least 1.57 inches or 39.9 millimeters high for easy readability from the back of the room. Additionally, the resolution quality of the projected map is lower than for the print map, so the print symbol can be slightly smaller (1.15 inches or 29.3 millimeters).

For map type to be recognized from a distance, it must be seen at a one-minute or greater visual angle. The required minimum type sizes for different viewing distances listed in table 7.2 follow the guidelines for minimum symbol size shown in table 7.1 but are about 50 percent larger because of the increased visual complexity of type. Because letter forms are more complex, a general rule of thumb is that any type on a print map that is read at a normal viewing distance (about 18 inches or 46 centimeters) should not be smaller than six points, or about 1/12 inch or 0.21 centimeters in size. Lack of readability becomes even more apparent when maps designed for viewing at one distance are used indiscriminately at another distance. For example, a person with 20/20 vision seated at the back of a 30-foot-long room cannot clearly see type projected on a screen at the front of the room unless it is 2 1/4 inches high or more on the desktop computer screen.

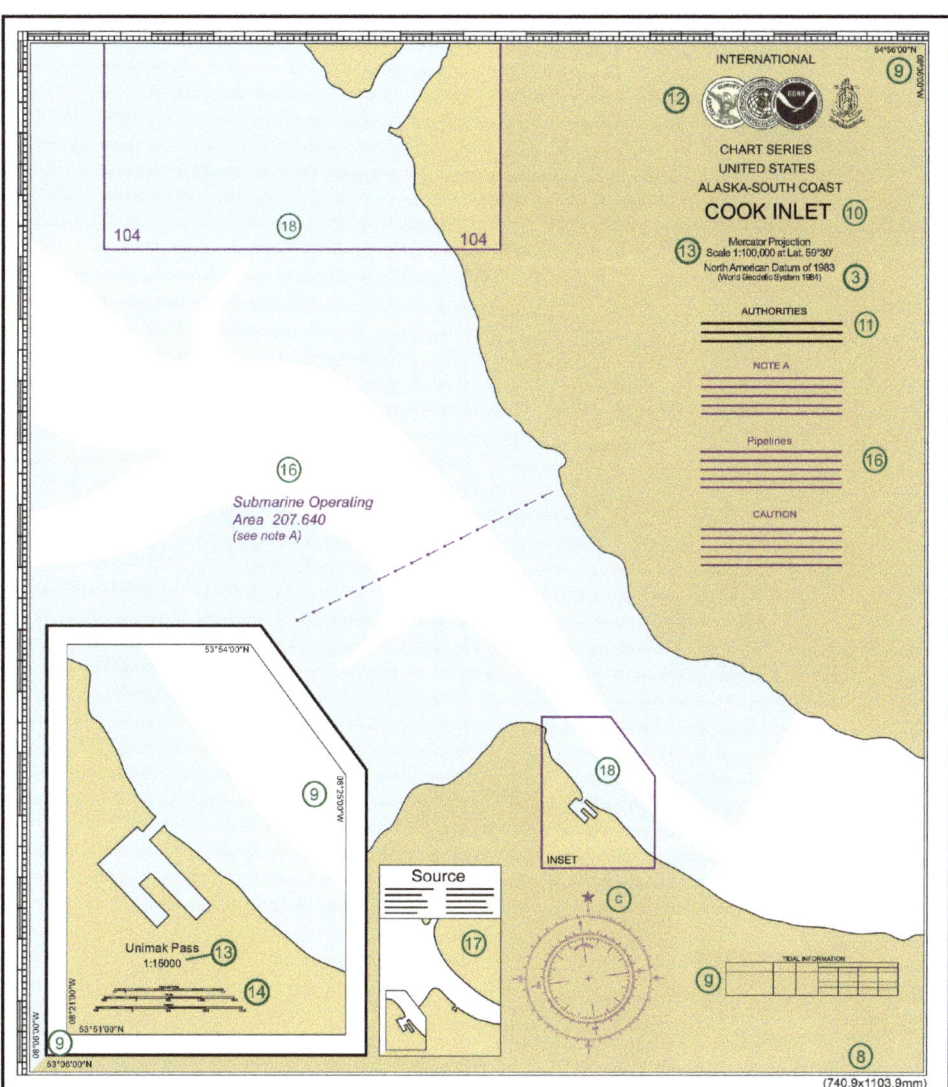

Figure 7.7. The layout for a NOAA nautical chart showing many of the common map elements. From NOAA Chart No. 1.

Map elements

Map layouts have two basic types of content: the map and the **map elements** (also referred to as **map surrounds** or **map marginalia**), which are additional objects on the layout that help explain or support the map. Examples of map elements include the neat line, title, legend, north arrow, and scale bar, as well as text blocks, graphs, charts, pictures, and even other maps.

Common map elements

To better understand some of the common map elements and how they might appear on a map, refer to figure 7.7, which shows the layout for a NOAA nautical chart. Figure 7.8 shows the guide for the elements on the layout. The following discussion of map elements refers to this layout, unless specified otherwise.

The **neat line** bounds the map extent. The neat line may be simple (see the right and bottom neat lines), decorative, or functional (see the top and left neat lines). A title (*10*) and sometimes a subtitle are included on almost all maps. The **legend** provides an explanation of the symbols and labels found on the map. It is usually shown in graphic form (see figure 7.2), although it is sometimes given as a word phrase, as on dot density maps, where you might see an explanation, such as "One dot equals 10,000 people" (see figure 9.19). For detailed maps with many features and symbols, the legend may be found on a separate page. The legend should explain any symbols that are unclear or confusing, especially if the map is for a general or international audience that has widely varying map reading experience. The legend title can also sometimes be used as a map title or subtitle.

Magnetic Features → B Tidal Data → H				
①	Chart number in national chart series	⑮	Linear border scale on large scale charts. On smaller scales use latitude borders for sea miles.
②	Chart number in international (INT) series (if any)	⑯	Cautionary notes (if any). Information on particular features, to be read before using chart.
③	Reference ellipsoid of the chart	⑰	Source Diagram (if any). Navigators should be cautious where surveys are inadequate.
④	Publication note (imprint)	⑱	Reference to a larger scale chart
⑤	Copyright note	⑲	Reference to an adjoining chart of similar scale
⑥	Date of current edition	ⓐ	Conversion scales
⑦	Notice to Mariners corrections	ⓑ	Reference to the units used for depth measurement
⑧	Dimensions of inner borders	ⓒ	Compass rose
⑨	Corner coordinates	ⓓ	Bar code and stock number
⑩	Chart title	ⓔ	Glossary: Translation of words on chart that are not in English
⑪	Explanatory notes on chart construction, etc. To be read before using chart.	ⓕ	Identification of a latticed chart (if any)
⑫	Seal(s)	ⓖ	Tidal and Tidal Stream information within the chart coverage
⑬	Scale of chart. Some charts have scale at a stated latitude.		
⑭	Linear scale on large scale charts		

Figure 7.8. The guide for the map elements on the layout for a NOAA nautical chart. From NOAA Chart No. 1.

A **scale indicator** (*13* and *14*) gives information about the map scale in one or more of the formats described in chapter 2. An **orientation indicator** helps map readers determine the direction to north on the map. It is often shown as a north arrow, although a **compass rose** is commonly used on charts (*c*). An orientation indicator is essential if the map is not in **normal orientation** (with north at the top of the layout), as with the map of Prince Edward Island in figure 7.4. An orientation indicator should not be used on maps in which the direction to north varies (see chapter 3). In these cases, the graticule can be used instead, or the orientation can be determined from features that align to the graticule, such as many administrative boundaries in the United States.

Metadata includes projection and map scale information (*13*) and, especially for maps that are used to make measurements, the datum (*3*) (see chapter 1). Text blocks can provide information about data sources (*17*), publication notes (*4*), the date of edition (*6*), explanatory notes (*11*), cautionary notes (*16*), and other pertinent information to help readers use the map and assess its accuracy and currency. Graphs, tables (*g*), photographs, logos or seals (*12*), and other ancillary elements can also be added to the layout to help explain the map and support its use.

Inset maps show areas that are geographically distant from the main map extent or are close-ups of areas that are too congested to easily see on the main map. The maps of Alaska and Hawaii on the layout in figure 7.10 are examples of inset maps showing distant areas. In figure 7.7, an inset map (*9*) is used to show the outlined area to the right that references a larger scale chart (*18*).

To familiarize readers with the location of the mapped area, a **locator map** can be used. This small-scale map provides context for a larger, more detailed map by showing

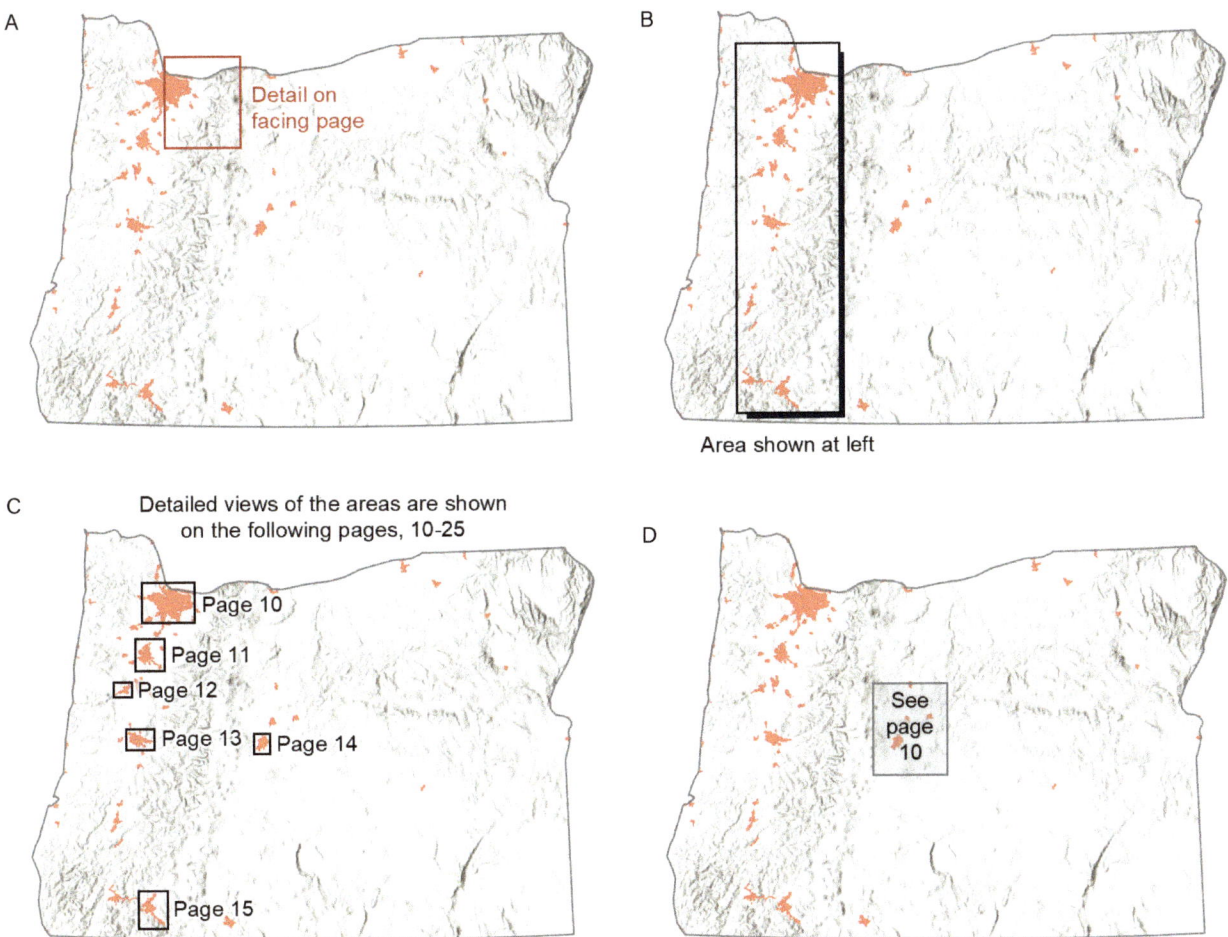

Figure 7.9. The extent of other maps can be shown on locator maps in several ways.

the location of the main area within a broader geographic region. A globe locator map (see figure 7.10) provides a quick and easy way for readers to understand the location of the mapped area in a global context.

Not all maps need all these elements, but some maps—such as reference maps and charts—include many of them. With the increased use of web maps, map elements can be shown in creative ways, or their need may be eliminated. For example, web maps can be panned and zoomed, so inset maps and locator maps may not be required. Metadata can be shown in pop-ups, and navigation tools can provide an indication of orientation.

Layout design

We now turn our attention to **layout design** to understand what makes the map and its elements come together as a cohesive whole as an aesthetically pleasing information product. To help them make good decisions, cartographers rely heavily on graphic design concepts that apply to almost any visual art. These include the concepts that relate to map organization (figure-ground, hierarchy, and contrast), as well as the golden ratio and proportion.

Golden ratio

When making decisions about where to place map elements on a layout, cartographers can rely on a principle of form and layout called the **golden ratio** (which is also known as the **golden section**, **golden mean**, **divine proportion**, or the Greek letter *Phi*). This ratio (which, for ease, can be expressed as 3:5) is considered to produce images that are pleasing to the eye. The golden ratio is approximately equal to 1.618, such that if one side of a rectangle is one unit, the other is 1.618 units. Figure 7.11 illustrates the **golden rectangle**—a rectangle based on the golden ratio that divides into a square and a smaller rectangle that also has a golden ratio. Drawing quarter circles within each square results in a curved line called the **golden spiral**—a pattern that can be found in nature in such things as sunflowers, pinecones, and seashells.

The golden ratio can be used to design the space on the layout and the dimensions of map elements and their parts.

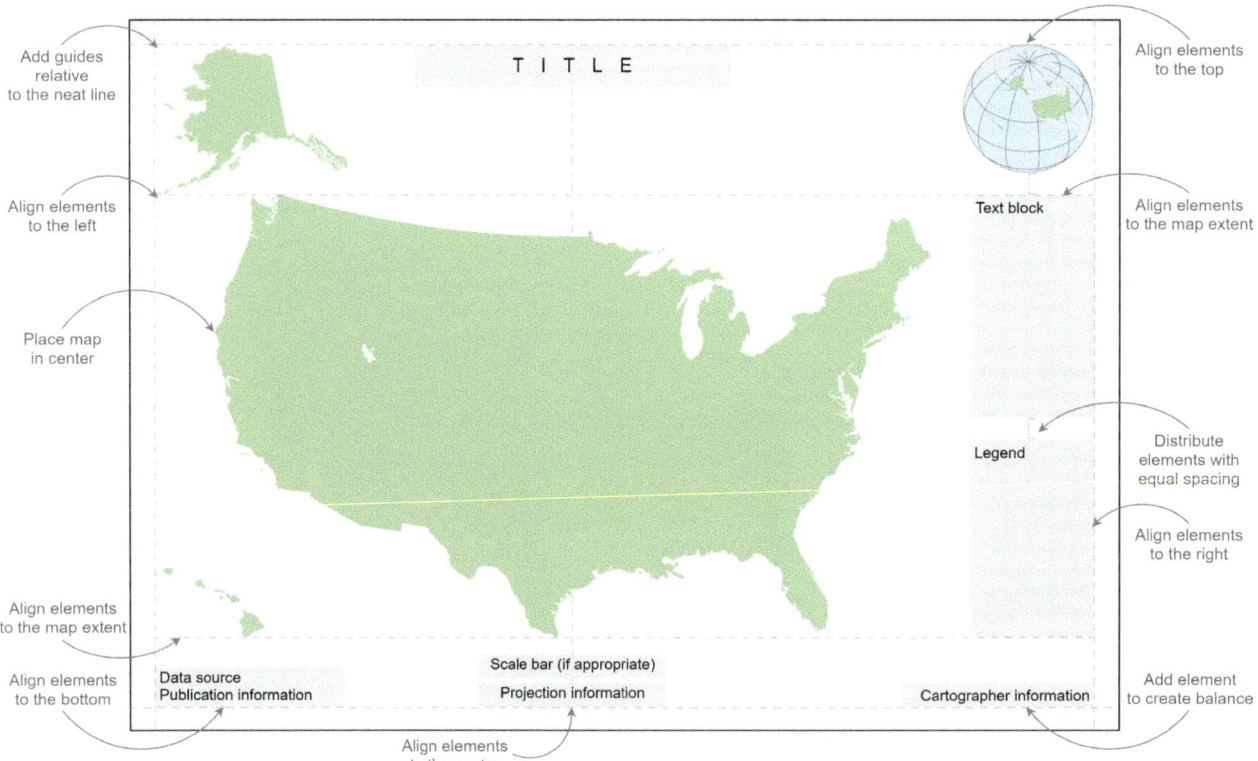

Figure 7.10. A layout for a map of the United States with many of the common map elements. Notes in the margins provide guidance for achieving an aesthetically pleasing layout.

Chapter 7: Map organization and layout **177**

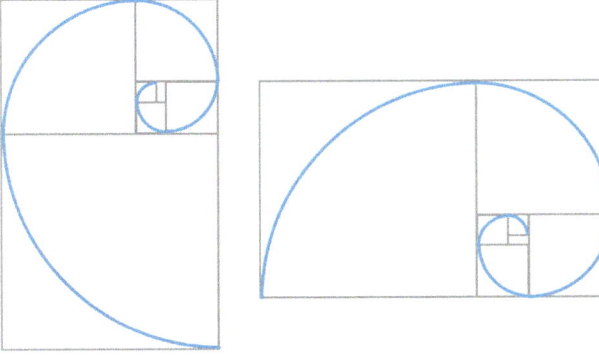

Figure 7.11. The geometry of the golden rectangle, in gray, and its relationship to the golden spiral, in blue.

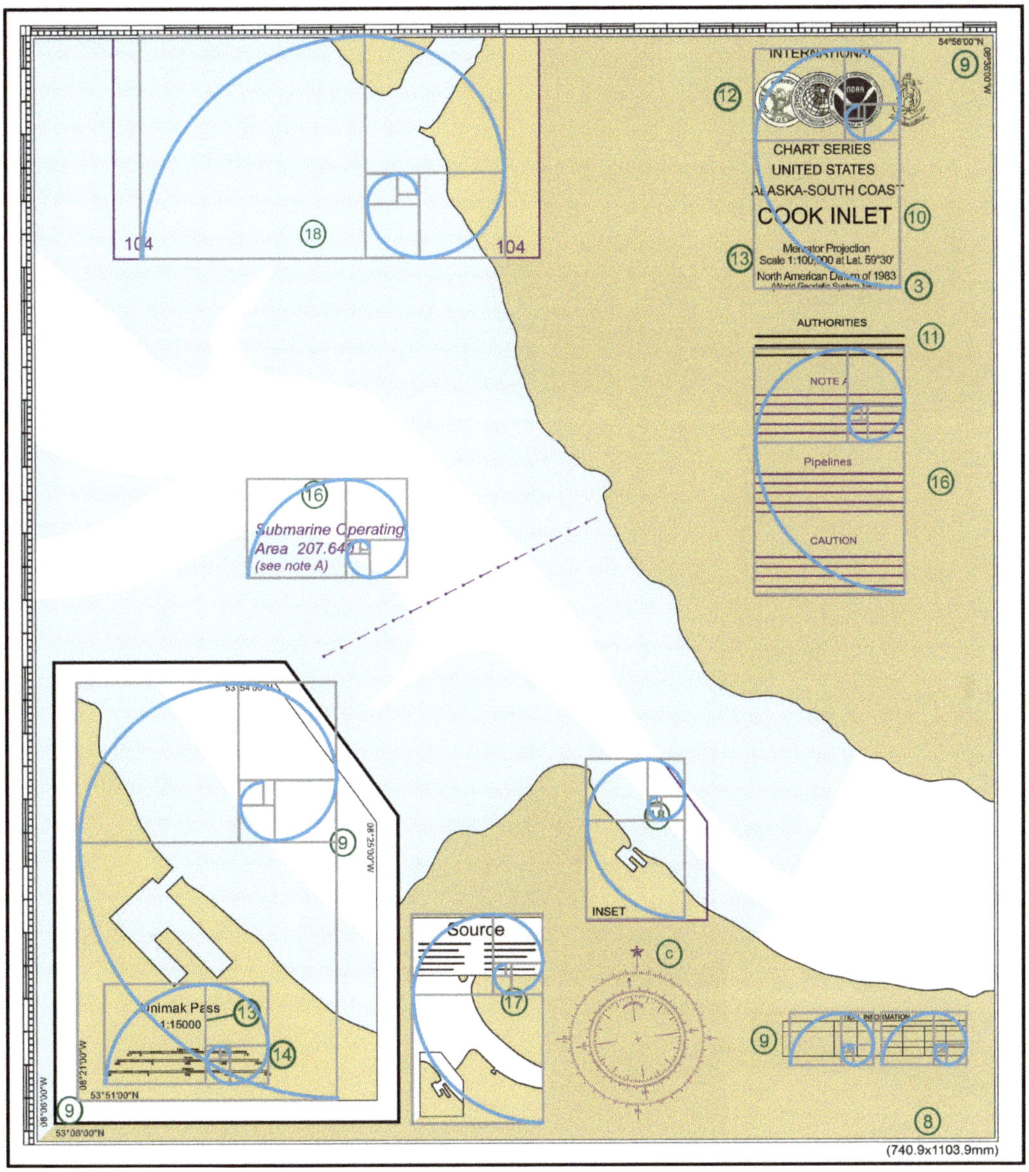

Figure 7.12. The golden rectangle overlaid on many of the map elements in the NOAA chart layout. Chart layout courtesy of the National Oceanic and Atmospheric Administration.

For example, a legend can be designed using a 3:5 ratio, as can the **legend patches** (the small, graphic examples that correspond to the symbols on the map). Many of the map elements in the NOAA chart layout have dimensions similar to those of the golden ratio (figure 7.12).

Proportion

Proportion is the relationship between one element on the map and another or the whole layout. Good proportion is achieved when a visually desirable relationship exists between the elements with respect to size, shape, scale, color, quantity, degree, visual weight, or setting. Design concepts that relate to proportion include **symmetry** (one side is a reflection of the other), balance, and harmony. Symmetry is hard to achieve in cartography because few geographic extents are as symmetrical as the rectangular state of Wyoming. Instead, we turn our attention to balance and harmony.

Balance

Balance is the distribution of visual weight within a layout. The objective is to create an impression of stability and order through **visual equilibrium** such that the layout is not off-balance in any direction. The impression of visual balance is controlled by the size, visual weight, and location of the symbols and labels on the map and the map

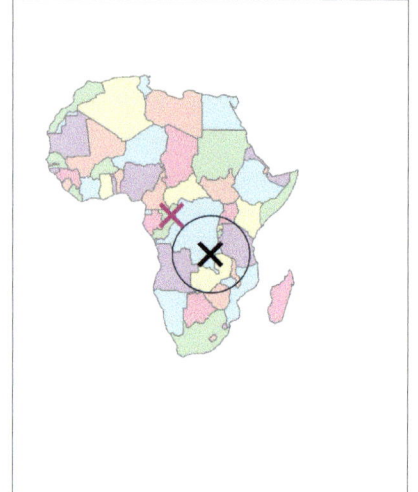
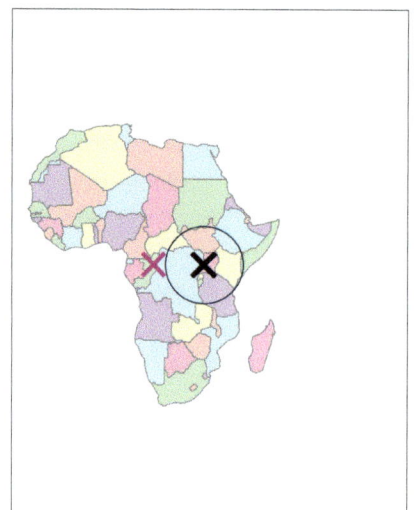
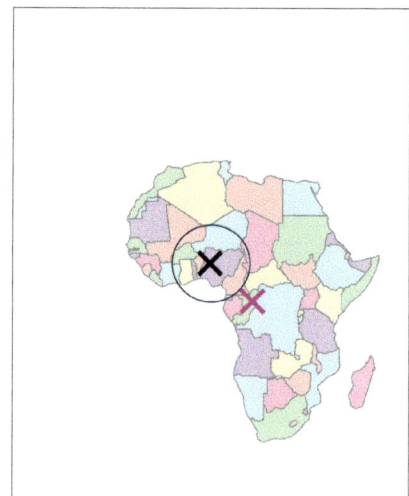
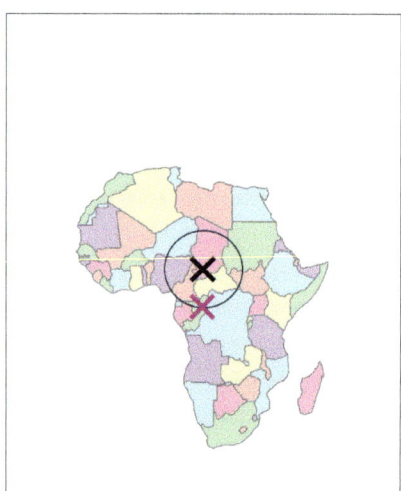
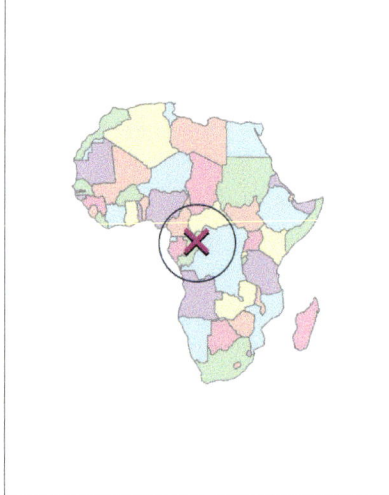
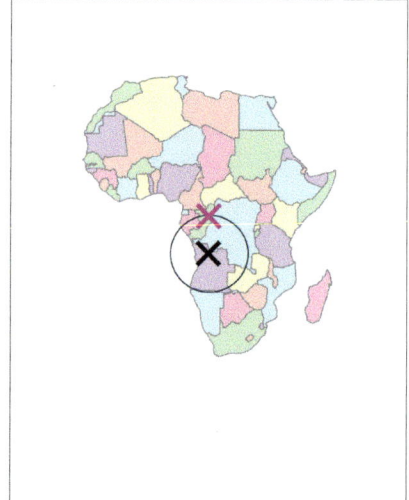

Figure 7.13. Which of the layouts seems most balanced? It should appear that the map at the bottom right has the best visual balance, achieved by placing the map's figure (Africa) slightly above center on the layout.

elements on the layout. One way to promote visual balance is to place the map's figure on the **visual center**, which is a point just above the geometric center of the layout. This is the point on which the eye first focuses, and it serves as the fulcrum, or balancing point, for the layout.

Most of the examples in figure 7.13 look off-balance, tipped toward the edge of the layout in one direction or another, except for the example at the bottom right. On this layout, visual balance is achieved by centering the figure (Africa) horizontally and shifting it up slightly to the visual center.

Visual balance is also affected by other elements on the layout. Figure 7.14 shows different ways of positioning the map and its elements to promote visual balance. The first two layouts (*A* and *B*) illustrate how placing the map's title, legend, and text blocks to fill **white space** (empty space on the map) can create what is called **informal balance**. **Formal balance** is achieved on the third layout (*C*) on which the title, map, and legend align with the vertical center of the layout. The combination of text blocks at the bottom can be considered a single element that is also centered on the layout.

Harmony

When a layout has **harmony**, the map and its elements have a cohesive arrangement and present a meaningful whole. Harmony can be promoted through the alignment and distribution of elements on the layout. **Alignment** is the arrangement—generally in an orthogonal direction relative to the layout—in a straight line or in an appropriate relative position (see figure 7.10). Proper alignment creates an ordered appearance by ensuring that the elements have an obvious and inherent relationship with each other. **Distribution** is the spacing of elements, parts of elements, or sets of elements so that they are equidistant. Distribution can be employed with features within elements, such as parts of a legend, or with a set of related elements, such as the locator map, text block, and legend in figure 7.10.

Web maps

Although we discuss online maps throughout this book, some unique characteristics of these types of maps deserve special consideration in terms of map organization and layout. **Web maps** are digital maps that are accessible in a web browser. The layout for a web map is often a **web app**—application software that runs on a remote server and is accessed in a web browser. Web apps allow users to view and interact with maps and related content, such as text, charts, photos, videos, social media data, and other web pages. For example, the wildfire web map in figure 7.15 is embedded in the Wildfire Aware app that also reports information about a selected fire, air quality, and weather conditions at the location, population and housing within the fire perimeter, and ecosystem characteristics, such as land cover, critical habitats, and protected areas.

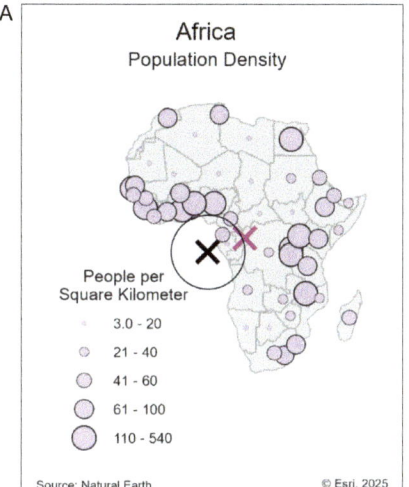
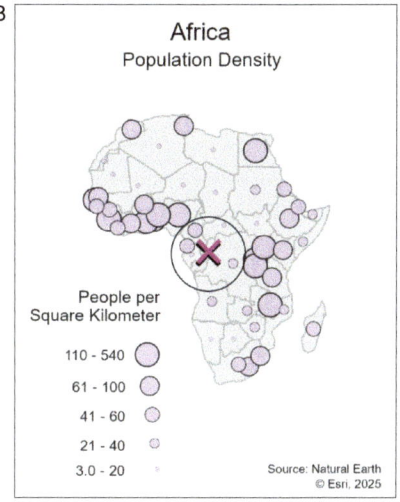
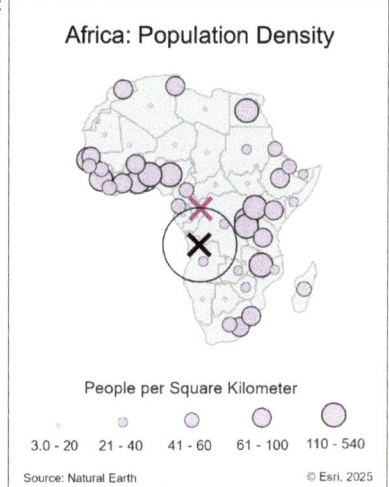

Figure 7.14. The first two layouts (*A* and *B*) are designed with informal balance. On the third layout (*C*), the map elements are arranged in a symmetrical manner to create formal balance.

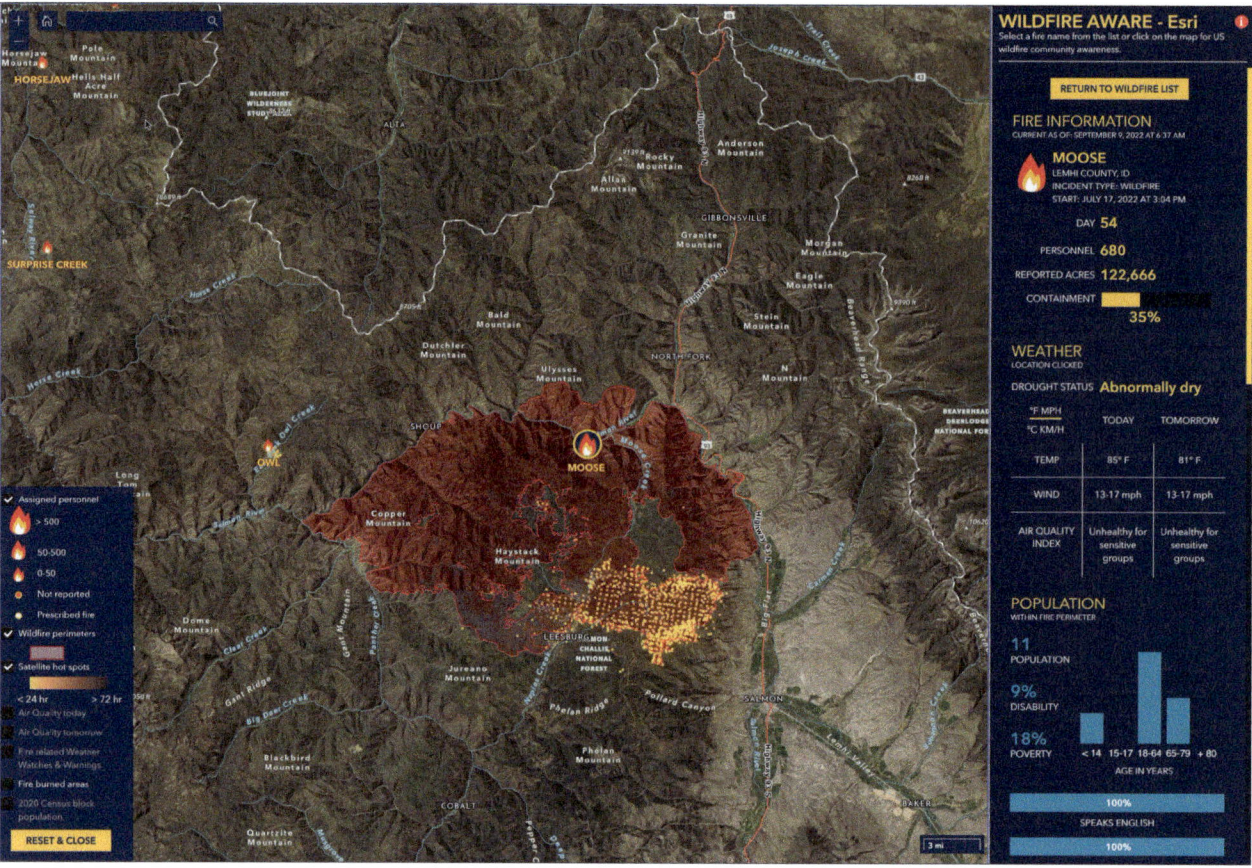

Figure 7.15. The Wildfire Aware app provides a broad understanding of the impact of wildfires on people, plants, and animals through its map and many map elements. Courtesy of Environmental Protection Agency, NASA, National Weather Service, US Department of Agriculture, US Census, USDA Forest Service, US Fish and Wildlife Service, US Geological Survey, National Interagency Fire Center, NatureServe, and Esri.

Web map organization

Web maps contain **web map layers**, or layers for short. A layer is the visual representation of a geographic dataset that relates to a theme. Conceptually, a layer is like a slice or stratum of the geographic environment. The web map in figure 7.15 contains multiple layers for such themes as the wildfire report locations, current fire perimeters, and thermal hot spots. Web maps are often created by overlaying operational and reference layers on a basemap (figure 7.16). A **basemap** is a digital representation of the earth that provides locational reference for a web map. A variety of basemaps have been created to support diverse web map uses and designs. A basemap can consist of a terrain surface (see chapter 10), an imagery layer (see chapter 11), or layers with features, such as parcels or roads, which are used for special purposes. **General-purpose basemaps** typically include layers for terrain, hydrography, transportation, administrative boundaries, and land cover. For certain basemaps, the administrative point features (such as cities), labels, and boundaries are on a separate **reference layer**, giving the mapmaker the choice to show these features or not.

Although it is only possible to turn the basemap and reference layers on and off, **operational layers** store data that users can often interact with. For example, clicking features on an operational layer may open a pop-up, symbol editor, or analysis pane.

Web map design

The design of web maps is often driven by the expectations of their users. Web map users expect the map data to be up-to-date and sometimes continuously updated, as on maps that show traffic, weather, or natural disasters. Web map users often expect a minimum level of interaction, such as zooming, which requires layers to be designed

Chapter 7: Map organization and layout   181

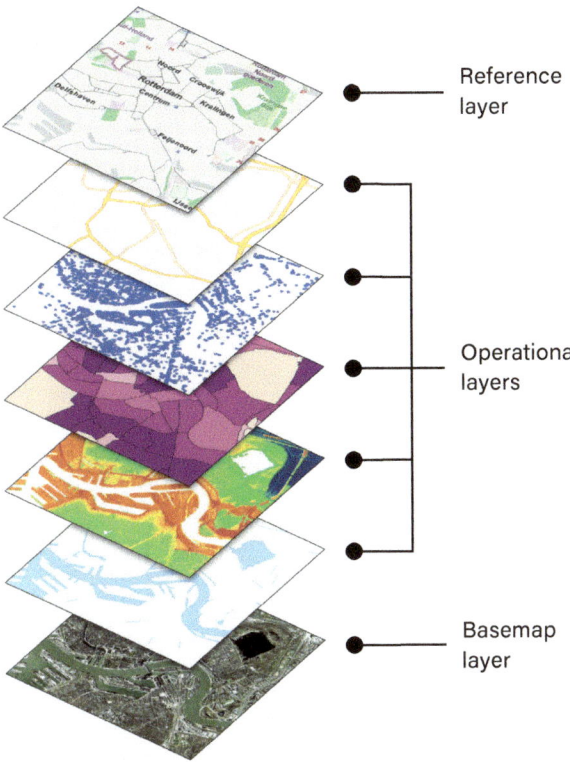

Figure 7.16. A web map is organized with a basemap as the foundational layer. Operational layers can be added, and a reference layer is sometimes overlaid on top.

for each zoom level (see chapter 2). Web map users may expect the map to be viewed on different devices, such as smartphones, tablets, and desktop computers. This requires **responsive design** so that the map automatically adjusts to different screen sizes and resolutions, ensuring a consistent and optimal user experience across all devices. Information that is normally displayed as map elements in the margin of a print map can be revealed on demand in a web map—for example, in pop-ups, tooltips, and expanding windows or panels. Web map users may even expect to be able to customize the map by adding or removing layers or changing symbology. This expectation creates a challenge for web map designers because they no longer control how the map will look in the end.

Digital twins

Perhaps the most challenging web map to design is a **digital twin**—a virtual representation of a physical geographic environment that mirrors the real-world landscape. It is created with advanced mapping technologies to simulate in a digital environment the characteristics, behavior, and interactions of its physical counterpart in the real environment. With these requirements for correspondence to reality and simulation of characteristics and behaviors, the power of the cartographer to abstract the complexity of the

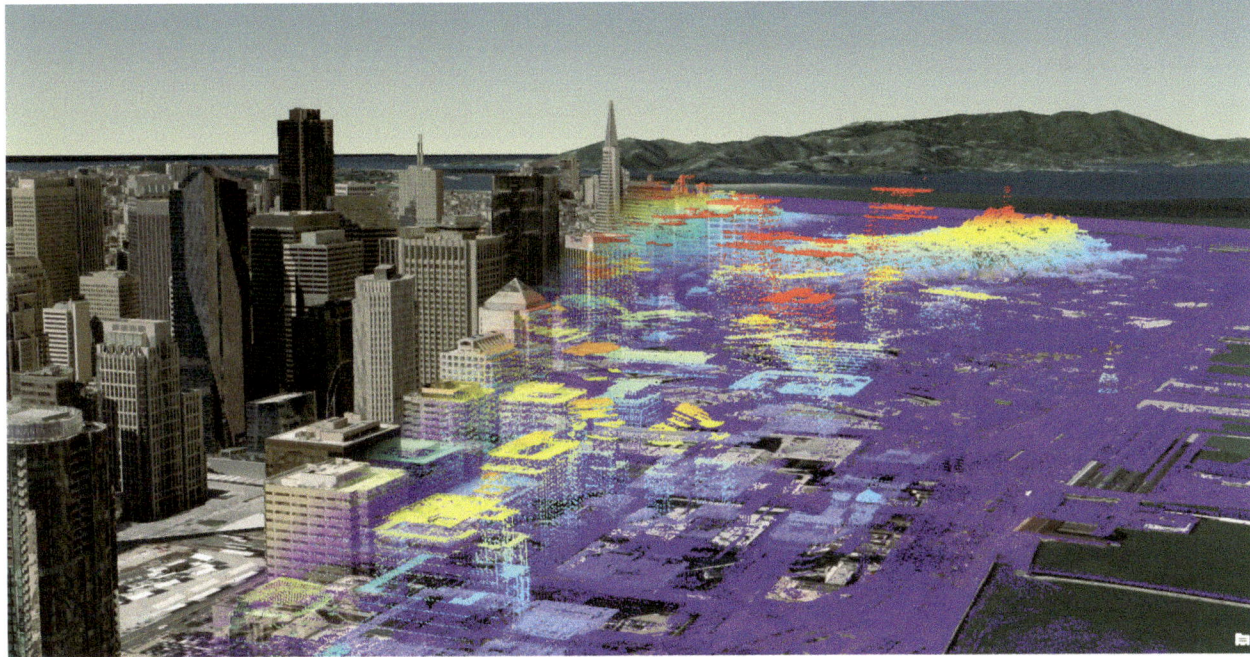

Figure 7.17. A digital twin may closely approximate the real environment, but it is still a map. Courtesy of Esri, EagleView Pictometry, and the National Oceanic and Atmospheric Administration.

real world (through selection, generalization, and classification) into a representation that is easier to understand (through symbolization, labeling and layout) is severely limited. A digital twin approximates the real environment with all its detail, intricacy, and noise, yet it is still a map. This conundrum of the map as reality and reality as a map is at the heart of chapter 12.

Selected readings

Battersby, S. E., M. P. Finn, E. L. Usery, and K. H. Yamamoto. 2014. "Implications of Web Mercator and Its Use in Online Mapping." *Cartographica* 49 (2): 85–101.

Bettencourt, L. M. A. 2024. "Recent Achievements and Conceptual Challenges for Urban Digital Twins." *Nature Computational Science* 4:150–53.

Konstantinou, E. N., A. Skopeliti, and B. Nakos. 2023. "POI Symbol Design in Web Cartography—A Comparative Study." *International Journal of Geo-Information* 12 (7): 1–33.

Ledermann, F. 2023. "Minimum Dimensions for Cartographic Symbology—History, Rationale, and Relevance in the Digital Age." *International Journal of Cartography* 9 (2): 319–41.

CHAPTER 8
Qualitative thematic maps

In the introduction, we explained that maps can be divided into three broad categories—reference maps, thematic maps, and charts—based on their purpose or function. In this chapter, we focus on qualitative thematic maps, which emphasize a single theme or a small set of related themes with nominal-level data. First, we look at the functions of qualitative point, line, and area graphic marks on maps. Then, we explore how qualitative data is symbolized on maps. We also look at examples of multivariate maps that show the relationships between two or more themes as well as maps that show change in feature locations or attributes. Finally, we explore dynamic qualitative thematic maps.

Qualitative information

The essence of mapping is to let something "stand" for something else. The graphic marks on the map stand for features in the environment, but if you do not understand what the marks mean, you cannot read, analyze, or interpret the map. We start by exploring how the marks for point, line, and area features function graphically. We also examine how qualitative geographic data is collected and represented as points, lines, and areas.

Point features

Point marks serve several functions on maps. They can be used to represent features that exist at points, features that are referenced to points, or features that are conceived of as points for mapping purposes.

The first function is to represent **point features**—zero-dimensional entities without width or area that are defined solely by their geographic location. For example, the volcanoes on the map in figure 8.1 are referenced to the locations of their summits. The summit may have been determined using surveying (see chapter 5), remotely sensing (see chapter 11), or some other method, but all capture the feature as an x,y (latitude, longitude) point location, so it is truly zero-dimensional in nature.

A second function of points is to represent features that are referenced to points. An example is the **mean center of population**—the balancing point of a population distribution on a map. Figure 8.2 uses a series of labeled points to show how the mean center of population for the United States has shifted westward from 1790 to 2020.

The third function of point features is to represent

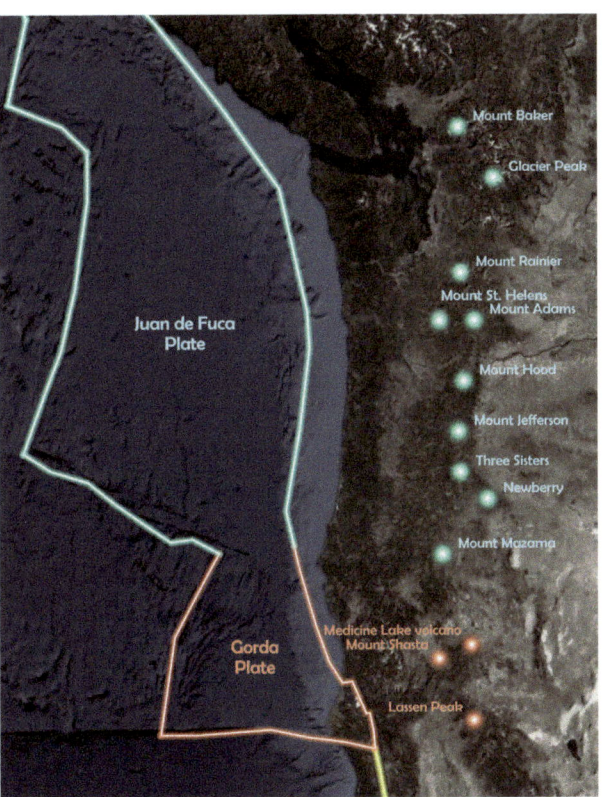

Figure 8.1. A map of volcanoes associated with tectonic movement of the Juan de Fuca (cyan) and Gorda (orange) plates.

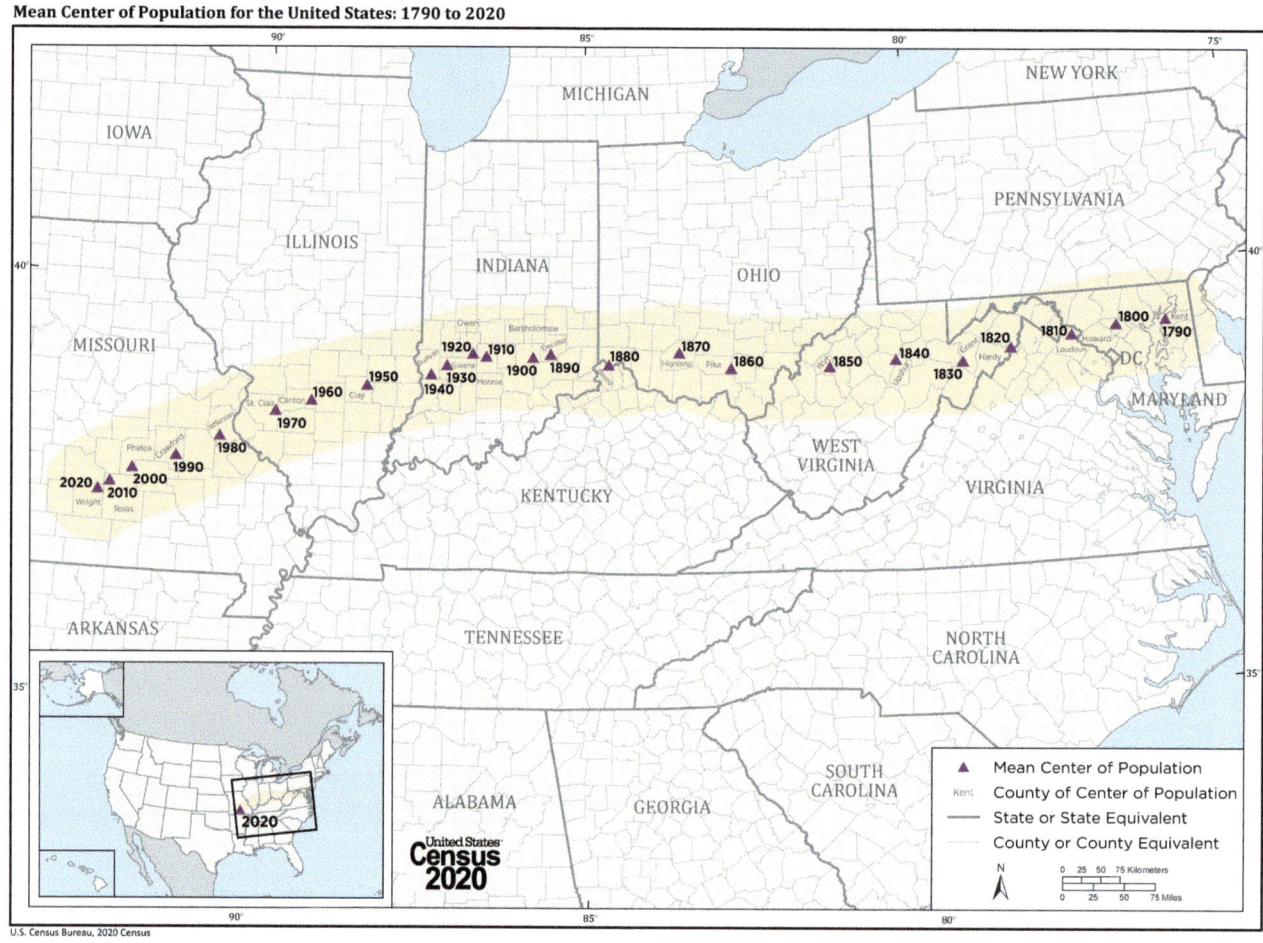

Figure 8.2. A map of the change in location over time of the center of population for the United States, from Maryland to Missouri. Courtesy of the US Census Bureau.

two-dimensional features that are conceived of as points for mapping purposes. In this case, features that are really areas on the ground are shown as points on a map. You have probably seen many uses of this mapping practice without realizing it. Think of how many times you have seen a map that shows cities as points, as on the segment of a map of Japan shown in figure 8.3. Large metropolitan areas such as Kyoto and Nagoya are symbolized with a light-red tint, but the cities labels relate to point symbols.

Data about the locations and attributes of point features is collected in a variety of ways. To illustrate this, we can think about how data for one theme might be captured and used for mapping. In our example, we consider how a city might collect data for all the trees that it manages. The city managers will want to know the location of each tree and certain tree characteristics (attributes). Qualitative attributes might include the tree species, heritage tree status (whether a tree is recognized for its unique age or historical significance), and site type (for example, a street median, the strip between a curb and a sidewalk, or a concrete cutout).

A fieldworker might use a GNSS-enabled mobile device to capture the locations of the tree trunks, and attribute data might be gathered through **field observation** (observing features in situ) or a **field sketch** (drawing of the sites). Another approach might be to **digitize** (convert into digital format) the tree location from aerial imagery (see chapter 11) as the tree canopy (the extent of the tree's branches and leaves) and record visible attributes. GIS can then be used to create a centroid for the canopy polygon. The map in figure 8.4 shows the GNSS-collected locations of tree trunks in orange. Tree canopy polygons digitized from imagery and their centroids generated by GIS are shown in green. Notice that the GNSS data and the canopy

Figure 8.3. On this map of Japan, city locations are shown with point symbols even though the city areas are much larger. Courtesy of Benchmark Maps, an imprint of East View Information Services.

Figure 8.4. Tree locations collected with GNSS data recorders are shown in orange. Tree canopies from aerial images and canopy centroids from GIS are shown in green.

centroids do not coincide because the tree trunks are not exactly centered under the canopies.

Line features

As with points, lines serve several functions on maps. They can represent truly one-dimensional features that have length and direction or features that are referenced to lines for mapping purposes. Surveyed lines such as property, political, and administrative boundaries are true one-dimensional features, as are the tectonic plate boundaries in figure 8.1. Lines also serve other graphic functions on maps, including representing paths of movement and the structure of linear networks.

Many lines that you see on maps represent features that have width but are portrayed graphically as one-dimensional lines—that is, they represent features that are conceived of as lines for mapping purposes. The roads and stream centerlines in figure 8.5 are examples. These

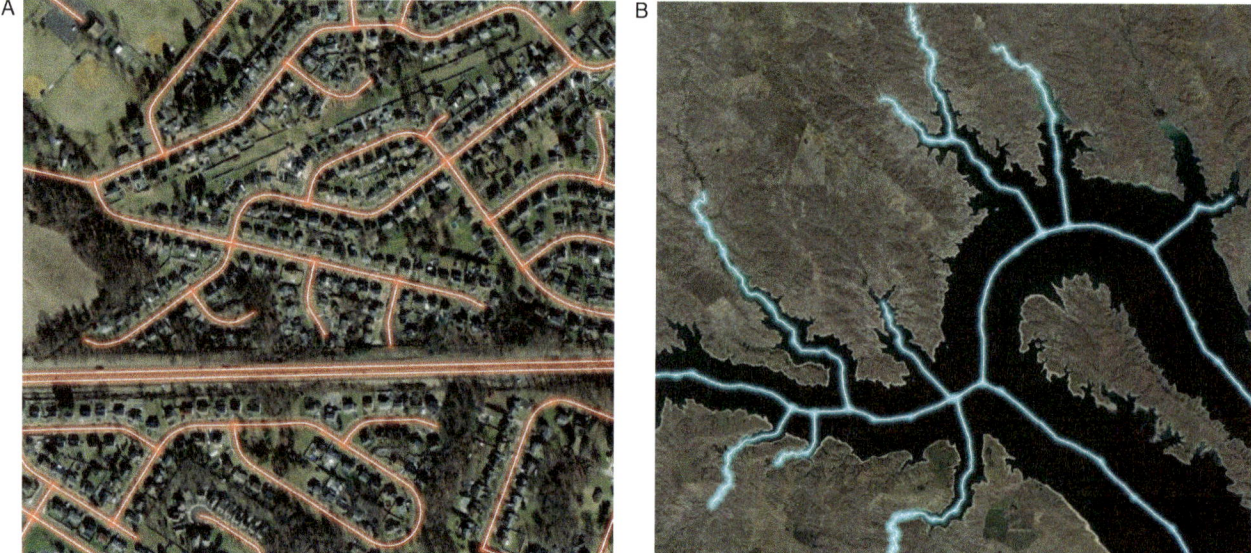

Figure 8.5. Areal features, such as roads (*A*) and streams (*B*), can be represented on maps as centerlines. Data courtesy of the USGS National Hydrography Dataset Plus High Resolution and Esri (*B*).

centerlines can be digitized from imagery or modeled with GIS. They can then be used to represent the transportation (orange) and hydrological (blue) networks.

As with point feature data, line feature information is collected through a variety of methods, including surveying, field sketching, GNSS data collection, and aerial or satellite image interpretation. For the roads and streams in figure 8.5, the features were digitized as **centerlines**—imaginary lines through the center and roughly equidistant from the sides of the features. Field observation, GIS or image analysis, interviews, or other methods can be used to collect the feature attributes. Examples of qualitative attributes are road type, such as expressway, secondary highway, or local road (see table 8.1), and stream type, such as perennial, intermittent, or ephemeral.

Area features

The graphic functions of areas on maps also include representing truly two-dimensional features, such as the light-red urban areas in figure 8.3, or features that are conceived of as areas for mapping purposes, such as the space between plate boundaries in figure 8.1. Mapmakers collect qualitative area feature data using the same methods as for points and lines. An additional method for gathering area data is a census.

Census data

A **census** is a systematic survey that collects data—within a defined area (or **data collection unit**) at a specified time—about all the members of a population. A variety of attributes are recorded for each member of the population. Federal, state, and local government agencies often conduct a **population census** to **enumerate** (count) the number of people and learn about their demographic, economic, and social characteristics. An **agricultural census** gathers information about the number of farms, types of crops grown, livestock, and farming practices. Other censuses can also be taken to focus on, for example, industry, transportation, wildlife management, or tourism.

When you read or make a map of census data, it is important to understand the geographic areas in which census data is collected. The data collection unit for a census is called an **enumeration area**, or sometimes **enumeration unit**. Several types of enumeration areas are used for data in the United States, including postal code areas, school districts, and urban areas, as well as those areas defined specifically for statistical surveys and census taking.

The **US Census Bureau**—the US federal agency that is responsible for producing data about the American people and economy—has developed enumeration areas for the United States. Figure 8.6 shows the enumeration areas for the state of Iowa. The state is divided into counties that contain census tracts consisting of census block groups.

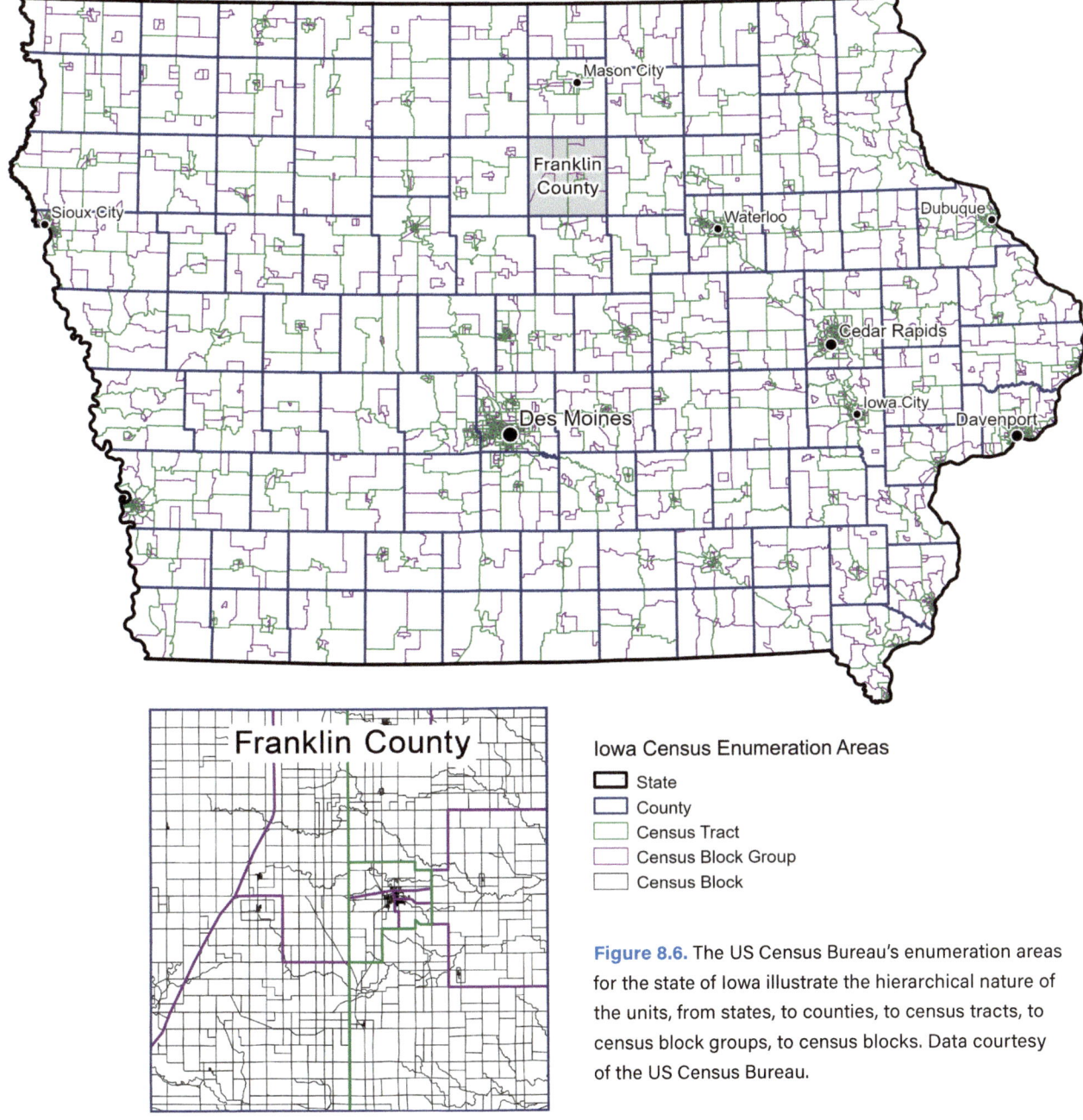

Figure 8.6. The US Census Bureau's enumeration areas for the state of Iowa illustrate the hierarchical nature of the units, from states, to counties, to census tracts, to census block groups, to census blocks. Data courtesy of the US Census Bureau.

These block groups are made up of the smallest enumeration area, census blocks. These areas are hierarchical so that the smaller units nest within larger units. With this approach, data for the larger units can be summarized from the smaller units.

Data accuracy

It seems logical that a map showing data collected using the methods we have described would be accurate. This assumption is true in most cases, but not always. A number of errors may reduce the quality of the data. The errors can come from instrumental, methodological, or human deficiencies during data gathering or analysis. Such errors are usually well disguised on maps, sometimes deliberately and sometimes inadvertently.

Consider the **Census of Population and Housing**, which is conducted for the United States each decade by the US Census Bureau. Because every household in the country is supposed to be surveyed, maps produced from the census are flawless…or are they? A full head count of more than 300 million people is an immense job. We will look at how error can creep in.

Once every 10 years, the US Census Bureau hires individuals, called **enumerators**, to collect data from the households that have not yet responded to mailed questionnaires. The enumerators are trained in about a week for a job that typically lasts several weeks and only pays a little above minimum wage. Enumerators are taught how to deal with reluctant respondents and those who refuse to participate in the survey. In the field, the enumerators must brave neighborhoods previously unknown to them, try to reach people who may not be home, and when they are home, may be hostile or let the enumerator know that their dog might be. The people interviewed are supposed to respond truthfully, even though this candor means sharing information that might get them into trouble with their landlords, the welfare office, or the authorities. Finally, after all the information is gathered, it still must be processed, analyzed, and mapped, which presents myriad opportunities to introduce bias or error into the final product.

You can see the potential for all three types of errors: instrumental (the questionnaire), methodological (the interview process), and human (the interviewer and the respondent, as well as the analysts handling the data). Yet the Census Bureau has maps and a website that, at any time, will give you the actual number of people in the United States. You must consider the potential for error when you judge the accuracy of census data and read census maps.

Aside from a census, other measurements are subject to these same sources of error. Imagine determining the number of fish in a stream by netting at several sample points. The net may not be fine enough to capture all species of fish (instrumental), the sample points may not be representative of the entire stream (methodological), and the person doing the netting may have problems performing the job thoroughly (human). Nonetheless, much time and effort are put into assuring that the data is collected in the most accurate yet efficient manner possible.

Homogeneity

The areal data collection units on maps are considered to be **homogeneous**, or uniform in structure or composition throughout. Consider mapping a stand of trees. The stand may prove to be entirely of the same species, in which case the stand is a completely homogeneous qualitative area feature because all its defining objects (trees) are of the same category (species) within the homogeneous area (stand).

The interesting thing about features mapped as though they are homogeneous is that most actually are not. In our tree example, although the trees within the stand may be the same species, there may be an understory of shrub, brush, or grass. At the edges of the stand, the trees may mix with other species to create a mixed-species transition zone between two single-species stands. Although neither of the stands are truly homogeneous area features, they are mapped as being so.

Mapmakers deal with this real-world problem in several ways. They may define the boundary between the two single-species stands as the middle of the transition zone. Or they may add a transition zone to the classification so that the map includes a mixed-species class as well as the two single-species classes. This solution increases the complexity of the map, but it allows the single-species stands and the transition zone to each be mapped as homogeneous areas that more accurately reflect what is on the ground. Alternatively, they can add the transition zone to one of the stands and add a note on the map explaining what they have done.

Homogeneity is also a consideration for line features. As an example, many streams may flow entirely through a natural channel. When these homogeneous features are shown on the map as natural features, the map agrees with what is on the ground. However, other streams may have reaches that flow naturally in some places and through canals or ditches, pipelines, or underground conduits in others. To show these variations correctly, the streams can be **segmented** by splitting them at locations where the stream type changes.

The feasibility of this approach relates to the map scale (chapter 2). For example, if you were planning a canoe trip, you would not use a small-scale map on which streams that flow through different types of watercourses are classified into one category. A larger-scale map showing different types of watercourses will help you avoid problems with planning and navigating your trip.

Single-theme maps

The simplest and most common qualitative thematic maps show a single theme. We now examine these qualitative single-theme maps for point, line, and area features.

Point feature maps

Point feature maps show point locations and their characteristics. The word *point* has a loose interpretation here because, as you saw earlier, points can be zero-dimensional features, the location to which something is referenced, or the location that represents a feature conceived of as a point for mapping purposes. The appropriate visual variables for showing qualitative data are shape, hue, orientation, and arrangement (see table 6.3). The map in figure 8.7A uses hue to distinguish the airfield and landside lights at San Francisco International Airport, and the map in figure 8.7B uses orientation to show the direction from which the navigational lights (shown in purple) on the

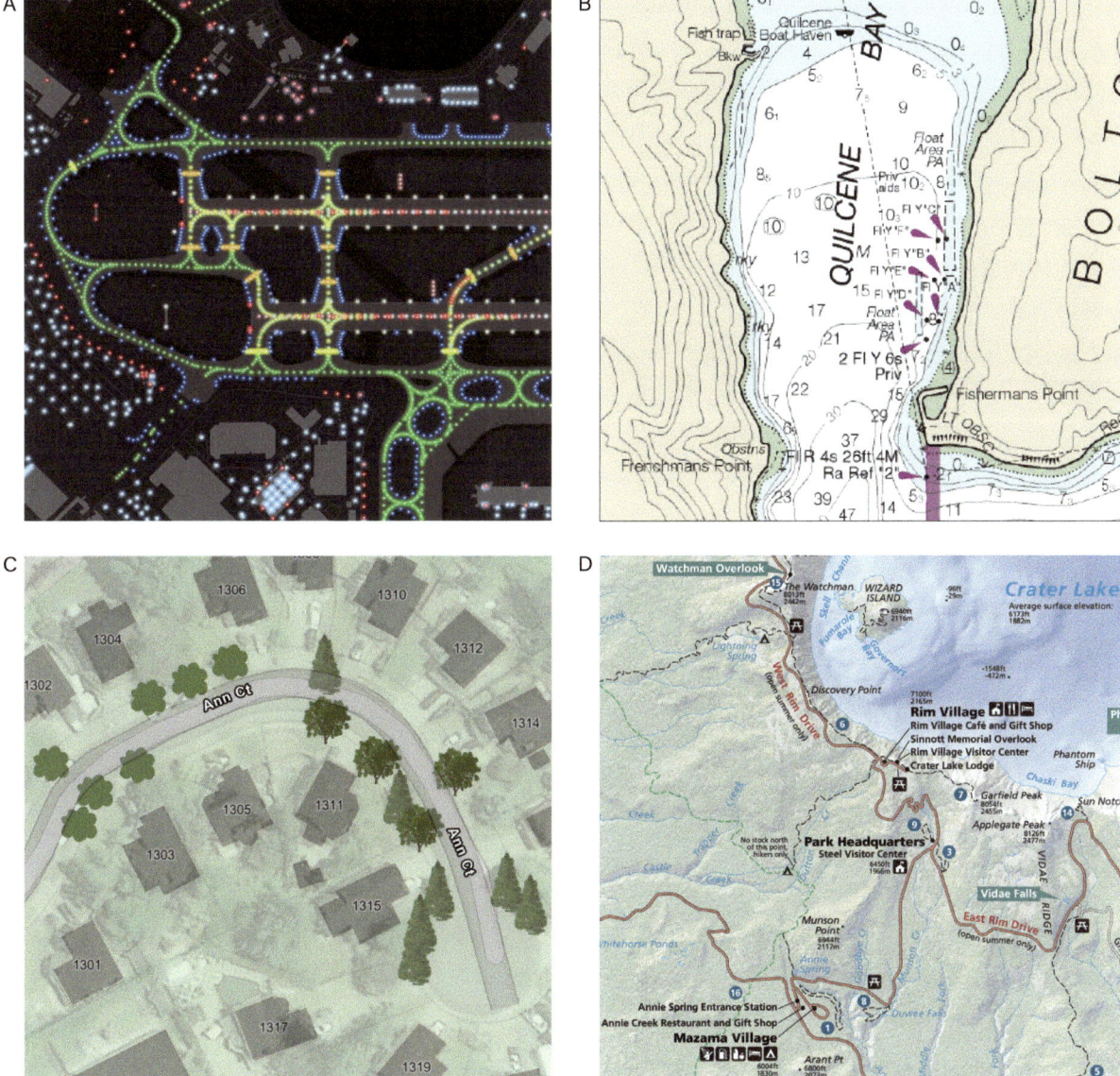

Figure 8.7. Qualitative point symbols can be symbolized with hue (*A*), orientation (*B*), or shape (*C*). The National Park Service map in *D* illustrates the use of standard symbols. Courtesy of San Francisco International Airport (*A*), National Oceanic and Atmospheric Administration (*B*), and National Park Service (*D*).

bathymetric chart are visible. Qualitative point features are usually represented by point symbols that are somewhere on a continuum between geometric and pictographic, as discussed in chapter 6. The maps in figure 8.7C illustrate how symbols for the tree data can vary from mimetic (*left*) to more pictographic (*right*).

A special type of point symbol is a **standard symbol**, which is used as a standard within a mapping company or agency or for a particular mapping product. For example, the US **National Park Service (NPS)**, which manages all national parks and other natural, historical, and recreational properties, uses its standard symbols to show transportation, recreation, and other activities and amenities on its maps. Examples are shown in figure 8.7D and figure 8.8D. Standard symbols often have a more professional look, but they suffer the same problems associated with

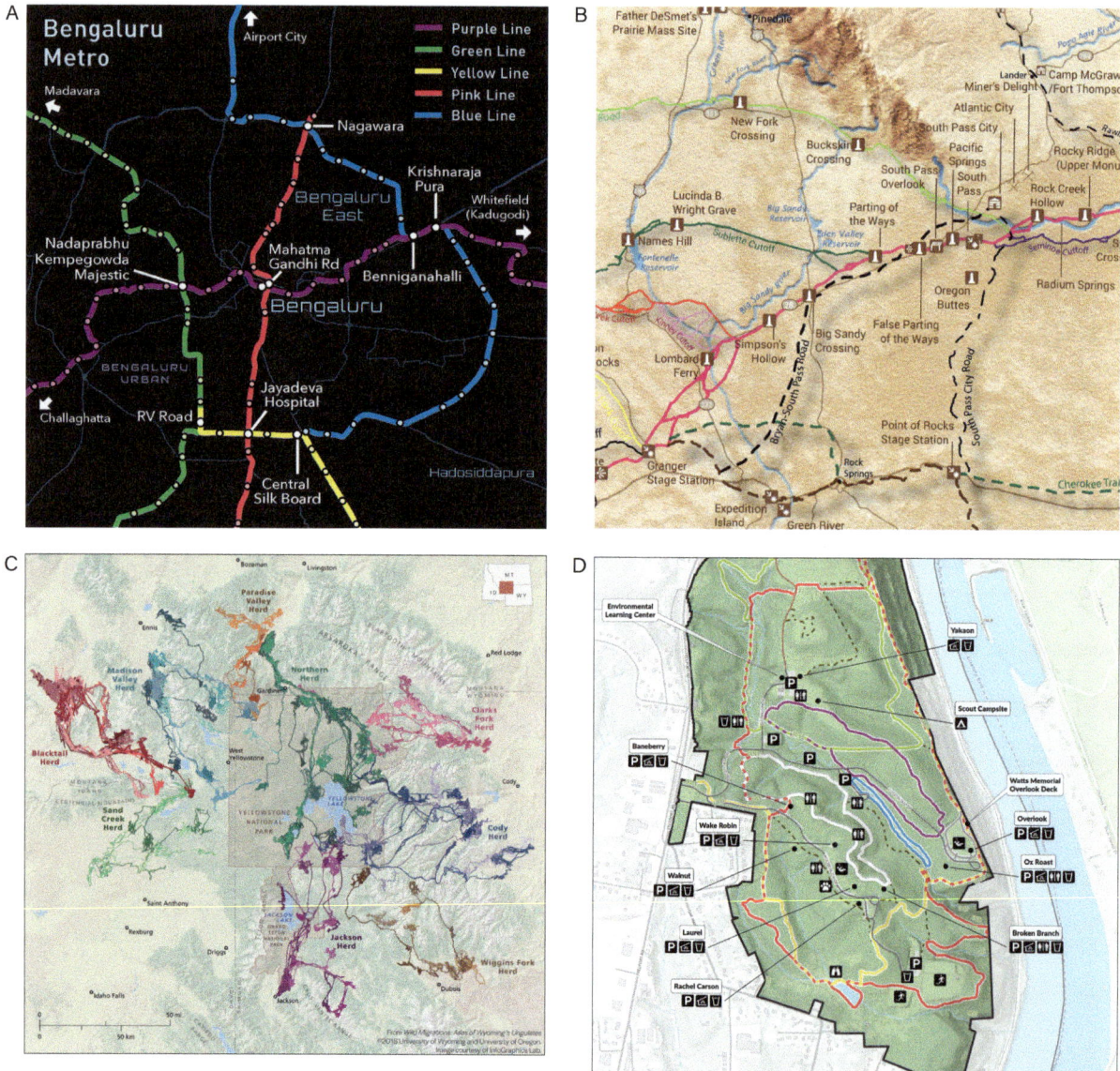

Figure 8.8. Qualitative data for lines can be shown with hue (*A*) or shape (*B*). Variations in a single hue can be used to show subsets within a category. Hue and shape can be used together, for example, to show overlapping types of features (*D*). Courtesy of the Bureau of Land Management, Wyoming State Office (*B*), from *Wild Migrations: Atlas of Wyoming's Ungulates* © 2018 University of Oregon and University of Wyoming, image courtesy of InfoGraphics Lab (*C*), and courtesy of Allegheny County, Pennsylvania (*D*).

pictographic symbols in general. They must be relatively large for details within the symbols to be legible in crowded spaces or at small sizes. Because readers can become confused if there are too many symbols or frustrated if they have to constantly refer to the legend, only a limited number of symbols can be used effectively. Although these symbols are intended to be intuitive at a glance, you will often have to check the map legend to determine what is being symbolized.

The drawbacks of more complex symbols for qualitative point features are overcome by using simple geometric shapes, such as circles, squares, triangles, and so on. Although these symbols are abstract, they can be correctly distinguished by the map user, even when small. They can also be used for categories with larger numbers of feature types. Their small size allows mapmakers to pack more information onto the map than they can with larger, more complex point symbols. Also, since the correspondence between real-world features and their geometric symbols is strictly arbitrary, any point feature can be represented this way. The greater level of abstraction embodied in geometric symbols increases their flexibility as symbols, but it also requires the reader to study the legend to understand what the symbols represent.

Table 8.1. Transportation symbols

Railroad features	
Railroad	┼┼┼┼┼┼┼┼┼┼
Road features	
Closed road	‒ ‒ ‒ ‒ ‒ ‒
Expressway	▬▬▬▬▬▬
Ferry	‒ ‒ ‒ ‒ ‒ ‒
Four-wheel drive	‒ ‒ ‒ ‒ ‒ ‒
Local connector	───────
Local road	───────
Ramp	───────
Secondary highway	───────
Tunnel	)‒ ‒ ‒ ‒(
Trails	
Snow	·‒·‒·‒·‒
Terra	·‒·‒·‒·‒
Water	·‒·‒·‒·‒

Source: US Geological Survey.

Line feature maps

Line feature maps show linear features and their characteristics. Examples of qualitative line feature maps are shown in figure 8.8. Hue is used in figure 8.8A used to symbolize the five metro lines in Bengaluru, India, and in figure 8.8C to show the paths of migration for different elk herds in Yellowstone National Park. On this map, each herd is shown with a different hue, and herds that migrate in similar areas have similar hues. On the map in figure 8.8B, hue and shape (dashed lines) show trails of historical significance in Wyoming. The map in figure 8.8D, which shows trails in a county park, uses arrangement to show where trails overlap. To read these kinds of maps correctly, you often must refer to the legend, unless the line types are labeled on the map.

As noted in chapter 6, shapes can be repeated along lines to create line patterns, which may have dashes, dots, hatches, and other shapes. For example, the common symbol for a railroad is a solid line with crosshatches that mimic the railroad ties (table 8.1). Using hue, line pattern, and size together allows cartographers to show a greater variety of lines. This is illustrated in table 8.1, which shows some of the transportation symbols used on USGS topographic maps.

Area feature maps

Area feature maps show the extent and characteristics of areas. The map in figure 8.9A uses hue to differentiate the second-most prevalent race or ethnic group in each US county, and in figure 8.9B, hue is used to shows lakes, forests, and urban areas on the Azerbaijan Peninsula. The nautical chart in figure 8.9C illustrates the use of area patterns to show lava flows (the elongated polygons), sandhills (shown with a stipple pattern), and steep terrain shown with a linear pattern in the steepest downslope direction. In figure 8.9D, ecoregions are shown with different hues, and the subregions are shown with variations of the hues.

The maps in figure 8.9A and 8.9D can be used to illustrate an important difference in the types of data collection units shown. The census data map (A) is based on federally defined data collection areas (counties). The areas are correctly portrayed as homogeneous because the calculation to determine the predominant race or ethnicity group was computed for the entire area. This type of map is often called a **categorical map**—a map with polygons that enclose areas assumed to be uniform or areas for which a single description can apply.

The ecoregion map (D) is a map of area features that

Race or ethnicity group

- White alone, non-Hispanic
- Black or African American alone, non-Hispanic
- American Indian and Alaska Native alone, non-Hispanic
- Asian alone, non-Hispanic
- Two or More Races, non-Hispanic
- Hispanic or Latino, of any race
- No single group was the second-most prevalent

Figure 8.9. Qualitative data for areas is often shown with hue (*A*) and (*B*). Using shape generally results in different types of area patterns, as in (*C*). As with lines, variations in hue can be used to show subsets within categories. Courtesy of the US Census Bureau (*A*), Azerbaijan National Aerospace Agency – Institute for Space Research and Natural Resources (*B*), National Oceanic and Atmospheric Administration (*C*), and US Environmental Protection Agency (*D*).

A
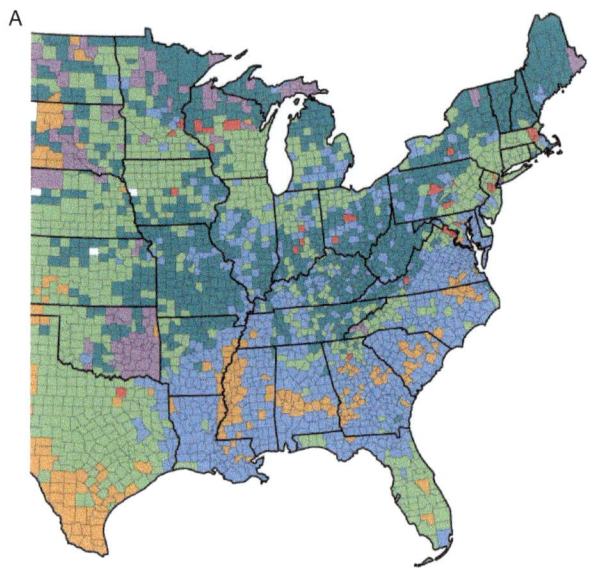

B
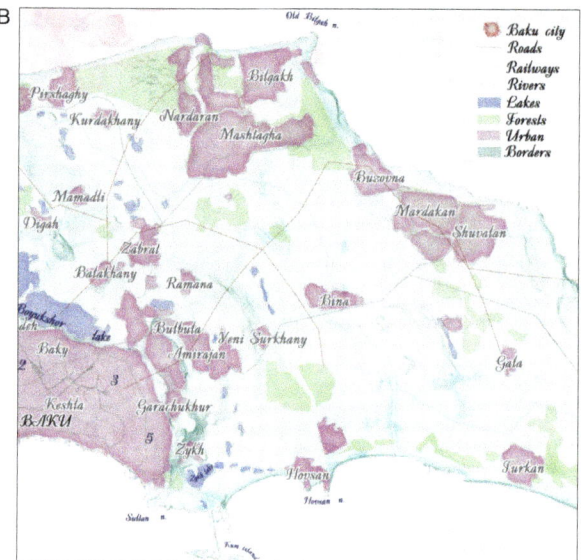

C

D
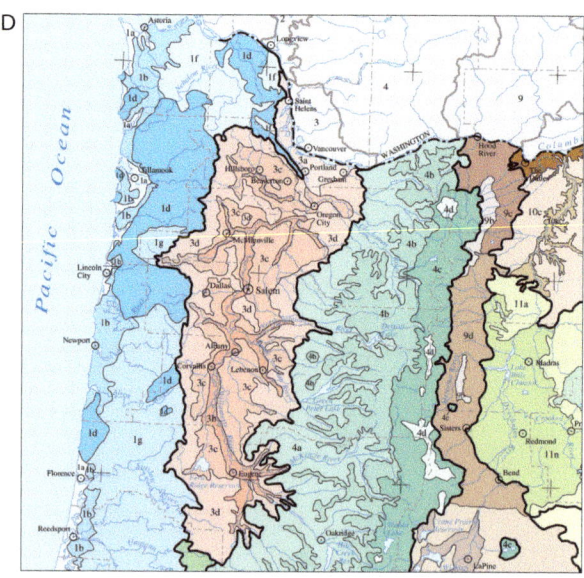

are inherently homogeneous in some way, such as having the same ecological characteristics. This map is compiled by letting the environmental data determine the boundary between categories—this differs markedly from a boundary that is compiled with no natural relation to the data.

Multivariate maps

Most qualitative thematic maps show a single attribute of a feature, but sometimes mapmakers show several feature attributes on a multivariate map (see chapter 6). Several methods can be used to symbolize qualitative multivariate information on maps. In this section, we look at such methods for point, line, and area features.

Multivariate points

In theory, mapmakers could show four different qualitative attributes at once by varying a point symbol's shape, hue, orientation, and arrangement (see table 6.3). In practice, point symbols that show two or three attributes may be difficult to understand even if they are carefully created. This difficulty is illustrated on the map in figure 8.10, which shows the age and height of lighthouses along the Eastern Seaboard, as well as their service status (retired or active). You will usually have little trouble telling when multivariate information is mapped, because the symbols appear more complex than those that show a single attribute. Thus, you will need to spend some time studying the legend to understand the symbols.

Multivariate lines

You will not likely find many examples of maps that show multivariate qualitative line features. One reason is that more complex line symbols are thicker, thus taking up more space on the map and potentially appearing as areas. The map in figure 8.11, which shows the advance

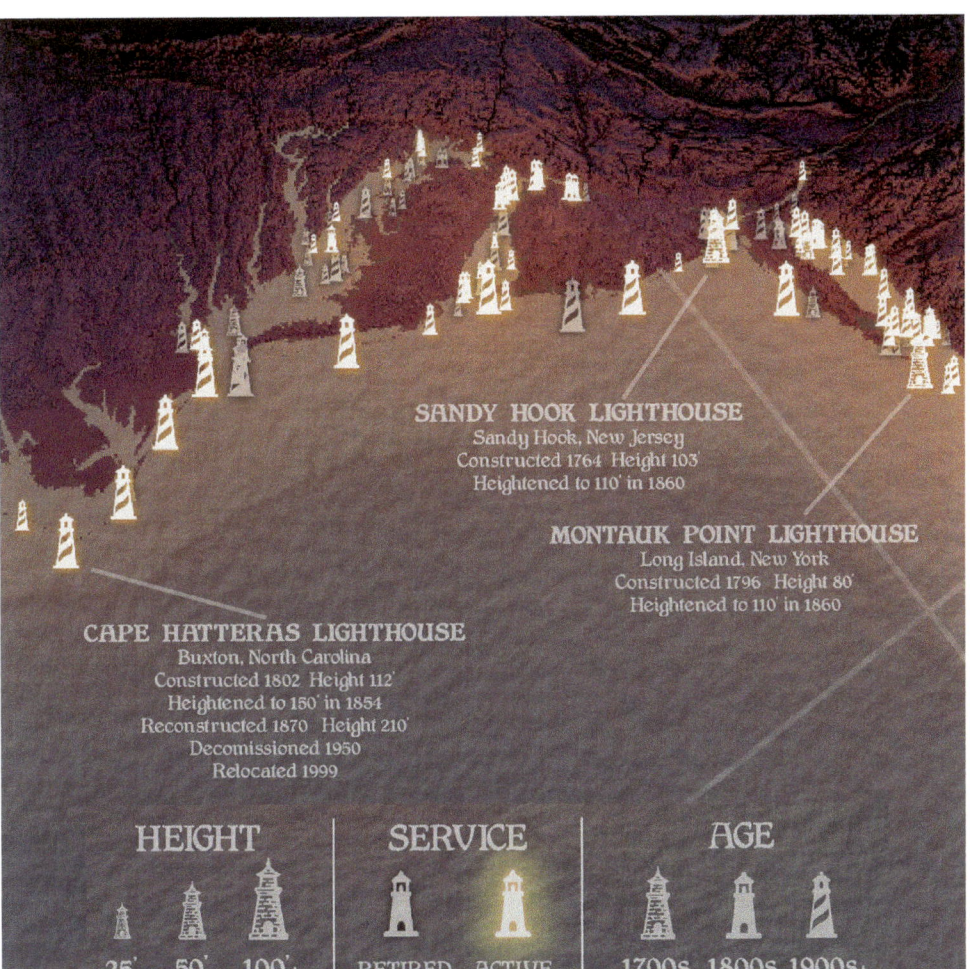

Figure 8.10. Multivariate point symbols are used to show three characteristics of lighthouses along the Atlantic Coast. Courtesy of Mike Huff and Ryan Huff, Burgess & Niple Inc.

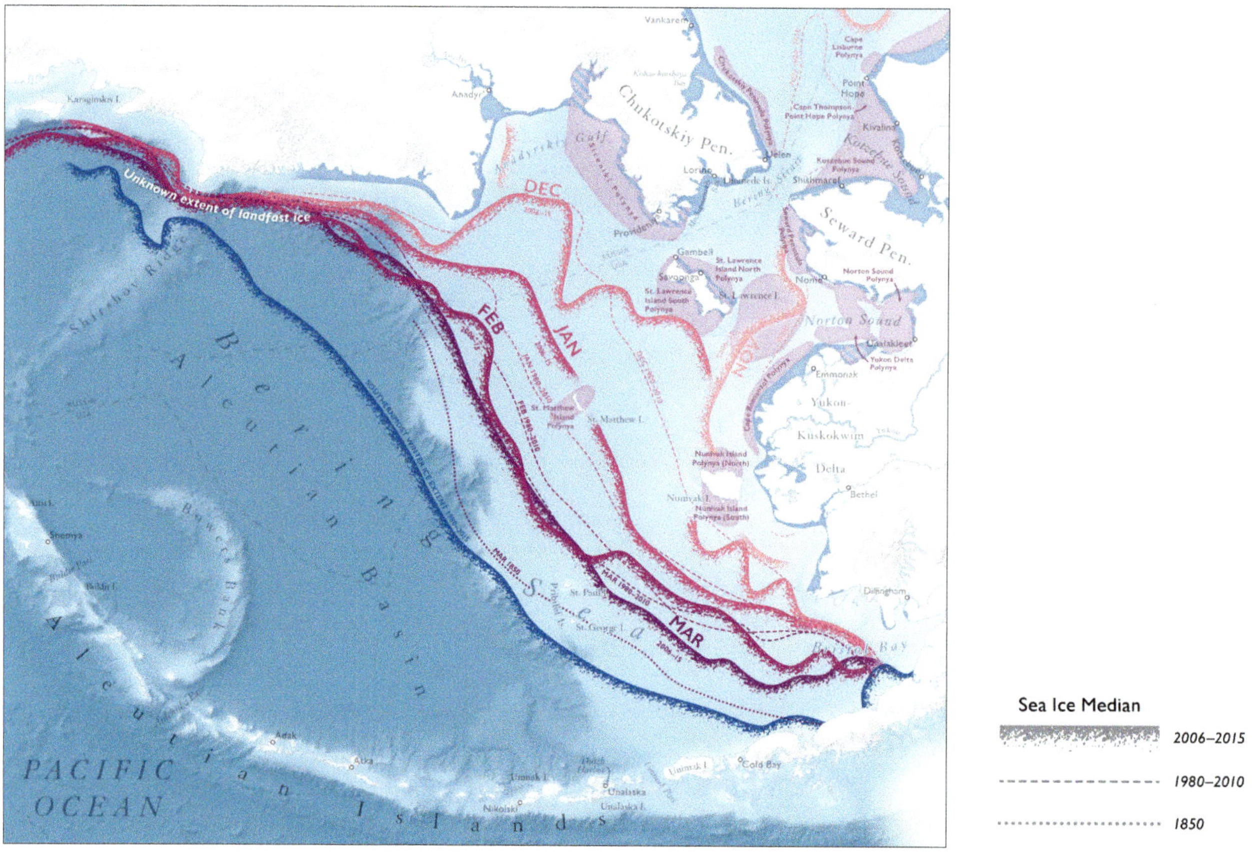

Figure 8.11. Multivariate line symbols can be used to show several attributes of the leading edge of sea ice. From *Ecological Atlas of the Bering, Chukchi, and Beaufort Seas*, © 2017 Audubon Alaska.

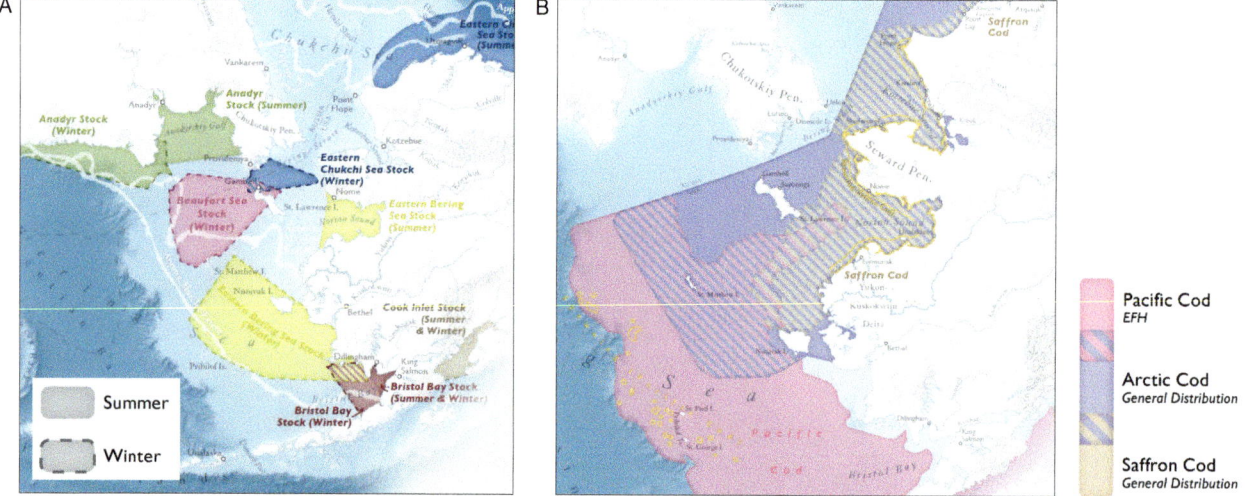

Figure 8.12. Multivariate area symbols are used to show two or more attributes on these maps of the summer and winter ranges of five stocks of beluga whales (*A*) and the general distribution of three species of North Pacific cod (*B*). From *Ecological Atlas of the Bering, Chukchi, and Beaufort Seas*, © 2017 Audubon Alaska.

of sea ice margins in the Arctic waters between Russia and the United States, illustrates how one clever cartographer addressed this challenge . The thick lines are solid toward the north of the map and stippled toward the south, symbolizing the southerly advance of sea ice from 2006 to 2015. Dashed lines show the maximum extent for the 1980 to 2010 period, and dotted lines show the 1850 margin. The pink hues relate to the months during which the sea ice advances, and the blue line shows the southernmost extent of winter ice from 1980 to 2015.

Multivariate areas

Mapmakers can make multivariate maps for qualitative area features in several ways. One common method is to combine two types of area symbols that are appropriate for nominal data. The most common visual variables used are hue for one attribute and area pattern for the other. You will commonly find these combinations on geology maps (see figure I.6C). Other examples are shown in figure 8.12. The first map (*A*) shows the range of five stocks of beluga whales as polygons with different hues. The polygon outlines differentiate the range of the summer stocks (shown with a solid line) from winter stocks (shown with a dashed line). On the second map (*B*), the cartographer has shown overlapping distributions of three species of Arctic cod by interlacing their symbols. Each species is shown with a different hue, either pink, purple, or yellow. Where two species overlap, alternating lines of the two hues are used to fill the area, and where three species overlap, a line pattern with all three hues fills the area.

Mapping change

Portraying changes in the environment has long challenged mapmakers. In the past, changing phenomena were not attractive candidates for mapmaking because the data was often difficult to find and the maps were more time consuming to create. Adding change to a map also makes it more difficult for the map reader to understand. But recent technological and graphic advancements have allowed mapmakers to better overcome these limitations, so you are now likely to find more **qualitative change maps** than in the past. The two types of qualitative change maps are those that show change in the location of qualitative features and those that show change in the nominal-level attributes of features.

Change in location

Change in geographic location is related to **movement**— the positional change of a feature from one point in space to another. Here, we look at how change in location can be shown for point, line, and area features.

Point feature locations

The changing positions of point features can be mapped by superimposing the points for each of the times on a single map. This is illustrated on the map of the changing position of the mean center of population in figure 8.2 and the map of migrating whales in figure 8.18. A map with superimposed features is particularly useful for tracking the movement or evolution of features to understand how they have shifted or progressed over time.

Line feature locations

As with points, the movement or change in the location of linear features, such as rivers, roads, or boundaries, can be shown by superimposing the line features for different times on one map. This is illustrated on the map of sea ice advance in figure 8.11. On this map, the features have little overlap, so each period is easily distinguishable. When there is more overlap, the mapmaker must use clearly distinguishable symbols so that each discrete period can be seen, thus allowing the reader to accurately study the historical pattern of change on the map. The maps in figure 8.13 use highly saturated colors to show changes in the meandering main Mississippi River channel from 5000 YBP to AD 1800.

Sometimes the movement of features is mapped as a **route line** that indicates a path or trail on a map. Significant events or locations along the route, such as stopping points, can be symbolized and annotated to explain the path taken and notable places along it. The map in figure 8.14 shows the routes taken by the expedition led by geologist Ferdinand Vandeveer Hayden in 1871 (red) and 1872 (gray) to survey the region of northwestern Wyoming that became Yellowstone National Park in 1872. The route is annotated with dates and directions of their travels.

Area feature locations

Change in the location of qualitative area features can be mapped as superimposed features, as with points and lines. This works well if there are not many features and if the features on top do not obscure the features on the bottom. The map in figure 8.15 illustrates how this method was

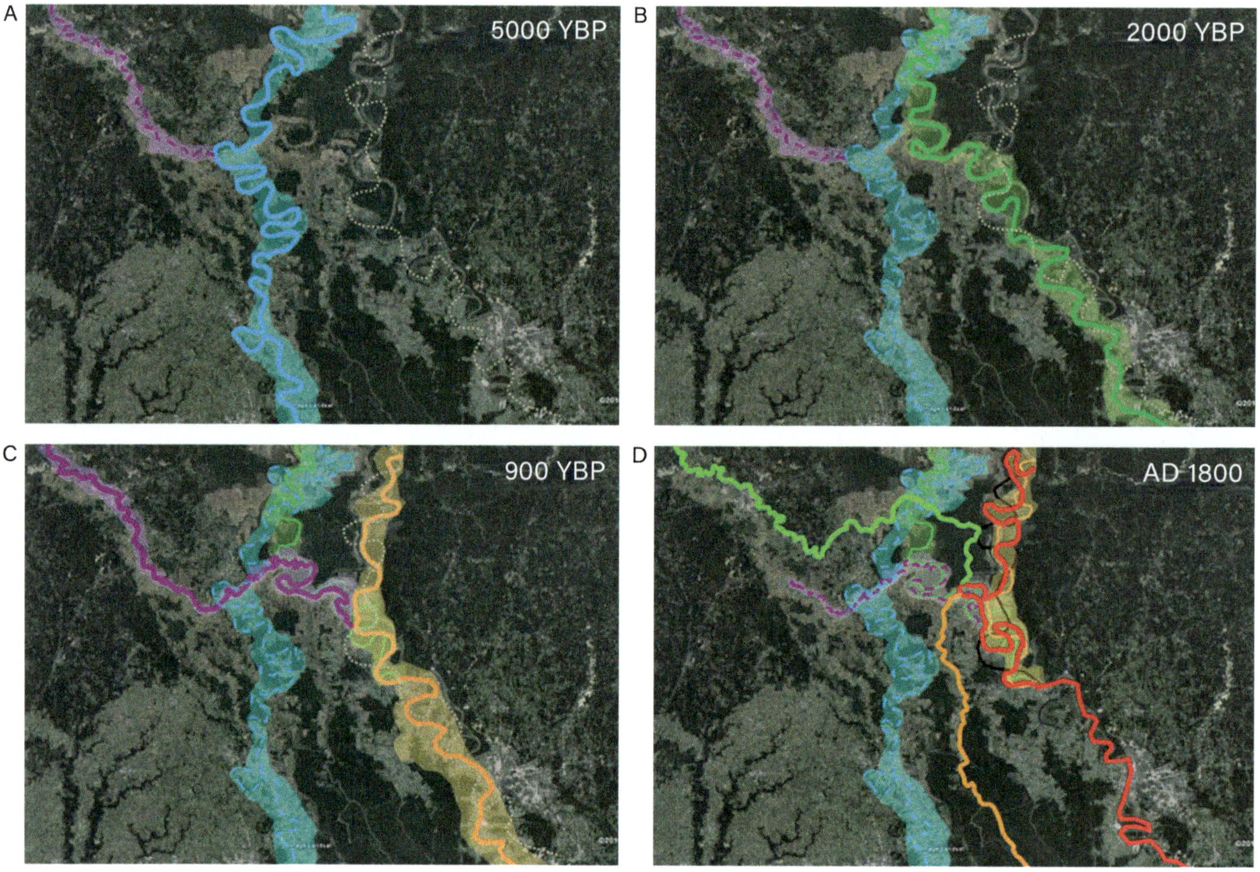

Figure 8.13. Changes in the location of the main river channel of the Mississippi River are illustrated in these maps spanning nearly 7,000 years. Courtesy of Dr. Andres Aslan, Colorado Mesa University.

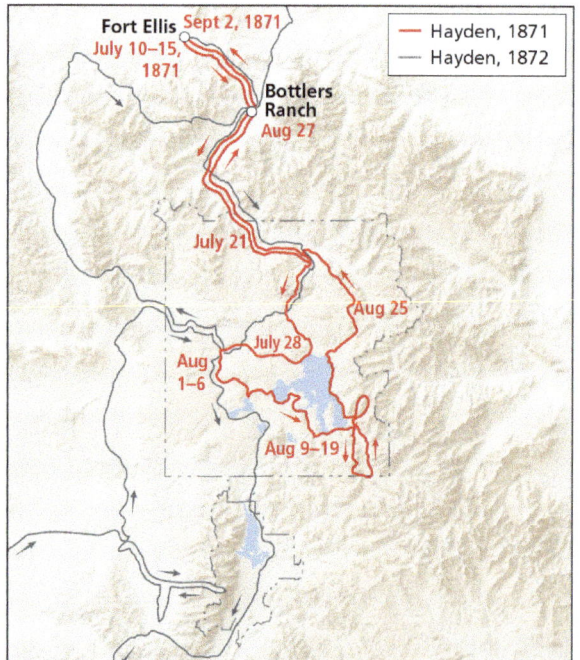

Figure 8.14. This map shows the routes taken by the Hayden expedition through northwestern Wyoming as they surveyed the region to help chart the western United States. From *Atlas of Yellowstone*, second edition, © 2022 University of Oregon. Image courtesy of InfoGraphics Lab.

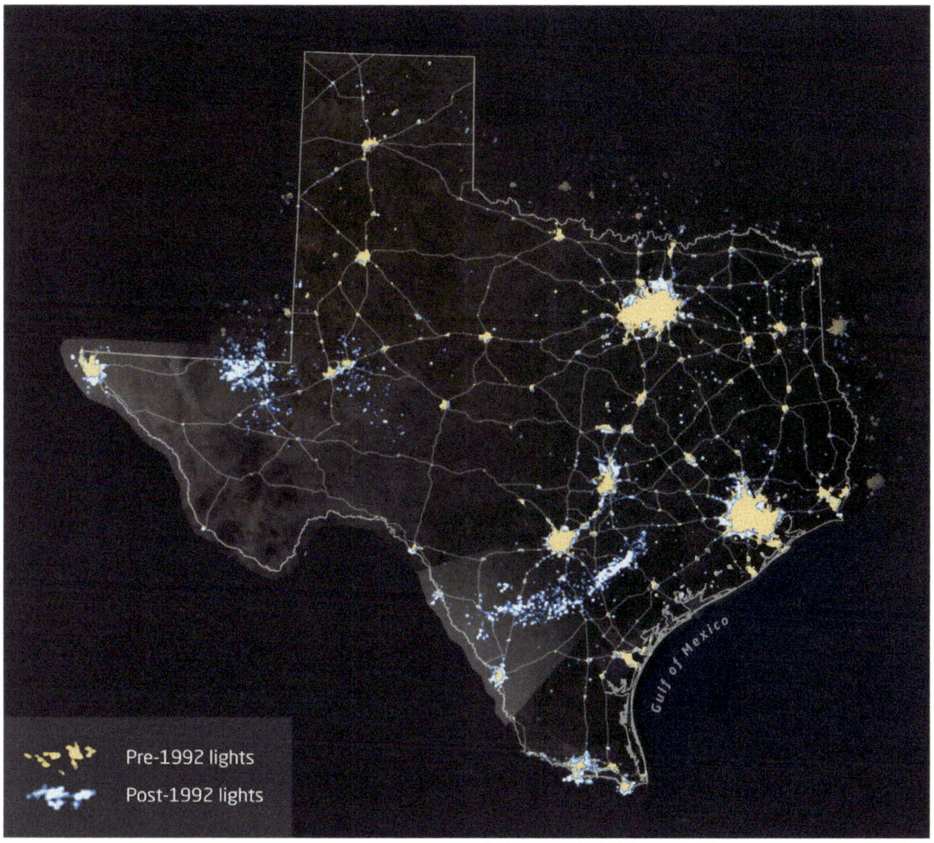

Figure 8.15. This map of Texas shows change in the extent of radiant lights from 1992 (yellow) to 2016 (blue). Map courtesy of the Texas A&M Natural Resources Institute.

used to show change in the extent of nighttime illumination from sources such as streetlights, industrial sites, or gas flares. By simply superimposing the data from two time periods, the map depicts the dramatic increase of light sources between 1992 and 2016.

Change in attributes

We now turn our attention to the second type of qualitative-change maps—those that show change in the attributes of features.

Change maps

A **change map** shows only those locations in which attributes have changed over time. A good example is the map in figure 8.16, which shows the change from one International Residential Code (IRC) to another across the conterminous United States. The IRC is a building code that specifies structural design requirements for family dwellings and townhouses based on their occupancy and the expected ground shaking from earthquakes. The IRC categories, A though F, have subcategories identified by numbers. For example, the legend for this map shows that category D is subcategorized into D0, D1, and D2. In 2024, the 2018 maps were updated using the most recent National Seismic Hazard Model for the United States. The areas where the codes have changed are shown on the map. Light red symbolizes where the category shifted to the following letter (for example, from A to B), and light blue shows shifts to the preceding letter. Darker red and blue areas show shifts of two letters.

It may seem that the data for this map is quantitative, with a code of B meaning something "more" than A. The categories for these codes are defined by a number of variables, some of which are quantitative (such as building height), whereas others are not (such as construction type), but, in essence, the building codes are categorical. Be sure to look for such subtle nuances when reading change maps to ensure that you understand the map correctly.

Small multiples

Rather than mapping change on a single map, mapmakers can use multiple maps. This mapping method is referred to as **small multiples**, in which multiple maps with the same basemap design and map scale are overlaid with the

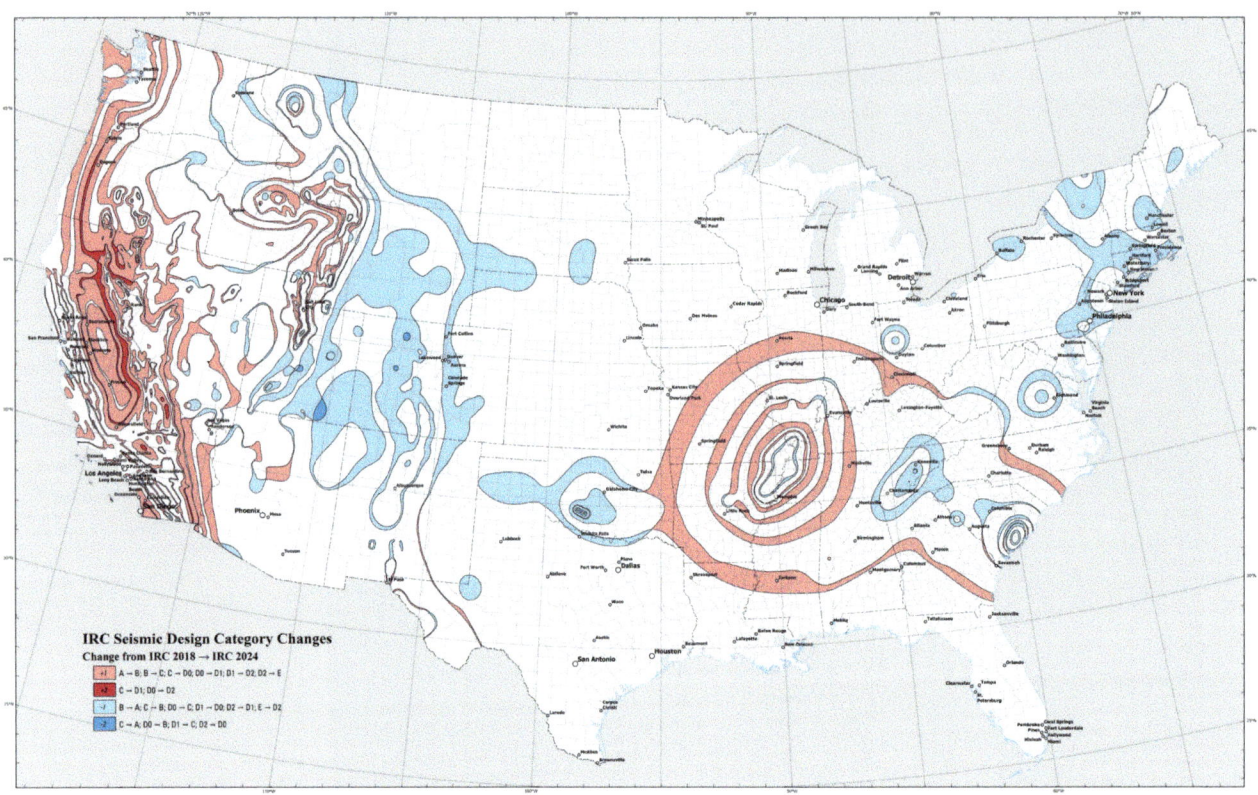

Figure 8.16. This map shows where the building codes based on occupancy and the expected ground shaking from earthquakes have changed from 2018 to 2024. Courtesy of the US Geological Survey.

changing data. With this method, the consistent design assures that the reader's attention is directed toward changes on the maps. This method can be used to map changes in location or attributes, and it can be used for point, line, or area features. It can also be used for quantitative data, which you will see in chapter 9.

The maps in figure 8.17 show the expansion of warehouse space from 1972 (*A*) to 2021 (*C*) in the Inland Empire, a region east of Los Angeles, California. Parcels with warehouses at each time period are shown with different hues. It is easy to see the pattern of growth—what began as primarily a couple of World War II military logistical sites has expanded to millions of square feet of warehouse space.

To read small multiple maps correctly, you must first be able to understand the basemap, then you need to be able to decipher the thematic data, and finally, you must be able to compare the maps to see the changes. The opportunity for misinterpretation increases with each step in the map reading process and every map added to the series.

You can imagine that the maps in small multiples are like frames in an animation. Indeed, the maps in figure 8.18 are from a larger set used to create an **animated GIF**—an image file format that contains multiple frames displayed in a sequence to create the illusion of movement or animation. This brings us to the final topic in this chapter—dynamic qualitative maps.

Dynamic maps

One disadvantage to all the change-related mapping approaches we have discussed is that the maps are static, whereas the phenomena they are designed to show are dynamic. A better solution might be to display the phenomena dynamically so that you can intuitively understand changes in the location or attributes of features. The map in figure 8.18 is an excellent example. The full-size map is animated to show blue whale migration data for each month of the year. The small multiples (*at right*) show six of the monthly frames for a smaller map area (outlined in white on the main map).

Animation gives life and visual appeal to our maps. It

Chapter 8: Qualitative thematic maps   199

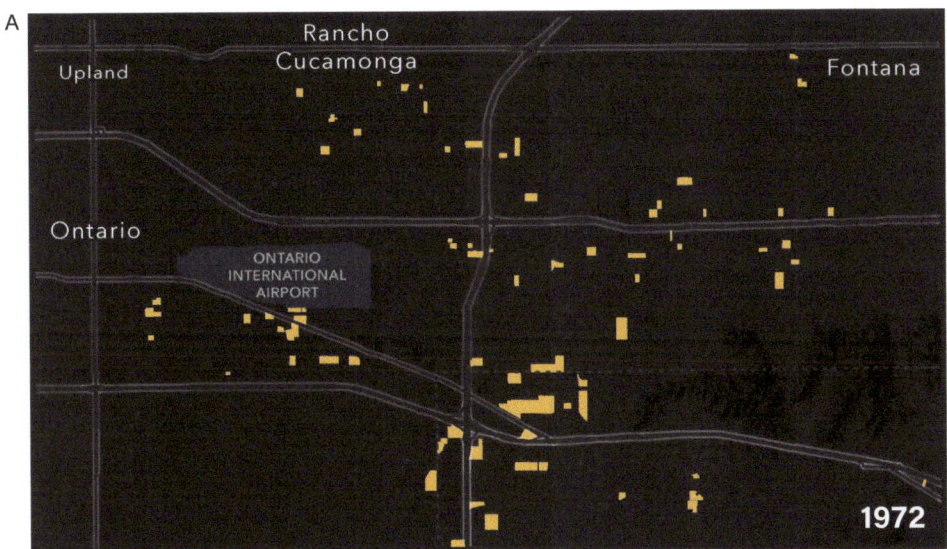

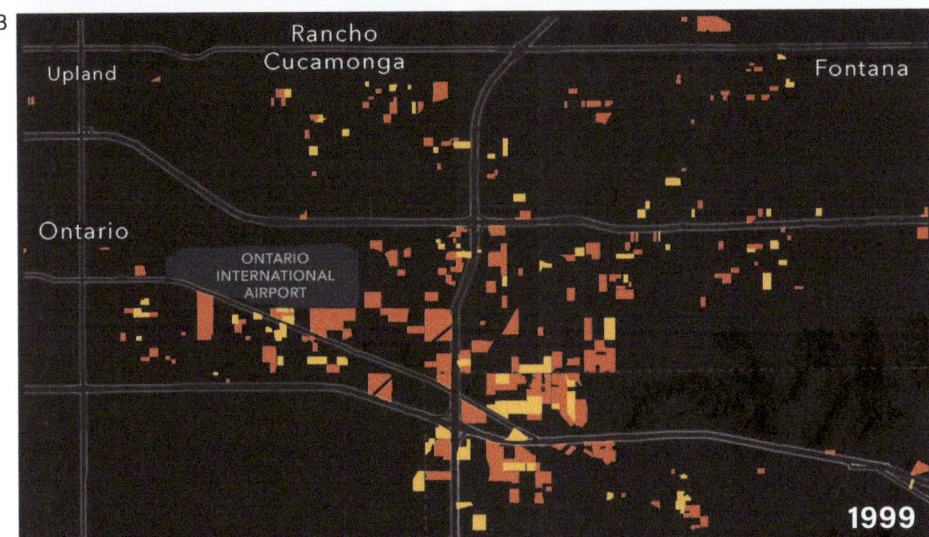

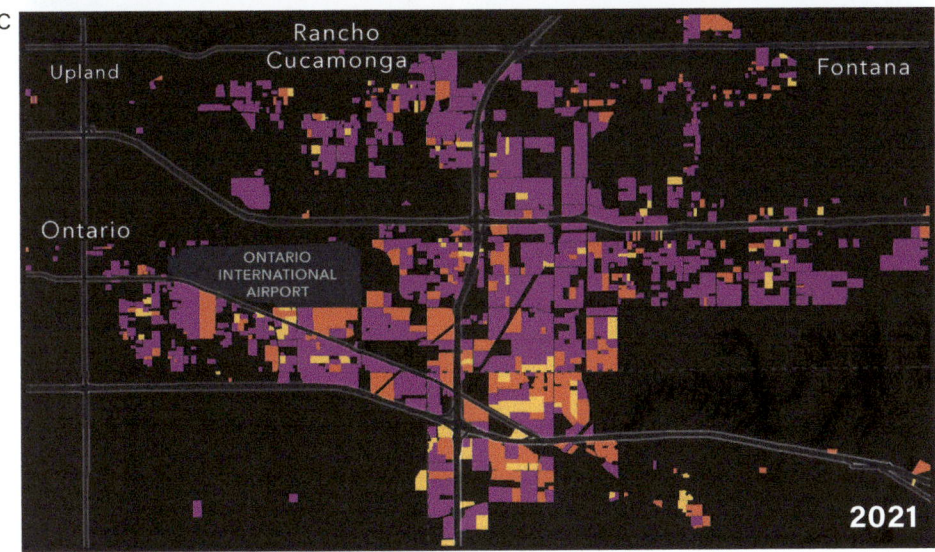

Figure 8.17. These small multiples show the expanding area covered by warehouses in Southern California east of Los Angeles.

Figure 8.18. The full-size map is animated to show the migration routes of northern Pacific blue whales for each month of the year. A smaller area of the full-size map is shown in the small multiples (*at right*) on which six of the twelve months of data are displayed.

can also help us better understand the dynamic nature of what has been mapped—think of a weather map that has clouds with rain and lightning strikes for thunderstorms or rotating **cyclone symbols** (circles with lines extending in the direction of movement) for advancing hurricanes. Nonetheless, you may find dynamic maps challenging to read, especially if individual scenes are not displayed slowly enough or if they are not repeated often enough. The optimal situation is for you to have interactive control over the display, which is also becoming more common.

Selected readings

Abdurakhmonov, S., K. Bekanov, S. Ochilov, et al. 2023. "Advances in Cartography: A Review on Employed Methods." *E3S Web of Conferences* 389: 1–8.

Aslan, A., W. Autin, and M. Blum. 2005. "Causes of River Avulsion: Insights from the Late Holocene Avulsion History of the Mississippi River, USA." *Journal of Sedimentary Research* 75 (4). https://www.researchgate.net/publication/229021598_Causes_of_River_Avulsion_Insights_from_the_Late_Holocene_Avulsion_History_of_the_Mississippi_River_USA.

Brehmer, M., B. Lee, P. Isenberg, et al. 2020. "A Comparative Evaluation of Animation and Small Multiples for Trend Visualization on Mobile Phones." *IEEE Transactions on Visualization and Computer Graphics* 26 (1): 364–74.

Calvo, L., F. Cucchietti, M. Pérez-Montoro, et al. 2023. "Measuring the Effectiveness of Static Maps to Communicate Changes over Time." *IEEE Transactions on Visualization and Computer Graphics* 29 (10): 4243–55.

Gil, T. L. 2021. "Taking Speed Seriously: Motion, Simultaneity, and Context in Mapmaking for Historical Analysis." *Cartography and Geographic Information Science* 48 (4): 320–37.

Klasen, V., E. P. Bogucka, L. Meng, et al. 2023. "How We See Time: The Evolution and Current State of Visualizations of Temporal Data." *International Journal of Cartography* 9 (2): 392–409.

Nelson, J. 2023. "How to Make This Animated Map of Blue Whale Migration." *ArcGIS Blog*, August 22, 2023. Accessed at https://www.esri.com/arcgis-blog/products/arcgis-pro/mapping/how-to-make-this-animated-map-of-blue-whale-migration.

Vlad, V., M. Toti, S. Dumitru, et al. 2021. "Developing Reliably Distinguishable Color Schemes for Legends of Natural Resource Taxonomy-Based Maps." *Cartography and Geographic Information Science* 48, no. 5: 393–416.

CHAPTER 9
Quantitative thematic maps

In chapter 8, you saw the methods to map qualitative information. In this chapter, we focus on thematic maps for quantitative information—numerical data at the ordinal, interval, or ratio level. First, we look at how quantitative data is collected. We examine the ways that quantitative data is normalized to adjust data values to a common scale and transformed to derive new values. Then, we explore the ways that quantitative themes are mapped for point, line, or area features. We end with an exploration of quantitative thematic maps that show multiple attributes, change, and animation.

Quantitative information

Sometimes you want to know not only what and where things are but also how much of something is at a location. To find out, you can turn to **quantitative thematic maps**, which show **quantitative information** that carries a magnitude message, such as a count or an amount (measurement). **Counts** are the total number of features, and **amounts** or measurements describe the dimensions, capacity, quantity, ranking, or magnitude associated with a feature. Next, we look at how counts and amounts can be collected for point, line, and area features.

As with qualitative data, quantitative data is collected using a variety of methods, such as interviews, field observation, GNSS data collection, surveying, map digitizing, aerial or satellite image interpretation, and GIS analysis. Quantitative data can also be collected using sensors that measure certain numeric characteristics of features. Two examples are earthquake magnitude (figure 9.1A), which is recorded automatically with **seismometers** (sensitive instruments that detect ground motion caused by seismic waves from earthquakes), and streamflow (figure 9.1B), which is measured with **stream gauges** (structures installed adjacent to a stream or river that contain equipment to measure and record the water level and other characteristics). Sensor data collection requires little human intervention, and the measurements taken can immediately be transmitted via satellites, cellular networks, or radio to a central processing system. When data is collected, processed, and made available immediately, it is called real-time data (see the introduction).

Spatial samples

Data can be collected for every feature within a region to obtain what is called a **population count**. Collecting population count data is possible when the number of features is fairly small or the data collection method can easily gather information about the entire population, as with the earthquake data mentioned in the preceding paragraph. Another approach is to collect data for a representative portion of the region or population to identify patterns or trends in the larger dataset—this is called taking a **spatial sample**. An example is the data collected with stream gauges at select locations along streams and rivers. The key difference between a population count and a sample is that the goal with a population count is to identify the characteristics of the population, whereas the goal with a sample is to make inferences about the characteristics of the population from which the sample was drawn.

The idea behind sampling is to use a smaller part of the population to find out what you want to know about the entire population. Samples to estimate what the entire population is like are used for several reasons. The time and cost involved in obtaining a full population count may be prohibitive, so only a sample is feasible. Samples are also used for **continuous phenomena** that exist everywhere and have values that change gradually over space or time. Examples include temperature and precipitation. It is impossible to measure these values constantly at every point on the earth, so a spatial sample of temperatures and

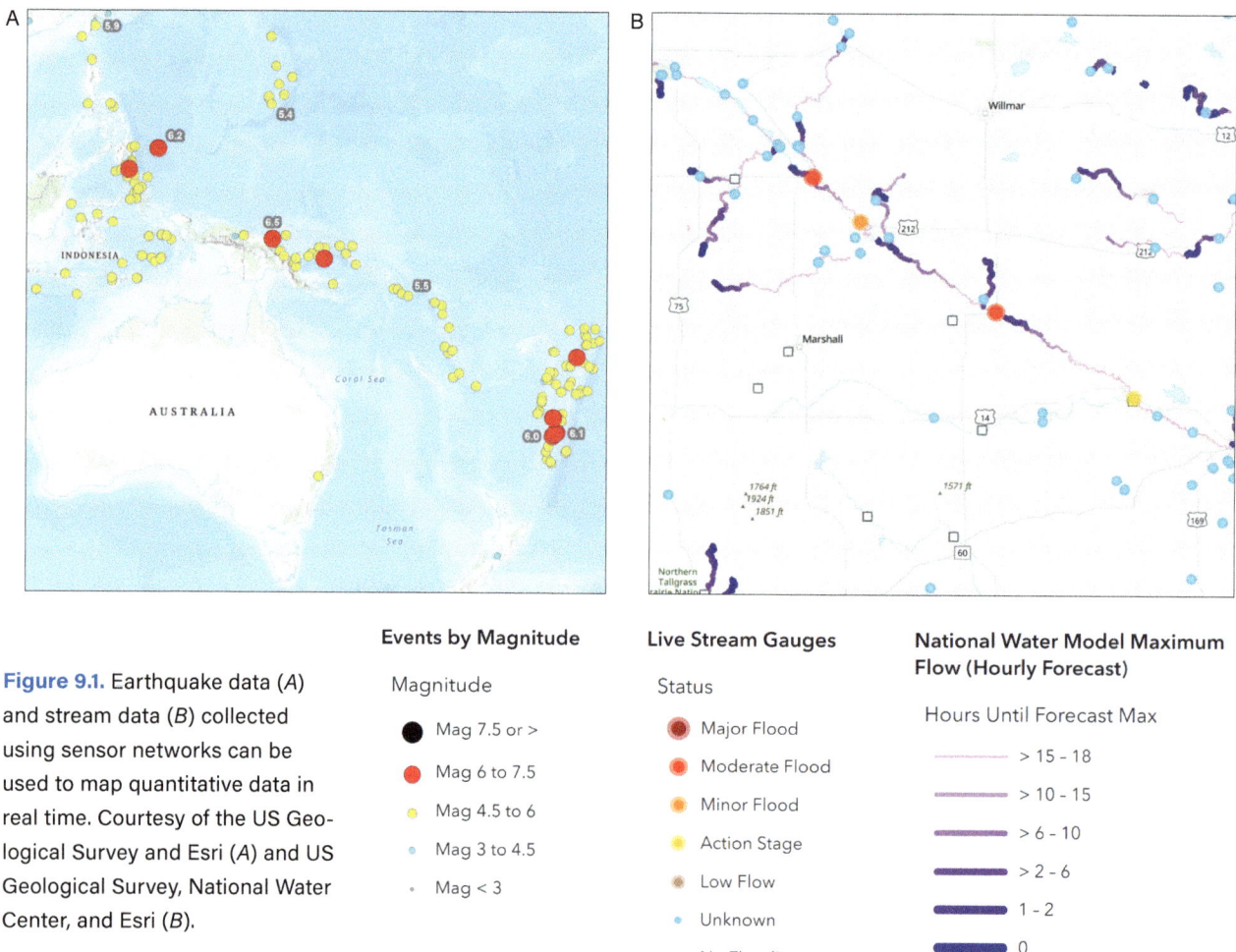

Figure 9.1. Earthquake data (*A*) and stream data (*B*) collected using sensor networks can be used to map quantitative data in real time. Courtesy of the US Geological Survey and Esri (*A*) and US Geological Survey, National Water Center, and Esri (*B*).

rainfall obtained from weather stations can be used to estimate the temperature or precipitation at any location over a selected period of time. Figure 9.2 shows a map of the total rainfall forecast for Florida for the next 72 hours—the estimates were made from data recorded at the many weather stations shown with small black circles.

There are several methods that can be used to gather sample data. To illustrate, we can consider how data for the average temperature of the lake in figure 9.3 may be collected. One approach is to travel by boat to several geographic positions and measure the temperature at each position, called a **sample point**. A second method is for the boat to follow and obtain measurements along **transect lines**, collecting data either continuously or at a constant time or distance interval. A third method is to navigate the boat to sampling areas predefined using **quadrats** (rectangular plots of land or areas of water) to do the sampling.

The arrangement of sample points, transect lines, and quadrats can be random or systematic, as shown in figure 9.3. With **random samples**, the locations are selected randomly so that all locations are given an equal probability of being included in the data sample. In contrast, **systematic samples** are arranged to collect data at regularly spaced distance intervals. With this approach, sampling is often done in a rectangular or triangular grid of sample points, along equally spaced parallel transect lines, or at equally spaced quadrats. In a **stratified sample**, known characteristics of the population's distribution are used to guide data collection. This type of sample is obtained by first subdividing the feature or region into zones, or **strata**, which possess distinctive traits that affect the value of individual samples. Random or systematic samples are then taken to reflect the importance of each stratum. In our lake example, the randomly selected sample points can be stratified by water depth because the shallow and deep portions of the lake are equally important to estimating the average temperature.

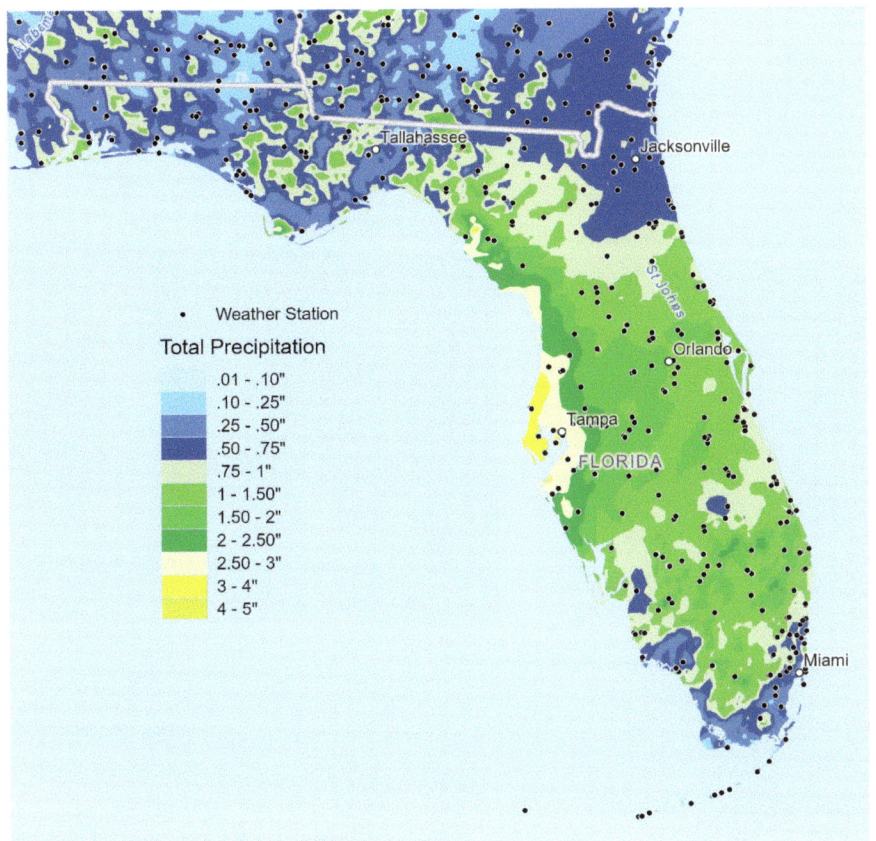

Figure 9.2. This map shows the quantitative precipitation forecast (QPF) for the next 72 hours produced with data updated hourly from the National Digital Forecast Database by the National Weather Service. Data courtesy of the National Weather Service.

The same number of sample points are randomly placed in the deep and shallow sections of the lake, reflecting their equal importance.

Statisticians may argue about which form of sampling gives the best data for estimating a true measurement of the population, but one thing is certain—a higher density of randomly or systematically placed sample points, transect lines, or quadrats should provide a better estimate, particularly if the sample is stratified appropriately.

Next, we look at how count and amount data is collected for point, line, and area features, as well as continuous surfaces.

Point features

Quantitative data for point features can be collected in the same ways that qualitative point feature data is collected; however, the **Internet of Things (IoT)**—the network of interconnected devices and systems that communicate with each other and exchange data over the internet—has enabled additional point data collection methods using, for example, crowdsourcing and location-based services.

Crowdsourcing is a way to collect data by soliciting contributions from a large group of people, typically over

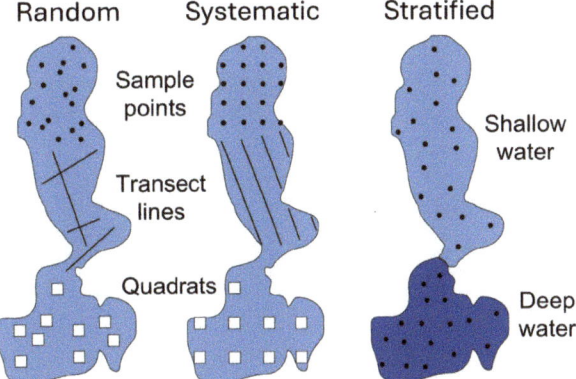

Figure 9.3. Different spatial sampling methods can be used to find the average temperature of a lake. Sampling points, transect lines, and quadrats can be randomly and systematically located across the lake, and a random point sample stratified by water depth can also be taken.

the internet. The map in figure 9.4A shows the locations of plant and animal species reported globally by a social network of naturalists, and the map in figure 9.4B shows the number of flood damage reports by local citizens in the

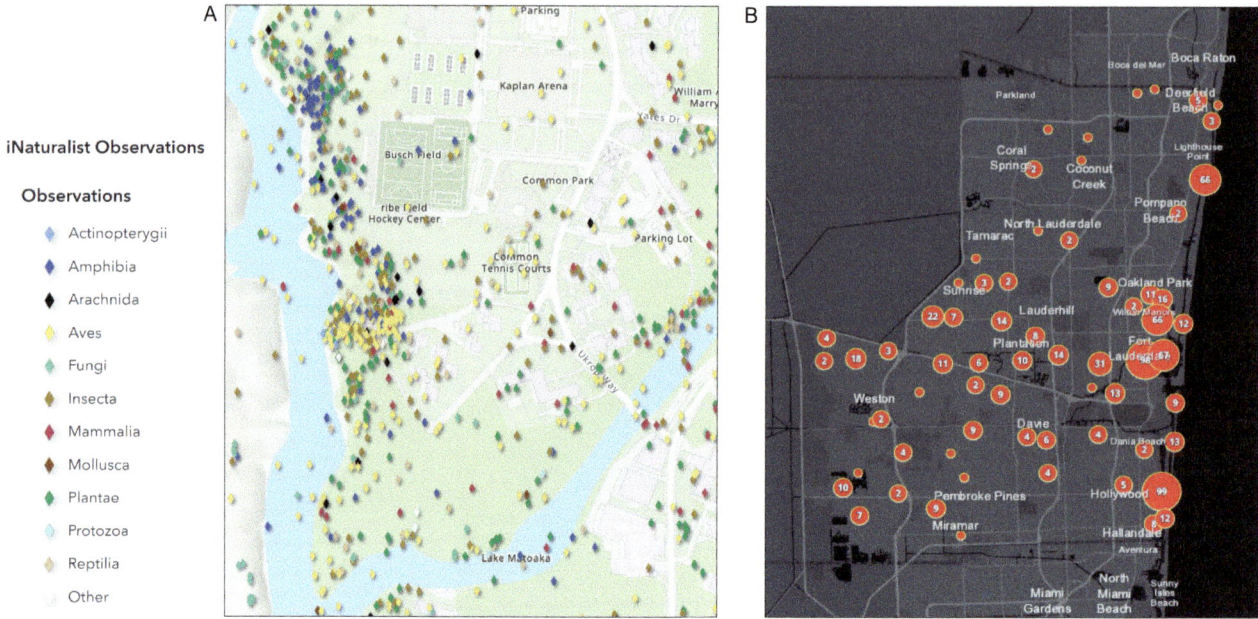

Figure 9.4. The first map (*A*) shows crowdsourced species observations contributed through the iNaturalist social network app and website. The second map (*B*) shows the count of flood damage reports by citizens of Broward County, Florida, in the aftermath of Hurricane Nicole in November 2022. Courtesy of iNaturalist and Esri (*A*) and Broward County, Florida, Resilience Unit (*B*).

aftermath of a hurricane. A **location-based service (LBS)** is also used to collect point feature data, with the point often being a person or, more specifically, a person's GNSS-enabled mobile device. LBSs can be used, for example, in emergencies to locate individuals in distress more quickly and accurately. Of course, the benefits of LBS must be weighed against concerns for personal privacy.

Line features

Often, data for line features is collected at points along the features. A good example is shown in figure 9.1B for stream gauges along rivers and streams. Water levels are used to compute the streamflow at each gauge, and this data is used to model the number of hours until the maximum flow for each **stream reach** (segment). This gives resource managers, disaster responders, and others time to prepare for high flows and potential risks, such as flooding.

Data for other linear features, such as roads, can also be recorded by sensors at points throughout the network. Vehicle counts can be taken from **automatic traffic recorders (ATRs)** at stations permanently installed throughout the road network. ATRs for the Portland, Oregon, area are shown in figure 9.5A, along with traffic flow amounts for segments of the highway network. More commonly, navigation apps, such as the one you probably have on your smartphone, provide real-time navigation and traffic information for your location using LBS data from the smart devices and navigation system in your vehicle and in those around you. Instantaneous processing of the data allows traffic flow and congestion levels to be updated immediately in real time. An example of a map made with this kind of data is shown in figure 9.5B. You can see that the primary visual variables on these maps is hue. You must be careful to note when color hue is used to show quantitative information because the tendency of your eye is to relate this visual variable to categorical, not numerical, data.

Area features

When you are more interested in the spatial distribution of features than their location, a map that shows data for areas or features within areas will better suit your needs. For these types of maps, it is important to understand the areas in which data is collected. In chapter 8, we noted that there are two types of data collection units for areas—those that are defined by natural phenomena and shaped by the environment, and those that are delineated by humans for administrative, political, or other purposes. Both are assumed to be inherently homogeneous in some way (see chapter 8), both are usually irregular in shape and size, and both have units that are hierarchical so that smaller units

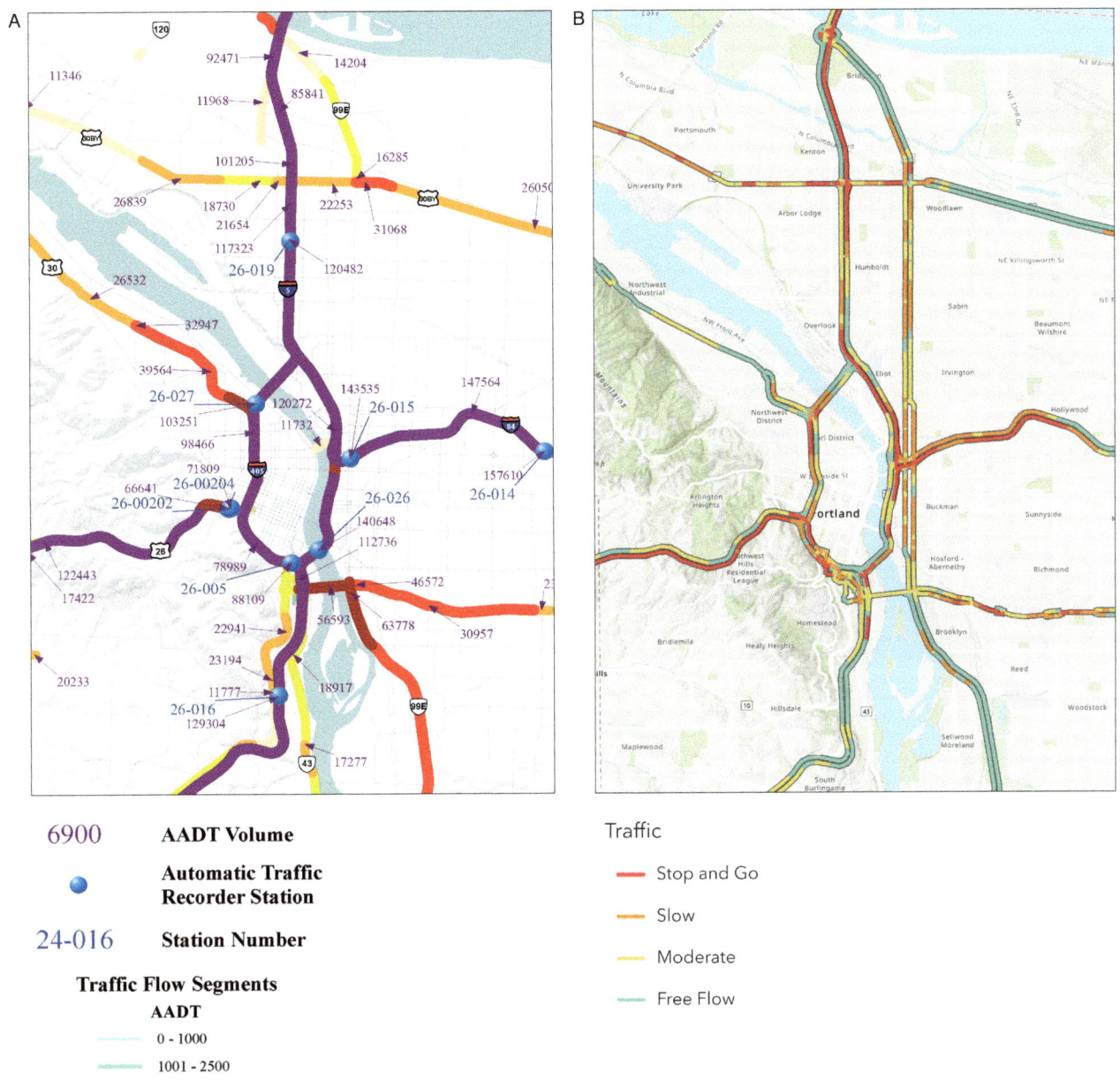

Figure 9.5. Traffic data can be collected using sensors called automatic traffic recorders or using crowdsourced data from the GNSS receivers in vehicles and on smart devices. Courtesy of Oregon Department of Transportation (A) and Esri and TomTom (B).

nest within larger units. In chapter 8, we looked at census enumeration as human-defined data collection units. Here, we examine watersheds as data collection units that are shaped by nature.

Watersheds are drainage basins in which all surface water flows to a common outlet, such as a stream, river, lake, or the ocean. For the United States, the USGS has defined four hierarchical levels of watersheds, which they call **hydrologic units**: regions, subregions, accounting units, and cataloging units. Hydrologic unit boundaries are determined based on topographic, hydrologic, and other relevant landscape characteristics without regard for administrative, political, or jurisdictional boundaries. In figure 9.6, the regional hydrologic units are shown with different hues, and the subregions are outlined in white.

USGS hydrologic units form a standardized system for organizing, collecting, managing, and reporting hydrologic information for the nation. The hydrologic units are used to summarize landscape characteristics and report on water resources, including water availability, water quality, point source or nonpoint source pollution, and more.

Figure 9.6. Except for the Great Basin, which drains internally, the regional-level hydrologic units in the conterminous United States flow directly into or through other watersheds to an ocean, the Gulf of Mexico, or the Hudson Bay. Subregions, outlined in white, nest within the regions. Data courtesy of the US Geological Survey in partnership with US EPA, USDA NRCS, and other state, federal, and local partners.

Continuous surfaces

So far, we have looked at how quantitative data can be collected for point, line, and area features, but such data can also be collected for **continuous surfaces**—surfaces that represent geographic phenomena that do not have clear, defined boundaries, exist over the entire region, and have values that vary gradually. You can think of the elevation of an area, the air quality over a region, and a precipitation forecast as continuous surfaces, because the phenomena exist at every location and their values vary gradually from place to place. This type of map is often made by interpolating data from points across the region of interest, such as the weather stations shown in figure 9.2.

Interpolation

Interpolation is a mathematical method to estimate unknown values that fall between known data points. For mapping, it can be used to create a continuous surface that is often represented graphically with lines that connect points of equal value, called **isolines**. You can consider the boundaries between different colors on the map in figure 9.2 as such boundaries.

The simplest method of interpolation is **linear interpolation** (figure 9.7), which assumes a linear relationship between known points to predict values at intermediate points along a straight line. The first step in linear interpolation is to identify a set of known data points. The range

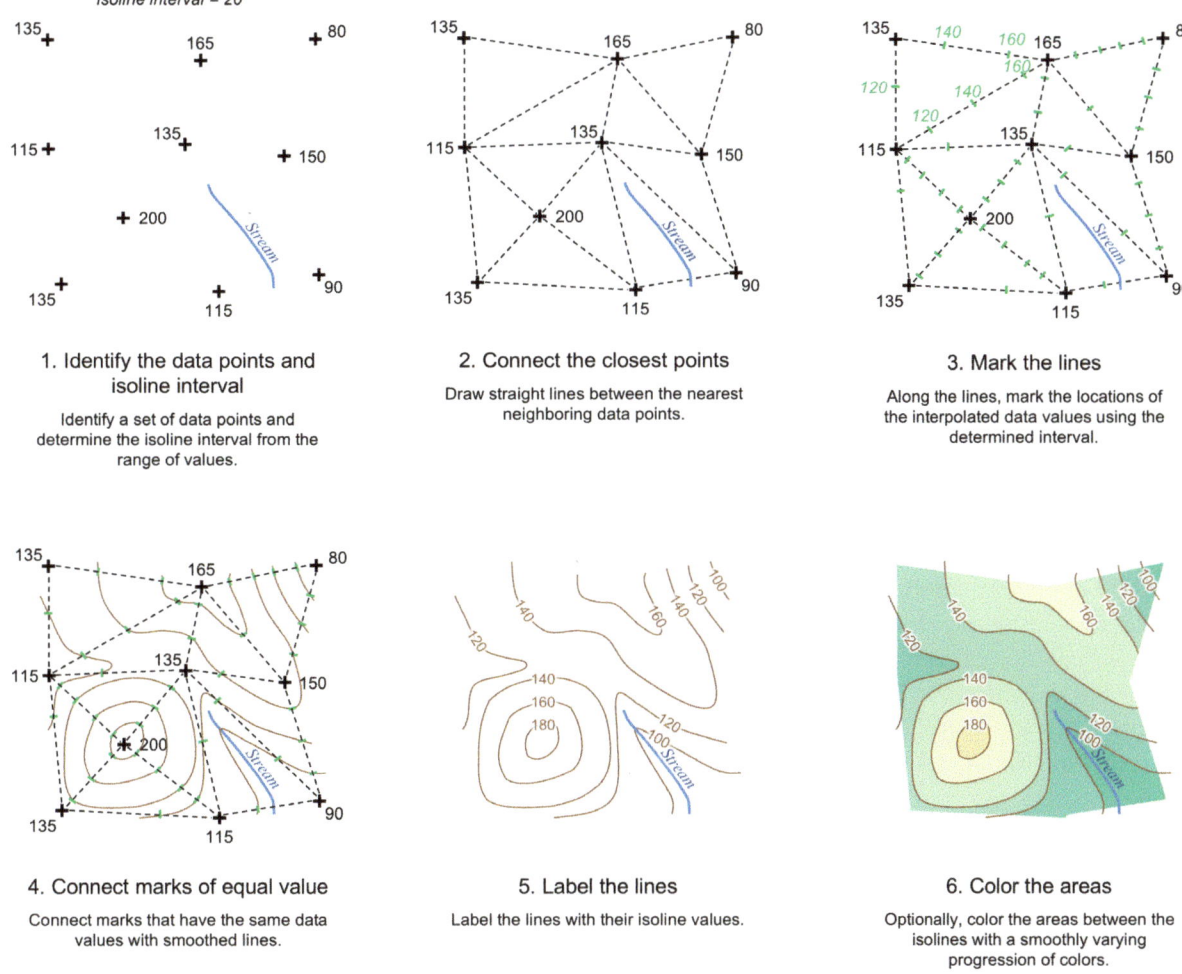

Figure 9.7. Linear interpolation starts with a set of known data points (*1*). The nearest points are then connected with straight lines (*2*). Marks are placed along the lines at the intervals (*3*) so that points with equal values can be connected (*4*). The lines and the areas between them can then be symbolized on maps (*5* and *6*).

of data values for all the points is used to determine the numeric interval of the lines that will connect points of equal value. The next step is to connect the nearest neighboring data points with straight lines. Linear interpolation is used to mark the locations of data values at the chosen interval along the straight lines. Lines are then drawn to connect points with the same values. Because the lines represent a continuous surface, the lines should be smoothed to remove sharp angles. For mapping purposes, the lines can be symbolized and labeled with their values. Optionally, the areas between the lines can be symbolized with a continuous color scheme (chapter 6). Today, linear and other interpolation methods are straightforward with the aid of GIS software, as is mapping the results.

Data normalization

A common mistake made by beginning mapmakers is to neglect normalizing the data, especially when it is population dependent. **Data normalization** standardizes and rescales the data values to a common scale that ensures that the data is consistent and comparable. When population data is not normalized, the resulting map often just reflects the total number of observations within units because where there are more observations, there is generally more of whatever the mapmaker is trying to show (figure 9.8A). This phenomenon is called **population dependence** because the thing of interest will vary depending on the size of the overall population.

A common way to normalize demographic data is to take the size of the data collection unit into account by dividing the count for each unit by its area to obtain a density (figure 9.8B). Other normalization approaches include dividing the count for one unit by the sum of counts for all units to obtain a proportion and obtaining a percentage by multiplying the proportion by 100. GIS can be used to easily normalize data for mapping using these simple, as well as more complex, approaches.

Single-theme maps

Now that we have seen how count and amount data for point, line, and area features, as well as continuous surfaces, can be collected, we will look at how the data can be shown on quantitative thematic maps. Cartographers have developed several common mapping methods for quantitative data. All rely on the use of the quantitative visual variables (size, value, saturation, or pattern texture) to portray variations in the counts or amounts of the features (see chapter 6).

Point feature maps

Cartographic research has shown that size is the most effective visual variable for showing quantitative variations in point features, so graduated and proportional symbols are often on quantitative point feature maps.

Graduated and proportional symbol maps

With **graduated symbols**, the quantitative range of values is classified and all the features within a class are shown with a symbol of the same size. The classes can be either ranked at the ordinal level (for example, low, medium, high) or numeric at the interval or ratio level (such as 1–10, 11–20, 21–30, and so on). With graduated symbols, you cannot tell the value of an individual feature, but you can tell that its value is within a certain range. **Proportional symbols**, on the other hand, are used to represent the unclassified data values by scaling the size of each symbol in exact proportion to its data value, as illustrated in figure 9.9.

The point symbols are usually geometric, such as circles, squares, or triangles. Circles are the most used geometric point symbol because they have the most compact geometric form, they are easy to scale, and they have more visual stability (that is, they cause less eye wandering).

Reading graduated or proportional symbols is not as straightforward as it might seem. The difficulty arises because of the way the human eye and brain work. The brain does not perceive the size of the symbols in proportion to their mathematically computed size. Instead, the size of symbols is progressively underestimated as the area or volume of the symbols increases. For this reason, cartographers sometimes use **appearance compensation**—an algorithm defined by the American cartographer **James Flannery**—for scaling up larger symbols to compensate for the discrepancy in perception. Because Flannery's research applies specifically to the perception of circles, appearance compensation should be used only with circular symbols.

Limiting the symbol sizes to a small number of classes (for graduated symbols) or a small range of values (for proportional symbols) helps map users correctly relate the symbols on the map to the values for the features. The goal of the mapmaker is to use symbols that are different enough in size that the eye can easily tell them apart. Although

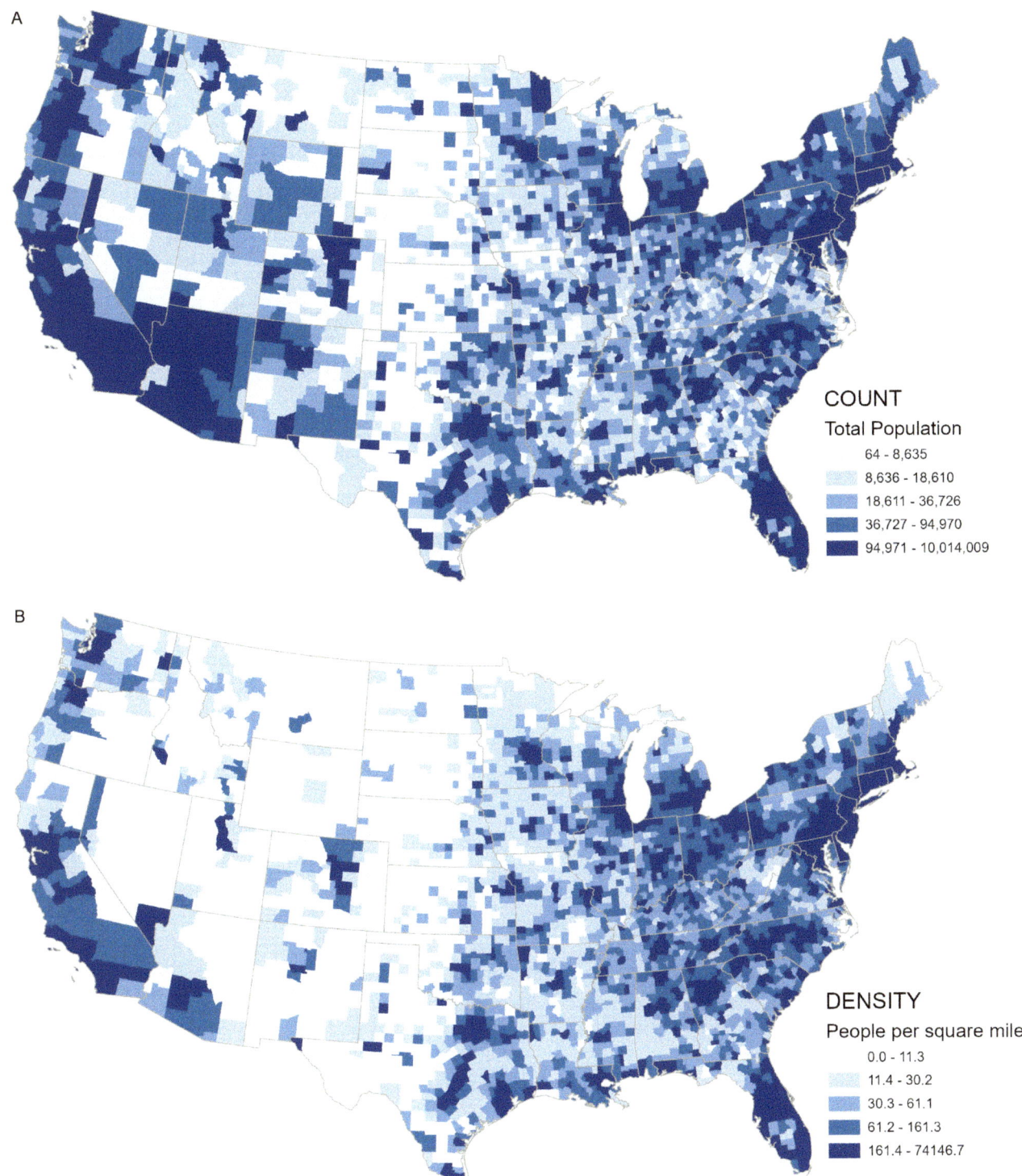

Figure 9.8. Data that has not been normalized does not account for the area of a unit or its total population, thus leading to population dependence. The resulting pattern (*A*) highlights counties with high population. Normalization leads to a more understandable map (*B*) because areas and their population density can now be compared. Data courtesy of the US Census Bureau.

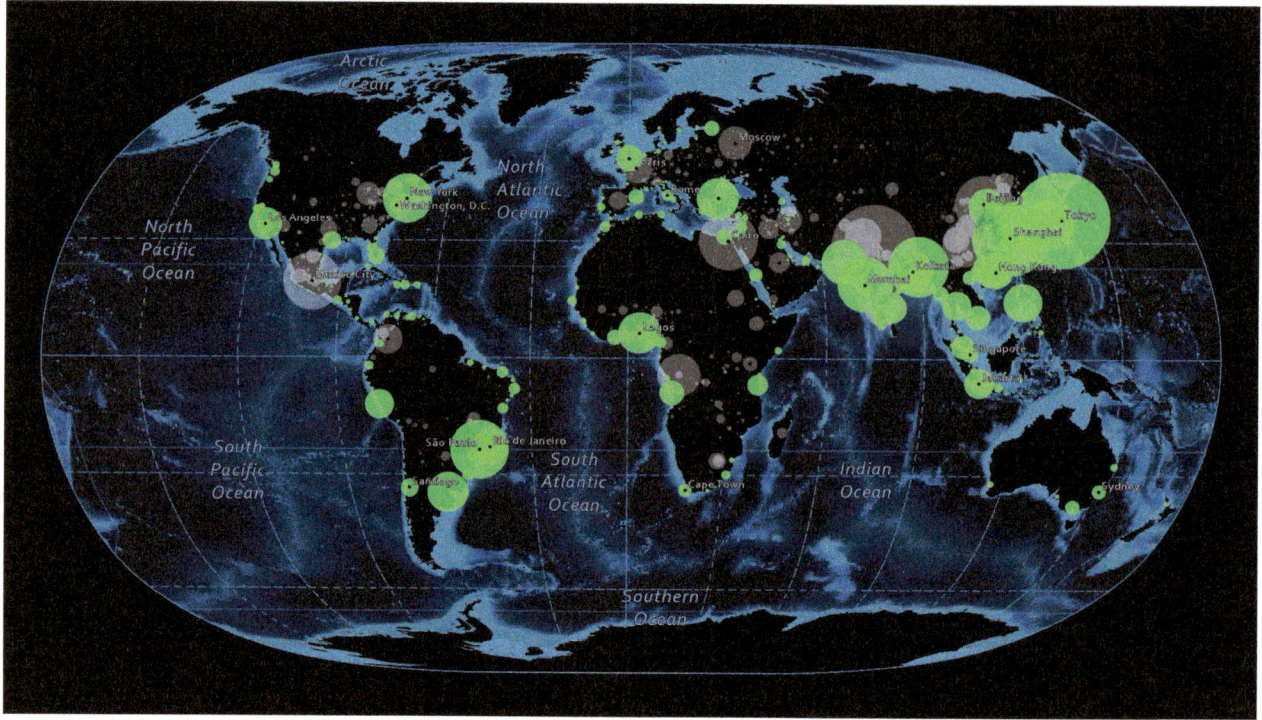

Figure 9.9. Urban areas of more than 300,000 people are shown with proportional circles on this map. Areas along the coasts are shown in green and inland areas are shown in gray, highlighting the fact that 40 percent of the earth's population lives within 100 kilometers of the coast.

information is lost by reducing the quantitative data to a few classes, graduated symbol maps are generally the easiest to read of the point symbol maps for a single attribute. Difficulties in reading proportional point symbols are largely avoided when symbols are also labeled, though this can make for a cluttered map.

Line feature maps

On line feature maps, mapmakers often show quantitative information associated with segments of the lines. As with points, quantitative line symbols can be graduated or proportional. On the traffic map in figure 9.10A, the graduated line widths change abruptly at some points, such as locations where roads cross or merge. On the streamflow map in figure 9.10B, proportional symbols show smoothly varying changes in values, as you would expect with the amount of flow in nature. With proportionally scaled line symbols, the magnitude at any point along a line can be difficult to interpret because the human eye is not sharp enough to discriminate between slight changes in width.

Flow maps

Flow maps represent the movement, distribution, or flow of objects, phenomena, or information within a geographic area or network. These maps use **flow lines** to depict the direction, volume, intensity, or relationships of flows between different locations or entities. To show the direction of flow, arrows can be added to one end of the flow line. On many flow maps, the mapmaker must decide where to place the lines so that they do not overlap other flow lines.

Four of the most common types of flow maps are illustrated in figure 9.11. **Radial flow maps** (*A*) are a type of visualization that represents the flow of information, resources, or relationships from a central point outward in a radial pattern. **Network flow maps** represent the movement of entities, resources, or information through a network, such as a transportation (*A*), hydrologic (*B*), or communication system. The map in figure 9.11B is a multivariate network flow map showing vessel traffic in marine shipping lanes. **Distributive flow maps** (*C*) illustrate the distribution or allocation of resources, goods, services, or information from a source to multiple destinations.

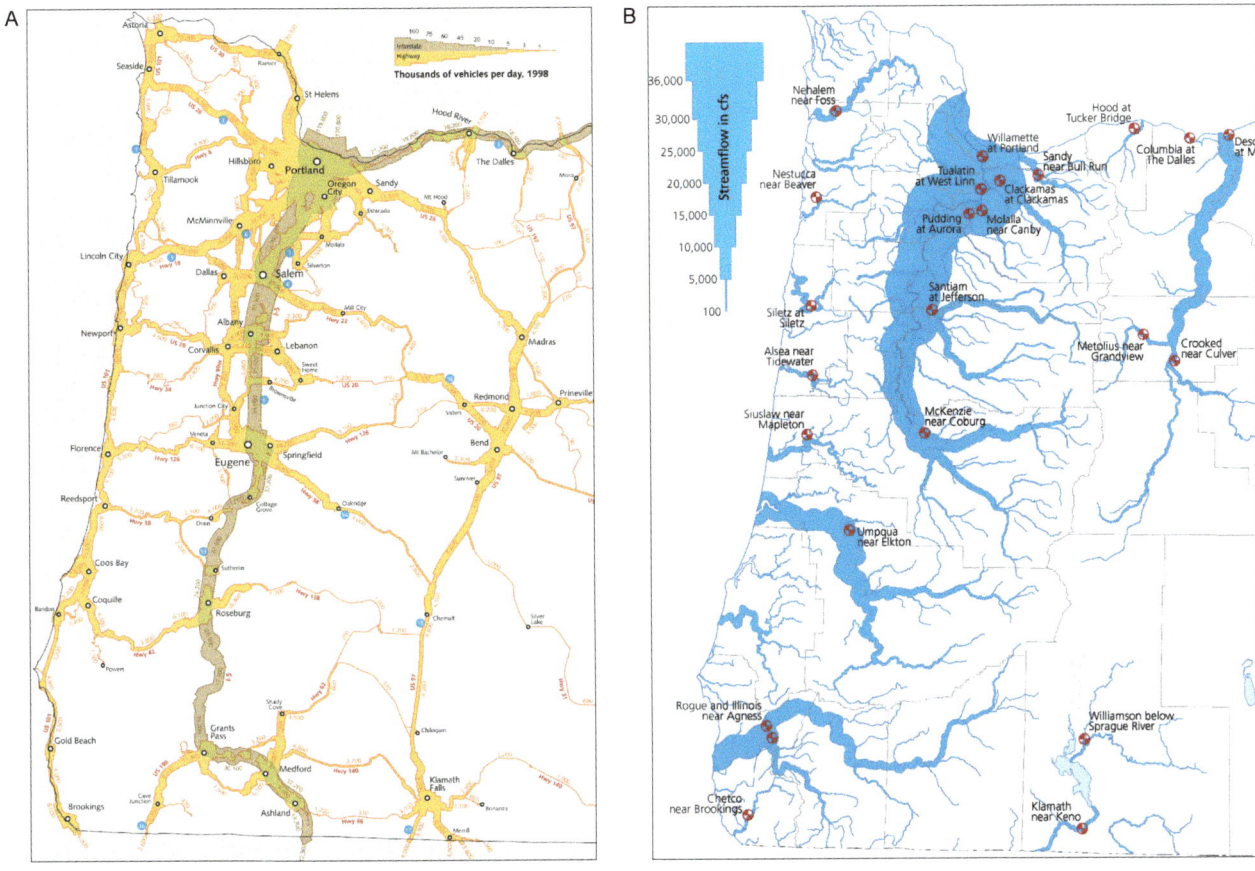

Figure 9.10. Graduated (*A*) and proportional (*B*) symbol maps for line features. From *Atlas of Oregon*, second edition, © 2001 University of Oregon. Images courtesy of InfoGraphics Lab.

When you read a flow line map, focus your attention on the magnitude information, not the flow line's precise location. Mapmakers often distort the path to avoid overlapping flowlines or the geography to accommodate wider flow lines. A flow line that shows the volume of ship traffic through the Strait of Gibraltar, for example, might be too wide to fit into the small space on the map between Spain and Morocco. Thus, the strait must be widened on the map; otherwise, it will look as if the ships are traveling over land (figure 9.11C).

Vector flow maps

Vector flow maps often use arrows or vectors to indicate the direction and magnitude of flow at different locations within a geographic area. In figure 9.11D, flowlines are used to depict daily wind speed and magnitude—faster winds are shown with thicker lines and areas with the highest speeds are shown with yellow lines against the dark background. When animated, these symbols allow readers to visualize fluid flow patterns in rivers, oceans, air currents, and other fluid systems.

Area feature maps

One of the more confusing types of quantitative thematic maps uses point symbols to represent quantities in areal data collection units. The problem is that you are led to think that you are looking at data for a point feature when the data is for an area. This is an entirely valid mapping approach though, because, as you saw in chapter 8, one of the graphic functions of points is to represent two-dimensional area features that are conceived of as points for mapping purposes. On **area point symbol maps**, the cartographer places the point symbol usually in the center of each data collection unit (at the centroid) and designs the point symbol using the same methods as for point features—that is, graduated or proportional symbols.

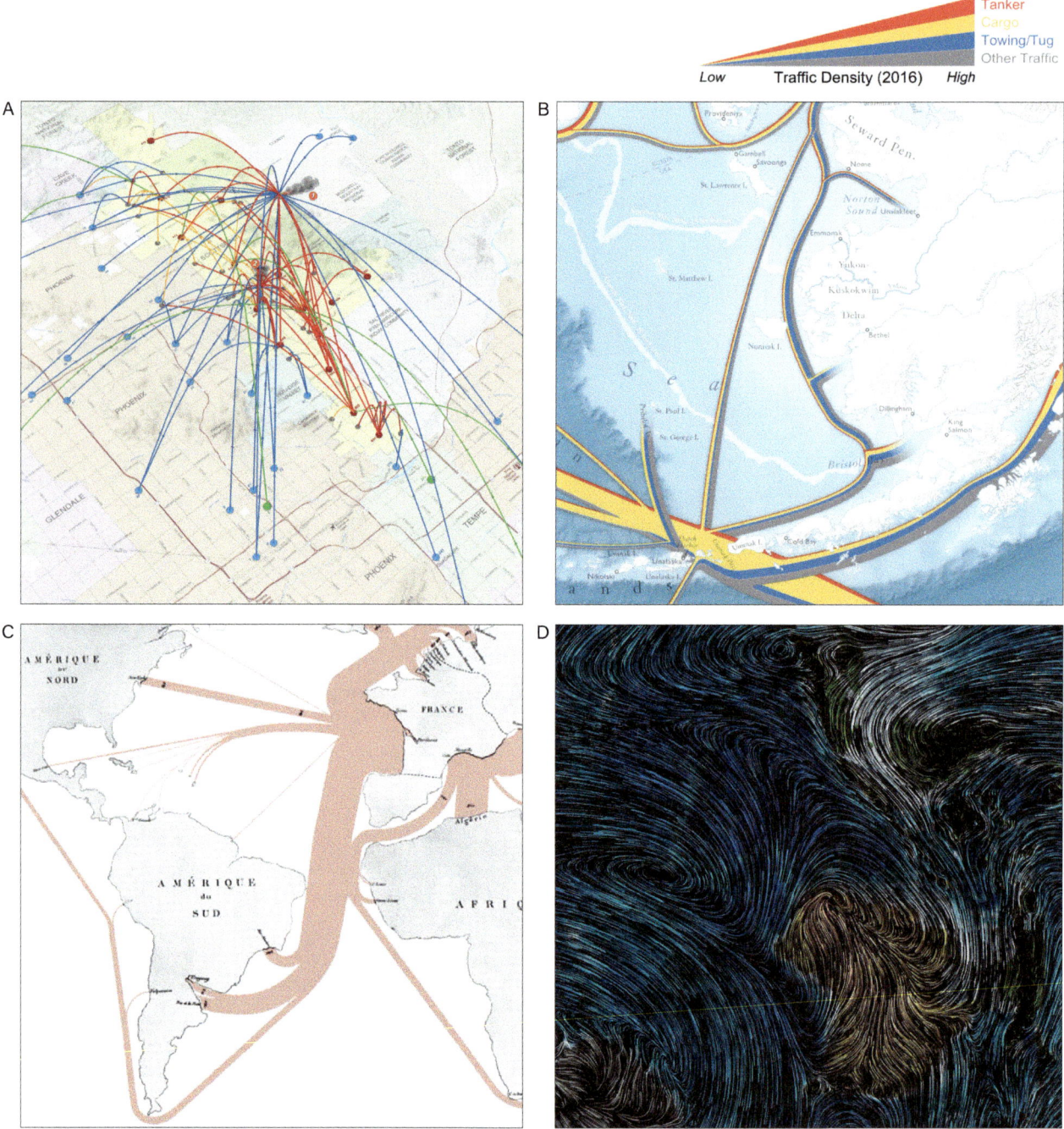

Figure 9.11. Flow mapping methods can be used to show the spokelike pattern of connections between origins and their destinations (*A*), the amount of flow along a path or network, such as shipping lanes (*B*), the distribution of goods or commodities, such as wine, from their distribution centers to their destination points (*C*), and fluid flow patterns in fluid systems, such as air currents (*D*). Courtesy of City of Scottsdale, Arizona (*A*), from Ecological Atlas of the Bering, Chukchi, and Beaufort Seas, © 2017 Audubon Alaska (*B*), and Charles Minard, public domain, via Wikimedia Commons (*C*).

Graduated and proportional symbols are also used to map clusters of points. **Clustering** groups point features that are within a certain distance of one another on the map into one symbol. Clusters are most often represented by proportional symbols based on the number of point features in the cluster. The number of features grouped into a cluster depends on the cluster radius, or the length of the radius from the centroid of the cluster symbol.

With online maps, clustering is applied dynamically based on the map scale, so as you zoom out, the points are aggregated into fewer clusters, and as you zoom in, more clusters are created. When you zoom to a level at which the clustering area around one point feature no longer contains other features, that point feature is displayed as a single feature. The single feature could represent a point feature, an areal feature, or the centroid of an areal feature. The challenge for map readers is to discover whether the map uses clustering, because at first glance, it will appear as though individual point features are being mapped (figure 9.12).

Choropleth maps

Many quantitative thematic maps of ordinal-, interval-, or ratio-level area data are made using the **choropleth mapping** method. The term comes from the Greek terms "choro," meaning area or region, and "plethos," meaning magnitude or multitude. With this method, each data collection area is given a particular color lightness, color saturation, or pattern texture depending on its magnitude. Areas are assumed to be homogeneous within, and abrupt changes can occur at boundaries between areas. The maps in figure 9.8 are examples, although only the bottom map correctly uses normalized data.

Dot maps

Whereas choropleth maps fill the data collection units with colours that vary in value or saturation, **dot maps** fill the units with dots that give the impression of "more" where the symbols coalesce. In this way, dot maps essentially use pattern texture as the visual variable. An advantage of this mapping method is that dots of different hues can be used to show different types of things. A challenge for the mapmaker is to assure that all the dots are visible, and a challenge for the reader is to decipher what different dot colour combinations mean. An example is shown in figure 9.13—in this map, each person is represented by a single dot whose hue depicts their religious allegiance.

Dasymetric maps

On **dasymetric maps**, the mapped areas are not political data collection units, such as counties, but rather areas of inherent homogeneity in the data. In dasymetric mapping, the mapmaker starts with a map of standard data collection units, such as the census tracts in figure 9.14A. Ancillary data, such as land cover (figure 9.14B), is used to adjust the unit boundaries to more realistically represent homogeneous areas. Then the original values (figure 9.14C) are recalculated to more accurately reflect the underlying data distribution (figure 9.14D). As you might imagine, having access to different layers of information and data analysis methods with GIS greatly aids dasymetric mapping.

In the Delaware County, Pennsylvania, example in figure 9.14, the rural part of the county population is reassigned to the developed areas, and the densities are recomputed. As a result, the population density in the west decreases in the agricultural and forest areas, and the density in the east decreases in the water and wetland areas. You can see that a dasymetric map with the homogeneous areas modified to reflect the data (*D*) should give a more faithful representation of population density than a choropleth map that shows the density class for each county (*C*). The challenge for the mapmaker is to find the right ancillary data to use, and to then correctly recompute the area values.

Cartograms

You may have seen maps that show an eye-catching portrayal of the relative magnitudes of area features rather than their exact spatial locations. At these times, you were most likely looking at an odd-looking map called a **cartogram** (figures 9.15B, 9.15C, and 9.15D). Mapmakers create cartograms by distorting the geographic size of data collection units in proportion to the numeric value being mapped. Although these maps look a little strange, they help overcome the problem of choropleth and area feature point symbol maps when the size of the data collection units varies widely.

When creating cartograms, mapmakers try to retain many spatial characteristics of conventional maps. These characteristics include shape, proximity, and contiguity. Preserving the shape of data collection units makes it easier to relate a cartogram to a conventional map. Preserving the **proximity** (nearness) and **contiguity** (boundary connectedness) to neighboring areas also makes the cartograms easier to read. But shapes cannot be preserved without modifying or losing the proximity and contiguity of neighboring areas,

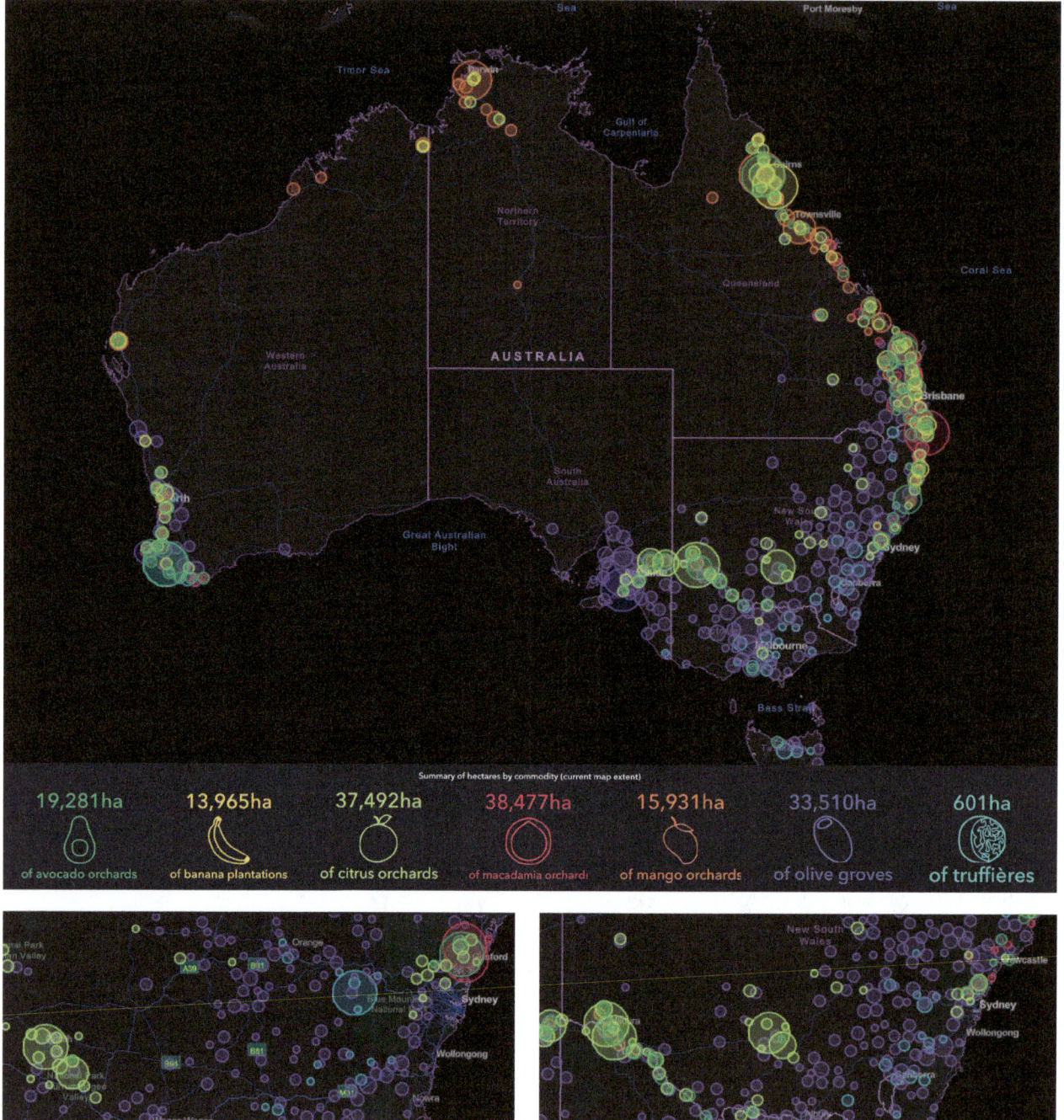

Figure 9.12. This map interactively clusters parcels with commercial horticulture tree crops in Australia. At the scale of the nation, more parcels are aggregated into fewer clusters. The number of clusters increases as you zoom in until the parcels are shown as individual areal features. Courtesy of the University of New England, Applied Agricultural Remote Sensing Centre.

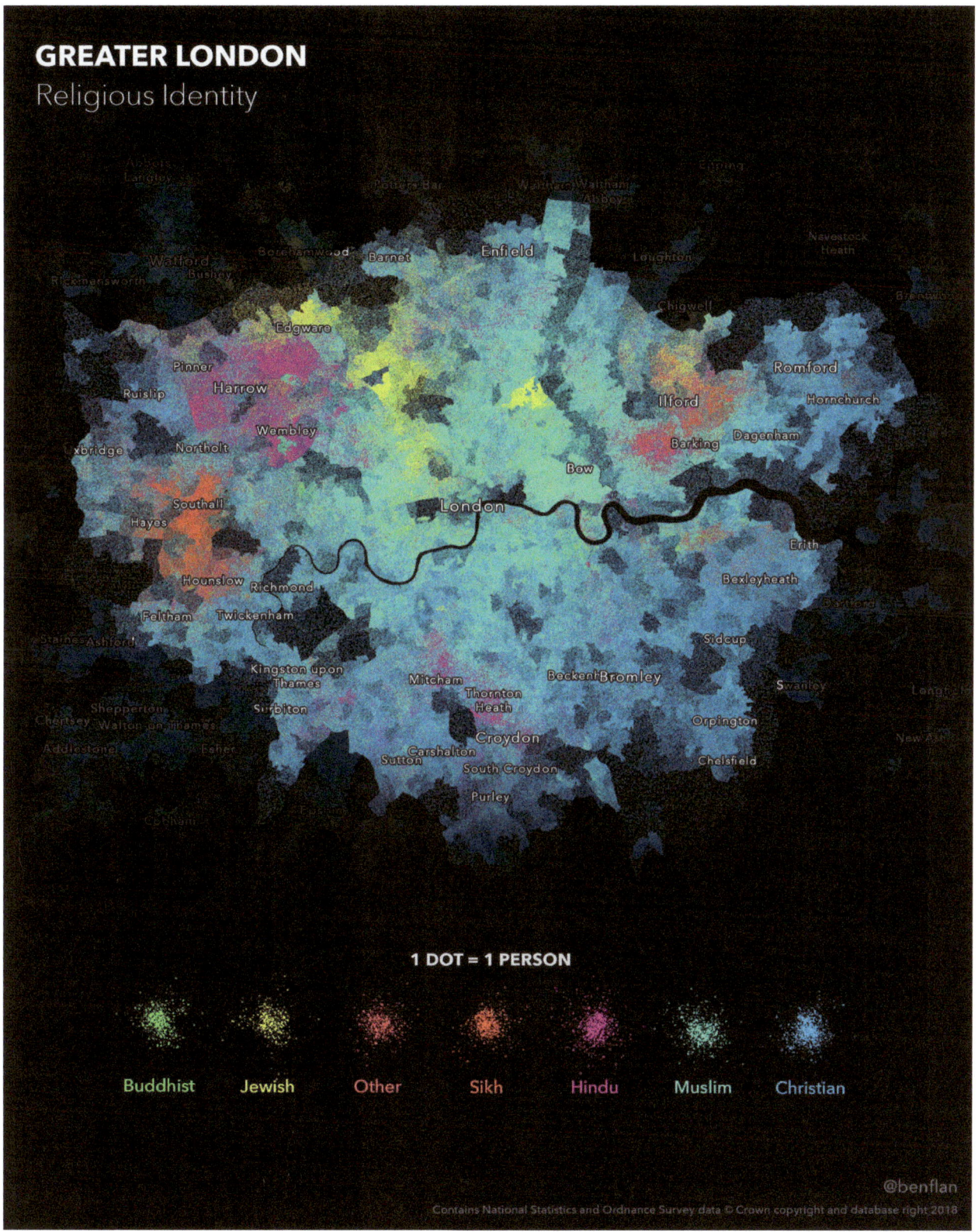

Figure 9.13. This dot map reveals people's religious identity across London.

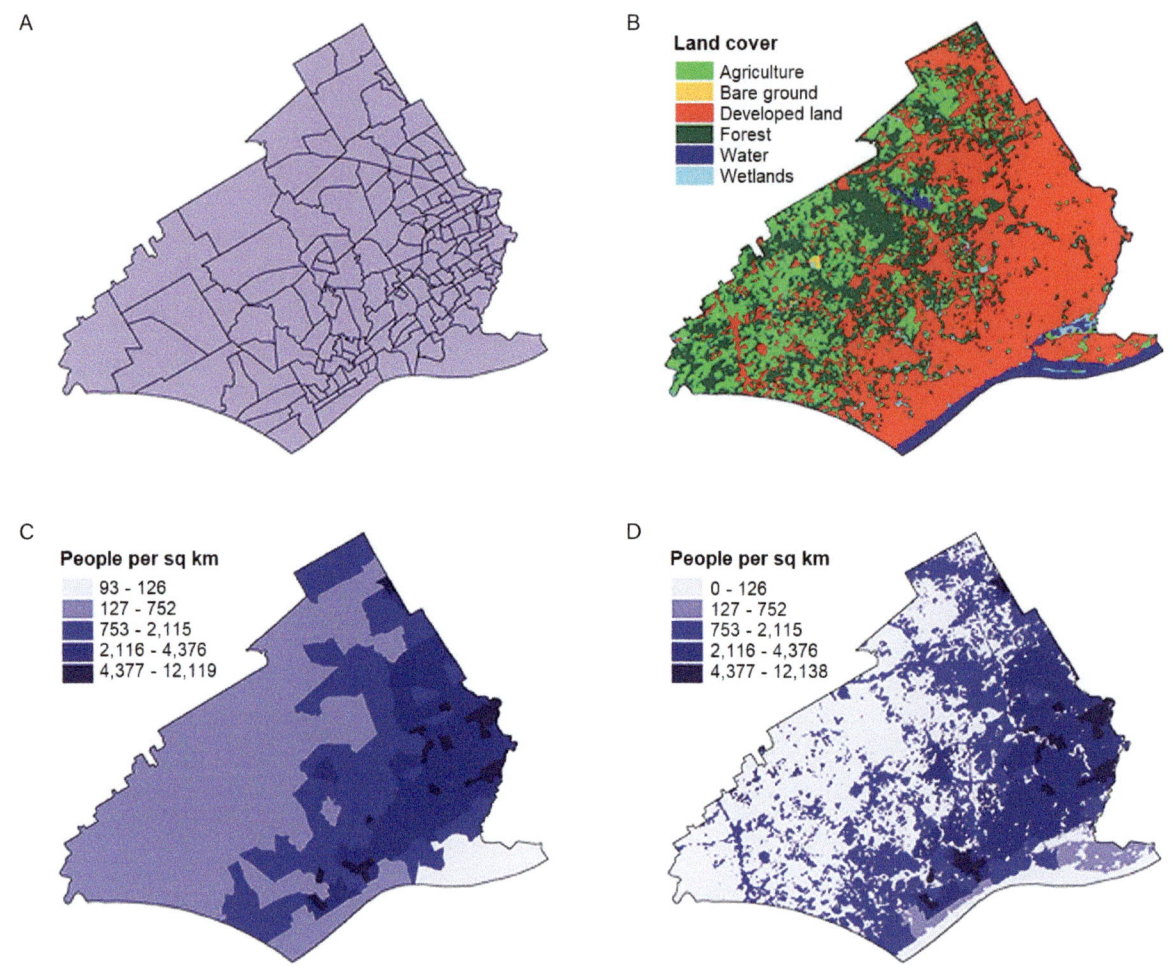

Figure 9.14. Dasymetric mapping is used to depict a more realistic representation of population density in Delaware County, Pennsylvania. Population data for landcover data (*B*) is used to redistribute the population to areas in which the people are more likely to be living. The densities are then recalculated, and the new areas are symbolized on the dasymetric map (*C*). Courtesy of Jeremy Mennis and Torrin Hultgren.

and vice versa. Next, we look at these characteristics for different types of cartograms.

Contiguous cartograms are by far the most interesting to look at because they maintain the proximity and contiguity of neighboring areas, although sometimes at the expense of extreme shape distortion. In the contiguous cartogram of California county populations in figure 9.15B, most counties do not look at all like themselves. Yet despite this shape distortion, the cartogram is effective—at only a glance, you can see precisely which counties have the greatest population. The variable quality of shape preservation from one region to the next is not always a major distraction. We appear able to accept a fair degree of shape distortion before areas become unrecognizable.

With **noncontiguous cartograms**, the shape of units is preserved at the expense of proximity and contiguity. These cartograms are created by scaling the size of the data collection unit relative to the magnitude of its attribute value, without regard to connectivity to adjacent areas. To create the noncontiguous cartogram in figure 9.15C, the mapmaker enlarged or reduced each county in proportion to its population. The modified data collection units were then placed as closely as possible to their geographic positions on the conventional equal area map in figure 9.15A map by positioning the centroid of the area on the cartogram in relation to the centroid of the corresponding area on the conventional map. Preserving the shapes of data collection units makes it easy to recognize and compare them with

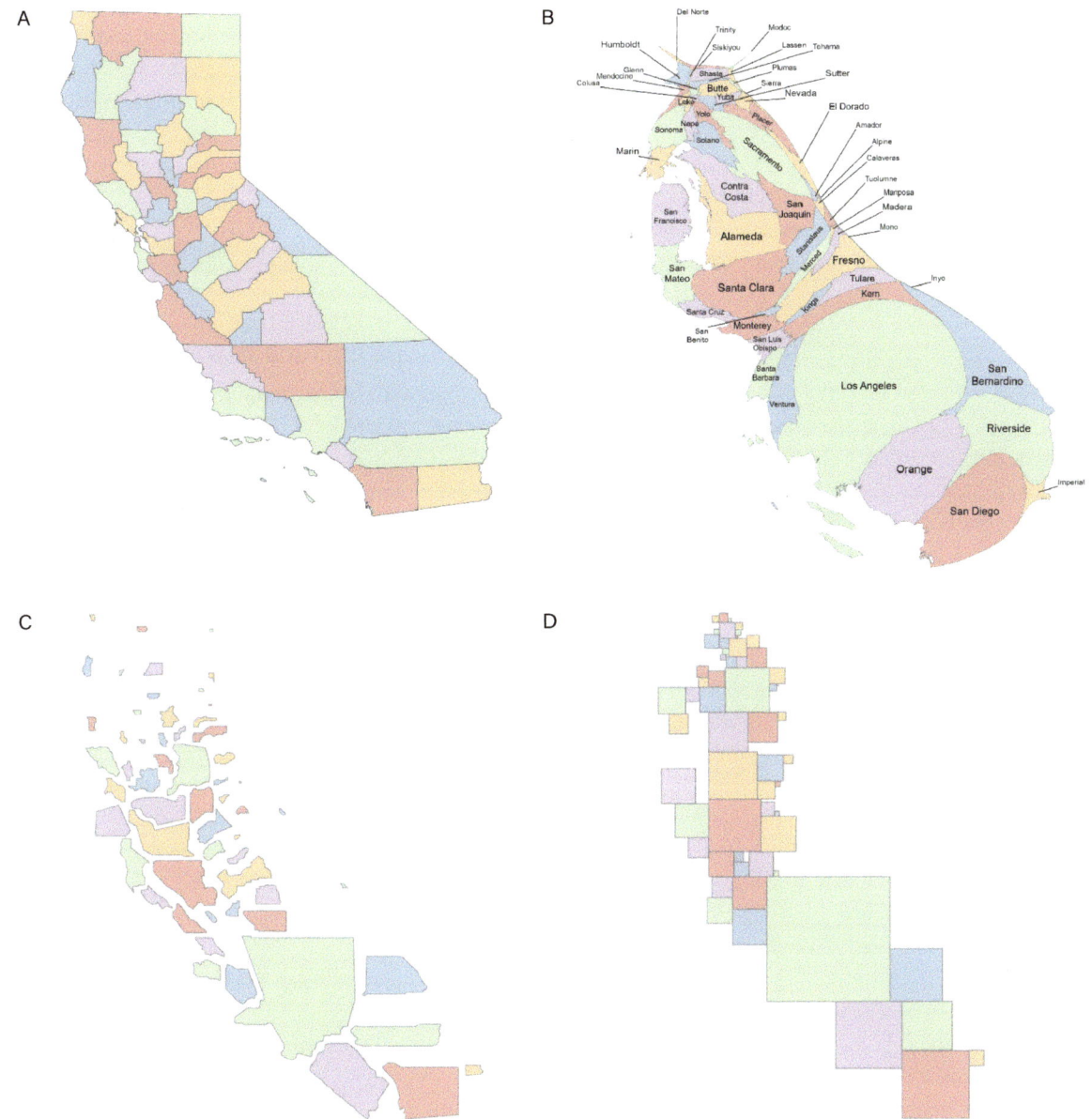

Figure 9.15. An equal-area map (*A*) can be used as reference to interpret contiguous (*B*), noncontiguous (*C*), and pseudocartograms (*D*) for California state population by county.

their counterparts on a conventional map. But as figure 9.15C also shows, proximity relations are only roughly maintained, and contiguity is completely sacrificed.

A **pseudocontiguous cartogram** is made by transforming each data collection unit into a simple geometric shape—squares or circles are most often used—that is proportional in size to the count or amount being shown (figure 9.15D). The simple shapes are then arranged in what resembles their relative geographic position on the conventional map with few gaps between the shapes. By sacrificing (geographic) shape and proximity, these cartograms can preserve a considerable degree of contiguity.

To read a cartogram effectively, you must be able to compare the sizes of data collection units on the cartogram with the same areas on a map with a conventional equal-area map projection either in front of you or more often in your mind. If the positions, shapes, and relative sizes of data collection units on the equal area map are not familiar to you, a cartogram can be difficult to use because you will not be able to compare the unfamiliar shapes on the cartogram with your mental image of the correct shapes. For this reason, cartograms are most successful when they are used to map data collection units that are familiar to the map reader, such as countries, states, and counties. Cartograms are also easier to read when there is a conventional map displayed along with the cartogram.

Prism maps

A **prism map** shows the magnitude of an attribute by extruding the feature to a height that is proportional to the attribute value. On the map of Fort Wayne, Indiana, in figure 9.16A, each city block is raised vertically (extruded) above the base level to a height that is proportional to its assessed value per acre. The map in figure 9.16B uses the same approach to show parcel values in New Zealand. Notice that map scale affects the appearance of prism maps. Larger-scale maps (figure 9.16A) have more variation in the area of units, but larger units can become overemphasized. As map scale decreases and the mapped area increases, the extruded features appear more like matchsticks of varying heights (figure 9.16B).

Prism maps are visually impressive, but they present several challenges to the map reader. One drawback is that, although highs and lows are apparent, the exact height of a given area is difficult to determine. Labeling is also a challenge for prism maps, but solutions can be found, as shown on the map in figure 9.16A. Another limitation is that they are viewed from a vantage point from the side at a height that is somewhere between horizontal and vertical. A poorly chosen vantage point may cause important units to be obscured from view. If the map can be viewed in an interactive environment, map readers can overcome this limitation by viewing the map from many vantage points.

Binned maps

Our discussion of quantitative mapping for area features ends with **binned maps**, which show aggregated point data in identically sized areas that visually summarize the count of features or their amounts. The bins are features that can tesselate to cover a region with a pattern of repeated shapes that fit inside each other with no overlap. Commonly used shapes that tesselate include squares, triangles, and hexagons—the latter two offer the advantage that all contiguous bins are equally spaced. When mapmakers use hexagonally shaped bins, they create a **hexmap** (figure 9.17).

One of the earliest uses of binned maps was for the US Environmental Protection Agency's Environmental Mapping and Analysis Program. The hexagonal grid developed by this book's coauthor, A. Jon Kimerling, for the 50 US states was then expanded to other countries to gather information about environmental factors, such as Ph, nitrogen in streams, and so on. GIS software can now easily create grids of squares or hexagons at any size for any geographic extent.

Continuous surface maps

A **continuous surface map** portrays the continuously changing amount of a phenomenon from one place to another. On these maps, variations from one location to another are gradual rather than abrupt. This is illustrated on the map in figure 9.18A, which shows a population density surface derived from 2020 US census data. Using the attribute for people per square mile and the same data collection units as for the choropleth maps in figure 9.8, a smooth surface was created and then symbolized with a continuous color scheme that varies from red in the high-density areas to blue in the lower-density regions.

The data for continuous surface maps may be collected as points, lines, areas, or in raster format, and several methods can be used to map a continuous surface. The most common methods produce isoline maps, heat maps, dot density maps, and 3D surface maps.

Isoline maps

Isolines are created by connecting points of equal value, as illustrated in figure 9.7. The prefix "iso" means equal, and the type of phenomena that can be shown with lines of equal value may be either physical or cultural. Different names are given to isolines according to the type of information they show. Although the word *contour* is used to describe lines of equal elevation value, other types of isolines typically use the "iso-" prefix. For example, **isochrons** are lines of equal time—they are often shown on travel time maps (figure 9.18B), and **isohyets** are lines of equal rainfall—you will often see these on weather maps. The isohyets can be inferred as the boundaries between different colors on the map in figure 9.18C.

The most common way to read an isoline map is to ignore individual isolines and focus on the overall pattern across the map to visualize the continuous surface. Where isolines are placed far apart, the surface varies only slightly; where isolines are dense, the surface varies significantly.

Chapter 9: Quantitative thematic maps   221

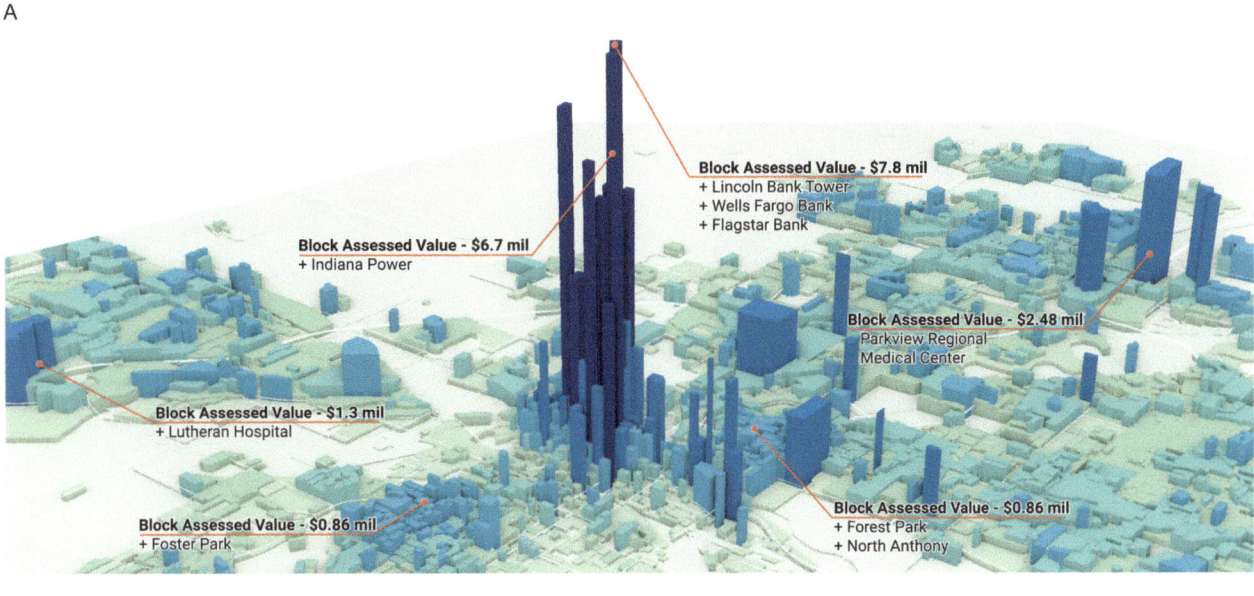

Figure 9.16. On the prism map of downtown Fort Wayne, Indiana (*A*), each city block is extruded to a height that is proportional to its assessed value per acre. On the map of downtown Auckland, New Zealand (*B*), the same approach is taken with parcel values. Courtesy of Houseal Levine Associates (*A*) and Urban3 (*B*).

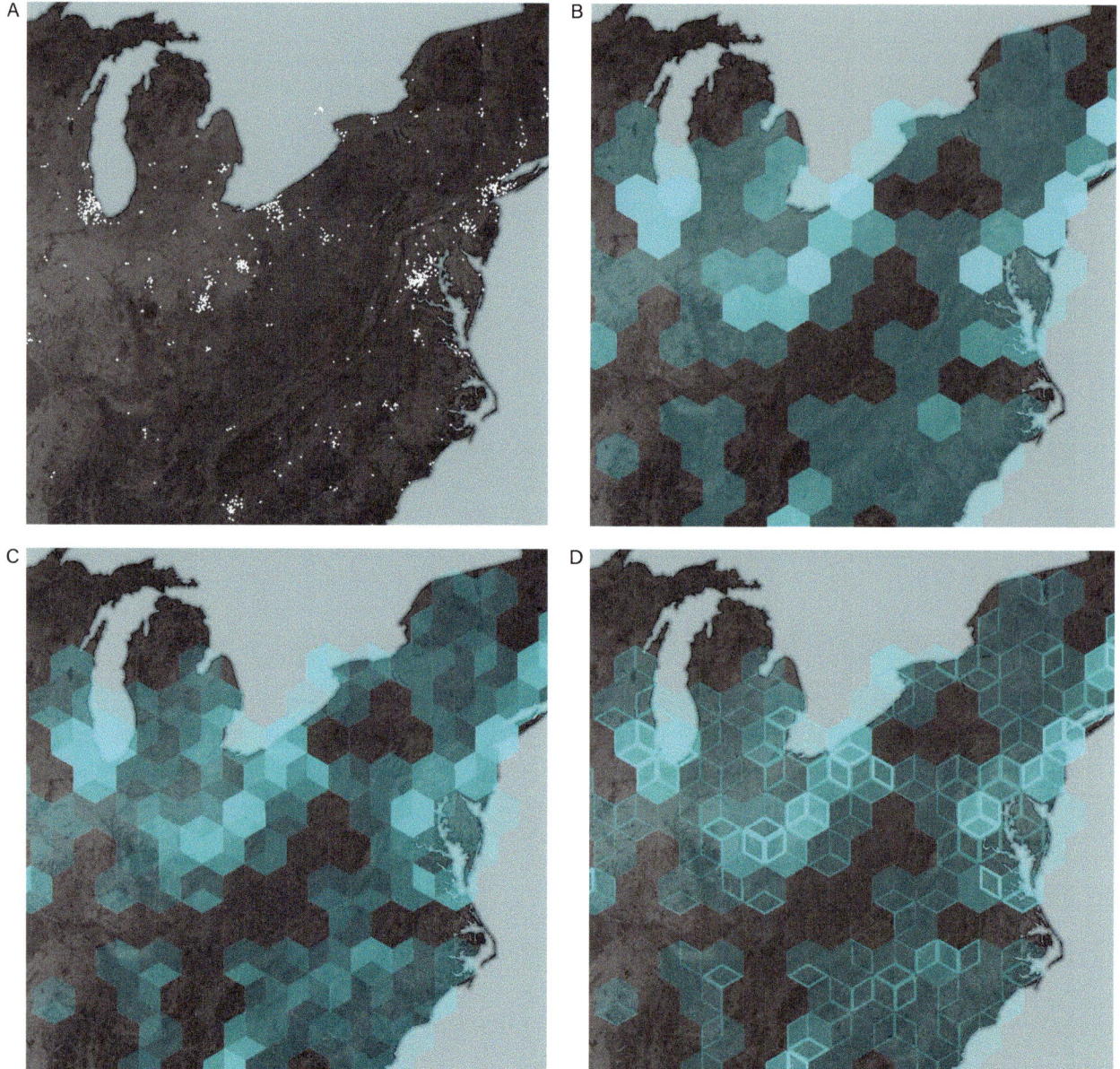

Figure 9.17. The data for a chain restaurant (*A*) can be binned into hexagons that appear either 2D (*B*) or 3D (*C* and *D*).

Some people find this hard to visualize, however. To help map users who have trouble focusing on the pattern rather than the separate isolines, filled isoline maps, which add color between the isolines, help map readers see a progression in the amount from low to high (figures 9.17B and 9.17C). The isolines, which serve as the boundary between areas with different amounts, often are not labeled, which further encourages readers to see the general pattern of surface highs and lows. If the lines are not labeled, numeric interval information for each area fill is found solely in the map legend.

Heat maps

Heat maps help readers see the concentration of many points that are close together or overlapping and cannot be easily distinguished. This mapping method works best for a large number of point features, especially if using individual point symbols would result in a lot of overlap. They are particularly useful for identifying areas of high and low data values or density of points, as well as clusters and outliers within a dataset. The map in figure 9.18D uses this method to show the concentration of fatal car crashes in Los Angeles, California.

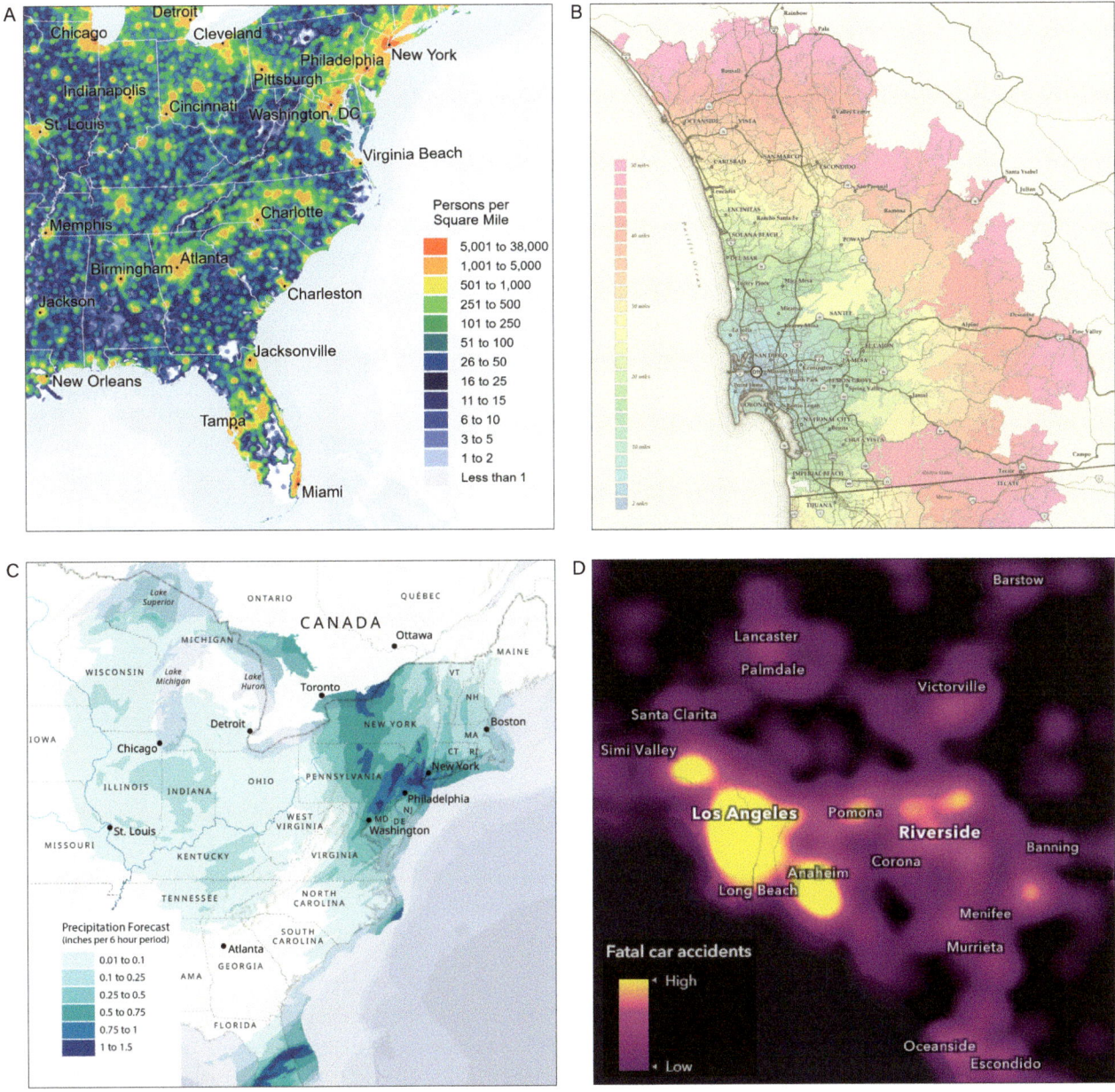

Figure 9.18. Different methods for mapping quantitative data for continuous surfaces are illustrated here. The continuous surface in A represents population density. Isochrons are used in B to show lines of equal travel time. Isohyets outline areas of equal rainfall in C. And a heat map highlights high concentrations of fatal car accidents in D. Courtesy of California Department of Transportation (B).

Dot density maps

Point symbols are the basis for one of the most effective (but not commonly seen) ways of showing variations in density across a surface. The method is called **dot density mapping**. Be careful not to confuse this mapping method with dot maps (in which each dot represents the location of a truly one-dimensional feature). What distinguishes dot density maps is that each dot represents more than one feature, and the density of dots gives the impression of a gradually varying continuous surface. This is illustrated in figure 9.19 using California population data. The value for each dot is indicated in the word legend: "Each dot represents 10,000 people." And the population density could be visualized as a 3D surface.

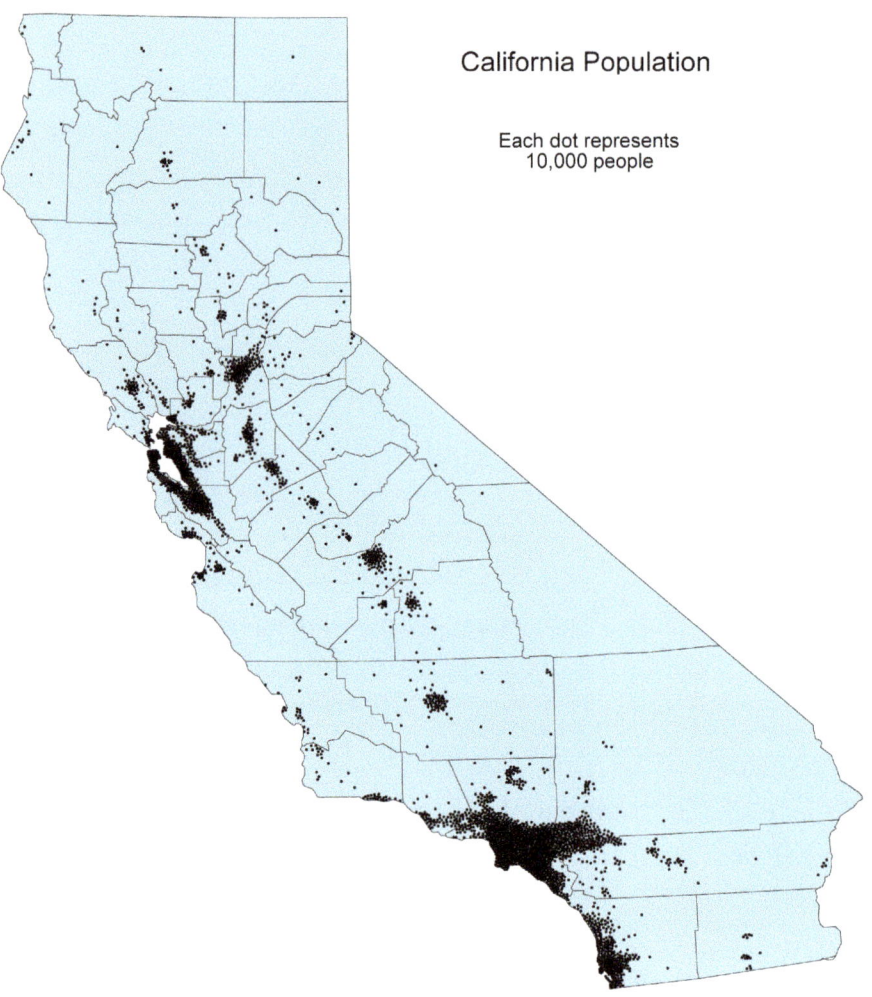

Figure 9.19. On this dot density map of California population, the greatest density is in Southern and Central California.

Although a circular dot is the most common symbol used in dot density mapping, the shape is irrelevant, because the symbol's meaning lies solely in the changes in symbol density produced by its repetition. However, as you saw with proportional and graduated symbol mapping, using circles as the symbol has clear advantages.

To create a dot density map, mapmakers first choose a **dot value** (the amount that each dot represents) and a **dot size** (the size of the dots they will put on the map). For example, they may decide, as they did for the map in figure 9.19, that each dot represents 10,000 people. They then divide the number of features (people, in this example) in the data collection unit (counties) by the dot value. The resulting number tells them how many dots to place within the data collection units. When producing a dot density map, the mapmaker places dots in a random-appearing fashion and blends the density of dots along the boundaries of the data collection units to show smooth variations in the dot density across the region.

Another distinguishing characteristic of dot density maps is that the dots are reallocated in areas within the original data collection units. Using secondary data, the areas within which the dots are placed are modified to assure that the phenomenon is shown where it is likely to exist, as with dasymetric maps. For example, more population is found in urban areas than rural areas, and little to no population is found in forests, parks, and water bodies. So, the mapmaker shifts the dots to the more populated areas by excluding or reducing no- or low-population areas from the data collection unit, recalculating the unit's area, coming up with a new density value, and adding the appropriate number of dots in the more populated areas. In practice, you will find that some mapmakers have not taken the steps to exclude areas and recalculate density values. Keep this in mind as you are reading these types of maps.

Dot density maps can be difficult to read if you want to determine exact density values for specific locations. As we have already discussed with proportional and graduated symbol maps, the human eye tends to underperceive magnitudes. Psychological experiments have shown that as the density of dots increases, our estimates tend to fall below the actual density at an increasing rate. The result is that people reading dot density maps typically get the impression that the range in dot density—and therefore the contrast in density from one region to another—is less than it really is. For this reason, when reading dot density maps, you may find it necessary to compensate mentally for underperception in your own density judgments to gain a true picture of the mapped distribution.

3D surface maps

A continuous surface can also be shown as a **3D surface map**. The map in figure 9.20 is the 3D equivalent of the 2D continuous surface map in figure 9.18A. As with prism maps, if the 3D surface map is viewed from the side, your ability to see all locations on the map is limited by the vantage point selected by the mapmaker. Interactive 3D perspective maps are ideal for viewing the details of the 3D surface because you can choose how to view the map.

Multivariate maps

The methods we discussed for showing multivariate qualitative data in chapter 8 can also be used for multivariate quantitative data. In addition to multivariate symbols, discussed next, cartographers also use composite indexes to map multiple attributes.

Multivariate symbols

Although they can be complex to read, multivariate symbols for quantitative data can carry a lot of information. A fairly simple approach is taken for the bivariate map showing total agricultural land with circles of different sizes and the proportion of land used for agriculture with color value (figure 9.21A). Nonetheless, readers must be

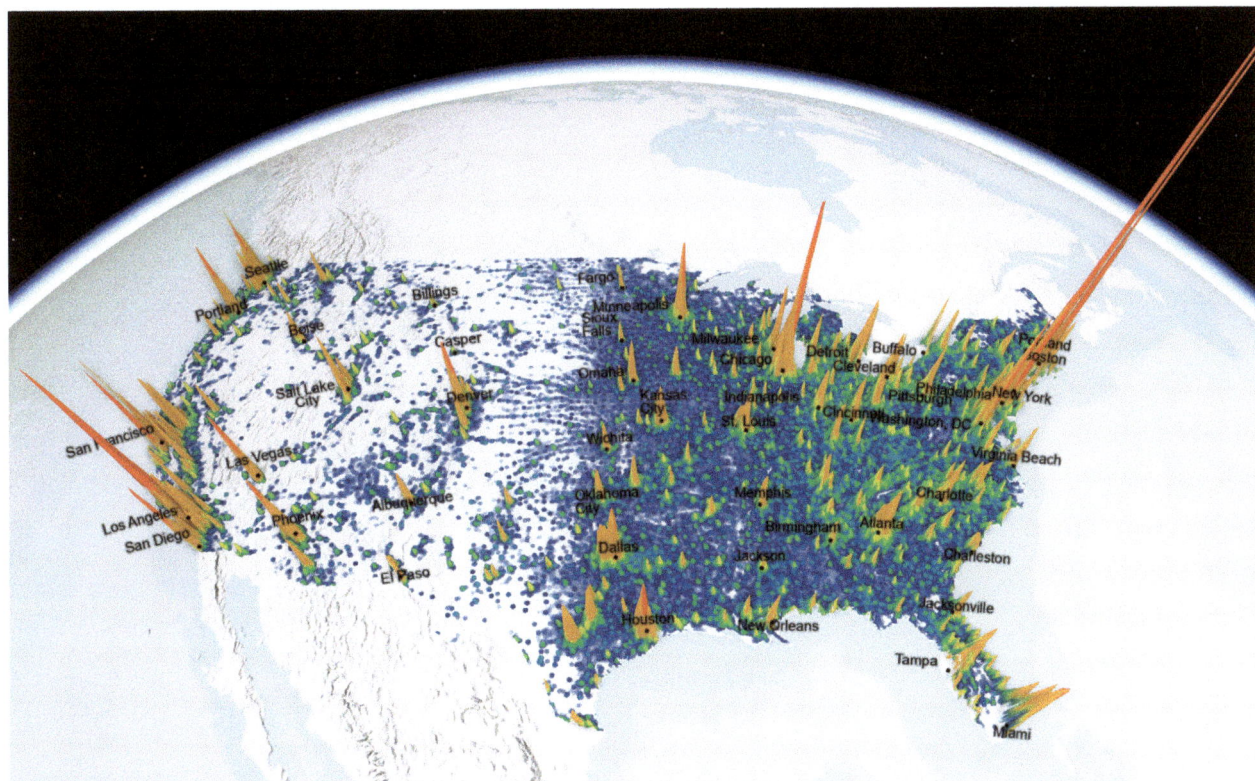

Figure 9.20. This 3D view of population density in the conterminous United States was created using the same data as in figure 9.18A.

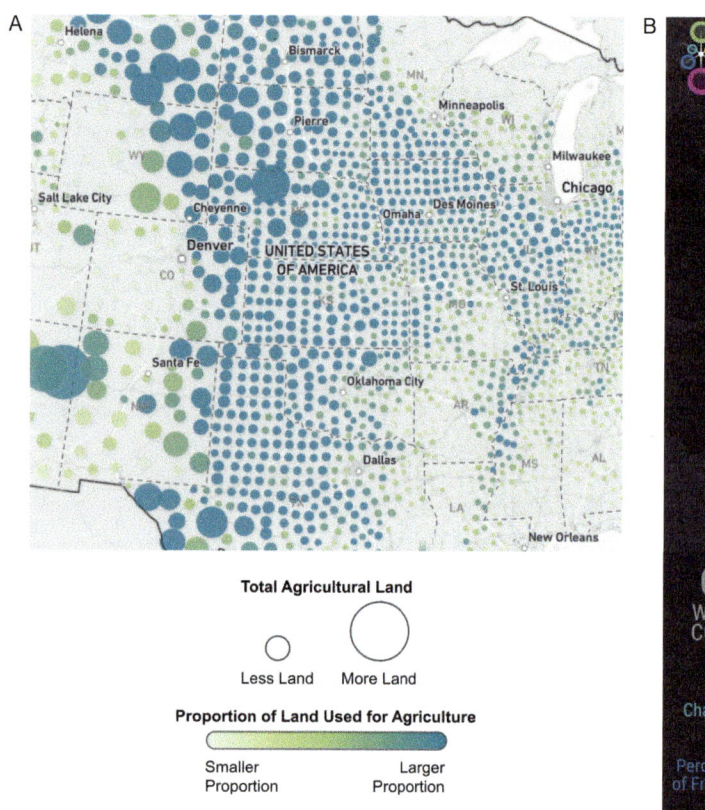
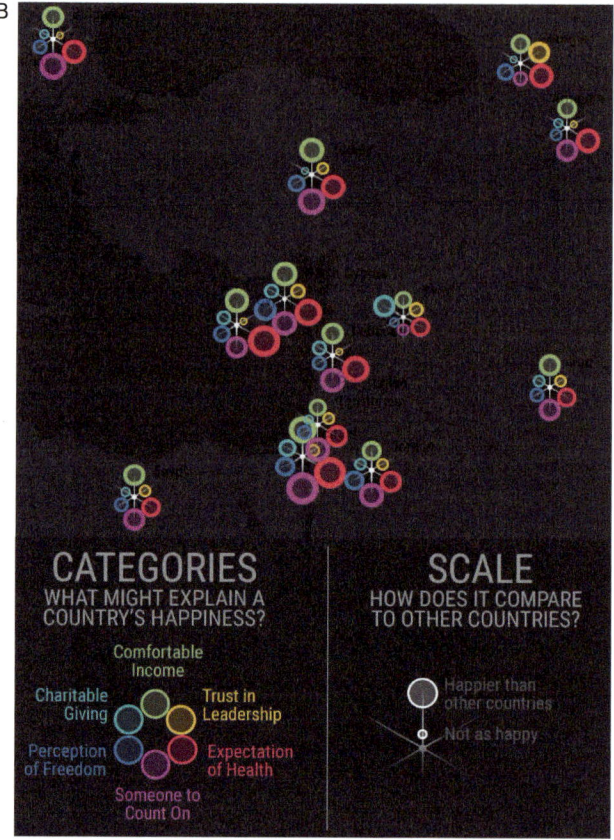
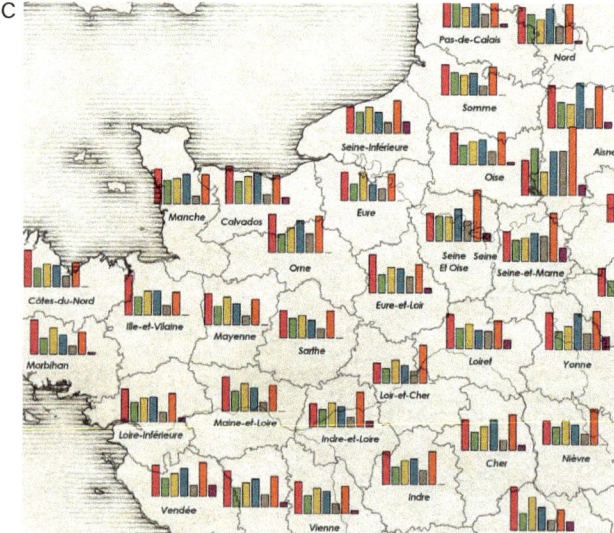
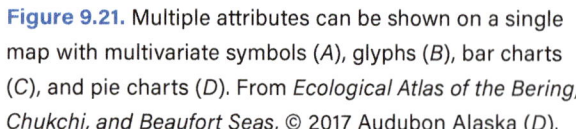
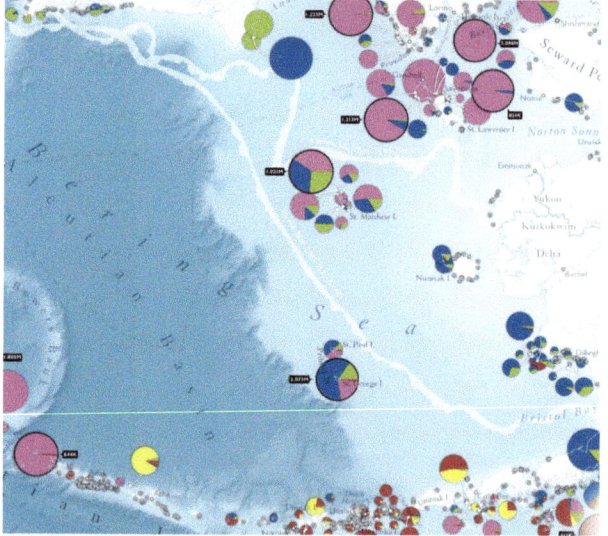
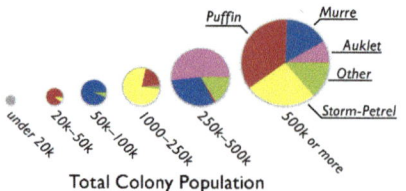
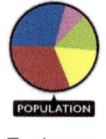

Figure 9.21. Multiple attributes can be shown on a single map with multivariate symbols (*A*), glyphs (*B*), bar charts (*C*), and pie charts (*D*). From *Ecological Atlas of the Bering, Chukchi, and Beaufort Seas*, © 2017 Audubon Alaska (*D*).

able to see the relationship between the two attributes on their mental map. The other maps in figure 9.21 use more complex multipart symbols called **glyphs**. Mapmakers use different parts of the glyph to show different attributes. For example, the map in 9.21B uses proportional circles with different hues arranged like petals around the central pistil of a flower to symbolize six attributes that relate to happiness. As the symbols get more complex, the reader's ability to decipher their meaning diminishes, so it sometimes helps to use glyphs that readers are familiar with, such as the bar charts in figure 21.C, showing seven traffic-related attributes for each of the metropolitan regions of France, and the pie charts in figure 9.21D, showing the size and composition of arctic marine bird colonies.

Composite variable maps

The combination of multiple quantitative attributes can be shown on **composite variable maps**. On these maps, the data for multiple variables or attributes is combined into a single numeric index, called a **composite index** (or **composite indicators**). These types of maps are used to simplify complex information by condensing multiple indicators into a single value, making it easier to interpret and compare different entities or scenarios. An example is the map in figure 9.22, which shows the effect that human activities are having on the ocean and the life it harbors based on the values of 12 human-made stressors, such as vessel traffic and sea temperature change.

Although it is easy to see areas of high and low impact on these types of maps, it is hard to determine the value at any specific location. Also, you do not know what attributes were used to calculate the index or how they were combined. Nonetheless, you will probably see more of these maps because GIS makes possible the analysis of large amounts of data to better understand complex environmental and human issues.

Mapping change

Quantitative change maps show change in the ordinal-, interval-, or ratio-level attributes of features over time. The mapping methods we explored in chapter 8 to show qualitative change, including superimposition, change mapping, and small multiples, can be used to show quantitative change as well. Here, we also explore time series maps and space-time cubes.

Figure 9.22. This map shows the estimated change in human impact on global marine ecosystems from 2003 to 2013. Bright-yellow areas represent increased human impact, whereas dark-blue areas represent minimal human impact.

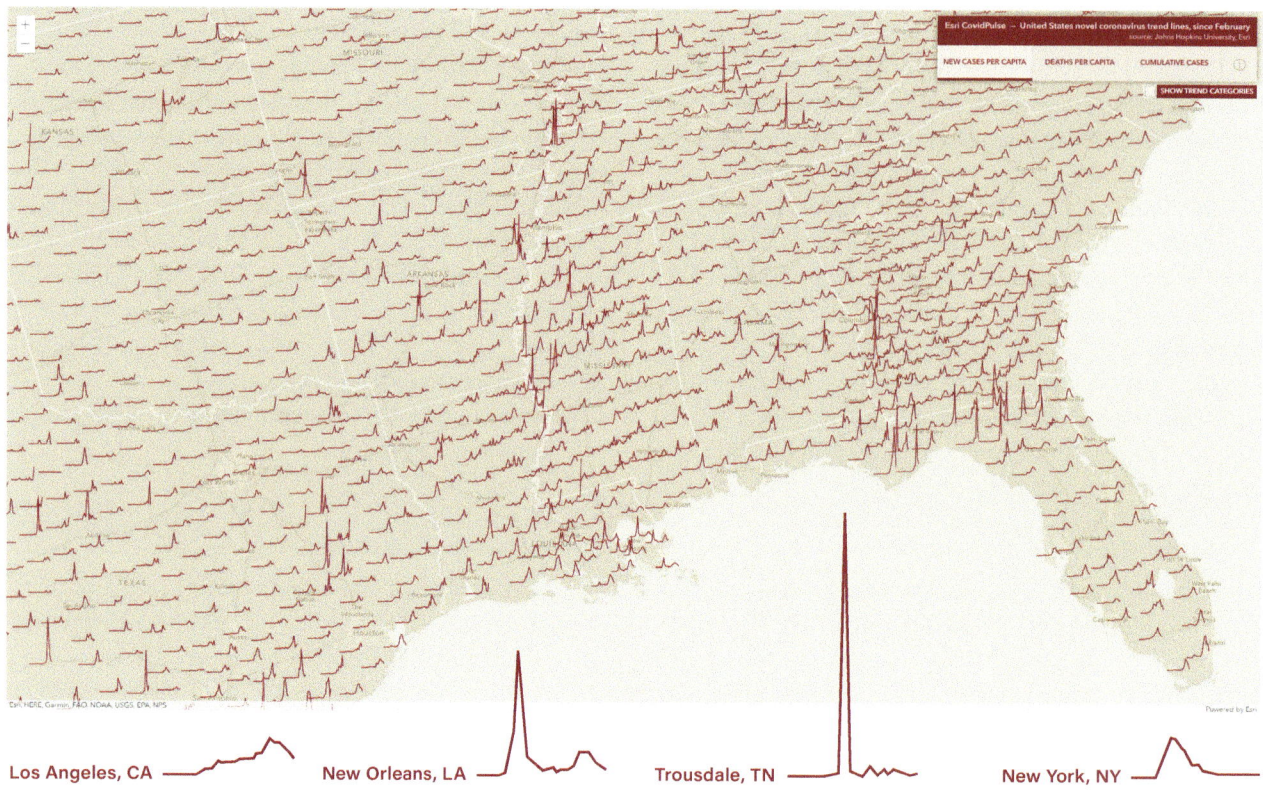

Figure 9.23. The online CovidPulse app uses sparklines to show the trend of COVID cases over time for counties in the United States.

Time series maps

Change over time can be shown on a **time series map**—a map that shows a sequence of data collected, recorded, or measured at successive points in time. Often line charts or bar charts are used as the symbol. A line chart example is shown in figure 9.23. On this map, the number of COVID cases by US county are symbolized in a miniature time line called a sparkline, introduced by the American statistician **Edward Tufte**. **Sparklines** are small graphics that show only the charted data with no labels or axes. The simplicity of the chart provides a quick visual summary of the life of a phenomenon or a trend in data values. The sparklines on the COVID map present six months' worth of daily data to show the trend of local rates of infection over time.

Space-time cubes

All the mapping methods for qualitative and quantitative data that we have looked at so far are used to show **spatiotemporal data**—information that is not only associated with a specific location in space but also varies over

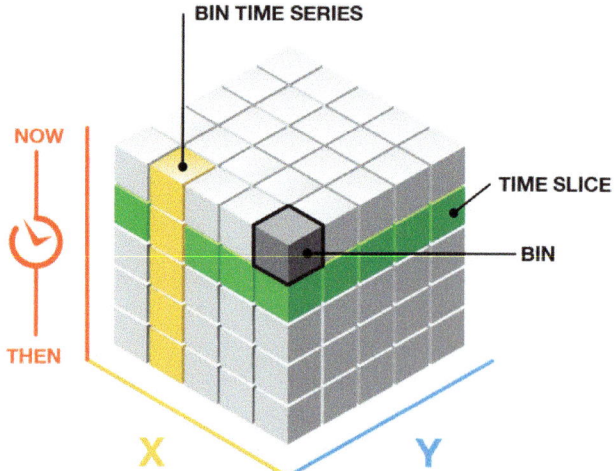

Figure 9.24. The structure of a space-time cube.

Chapter 9: Quantitative thematic maps   **229**

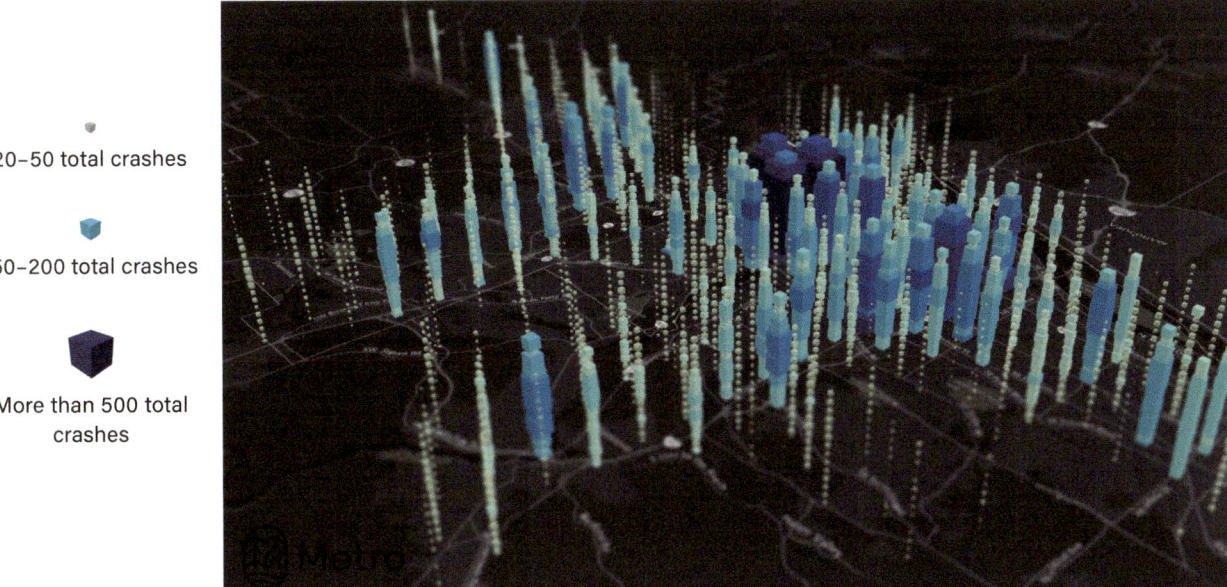

20–50 total crashes

150–200 total crashes

More than 500 total crashes

Figure 9.25. Select bin series within a space-time cube are revealed to show the number of vehicle crashes over time in Portland, Oregon. Courtesy of Matthew Hampton, Oregon Metro.

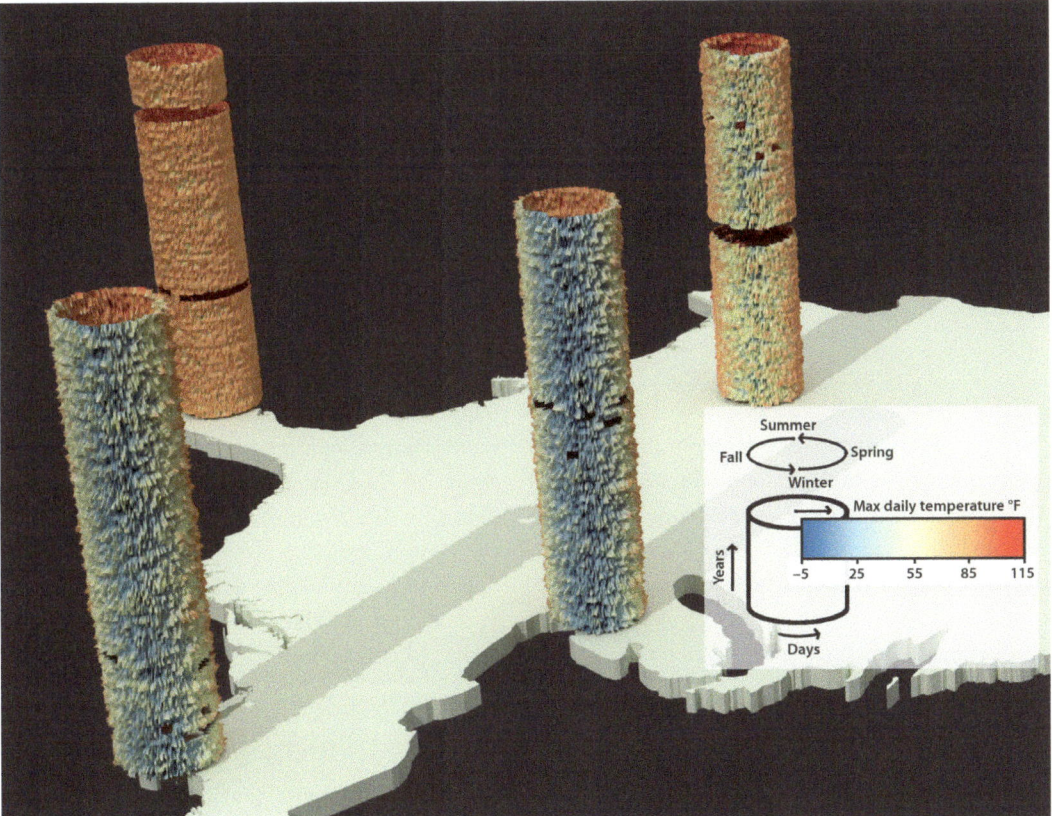

Figure 9.26. The temperature graphs on this map show the average monthly temperature range and 100 years of maximum temperature data for four weather stations in the United States. Courtesy of Drs. Christopher League and Patrick Kennelly, Long Island University.

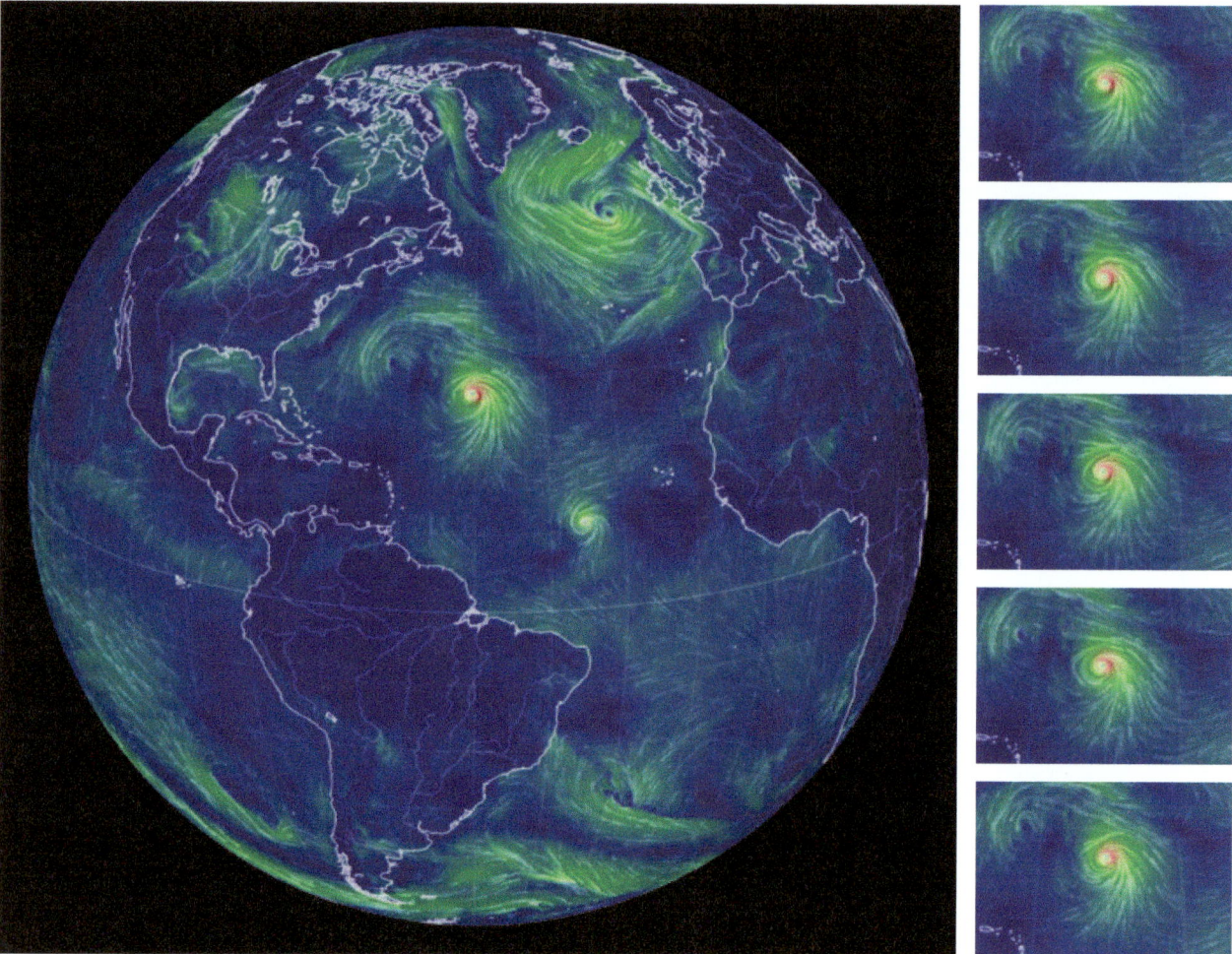

Figure 9.27. This animated map shows ocean currents as a dynamic phenomenon. © 2025 Nullschool Technologies Inc. Image courtesy of Cameron Beccario.

time. When that data is structured in a way that allows for the representation of information along multiple axes or dimensions, we call it **multidimensional data**. This type of data can be conceived of as a **space-time cube**—a three-dimensional cube composed of **bins**, with the x and y dimensions representing space and the t dimension representing time.

Maps can be created to show all the space-time bins, with controls for the user to "slice through" the map, or a show a bin time series for a certain location. The second option is illustrated in figure 9.25 which shows 15 years of total vehicle crashes for select locations in Portland, Oregon. Viewed online, the map can be examined from different vantage points and at different zoom levels.

Cyclical phenomena (phenomena that recur regularly) can also be shown with time in the vertical dimension. The temperature and seasonality over a region, for example, can be illustrated on a map consisting of annual graphs that show the yearly cycle of temperature at different locations (figure 9.26). The temperature graphs on this map show the average monthly temperature range and 100 years of maximum temperature data for weather stations at Plymouth-Kingston, Massachusetts; Fort Myers, Florida; Ann Arbor, Michigan; and Paris, Texas.

Dynamic maps

Mapmakers are now creating dynamic quantitative change maps as 3D displays, with time as the fourth dimension. These maps are often interactive, allowing users to change the vantage point and other display parameters through the user interface. The map of ocean currents in figure 9.27

is an example. It shows speed and direction of currents across the entire globe. Users can spin the globe to whatever point they want, and the animated display of ocean currents begins again. The small multiples on the right of the figure give you an idea of what the animation looks like at select points in time.

Selected readings

Baynes, J., A. Neale, and T. Hultgren. 2022. "Improving Intelligent Dasymetric Mapping Population Density Estimates at 30 m Resolution for the Conterminous United States by Excluding Uninhabited Areas." *Earth System Science Data* 14 (6): 2833–49.

Beccario, C. 2023. "Nullschool Earth." Website accessed at https://earth.nullschool.net/about.html.

Beconytė, G., A. Balčiūnas, A. Šturaitė, et al. 2022. "Where Maps Lie: Visualization of Perceptual Fallacy in Choropleth Maps at Different Levels of Aggregation." *International Journal of Geo-Information* 11, no. 1: 1–13.

Calka, B. 2021. "Bivariate Choropleth Map Documenting Land Cover Intensity and Population Growth in Poland 2006–2018." *Journal of Maps* 17, no. 1: 162–68.

Calvo, L., F. Cucchietti, and M. Pérez-Montoro. 2023. "Measuring the Effectiveness of Static Maps to Communicate Changes over Time." *IEEE Transactions on Visualization and Computer Graphics* 29, no. 10: 4243–55.

Cheshire, J., and A. J. Kent. 2023. "Getting to the Point? Rethinking Arrows on Maps." *The Cartographic Journal*. https://doi.org/10.1080/00087041.2023.2178134.

Cybulski, P. 2022. "An Empirical Study on the Effects of Temporal Trends in Spatial Patterns on Animated Choropleth Maps." *International Journal of Geo-Information* 11, no. 5: 1–15.

Cybulski, P., and V. Krassanakis. 2021. "The Role of the Magnitude of Change in Detecting Fixed Enumeration Units on Dynamic Choropleth Maps." *The Cartographic Journal* 58, no. 3: 251–67.

Dent, B. D. 1996. *Cartography: Thematic Map Design*, 4th ed. Wm. C. Brown.

Duncan, I. K., S. Tingsheng, Perrault, S. T., et al. 2021. "Task-Based Effectiveness of Interactive Contiguous Area Cartograms." *IEEE Transactions on Visualization and Computer Graphics* 27, no. 3: 2136–52.

Knura, M., and J. Schiewe. 2022. "Analysis of User Behaviour While Interpreting Spatial Patterns in Point Data Sets." *KN: Journal of Cartography and Geographic Information* 72 (3): 229–42.

League, C., and P. Kennelly. 2019. "Cartographic Symbol Design Considerations for the Space–Time Cube." *The Cartographic Journal* 56 (2): 117–33. https://doi.org/10.1080/00087041.2018.1533291.

Mennis, J., and T. Hultgren. 2005. "Dasymetric Mapping for Disaggregating Coarse-Resolution Population." *Proceedings of the 22nd International Cartographic Conference*, July 11–16, A Coruña, Spain.

Petrov, A. 2012. "One Hundred Years of Dasymetric Mapping: Back to the Origin." *The Cartographic Journal* 49 (3): 256–64.

Seipel, S., M. Andrée, K. Larsson, et al. 2020. "Visualization of 3D Property Data and Assessment of the Impact of Rendering Attributes." *Journal of Geovisualization and Spatial Analysis* 4 (2): 1–17.

Strode, G., V. Mesev, S. Bleisch, et al. 2020. "Exploratory Bivariate and Multivariate Geovisualizations of a Social Vulnerability Index." *Cartographic Perspectives* 95:5–23.

Tufte, E. 2001. *The Visual Display of Quantitative Information*. Graphics Press.

CHAPTER 10
Relief portrayal

The earth's surface provides the foundation on which we live, from strolling along the ocean near sea level to catching our breath at high elevations. It also can change quickly, as we ski down a mountain, pedal harder to bike over a hill, and generally play out our lives. Nothing in the environment is immune from the elevation at which we find ourselves. Our mobility, sense of direction, and environmental understanding are all affected by the vertical differences on the earth's surface, or **relief**. Yet it is easy to forget the significance of relief in our environment because many of us live in an essentially flat world. The floors of our homes, schools, and offices are flat; yards, parks, and sport grounds are often flat; and highways, railroads, and streets are designed to be as flat as possible. Thus, we tend to think of geographic location in purely horizontal terms. This limited thinking is often a perfectly acceptable way to simplify our world, but ignoring or misunderstanding relief can have severe consequences. Ships run aground, airplanes narrowly avoid mountainsides, and hikers lose their way—all because relief was not understood properly.

Relief relates to differences in height from place to place on the earth's **terrain surface**—a three-dimensional representation of the elevations of the physical environment at the earth's surface on land and below water. The term **hypsometry** is used to describe relief relative to sea level, whereas **bathymetry** refers to underwater relief. Surveyors, earth scientists, and others tend to focus on measuring and using elevation values on maps to study spatial relationships with other geographic phenomena, such as parcel boundaries, patterns of vegetation and rainfall, seabed variations, and so on. By contrast, cartographers often represent the terrain surface to help map readers visualize changes in relief. Different kinds of maps address these varying needs.

Elevation data

Today, many relief portrayal methods require data in the form of a **digital elevation model**, or **DEM**, which consists of cells (or pixels) arrayed in rows and columns (a grid), each containing a value that represents elevation. DEM data represents a continuous elevation surface, and all cells in the grid contain elevation values that are referenced to a common datum.

A **digital terrain model (DTM)** represents the elevation of the bare earth stripped of natural and cultural features (figure 10.1B). A **digital surface model (DSM)** is a model that represents the elevation of the terrain surface and all objects on it, such as vegetation, buildings, and other structures (figure 10.1A). DSMs can be processed to create DTMs by removing non-ground elevations. Because DSMs can be used to determine the presence and height of features on the terrain surface, such as the footprints and heights of buildings, they are used in applications such as land use planning, environmental modeling, and engineering projects. As a representation of the terrain surface, DTMs can be used to model and analyze landscape characteristics, such as slope, gradient, aspect, and curvature.

Spatial resolution

The level of detail in a DEM (and other raster data) largely depends on the cell size, or spatial resolution, of the data. **Spatial resolution** refers to the distance measured on the ground between cell centers. For example, if the distance between the centroids is two meters, the cell covers a 2 × 2 meter area, and the dataset is said to have two-meter resolution. The higher the resolution, the smaller the cell size and, thus, the greater the amount of detail. These relationships are illustrated in figure 10.2.

The cell must be small enough to capture the required detail but large enough that computer storage and analysis

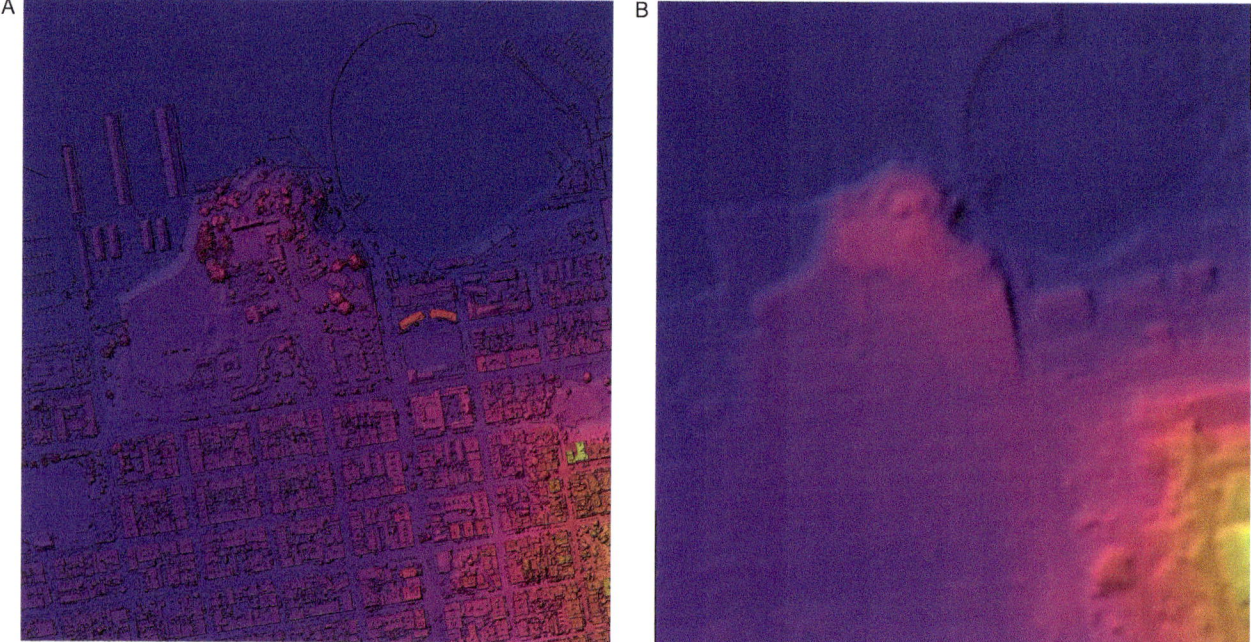

Figure 10.1. Digital surface models (*A*) represent the height of the earth's surface and all objects on it, such as buildings, vegetation, and above-ground structures. Digital terrain models (*B*) only include the elevation values of the land surface.

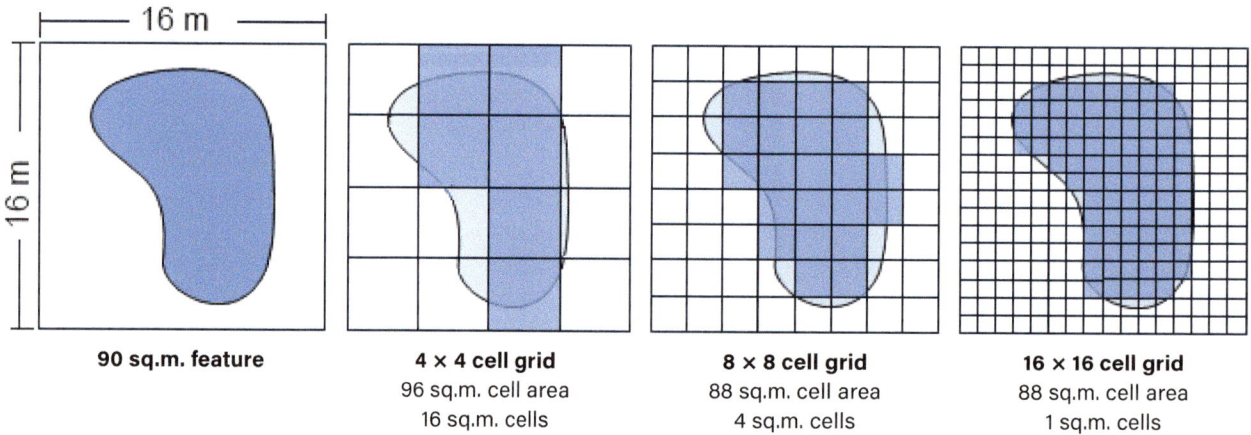

90 sq.m. feature

4 × 4 cell grid
96 sq.m. cell area
16 sq.m. cells

8 × 8 cell grid
88 sq.m. cell area
4 sq.m. cells

16 × 16 cell grid
88 sq.m. cell area
1 sq.m. cells

Figure 10.2. With raster data, more detail can be represented with smaller cell sizes at higher resolution.

Table 10.1. Differences between larger and smaller cell sizes

Larger size	Smaller size
Lower resolution	Higher resolution
Less precise feature representation	More precise feature representation
Smaller file size	Larger file size
Faster processing time	Slower processing time
Faster display time	Slower display time

can be performed efficiently. Smaller features or greater detail in features can be represented by a raster with a smaller cell size but at the cost of greater storage space and longer processing or display time. For example, for a given area, changing cells to one-half the current size requires as much as four times the storage space, depending on the type of data and data storage techniques used. These relationships are summarized in table 10.1.

Mapmakers create relief portrayal maps from data at the same or higher spatial resolution than the ground resolution of the map display pixels. The ground resolution of the map display pixels depends on the scale of the map. An equation you can use that relates optimal map scale expressed as a representative fraction ($1/x$), DEM cell resolution, and map display pixel resolution is given in equation (10.1):

$$\frac{1}{x} = \frac{1}{\left(\text{DEM cell ground size} \left[\frac{\text{cm}}{\text{cell}}\right] \times \text{map pixel density} \left[\frac{\text{pixels}}{\text{cm}}\right] \right)}$$

(10.1)

Here, one centimeter on the map represents x centimeters on the ground. For example, the optimal map scale for a 30-meter (3,000-centimeter) DEM mapped at 100 pixels/cm is 1/300,000.

DEM data collection

Before we had today's methods, DEM data was collected in the field by a team of land surveyors. The survey team set out a grid with x,y locations every 50 meters or more. Starting from a benchmark, usually a USGS or National Geodetic Survey (NGS) monument, the team captured the grid point elevations using **differential leveling**—a surveying technique that determines the difference in elevation between two points by measuring, at each point, the height of a leveling rod relative to a benchmark. If the survey team noticed that an area with rapid change in elevation, such as a hill or valley, was not covered by points on the initial grid, they added more points in the area. An elevation model was created by interpolating between the grid points to produce a smooth, continuous surface (see chapter 9 for a description of interpolation). This method to collect DEM data has changed over time, and today's methods most often use remote sensing to acquire the data (see chapter 11). We describe one such method next.

Lidar

Light detection and ranging (**lidar**) is a remote sensing system that uses rapid pulses of laser light striking the surface of the earth and features on its surface to determine elevations. The lidar sensor is mounted in the bottom of an airborne platform, usually an airplane or drone, with a

Figure 10.3. A lidar point cloud for an area along the East River near the Robert F. Kennedy Bridge in New York City. Courtesy of Jarlath O'Neil-Dunne, UVM Spatial Analysis Lab.

GNSS receiver used to determine the aircraft's horizontal position and an altimeter to record the aircraft's altitude. Some or all of each laser light pulse is reflected or scattered back to the sensor as a **lidar return**. The time it takes for a pulse to return to the sensor is proportional to the distance between the sensor and the reflected surface, which can then be used to determine the elevation of the surface. The elevations of the lidar returns (often shortened to **returns**) are used to create elevation models.

Lidar systems can record the distances for more than one return per pulse if the pulse is able to continue through a feature, such as vegetation. This characteristic of lidar datasets, called **point clouds**, coupled with the high resolution at which they are often collected, can result in hundreds of thousands or even millions of returns (figure 10.3). The first return captures the first object the pulse hits, which could be the tree canopy, buildings, power lines, bridges, and even birds flying in the area. In a vegetated area, the first, second, and third returns capture the canopy, understory, and bare ground, respectively.

Lidar data is often used to create high-resolution DSMs and DTMs. Figure 10.4 shows the point cloud (B) for a site (A) in profile view. The first return measures the elevations of surface structures, tree canopies, and the bare earth (C) and produces a DSM. The last return is used to model the elevation of the terrain surface. A DSM is created by subtracting the non-ground elevations from the surface elevations (D).

DEMs, DTMs, and DSMs have become key elevation models, and they are often available at no or low cost. Cartographers use them to create 3D maps and scenes styled in artistic and sometimes dramatic ways, and they can be draped with imagery or scanned maps, to create realistic world views and digital twins.

Vertical exaggeration

The appearance of relief in an elevation model can be enhanced using **vertical exaggeration**, abbreviated as **VE**, which is the amount that the vertical scale has been increased while keeping the horizontal scale unchanged. If you see on the map that vertical exaggeration is $5x$, or 'VE=$5x$", it means the vertical scale has been increased five times with respect to the horizontal scale. This, of course, also exaggerates the positional displacement of terrain features, but the exaggeration may serve the mapmaker well by giving the visual impression they want for the map.

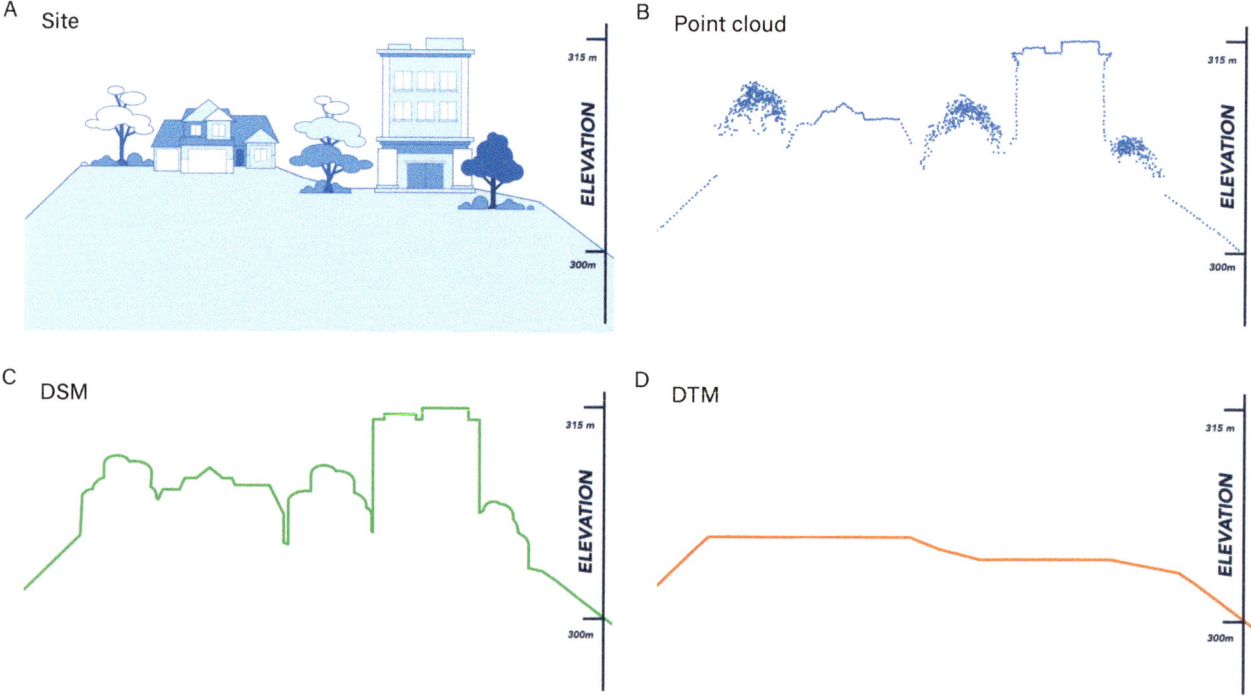

Figure 10.4. For a site (A), lidar data that is captured as a point cloud (B) is used to derive a DSM (C). Removing the non-ground elevations produces a DTM (D).

Map perspective

Many maps portray relief from an overhead viewpoint. These **planimetric** or **plan perspective maps** are useful when the area is essentially flat or when an area's relief is not of primary visual importance, especially when terrain information risks cluttering the map with unnecessary detail. Planimetric maps are useful for depicting precise elevation information as points, lines, and areas, but for relief portrayal, these maps are always deficient to some degree because they attempt to show a 3D surface in 2D.

An alternative to a planimetric map is an **oblique-perspective map**, which shows an area from a viewpoint somewhere between vertical and horizontal, like looking out the window of an airplane. Oblique-perspective maps have a more 3D appearance than planimetric maps, but showing exact elevation values at points is difficult. With their bird's-eye view of the landscape, oblique-perspective maps have strong intuitive appeal, accounting in part for their popularity, but there is a problem. On these maps, features are often displaced from their true horizontal locations. Positional displacement of a terrain feature occurs in direct proportion to its height on the terrain surface. The shift in horizontal position between the bottom and top of a tall mountain is much greater than for a small hill. The top, bottom, or middle of a hill can be placed in its correct horizontal position on the map, but not all three at once, so it is impossible for an oblique-perspective map to be planimetrically correct.

Examples of relief portrayal on planimetric (*A* and *B*) and oblique-perspective (*C* and *D*) maps are shown in figure 10.5. At first glance, relief portrayal may simply seem like an attractive way to display the earth's surface. However, to the eye of an experienced map user, it can help them better understand land use and settlement patterns, explore underwater landscapes, and identify areas at greater risk of natural disasters, such as flooding or landslides.

A **varied perspective map**, such as a globe, does not have a fixed viewpoint because the map user can look down on the map from a planimetric perspective or view it from the side in oblique perspective. These maps are created as solid physical models, sometimes with great vertical exaggeration, so that as the map user's vantage point moves from vertical (overhead) to horizontal (profile), the 3D relative relief is enhanced. These characteristics of varied perspective maps help explain their popularity.

We now turn our attention to the portrayal of absolute versus relative relief on maps. For relative-relief mapping, it helps to categorize the methods by their map perspective.

Absolute-relief maps

When accurate relief information is important, it is often best to turn to topographic and other planimetric maps that show **absolute relief**, or precise elevation values at specific locations on the earth's surface. **Absolute-relief maps** provide numerical elevation or water depth information at points, along lines, or within areas. These types of maps are used by engineers, surveyors, navigators, and others who require meticulous detail about the earth's 3D terrain surface.

Absolute land elevation and water depth values can be mapped in several ways. One approach is to measure and label individual points with their elevation or depth values. Surveyed elevation or depth values at specific points are often the most accurate terrain surface information shown on maps. Another mapping technique is to trace and label lines connecting points of the same elevation, resulting in isolines known as contours (or in the case of the underwater terrain, isobaths). A third method is to add colors to show elevation variations over the terrain surface using a technique called hypsometric tinting. Next, we look at each of these mapping methods in more detail.

Spot elevations, benchmarks, and soundings

On aeronautical and nautical charts, topographic maps, cadastral maps, and other navigational and large-scale maps, the elevation of the terrain surface is given numerically at individual points. These elevation values, relative to the mean sea level (MSL) datum (see chapter 1) are called **spot elevations**. Spot elevations are often located at control points or on features that are **recoverable survey marks**—that is, they can be easily found again. Searching for survey marks, or **mark recovery**, at permanent features in the landscape, such as road intersections or mountain summits, is much easier than at changeable locations, such as stream confluences or edges of vegetation patches. As shown in figure 10.6, spot elevations on historical USGS topographic maps are symbolized with a small *x*, followed by the elevation value above MSL. When spot elevations are located at a road intersection, the symbol is omitted, because the location of the survey mark is known. For survey marks that are also benchmarks (see chapter 1), the identifier "BM" precedes the elevation.

Soundings are the depths of waterbodies, such as lakes, rivers, and oceans, at specific points. For thousands of years, mariners have measured shallow water depths using a calibrated **lead line** (a line with a heavy lead weight tied to one

Figure 10.5. Segments of planimetric maps show the topography of Utah (*A*) and Mount Usu volcano in Japan (*B*). Oblique-perspective maps show Mount Fuji, Japan, with Sentinel-2 10-meter land cover data draped on the terrain (*C*) and the bathymetry of Challenger Deep in the Pacific Ocean near Guam (*D*). Courtesy of Lynn Roth, Utah BLM (*A*) and Asia Air Surveys Co. Ltd. (*B*).

end that is lowered into the water to determine depth by markings on the line) or a **sounding pole** (a pole marked with water depth values). Today, soundings are obtained by electronic depth-measuring instruments, such as **depth sounders** and sonar systems, that send an acoustic pulse to the bed and calculate the depth based on the time it takes for the sound to bounce back.

Soundings are not relative to MSL, but rather to specific definitions of low water. The two datums (see chapter 1) used in North America to define low water are the arithmetic average of all the low-tide levels recorded over a 19-year period, called **mean low water** (**MLW**), and the arithmetic average of the lower of the two daily low tides recorded over the same 19-year period, called **mean lower**

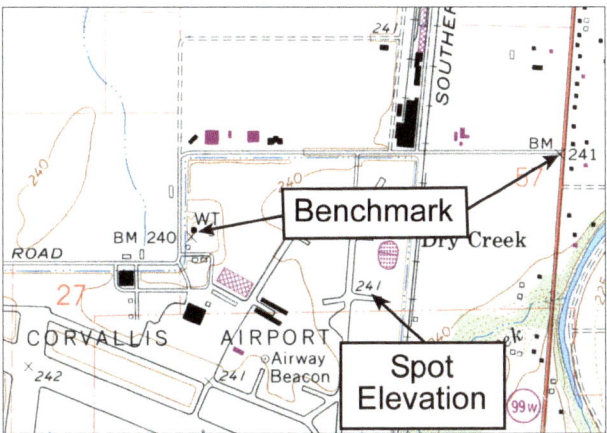

Figure 10.6. Spot elevations and benchmarks as portrayed on a historical 1:24,000-scale USGS topographic map. Map courtesy of the US Geological Survey.

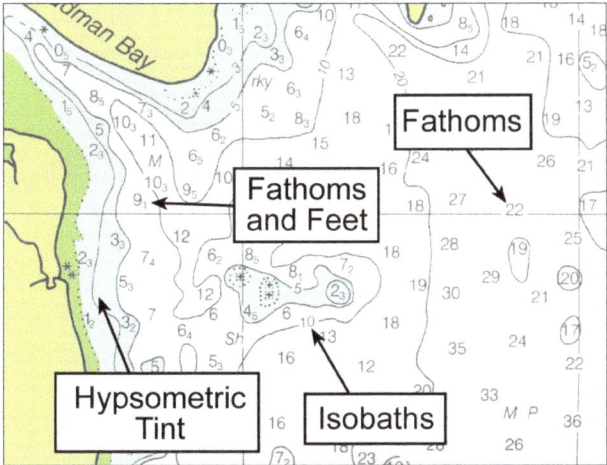

Figure 10.7. Soundings on US nautical charts, especially older ones, are usually shown in fathoms for depths greater than 11 fathoms. Fathoms and feet are used together for shallower areas. Chart courtesy of NOAA National Ocean Service.

low water (**MLLW**). Canadian nautical charts use MLW, but in the United States, MLLW is the official nautical chart datum of the US **National Ocean Service**, the agency responsible for preserving and enhancing the nation's coastal resources and ecosystems. The datum used to define high water to support harbor and river navigation is **mean high water** (**MHW**), the average height of all high tides over a 19-year period.

The MLLW datum is used on nautical charts for water depths because ship captains must decide whether there is enough clearance for their vessels between the changing water surface and a fixed submerged obstacle, such as the harbor bottom or a wreck at low tide. To do so, they must determine the minimum depth likely to be encountered at a given position. If overhead clearance is the concern, as when moving under a bridge at high tide, the situation is reversed, and the MHW datum is more useful.

Soundings have been printed on nautical charts since the early 1600s. Charts that cover the coastline of the United States traditionally use the **fathom** (six feet) as the unit of measurement for depth (see figure 10.7). In areas shallower than 11 fathoms (66 feet) in which more exact depths are required for safe navigation, fathoms and feet are used together. The value 9_1 in figure 10.7, for example, is read as nine fathoms and one foot, or 55 feet. Nautical charts made in most other nations show depths in meters, as do more recent US charts.

Contours

Contours are lines of equal elevation above a datum. If a contour were actually drawn on the earth, it would trace a horizontal path that is constant in elevation. It is common to use contours on topographic maps because variations in relief are shown by the spacing of contours, and landform features such as hills and valleys are shown by their shape. The segment of the historical 1:24,000-scale USGS topographic map shown in figure 10.8 contains several types of contours found on topographic maps. All but the approximate or indefinite contours are shown on current USGS topographic maps.

To understand the logic behind contours, imagine that you are on a small island in the ocean. It has two hills, one that is just over 700 feet (213 meters) high and another that is just over 500 feet (152 meters) high (figure 10.9). If you walked around the island at the shoreline when the tide was at MSL, you would trace the **zero contour**, or **datum contour**, that directly overlies the shoreline and return precisely to your starting point. All contours eventually close like this, although the closed curve cannot always be seen on a single map sheet.

Now suppose that you start climbing up the highest hill until you reach an elevation of 100 feet above MSL. If you again followed a path around the island at this constant elevation, your footsteps would trace a new contour, this time the 100-foot contour, and again you would return to your starting point. The effect is the same as if you walked along the shoreline after the ocean level is raised 100 feet. If you did the same thing for elevations of 200 feet, 300 feet, and so on, and the paths you walked were projected

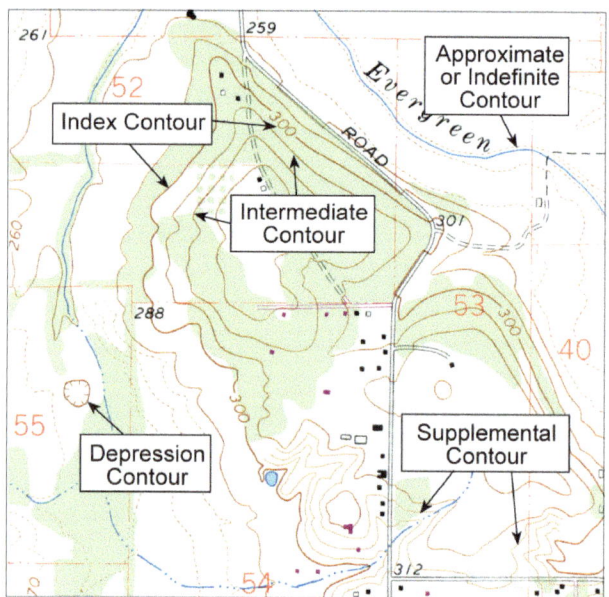

Figure 10.8. A segment of a historical 1:24,000-scale USGS topographic map for Corvallis, Oregon, with a 20-foot contour interval. Contours of different types are used to portray special landform features and to make the map easier to read and analyze. Map courtesy of the US Geological Survey.

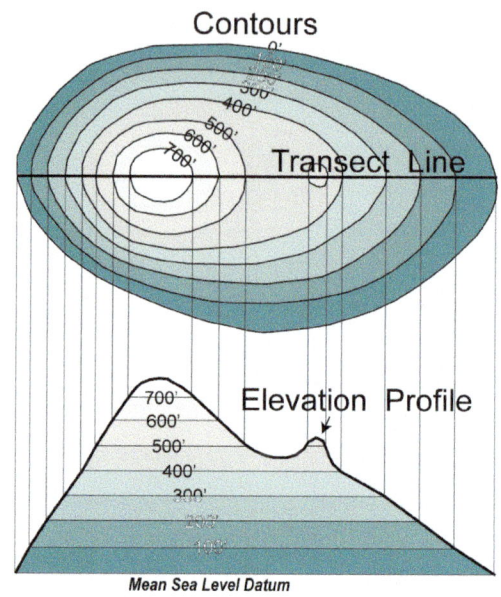

Figure 10.9. The logic of the contour method of relief portrayal is illustrated here by the relationship of the contours that intersect the horizontal transect line on the map (*top*) to their corresponding locations on the elevation profile (*bottom*).

vertically onto a flat map, the result would resemble the map at the top of figure 10.9. If the island is viewed in profile, it would look as if it were sliced into layers of equal thickness by imaginary horizontal planes, as in the **elevation profile**, or **terrain profile**, at the bottom of figure 10.9.

In our island example, the vertical difference in elevation between adjacent contours, called the **contour interval**, is 100 feet. The mapmaker, of course, may select a contour interval of 20 feet, 50 feet, or whatever seems appropriate, considering the extent of the area being mapped and the amount of relief in that area. The smaller the interval used, the more detailed the relief portrayal. Once you know the contour interval, estimating the elevation anywhere is a simple matter of "reading between the lines" on the map. If the spot you are interested in lies halfway between the 500-foot and 600-foot contours, you can conclude that it is around 550 feet high.

On most maps that show contours, the MSL datum and a constant contour interval are used. You are usually safe in treating these factors as constants so only the size of the interval varies from one map to another. In general, the greater the relative relief on a map, the larger the contour interval. Mapmakers use this relationship between relief and contour interval to keep contours from becoming too dense in areas of high relief and too sparse in areas of low relief.

When relief changes markedly from one part of the map to another, it is possible that several different but constant intervals are used on the same map. For example, this **variable contour density** might occur when the map sheets span high to low relief. Without variable density, closely spaced contours clutter the map and become difficult to differentiate, and widely spaced contours make it hard to determine slopes and predict elevations at points that fall between the lines.

Types of contours

On USGS and other topographic maps, you are likely to see a variety of **topographic contours** to show the shape or details of the land surface, including index, intermediate, supplemental, and depression contours (table 10.2).

Not every contour is marked with its elevation value; instead, every usually fourth or fifth line is labeled, depending on the scale and contour interval of the map. To help map readers more easily determine elevations, these labeled **index contours** are drawn with a thicker line (see figure

Table 10.2. Contours and their symbology

Contour	Symbol
Index	~~~8000~~~
Intermediate	~~~
Supplemental	~~~
Depression index	~~~4000~~~
Depression intermediate	~~~
Depression supplemental	~~~

Note: Contour symbology is taken from the *Topographic Map Symbols* document for 1:24,000-scale USGS topographic maps and doubled in line width for this figure.

Source: US Geological Survey.

10.8 and table 10.2). **Intermediate contours** are drawn with a thinner line between the index contours. **Supplemental contours** (table 10.2) are sometimes shown to enhance the depiction of relief in areas where elevation change is minimal. They almost always represent half the contour interval for the rest of the map.

To aid in identifying closed **depressions** (areas lower than the surrounding surface in all directions), small right-angle ticks may be added to the downslope (or inside) of contours (see figure 10.8 and table 10.2). These **depression contours** help focus attention when the depression is small relative to the size of the area mapped, which is often the case. One example would be an area pockmarked with sinkholes resulting from the natural collapse of underground caverns. Depression contours are especially useful in distinguishing depressions from small hills because there is seldom enough space for the mapmaker to label the contours of these features with their elevation values.

Isobaths

If a contour shows the land that lies underwater, it is an **isobath** or **bathymetric contour** (also called a **depth contour** or **depth curve**). Isobaths are lines of equal water depth below the MSL or MLLW datum. Isobaths are found on nautical charts and other **bathymetric maps** that show water depths. The first isobaths appeared on European charts of river estuaries in the late 1500s, demonstrating how easy it was to measure below a water surface with a lead line compared with measuring elevation above a datum with yet-to-be-developed surveying tools. By the end of the 18th century, enough soundings were taken that interpolation could be used to draw isobaths on many nautical charts, especially in shallower areas (see chapter 9 for more on interpolation). Today, electronic depth sounders provide the detailed information required to draw isobaths on virtually all maps and nautical charts of inland waters, the ocean, and other deep-water bodies.

Isobaths on nautical charts differ from contours on topographic maps in that isobaths use uneven depth intervals as opposed to a constant contour interval. For instance, charts produced by the US National Ocean Service show isobaths for 1, 2, 3, 5, and 10 fathoms, with greater depths shown in multiples of 10 fathoms, typically 20, 50, and 100. Shallow-water isobaths appear on charts for the same reason that shallow-water depths are shown in fathoms and feet—mariners need this information for safe coastal and harbor navigation.

Hypsometric tints

Hypsometric tints (also called **elevation tints** or **layer tints**) are symbols used to "color between contour lines," using colors to represent different elevation ranges. With hypsometric tinting, the elevation ranges can have a stepped appearance, much like a layer cake (figure 10.10). The elevation ranges between contours are given distinct colors, or sometimes gray tones, called **discrete hypsometric tints** or **discrete layer tints**. This mapping method allows you to determine the elevation range of any location but not necessarily the elevation at specific points. This approach can also be used to map depths of underwater surfaces.

For the terrain surface, cartographers often try to use colors that relate to the type of land cover found at different elevations, such as green for low-lying vegetated valleys; brown for high-elevation treeless areas; and white for snow-capped peaks. Between isobaths, depth ranges are often shown with a progression of blue tints, with darker blues generally used for deeper waters (figure 10.10). The simplest discrete hypsometric tinting is on US nautical charts, in which a light-blue tint is added to all water areas between land and the five-fathom isobath (see figure 10.7).

With today's mapping capabilities, the abrupt change between discrete hypsometric tints can be eliminated using **continuous hypsometric tints**, which gradually merge one tint into the next, giving a smooth appearance to the gradation of colors (figure 10.11). This mapping symbology is readily achieved using computer software and DEM data.

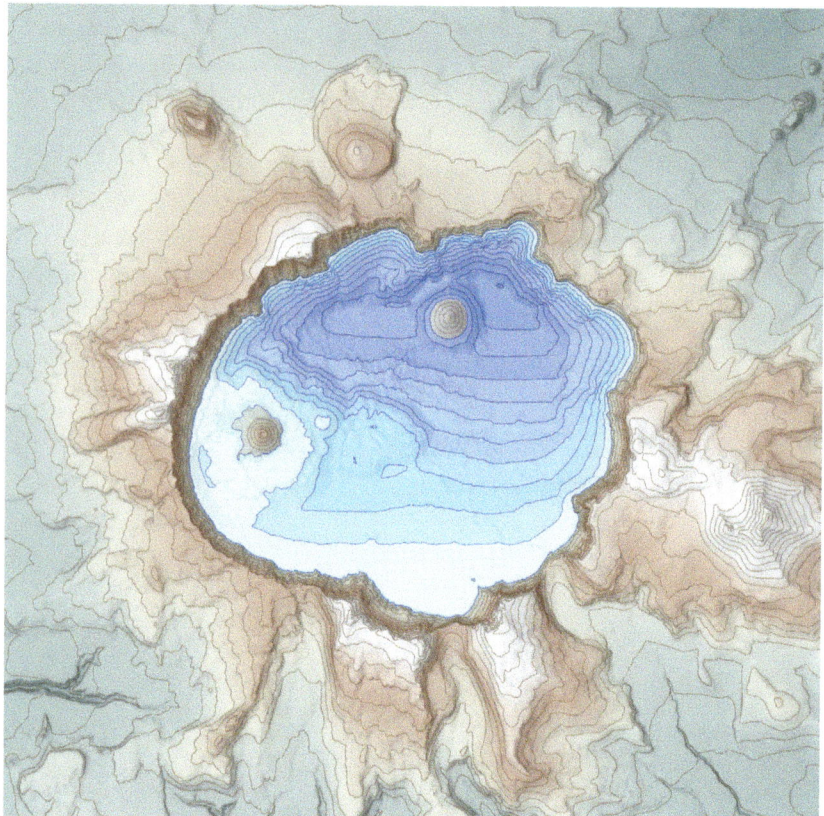

Figure 10.10. Green to brown to white discrete hypsometric tinting used to symbolize land and blue tinting used for water visually enhance land elevations and water depths for the contours and isobaths in the Crater Lake area in Oregon.

Figure 10.11. Continuous hypsometric tinting creates a smooth progression of tints for water depths and the land surface form.

The key advantage of this method is that each elevation value of the DEM is assigned a color that corresponds to its exact elevation. However, the colors vary so gradually that you cannot see the elevation ranges that are clearly visible with discrete hypsometric tinting. For this reason, it is harder to determine by eye an exact elevation or elevation range at any location. To find this information, you must rely on contours, if they are also shown.

Some mapmakers have combined classical hypsometric colors, which traditionally sought to mimic landscape colors of the Swiss Alps, with a wider palate of landscape colors found throughout the world. One approach uses **cross-blended hypsometric tints** to assign significantly different color ramps to arid and humid regions to create maps like the one shown in figure 10.12.

Relative-relief maps

In our day-to-day lives, we often think of relief in terms of the 3D features we can see in the landscape, such as hills and plains, mountains, and valleys. We are not as concerned about the actual heights of these features as we are the variations in their elevations. **Relative relief** is the difference in elevation between the highest and lowest points within a specific region. It helps us understand the topographic variations within an area and provides us with a sense of the terrain's ruggedness or flatness. The relative relief map in figure 10.13 clearly depicts the steep slopes, deep canyons, and rocky outcrops in the mountains in the Wasatch Range, whose front rises sharply from the gently sloping valley floor in northern Utah.

We now turn our attention to methods that mapmakers have developed to show relative relief. One way to categorize these methods is by map perspective, or the viewpoint from which the map is seen. As we noted, plan perspective maps are designed for top-down viewing, and oblique-perspective maps are designed for a bird's-eye view. But some maps can be viewed from all angles, or a variable perspective.

Variable perspective maps

A type of relief portrayal that is a special case in terms of mapping perspective and vertical exaggeration is a physical model of the terrain. Solid models of the landscape are unusual in that they typically vertically exaggerate elevations to create map displays that not only look three-dimensional, but whose bumps and indentations can be felt with the fingertips. Also, such models do not have a fixed viewing perspective, as the map user can look directly down on the model or view it from an oblique perspective. All of this helps to explain the popularity of the physical terrain models described in the following section.

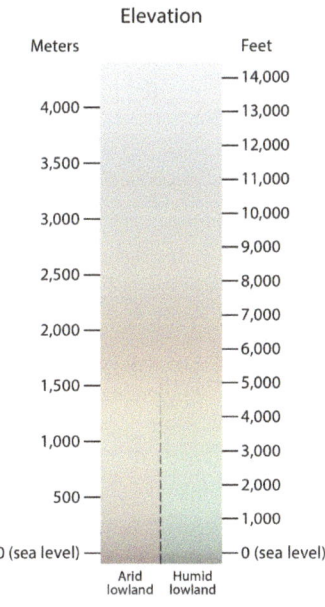

Figure 10.12. A map centered on the conterminous United States showing elevation with cross-blended hypsometric tints, which use different continuous hypsometric tints for low-lying arid and humid areas. Courtesy of Tom Patterson and Bernie Jenny.

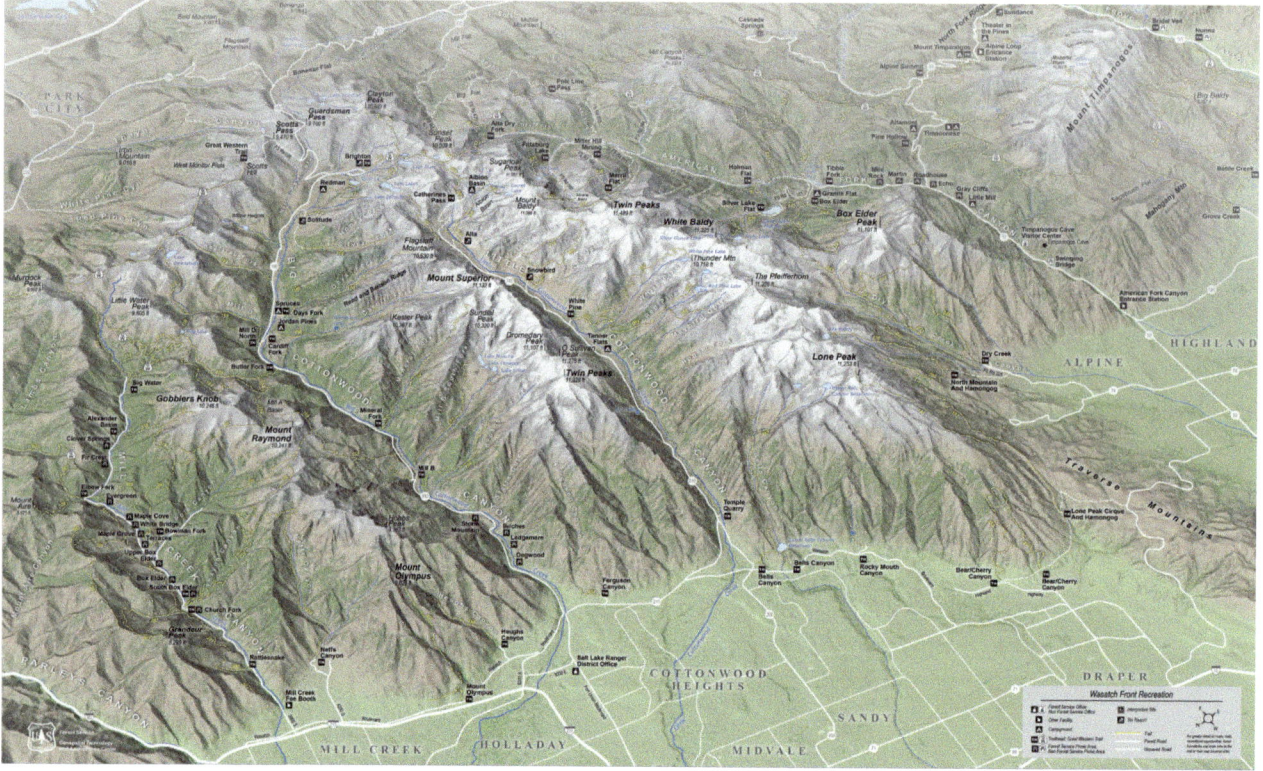

Figure 10.13. An oblique-perspective recreation map of the Wasatch Front, Utah. Courtesy of Andrew Keske, US Forest Service.

Raised-relief globes

A **raised-relief globe** is a globe whose surface features are raised and exaggerated to highlight the earth's relief (figure 10.14). Unlike traditional flat maps or globes that depict the earth's surface in a 2D format, raised-relief globes provide a tactile and visual representation of the planet's topography.

Because globes present the truest picture of the earth as a whole, you might conclude that raised-relief globes would provide the most realistic and useful portrayal of the vertical dimension. Considering the horizontal scale of globes with vast regions extending over extensive mountain ranges, mapmakers greatly exaggerate the relief on this form of terrain model. This is needed because, as you saw in chapter 1, if the earth were reduced to the size of a bowling ball (a common globe size), the earth would be smoother than the ball. In practice, a VE of about $20x$ is typical on raised-relief globes of a 24-inch (60-centimeter) diameter. Therefore, in addition to the highly generalized landforms because of their small scale and the handling and storage inconvenience of globes, users of raised-relief

Figure 10.14. Elevation variations on a raised-relief globe must be greatly exaggerated to give a reasonable impression of the earth's landforms. Courtesy of RäthGloben, Germany.

globes also face the problem of large VE values greatly distorting the actual vertical surface variation.

Raised-relief maps

A **raised-relief map** is a type of physical terrain model that partially sidesteps the problems of weight and high cost. It is made by printing a plan perspective map, often a topographic map, on a sheet of plastic and using heat to mold the plastic over a 3D model of the terrain surface (figure 10.15). The molded plastic map has the advantages of a planimetric map with the added enhancement of a 3D landform surface.

Raised-relief maps cost about 10 times more than a conventional map of the same size and area, and they are relatively fragile, prone to cracking, and difficult to store. Because the flat plastic sheet must be stretched during construction, some positional displacement of features on the raised surface is also common. This misrepresentation is inevitable and often compounded by the fact that these maps are usually produced at medium to small scales for wider sales appeal. With small-scale maps, a small stretching of the plastic gives greater distortion than the same amount of stretching for a large-scale model. But the realistic impression of relief often compensates for these drawbacks.

Terrain models

The problem of large VE can be minimized using physical terrain models that cover more limited areas. You can think of **terrain models** as blocks cut out of the surface of a giant raised-relief globe. They let you focus on a smaller portion of the earth at a larger scale than a raised-relief globe, allowing relief to be shown with less VE. **Stepped terrain models** are constructed with discrete elevation layers, much like discrete hypsometric tinting, but the areas between contours are flat planes at the same elevation (figure 10.16). The stepped effect is almost always a result of the mapping technique, not the nature of the terrain surface, although layered landforms do occur in the rare regions of fine terraced agriculture, open-pit mining, and horizontal sedimentary beds with varying resistance to erosion.

Smooth terrain models without stepped artifacts show a more realistic continuous terrain surface, although the relief is still typically vertically exaggerated to make the relief more apparent. The terrain model in figure 10.17 is an example. It was created using DEM data to guide a large computer-controlled milling machine to cut the terrain surface into a thick block of polyurethane foam. A custom 3D ink-jet printer then printed a satellite image on the model to add realism.

As with all physical models, terrain models have the disadvantages of bulk, weight, and a high cost of production. For these reasons, they are normally used only in permanent or semipermanent displays, such as national park or natural history museum exhibits; the lobbies of government and private agencies that deal with environmental problems; city or regional planning exhibits; and university geology, geography, and landscape architecture departments. They are also popular for visualizing promotional schemes or research projects, such as urban renewal proposals, mall developments, and dam sites. For these uses, the inconvenience of constructing the physical models is offset by the true-to-life impression of the landscape they provide.

Figure 10.15. A raised-relief map of Japan printed on a sheet of plastic molded to form a 3D surface. Map courtesy of Benchmark Maps, an imprint of East View Information Services. Photo by Esri.

Figure 10.16. The layer-by-layer construction of some terrain models often leaves them with a stepped surface. Courtesy of HowardModels.com.

Plan perspective maps

We now turn to ways that relative relief can be shown on plan perspective maps, focusing on relief shading and enhancements to this mapping method.

Relief shading

Relief shading, also called **hillshading** or **shaded relief**, sidesteps the problems of weight and cost associated with portraying the 3D appearance of the terrain surface by quickly and easily simulating the effects of light and shadow to create a sense of depth and elevation on a flat surface. You know from your experience looking out at the landscape from an airplane that patterns of lighter and darker shading on hills give the strongest impression of relief. Relief shading re-creates these tonal variations on maps to give the same 3D effect.

The principle underlying relief shading is that terrain features appear 3D to us when we use an imaginary light source from the northwest (top-left corner of the map) to create the relief depiction (figure 10.18A). The lightest gray is on the steepest, northwest-facing slopes, which are perpendicular to the imaginary rays of the light source.

The darkest gray is on the steepest, southeast-facing slopes, which are on the opposite side of the terrain surface facing the light rays. With this method, lightness and darkness are determined not only by slope steepness but also by orientation of the terrain surface with respect to the light source. If the light source comes from the opposite direction, the southwest (bottom-right corner of the map), the terrain surface appears inverted (figure 10.18B) resulting in **relief reversal**, which is exactly the opposite effect from what is intended. With relief reversal, hills look like valleys, and valley bottoms look like ridge tops to most users.

Multidirectional relief shading

Another challenge with relief shading occurs when features are oriented roughly parallel to the imaginary light rays because the terrain surface is not shaded enough to give the visual cues needed to see the surface in 3D. To circumvent this problem, mapmakers use **multidirectional relief shading**, which adds multiple imaginary light sources to the map's west and north sides so that a terrain surface with north–south and east–west orientation also receives sufficient shading. This use of multiple light sources customized

Figure 10.17. Side view from the bottom-right corner of a 24-inch-wide-by-36-inch-high terrain model of an area that extends from the Seattle, Washington, metropolitan area southeast to Mount Rainier. The map is carved from a sheet of polyurethane foam that is one to two inches thick. Courtesy of General Plastics Manufacturing Company.

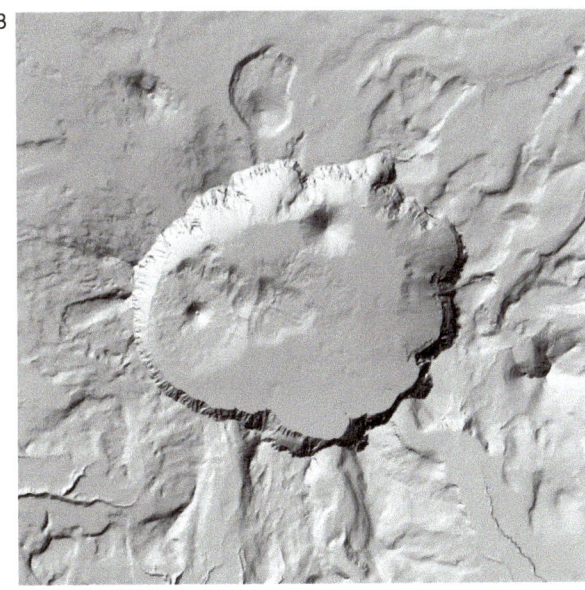

Figure 10.18. Shaded-relief maps of Crater Lake, Oregon, using oblique illumination with the light source from the northwest (*A*) and southeast (*B*). Relief reversal is evident on map *B* because hills look like valleys and valleys look like hills. Data courtesy of the US Geological Survey.

for individual landforms was the gold standard of Swiss and other expert cartographers to show relative relief on a flat map. The example shown in figure 10.19 looks as real as a physical terrain model of the Matterhorn in the Pennine Alps between Switzerland and Italy.

Relief shading used to be a laborious manual chore that required artistic skill and a thorough understanding of **geomorphology** (the study of the origin, evolution, and dynamics of landforms). The expense involved in producing shaded-relief maps by hand could rarely be justified. Thus, despite the dramatic visualization that relief shading afforded, for practical reasons, not many such maps were made. This situation has changed dramatically, because relief shading can be produced in seconds using computers and DEM data is plentiful. So, you will often find a shaded-relief surface as the basemap for many maps.

Relief shading and shadowing

Historically, mapmakers used shadows sparingly in artistic renderings, and this practice carried over to the early years of computer automation. This was because shadowed areas were typically shown in black and did not allow the relief beneath the shadows to be seen. This is a shortcoming of all the imaginary light rays coming

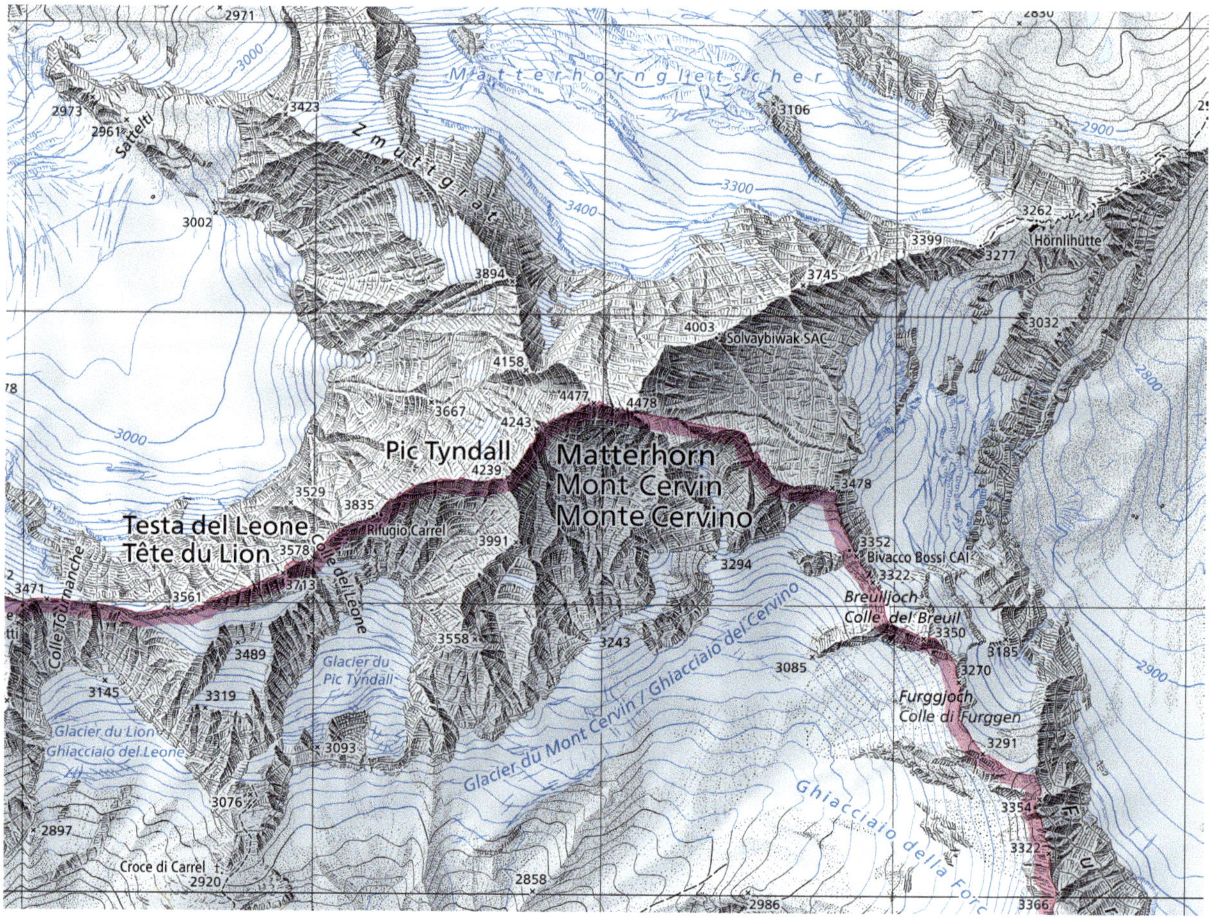

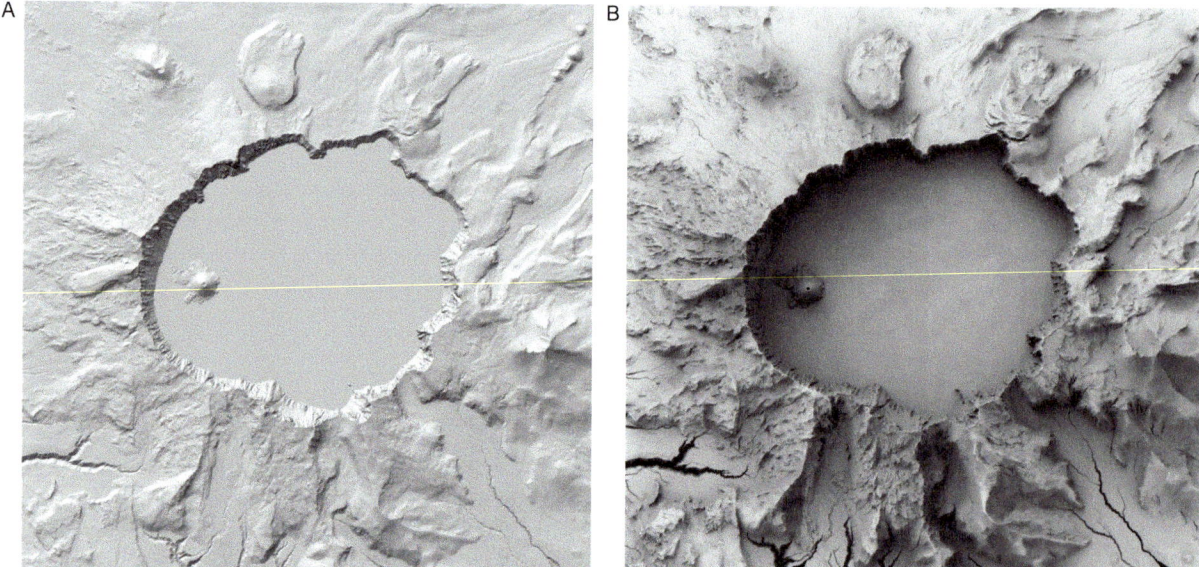

Figure 10.19. The use of artistic shading enhances the impression of the landform surface on this Swiss 1:100,000-scale topographic map. Rock outcrop drawings also enhance visual realism. © swisstopo.

Figure 10.20. Maps of Crater Lake, Oregon, rendered with (*A*) default relief shading and (*B*) a sky model with the lighting conditions of a clear day. Data courtesy of the US Geological Survey.

from only one northwestern direction. But this problem is addressed by more-advanced lighting models that create the types of soft shadows we see in nature on a cloudy or overcast day. **Sky models** combine relief shading and shadowing from hundreds of directions to approximate relief displayed under various natural lighting conditions. Advantages of this technique can be seen by comparing the maps in figure 10.20. Figure 10.20A was produced with the default hillshading method that uses a light source at an azimuth of 315° (from the northwest) and an altitude of 45° (halfway between the horizon and directly overhead). Figure 10.20B was produced with a sky model to create a map with a softer shading effect, enhanced relief, and variable patterns of shadowing, even on the water of Crater Lake.

Relief shading and hypsometric tints

It is common to see maps that combine both relief shading and hypsometric tints (figure 10.21). Indeed, it is unusual to find a map today that has hypsometric tinting without

Figure 10.21. Both relief shading and hypsometric tinting are shown on this section of a landform map of the United States. Courtesy of Tom Patterson and Bernie Jenny.

relief shading, primarily because both techniques can be easily created and combined using computers and DEM data. If the mapmaker can create one effect, it is easy to add the other, with the biggest challenge being to effectively assign the hypsometric tints to the elevation values.

Relief shading and contours

A skillful combination of contours, spot elevations, and relief shading is one of the most effective ways to portray absolute relief as well as the relative relief of the terrain surface. Large-scale Swiss topographic maps, such as the map shown in figure 10.19, are noted for combining absolute and relative relief mapping methods with clarity and beauty. These maps are works of art that combine sophisticated hillshading techniques with masterfully designed hypsometric tints and rock outcrop drawings, yet elevations can easily be determined from the contours, as well as spot elevations and benchmarks.

Oblique-perspective maps

Although an oblique-perspective map is unable to show the proper planimetric locations of elevation values on the terrain surface, such maps open a world of possibilities for mapmakers. Unlike varied perspective maps, oblique-perspective maps use a fixed viewing direction. Mapmakers have created oblique-perspective views in a variety of ways

Figure 10.22. Hillsigns and hachures on Jos Murer's 1566 map of Zürich Canton in Switzerland. Map scale is approximately 1:56,000.

over the centuries, so we start here with early attempts to provide an oblique view of the terrain surface on maps.

Hillsigns and hachures

The stylized drawing of hills from an oblique perspective is one of the oldest ways to show relief, and you will still see this mapping technique used today. From the Western European Middle Ages through the Renaissance and up to the mid-1800s, hills and mountain ranges on both large- and small-scale maps were represented by line drawings, sketches, and paintings of stylized hills called **hillsigns**. Some drawings look like conical mole hills that bear little resemblance to the actual features symbolized, but other artists mastered techniques to show the continuity of mountain chains and the relative heights of peaks with shading or linework. Jost Murer's map of Zürich Canton from 1566 demonstrates how hillsigns can be portrayed skillfully with the form lines of ridges and the depiction of slopes with **hachures**, or short parallel lines (strokes) drawn in the direction of the steepest slope (figure 10.22). Steeper slopes are represented by shorter, thicker **strokes**.

Hillsigns are commonly found on fantasy, recreational, and advertising maps, usually designed for the public, to give a general impression of the character of terrain features. Perhaps the most well-known use of hillsigns is on the fantasy maps of Middle-earth in J. R. R. Tolkien's books. This style has been adapted by mapmakers to create interesting portrayals of imagined and real places, as illustrated in figure 10.23 for the area around the Mediterranean Sea.

Block diagrams

A **block diagram** is a schematic representation of a portion of the earth's surface using simple symbology to represent landscape components and their relationships (figure 10.24). The vertical sides of the block diagram also allow the underlying rock formations, mining operations, or other subsurface features to be shown. Scientific diagrams, popular geographic magazines, and even advertisements take advantage of the x-ray perspective of block diagrams. Illustrations in geology, physical geography, and other environmental textbooks often use block diagrams to teach concepts about what is occurring beneath the ground compared with what we are seeing or experiencing on the surface.

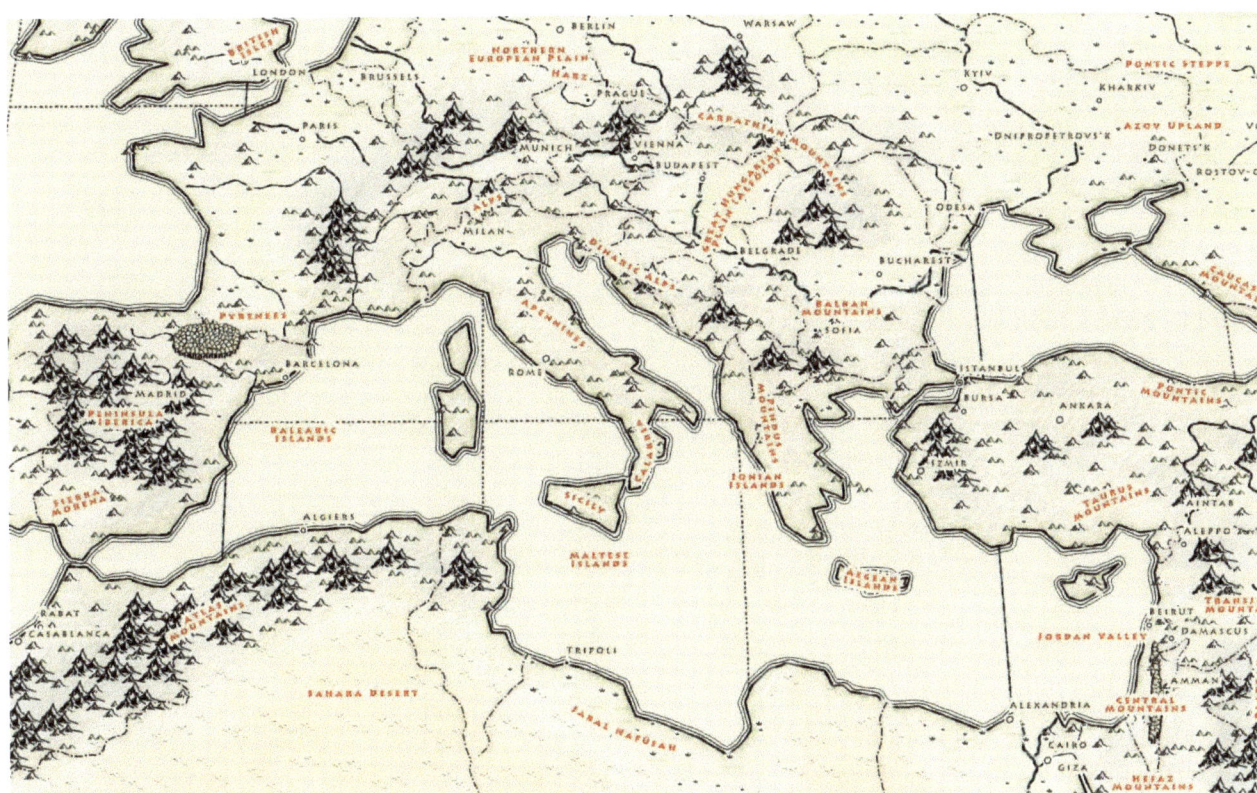

Figure 10.23. A modern version of stylized hillsigns on a map centered on the Mediterranean Sea.

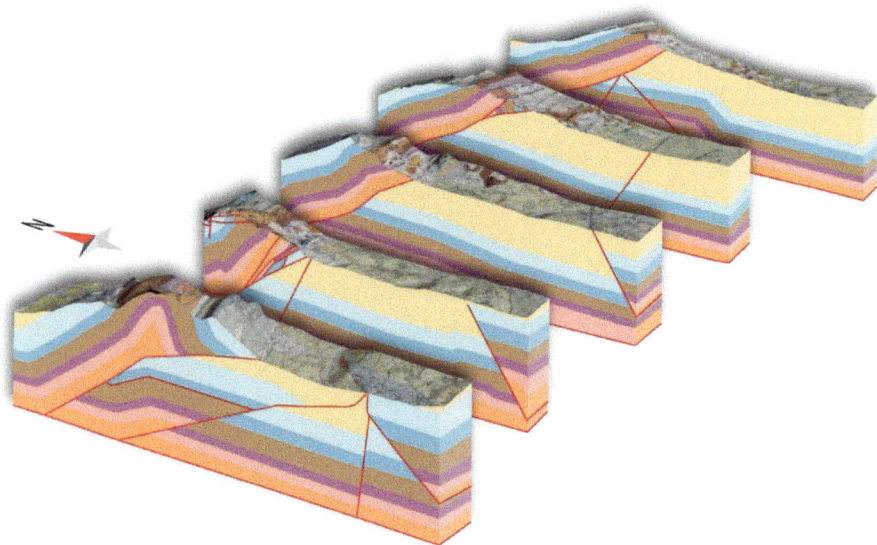

Figure 10.24. These block diagrams, which show the nature and structure of the subsurface, are an important tool for gaining detailed knowledge that relates to subsurface planning. © swisstopo.

Diorama maps

A **diorama map** is a 3D representation of a landscape used to create a realistic depiction of a specific location, phenomenon, or event. Diorama maps focus on detail and realism to immerse readers in the depicted environment as seen from a specific perspective or viewpoint. Diorama maps are used in museums, educational settings, and tourist exhibits to provide a realistic representation of the landscape of a region or an event in a locale. A striking example is shown in figure 10.25 of the Straits of Mackinac between Michigan's Upper and Lower Peninsulas.

Landscape drawings

Artistically rendered, oblique-perspective illustrations, called **landscape drawings**, can provide nearly photorealistic relief representations. Because of the immense amount of hand labor involved in their construction, the best examples are found for regions of special interest, such as national parks and popular tourist areas. Landscape drawings, such as the ski-area poster by Pierre Novat of France's Grand Massif shown in figure 10.26, are also popular as wall art.

True 3D

True 3D refers to a 3D image of a scene that accurately simulates depth perception, allowing the viewer to perceive the spatial relationships between different elements in the scene. Unlike 2D or 3D maps viewed traditionally, **true 3D maps** provide a sense of depth, volume, and perspective, making the visual image appear more realistic and immersive. A true 3D map provides a more comprehensive and realistic representation of landscapes by incorporating depth, thus enhancing the viewer's perception and understanding of the 3D nature of the visual scene.

True 3D has long held the interest of mapmakers because of its ability to add a strong impression of the third dimension to an otherwise flat map. Various techniques and technologies are used to create true 3D maps, such as stereoscopic views, anaglyphs, and ChromaDepth images, discussed next.

Stereoscopic views

A common way to see in true 3D is with a **stereoscopic view**—a 3D image or visual representation that simulates depth perception by presenting two offset images separately to each eye. When these two images are viewed simultaneously, the brain combines them to create the illusion of depth, making the image appear three-dimensional. Stereoscopic views are commonly created using **stereoscopic pairs** of images, in which each image is slightly offset to mimic the perspective seen by each eye. The brain combines these two offset images to perceive depth and 3D space. This visual trick replicates the way human vision works, in which each eye sees a view of the same scene from a slightly different vantage point. You can test this if you close first one eye and then the other—the scene you see shifts slightly. This phenomenon, in which the apparent position of an object seems to change when viewed from different perspectives, is called **parallax**.

Chapter 10: Relief portrayal 253

Figure 10.25. A diorama map of the Straits of Mackinac connecting Lake Huron to Lake Michigan.

Figure 10.26. A landscape painting of Grand Massif, France, by Pierre Novat. © 1986 Atelier Pierre Novat.

Stereoscopic views were traditionally observed with a **stereoscope** (figure 10.27). Today, stereopairs are viewed using a variety of modern techniques and devices that have evolved from traditional stereoscopes, including 3D glasses, 3D TVs and monitors, virtual reality (VR) headsets, and stereoscopic apps and software. With any of these, the vertical exaggeration of the view is subjective but will increase as the difference between the vantage points of the stereopairs increases. Despite their variations, all of the methods are based on the way we naturally see in 3D—with two eyes that have slightly offset vantage points.

Anaglyphs

An **anaglyph** is a type of stereoscopic view created by combining two slightly offset images, one in red and the other in cyan (or blue), that appear 3D when viewed through anaglyph glasses. The glasses have red and cyan (or blue) filters so that you see the red map with one eye and the blue map with the other (figure 10.28). Your brain combines the slightly offset images so you see the map stereoscopically and the terrain surface appears three-dimensional.

ChromaDepth maps

ChromaDepth maps, such as the one shown in figure 10.29, rely on a method for creating 3D images that is based on color separation. With ChromaDepth, different colors are assigned different depth values. The colors are arranged in a specific order to create a sense of depth perception when viewed through specially designed ChromaDepth glasses. With the glasses, certain colors in the visible spectrum appear closer to and others appear farther from your eyes, with red in the foreground, blue in the background, and orange, yellow, and green in intermediate positions. Only one image is required to see true 3D on a ChromaDepth map, and the perception of vertical exaggeration can be greatly enhanced by viewing the map from a distance.

Dynamic maps

Computer technology makes it possible to put mapped relief in motion in dynamic maps of relief portrayal. Dynamic relief maps let you explore the terrain surface from multiple vantage points that you can often choose yourself. With computers and sufficient data representing the terrain surface, images can be created from the digital database through a series of calculations that consider the vantage point of the observer, the orientation of the terrain surface, the amount of vertical exaggeration, and the position of the illumination source. The images are then displayed on-screen and recorded digitally for subsequent playback in an animation called a **fly-through** that gives the impression of a camera flying over and around a region.

The Swiss topographic mapping agency, swisstopo, has created a 3D virtual model of Switzerland and the Principality of Liechtenstein that shows every building, bridges, cable cars, forests, individual trees, and geographic names

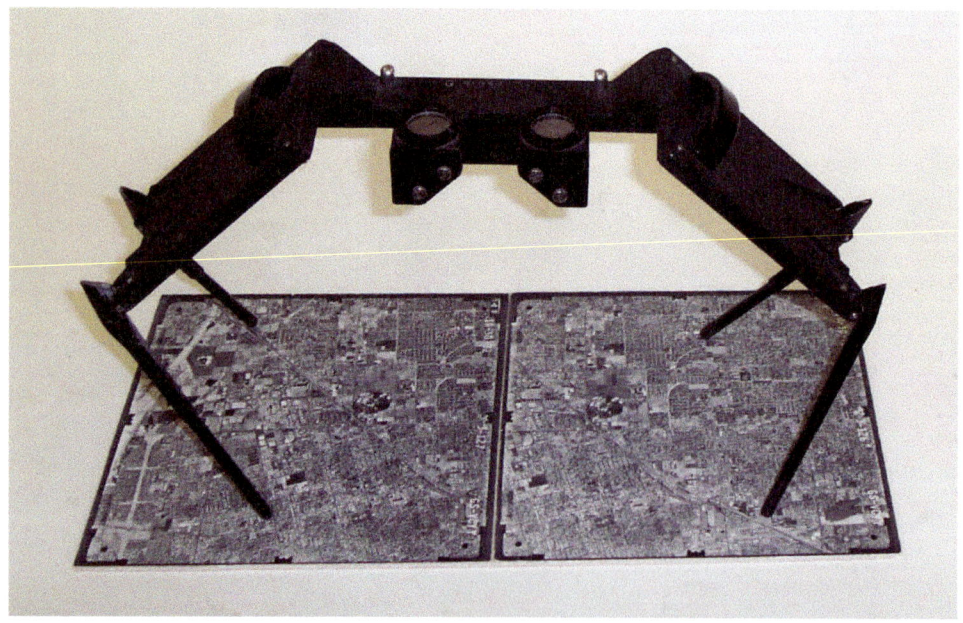

Figure 10.27. A stereoscope positioned above a stereopair. The aerial images, with areas of overlap near the middle, were taken from two slightly different vantage points. Courtesy of the US Geological Survey.

Chapter 10: Relief portrayal   255

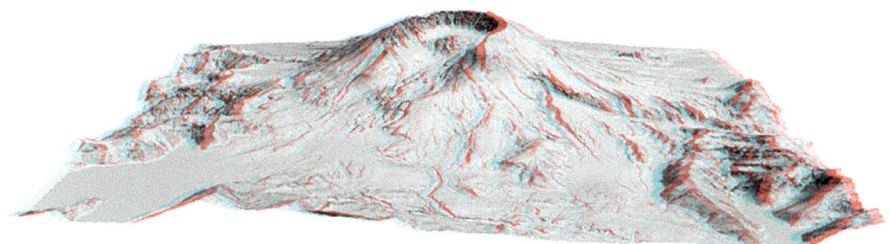

Figure 10.28. An anaglyph of a perspective-view map of Mount St. Helens, Washington.

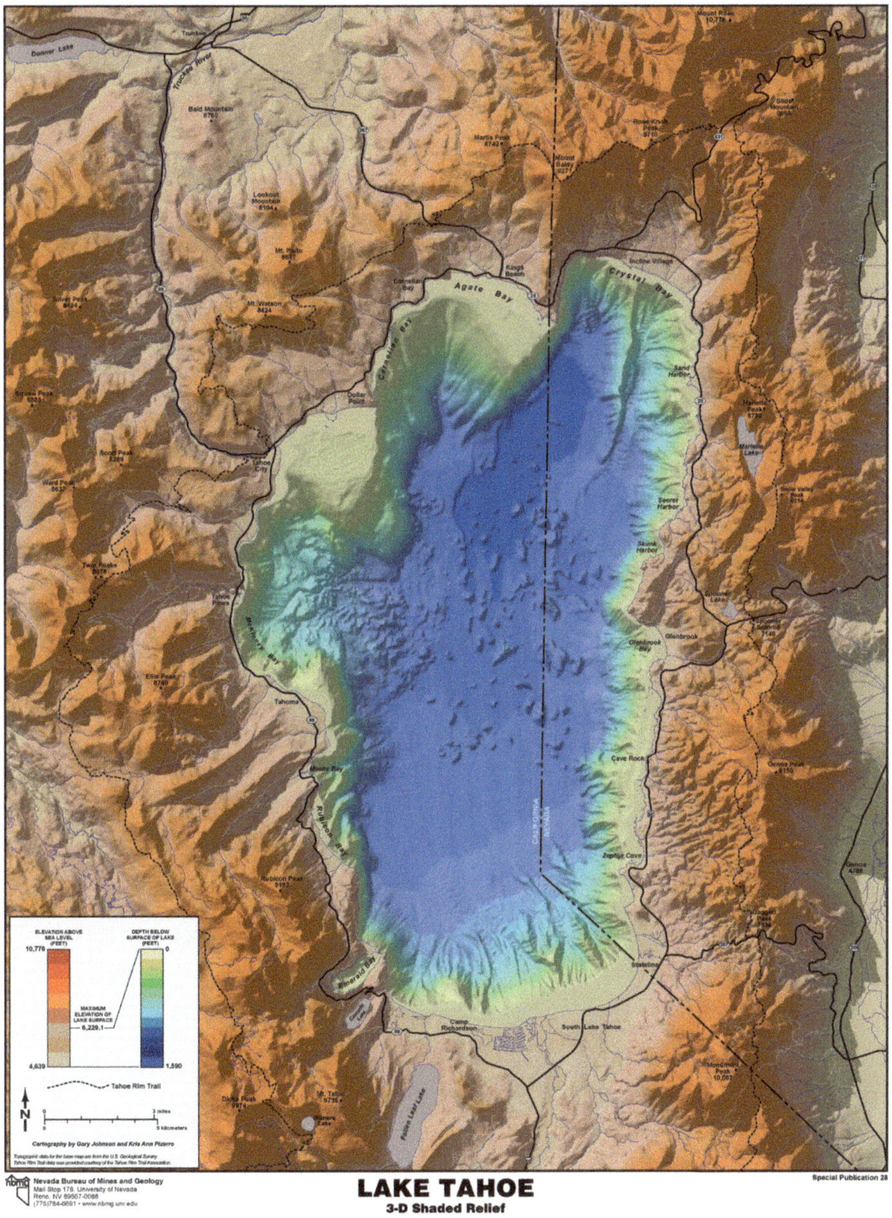

Figure 10.29. The relief in this planimetric ChromaDepth map of Lake Tahoe stands out vividly when viewed through ChromaDepth glasses. Courtesy of the Nevada Bureau of Mines and Geology.

Figure 10.30. The Swiss federal map viewer has a 3D mode that allows you to fly over the Alps, plan your next hike, or explore every building in a city. The map is built on a high-resolution 3D digital topographic basemap. © swisstopo.

in 3D (figure 10.30). These features rest on a DTM that represents the complex topography of the region. Viewers move around the 3D model using a mouse, touchpad, touchscreen, or icons on the interface. Through interactivity, viewers have control of what they see.

Examples of dynamic terrain maps are varied, and they continue to grow and evolve. When we encounter them, we are prone to fascination and exploration. It is easy to imagine getting lost in the landscapes they share with us and sometimes allow us to control. For example, the augmented reality sandbox shown in figure 10.31 allows users to sculpt sand that projects contours and hypsometric tints onto the modeled landforms.

To summarize, relief maps not only fascinate us but help us move beyond the fundamental 2D nature of most maps. Because we experience relief in our environment every day, relief portrayal on maps is something that captures our attention and imagination. Our understanding of the world around us and even other worlds (figure 10.32) is expanding through the techniques that mapmakers use to understand elevation and its variation on maps. With a greater understanding of these techniques, you can better appreciate and understand relief maps. And with today's mapping methods, you are in a better position than previous generations to explore the 3D nature of our world.

Selected readings

Arundel, L., A. Phillips, A. J. Lowe, et al. 2015. "Preparing the National Map for the 3D Elevation Program—Products, Process, and Research." *Cartography and Geographic Information Science* 42 (supplement 1): 40–53.

Barnes, D. 2002. "Using ArcMap to Enhance Topographic Presentation." *Cartographic Perspectives* 42 (Spring).

Eyton, R. J. 2005. "Unusual Display of DEMs." *Cartographic Perspectives* 50: 7–23.

Hurni, L., B. Jenny, T. Dahinden, and E. Hutzler. 2001. "Interactive Analytical Shading and Cliff Drawing: Advances in Digital Relief Presentation for Topographic Mountain Maps." *Proceedings of the 20th International Cartographic Conference*, ICC, Beijing, China.

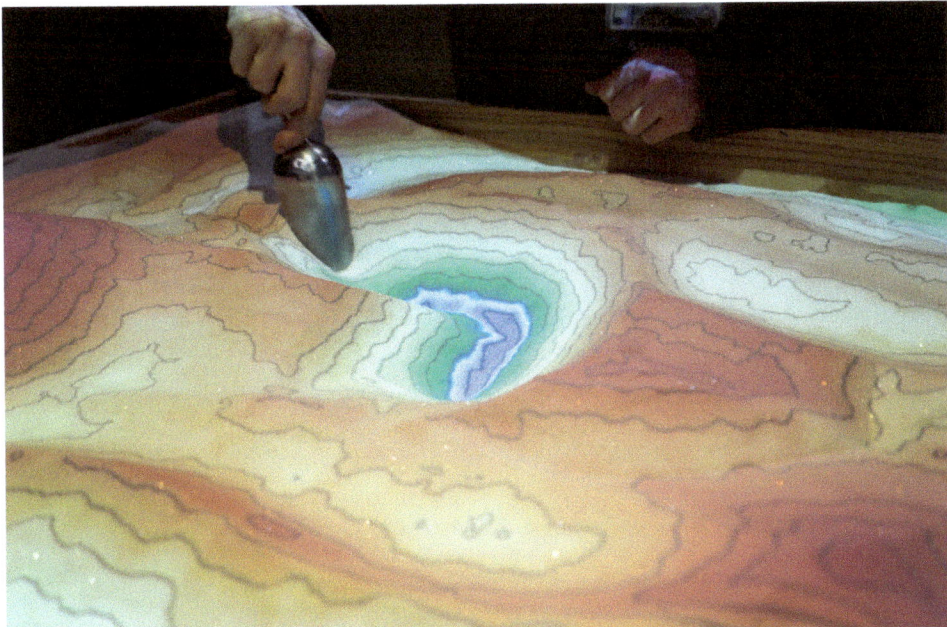

Figure 10.31. As the landforms in this augmented reality sandbox are modified, the resulting contours and hypsometric tints on the terrain surface are updated interactively. Courtesy of the US Geological Survey.

Figure 10.32. A relief map of the moon. Courtesy of the US Geological Survey.

Imhof, E. 2007. *Cartographic Relief Presentation*, translated and edited by H. J. Steward. Esri Press.

Jenny, B. 2001. "An Interactive Approach to Analytical Relief Shading." *Cartographica* 38 (1–2): 67–75.

Kennelly, P. 2017. "Terrain Representation." *The Geographic Information Science & Technology Body of Knowledge.* Edited by John P. Wilson. https://gistbok-topics.ucgis.org/CV-04-014.

Kennelly, P., and A. J. Stewart. 2006. "A Uniform Sky Illumination Model to Enhance Shading of Terrain and Urban Areas." *Cartography and Geographic Information Science* 33 (1): 21–36.

Kennelly, P., and A. J. Stewart. 2014. "General Sky Models for Illuminating Terrains." *International Journal of Geographical Information Science* 28 (2): 383–406.

Patterson, T., and B. Jenny. 2011. "The Development and Rationale of Cross-Blended Hypsometric Tints." *Cartographic Perspectives* 69: 31–46.

Price, W. 2001. "Relief Presentation: Manual Airbrushing Combined with Computer Technology." *The Cartographic Journal* 38 (1): 107–12.

Samsonov, T., S. Koshel, D. Walther, and B. Jenny. 2019. "Automated Placement of Supplementary Contour Lines." *International Journal of Geographical Information Science* 33 (10): 2072–93.

Tait, A. 2002. "Photoshop 6 Tutorial: How to Create Basic Colored Shaded Relief." *Cartographic Perspectives* 42: 12–17.

Trantham, G., and P. Kennelly. 2022. "Terrain Representation Using Orientation." *Cartography and Geographic Information Science* 49 (6): 479–91.

CHAPTER 11
Imagery and maps

Although we primarily rely on our own eyes to learn about the environment, indirect experience is also important. People have shared stories of distant places by word of mouth since the beginning of time. Explorers traveling to faraway lands wrote and sketched in personal travel logs. Scientists on explorations to remote areas kept detailed notes, recorded measurements, and took photographs. All of these provide valuable information about our environment, but they only detail specific places. With our expanding view of the world comes a demand for a more holistic approach to gathering information about our environment. In the last century, the significance of collecting data and images of the earth (and other planetary bodies) from a distance, called **remote sensing**, has grown to the point that we are now inundated with massive volumes of remotely sensed data about our surroundings.

Remote sensing allows us to observe features in the environment using cameras or other electronic imaging instruments (**sensors**) that are sensitive to the energy emitted or reflected from objects. With a camera, light from a subject or scene enters the camera through a lens that focuses the light onto an **image sensor** that converts it into an **analog image** recorded on film or into a **digital image** recorded in the camera's memory. Memory devices for digital images can store gigabytes of data for thousands of digital images. Composed of pixels that have integer values representing gray tones or colors, digital images have higher resolution (see chapter 10) than analog images on photographic film. Digital images can also be created by noncamera sensors, which are generally scanners that build up an image line by line instead of instantaneously.

Although remote sensing images are excellent for showing many aspects of the environment, they may fail to capture others. Invisible features, such as political boundaries, are not picked up on images unless they happen to align with physical features. Map symbols and labels are absent, as are useful map elements such as reference grids and graticules. And features on images are not identified in a legend or explained with notes. For all these reasons, remote sensing images are often made more interpretable and useful through cartographic enhancement, with symbols for point, line, and area features and labels for identification or description, producing what is called an **image map**. Image maps have many uses and capabilities, from serving as detailed basemaps to being draped over terrain models for realistic 3D displays.

One way to distinguish remotely sensed images is by the type of sensor used to collect the images or by the sensor **platform**—the vessel, craft, or instrument that the sensor is carried on. Historically, fixed-wing aircraft, such as airplanes and gliders, and even balloons, kites, and pigeons, were used to carry cameras for imaging. Today, rotating-wing aircraft, such as helicopters and some types of drones, have been added to the platform list, along with fixed-wing drones, jet aircraft, and orbiting satellites (figure 11.1). Although other platforms on the surface of the earth can also be used to image our environment, in this chapter, we focus on earth images from above the surface. We also refer to two primary types of sensor platforms: those on aircraft that operate within the earth's atmosphere (up to about 10 miles or 16 kilometers above the earth) and those on spacecraft in orbits from about 300 to 22,000 miles (500 to 36,000 kilometers) above the earth.

In this chapter, we will explore both **aerial imagery** (images of the ground taken with a camera on an aircraft) and **satellite imagery** (images from satellites and other spacecraft), but we start by looking at factors that affect the quality of the images.

Figure 11.1. Modern earth imagery is captured from sensors on a variety of platforms at a broad range of altitudes starting from ground level to more than 22,000 miles above earth. Although not meant to be an exhaustive inventory, this diagram illustrates some of the most common platforms.

Syncom

Landsat 8

SR71 Blackbird

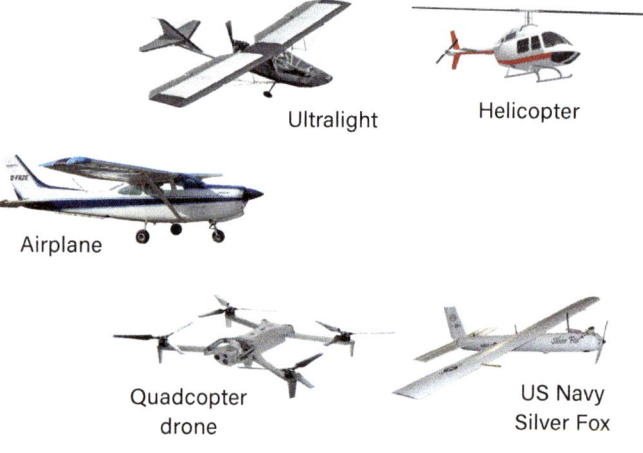
Ultralight Helicopter
Airplane
Quadcopter drone US Navy Silver Fox

Handheld spectrometer
Street-level mapping car
Doppler-on-wheels
Smartphone

Geosynchronous   22,236 miles

Satellites that match Earth's rotation appear stationary in the sky to ground observers. Although most commonly used for communications, geosynchronous orbiting satellites, such as the hyper-spectral geosynchronous imaging Fourier transform spectrometer (GIFTS) craft, are also useful for monitoring changing phenomena such as weather conditions.

Sun synchronous   375–500 miles

Satellites in this orbit keep the angle of sunlight on Earth's surface as consistent as possible, which means that scientists can compare the images of Earth-observing sensors over several years.

Jet aircraft   30,000–90,000 feet

Jet aircraft flying at 30,000 feet and higher can be flown over disaster areas in a short time, making them a good platform for certain types of optical and multispectral image applications.

General aviation aircraft   100–10,000 feet

Small aircraft able to fly at relatively low speeds and altitudes have long been an effective way to capture high-quality aerial photography and orthophotography. From airplanes to ultralights to helicopters, these are the workhorses of urban optical imagery.

Drones   100–500 feet

Drones are now commonplace for commercial, military, conservation, law enforcement, and recreational uses, such as amateur photography. Their ability to fly low, hover, and be remotely controlled offers advantages for aerial photography, with resolution down to sub-1 inch. Military unmanned aerial vehicles (UAVs) range from tiny drones to full-sized aircraft.

Ground-based/handheld   Ground level

Increasingly, imagery taken at ground level is finding its way into GIS workflows such as Google Street View, HERE street-level imagery, and Mapillary. Handheld multispectral imagers and other terrestrial sensors are finding applications in areas such as pipelines, security, tourism, real estate, natural resources, and entertainment.

Image quality

Many things can influence the quality of remotely sensed images, including the sensor's sensitivity to different wavelengths of electromagnetic radiation, the image's spatial resolution, the sensor's height, and relief displacement. Image quality can also be affected by atmospheric conditions and digital manipulation of the images.

Spectral sensitivity

Barring any visual impairments, when we look at the world with our eyes, we see a rainbow of colors representing a small portion of the electromagnetic spectrum. Much of our environment can be imaged in the same way that we see, by sensing light in the visible portion of the electromagnetic spectrum (figure 11.2). More recently, imaging capabilities have been extended outside the visible light spectrum at both ends of the eye's sensitivity range. Using filters to block unwanted ranges of wavelengths, it is possible to image specific bands of **visible light** energy, such as blue, green, and red, as well as **near-visible light** energy, such as **ultraviolet (UV)** and **near-infrared (near-IR)**. Non-photographic sensors expand our imaging capability into the **mid-infrared (mid-IR)** and other **thermal-infrared (thermal-IR)** ranges. Images can be obtained simultaneously in

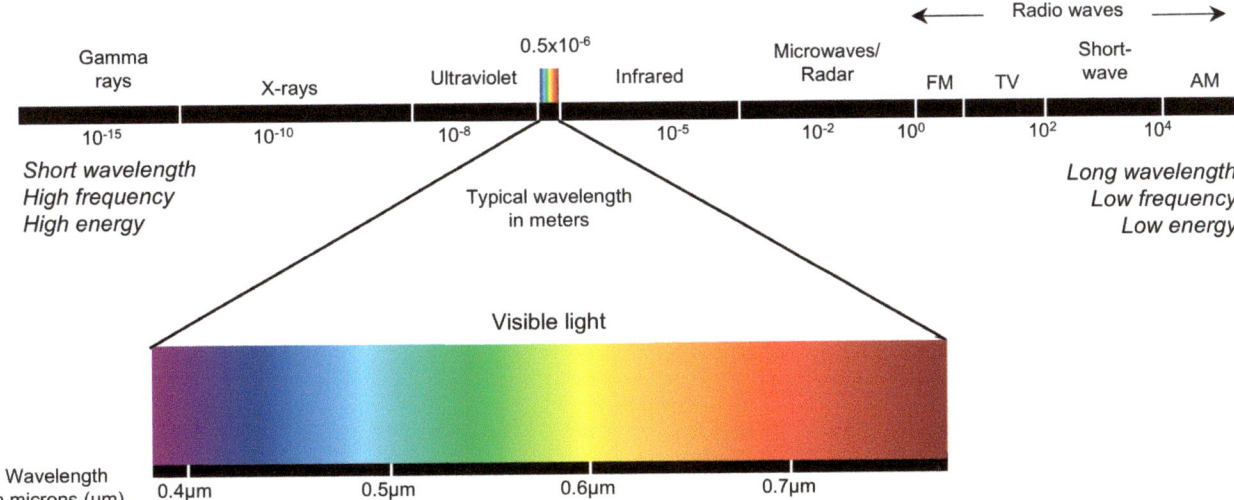

Figure 11.2. Wavelength range (in microns) for the near-ultraviolet to near-infrared portion of the electromagnetic spectrum.

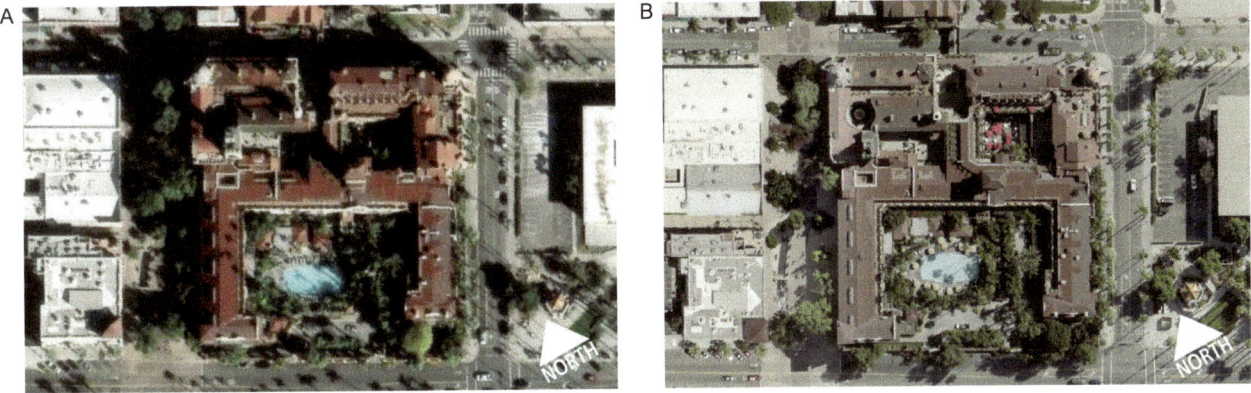

Figure 11.3. The Mission Inn in Riverside, California, is shown in an image with 30 cm resolution (*A*). The same area is also shown in an image with 8 cm resolution, making it easier to see details and resolve features (*B*). Courtesy of City of Riverside, California; Microsoft; and Esri.

different portions of the electromagnetic spectrum, called **spectral bands**, each of which provides valuable insights into the nature of our surroundings.

Spatial resolution

As we described in chapter 10 (for DEM data), the distance measured on the ground between pixel centers on the digital image defines the image's spatial resolution, also called **ground sample distance (GSD)**. For example, on a digital image with a 30-centimeter GSD, adjacent pixels on the image are 30 centimeters on the ground (figure 11.3A). The closer the spacing, the higher the resolution.

Sensor height

Detailed **low-altitude images** are taken anywhere from just above the ground to an altitude of around 1,500 feet (500 meters). These low-altitude images (figure 11.4) usually cover a small ground area at a larger map scale (see chapter 2).

Acquiring images at higher altitudes, from 1,500 to 10,000 feet (500 to 3,000 meters), provides less detail because the spatial resolution is lower, but more ground area is covered. The image in figure 11.5B was taken from an altitude of about 10,000 feet. Even higher altitude

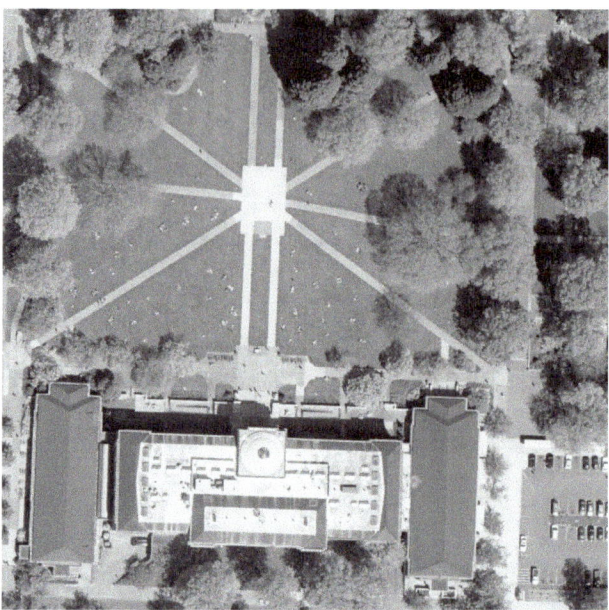

Figure 11.4. A low-altitude black-and-white image of the Memorial Union on the Oregon State University campus, Corvallis, taken from an altitude of around 1,000 feet (300 meters), on which you can see students on the grass in front of the building. Courtesy of City of Corvallis Public Works Department.

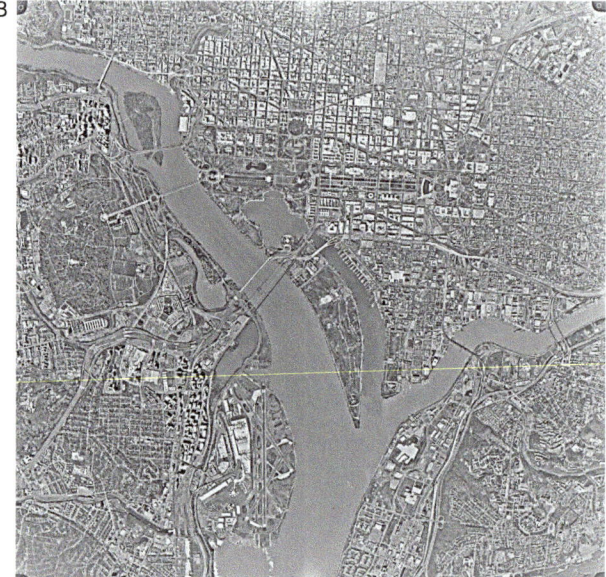

Figure 11.5. Portions of high-altitude (*A*) and low-altitude (*B*) black-and-white photographic images of Washington, DC (both reduced in scale from the original), taken at altitudes of around 10,000 and 40,000 feet (3,000 and 12,000 meters), respectively. The white box drawn on the high-altitude image outlines the area covered by the low-altitude image. Courtesy of the National High Altitude Photography Program (*A*) and the National Aerial Photography Program (*B*).

images (figure 11.5A) are taken from an altitude of about 40,000 feet (about 12,000 meters).

The advantage of high-altitude imagery is that a single image can cover a large ground area. This type of image coverage has many mapping applications. For instance, the damage caused by earthquakes, floods, and droughts can be quickly monitored and assessed. Forest resources, snow cover, crop yields, and many other environmental features can also be studied over seasonal, annual, and longer periods.

Relief displacement

On **vertical images** captured with a camera vertical to the ground, the scale of the image most likely varies radially away from its center. The primary reason for paying attention to radial **scale variation** on vertical images is that objects of different heights are displaced radially away from the **nadir** (the point on the ground vertically beneath the center of the camera lens). The camera lens creates a central perspective view of the ground, as you would see if you looked straight down at the ground from the sensor platform. **Relief displacement** is the apparent "leaning out" of the top of a higher feature on a vertical image. If the top of a feature is higher than the elevation of the nadir, it will be displaced outward and imaged at a slightly larger scale. For example, in figure 11.6, the sides and tops of tall buildings in downtown Chicago are displaced outward and radially away from the nadir of the image in the bottom left.

The geometry underlying this radially outward pattern of relief displacement is illustrated in figure 11.7. Notice the relation between the horizontal position of the six buildings on the ground at different distances from the nadir and their associated appearance in the image. The greater the height of the buildings relative to the ground

Figure 11.6. This northeast quarter of a high-altitude vertical aerial image of Chicago, Illinois, shows relief displacement, apparent by the "leaning out" of buildings away from the nadir (*bottom left*). Courtesy of the National Aerial Photography Program.

and the farther the features are from the nadir, the greater the relief displacement. This pattern of relief displacement also holds true for hills, valleys, and any other natural or cultural feature that varies in elevation.

The vertical rays in figure 11.7 indicate that there is no relief displacement at the nadir irrespective of how high or low the camera altitude is. The slanted rays show that relief displacement is related to an angle between lines from the lens to the nadir and from the lens to a feature on the ground. If the camera is vertical to the ground, relief displacement also depends on the camera's **focal length** (the distance between the optical center of the lens and the camera sensor or film plane when focused at infinity). A shorter focal length results in wider ground coverage and larger angles from the nadir, so there is more displacement. Finally, taking a digital image at a higher altitude will result in less relief displacement for locations at a set distance from the nadir. Thus, vertical images taken with cameras by space shuttle astronauts exhibited so little relief displacement and scale variation that they could be overlaid on topographic maps, with the only geometric discrepancies coming from the earth's curvature.

Aerial imagery

Aerial imagery refers to any image captured from an airborne sensor. The most common type is an **aerial photograph**, or **air photo**—a photographic image captured from an aircraft or other airborne platform that provides a top-down view of the landscape. For areas in the United States, aerial (and other) imagery from the Earth Resources Observation and Science (EROS) archives of the USGS is made available through the USGS EarthExplorer website. The US Department of Agriculture (USDA) Farm Services Agency acquires and makes available high-resolution aerial imagery through its **National Agriculture Imagery Program** (**NAIP**). NAIP aerial imagery, acquired during the agricultural growing seasons, is typically captured at a high resolution of one meter or less and made available as natural-color or color-infrared images.

Panchromatic imagery

Photographic media have been used for aerial reconnaissance since the middle of the 1860s. Traditional black-and-white air photos were created using a film emulsion sensitive to electromagnetic radiation in the 0.3–0.7 μm band covering most of the visible spectrum. Because of its sensitivity to visible light, this black-and-white film was

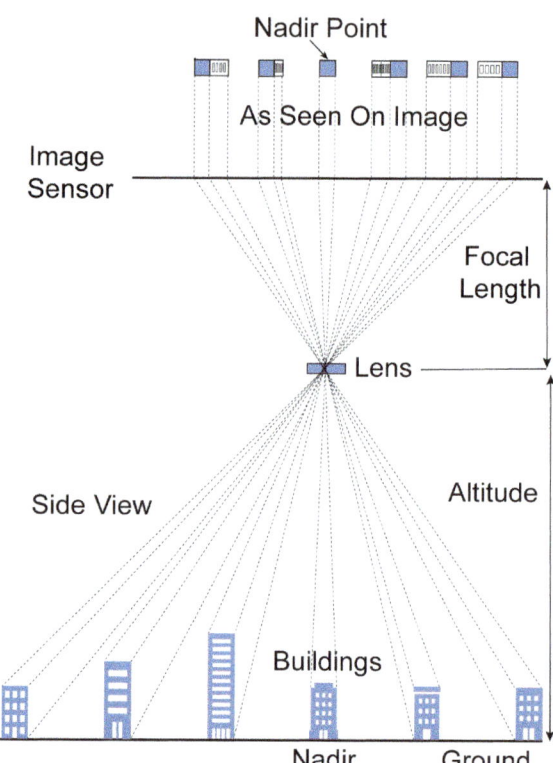

Figure 11.7. The relief displacement of features in a vertical image with central perspective can be seen in this side view of buildings shown horizontally on the ground relative to the nadir and their appearance in the image.

called **panchromatic** (meaning "all colors") **film**. Positive photographic prints were made from the film negatives. Today, these older prints are scanned and shared digitally, and more recent images are captured digitally (figure 11.8).

The greater the amount of visible light reflected from an object and gathered by the camera or sensor, the higher its numeric pixel value in a scanned or digital aerial image. Soil moisture content, surface roughness, and a feature's natural color all influence its tone on a panchromatic image because the range of low to high pixel values on digital images is usually displayed as a range of dark- to light-gray tones. Moist soils, marshlands, and newly plowed fields tend to be darker than surrounding features, whereas built features such as roads and buildings tend to appear lighter.

Natural-color imagery

For a long time after its development in the 1930s, natural-color air photos were little used in mapping. Color film was more expensive than conventional black-and-white film, partly because color film contains three separate emulsion

Figure 11.8. A panchromatic aerial image of the US Capitol building in Washington, DC. Courtesy of the US Geological Survey.

Figure 11.9. A natural-color aerial image of the US Capitol building in Washington, DC. Courtesy of Esri and the USDA Farm Production and Conservation Business Center.

layers sensitive to the blue, green, and red portions of the visible spectrum. Special film-processing requirements added further to the cost. In addition, early color film had poor resolution relative to its panchromatic counterpart. Color film came into widespread use in the 1950s and has since been used to create **natural-color images** (also called **true-color images**).

With digital cameras, the improved resolution and clarity of natural-color images makes them easier to read than their panchromatic counterparts because they capture the colors associated with landscape features, such as the green grassy areas and white rooftops in figure 11.9. To our eyes, it is easier to distinguish subtle differences in color rather than shades of gray. Color is especially useful in revealing the condition of features, such as the stage of a crop in its maturation (phenological) cycle.

For applications such as vegetation and soil classification, geologic mapping, and surface water studies, using color images is more informative and accurate than working from the equivalent panchromatic images.

Color-infrared imagery

Color-infrared (CIR) film, which records, in part, energy from portions of the electromagnetic spectrum invisible to the human eye, was developed to detect camouflaged military objects in the 1940s. CIR film is sensitive to near-IR wavelengths (0.7–0.9 μm), as well as to green and red visible light. In a **color infrared image** (now also known as a **false-color image**), near-IR light reflected from a scene appears as red, red appears as green, green as blue, and blue as black.

You can see in figure 11.10 that an environmental feature that absorbs a lot of near-IR energy, such as clear water in ponds, appears dark blue to black, whereas features that absorb less near-IR energy, such as building rooftops, appear white or tan. The most obvious feature, aside from water, which is black, on CIR imagery is healthy vegetation, which appears bright magenta or red because of its high near-IR reflectance.

CIR film is useful for distinguishing between healthy and diseased vegetation, delineating bodies of water, and penetrating atmospheric haze. In vegetation studies, CIR images are used for crop inspection, tree growth inventories, and damage assessment of diseased flora, especially as diseased plants can often be detected on CIR images well before they become visible to the unaided eye. Geologists use CIR images to map near-surface structural features, such as faults, fractures, and joints, which often collect water, encouraging lusher vegetation growth. Boundaries between soil and vegetation and between land and water can also be detected in CIR images. In urban mapping, the near-IR wavelengths can penetrate smog easily.

Aerial imagery capture

Aerial images are captured by flying the sensor on an aircraft or other airborne platform above the ground while taking the images. To systematically capture these images, the aircraft flies along **flight lines** (paths that the aircraft follows), typically in the north–south or east–west direction (figure 11.11). Images are taken along the flight line so that there is **overlap** (a duplicated image of the ground in two successive aerial images). Typically, more than a single flight line is required to cover the area to be mapped, and adjacent flight lines are planned with **sidelap** to ensure that there are no gaps in the coverage. The **forward overlap**, or **front overlap**, along the flight lines, is usually between 60 and 80 percent, and the side overlap, also called **sidelap**, between the flight lines is between 20 and 30 percent. A similar pattern can also be used to acquire imagery with drones.

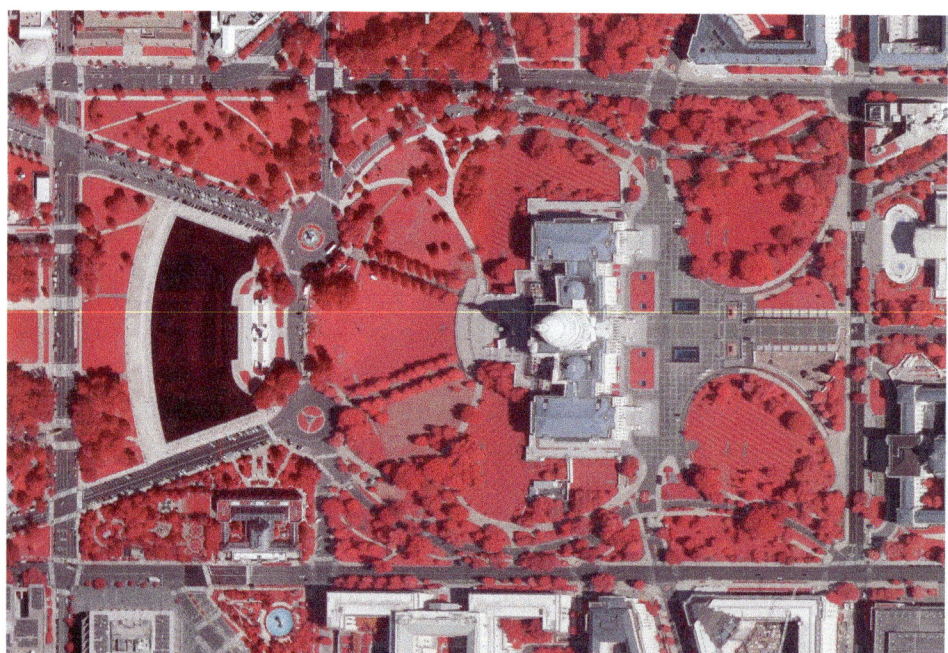

Figure 11.10. A CIR aerial image of the US Capitol building in Washington, DC. Courtesy of Esri and the USDA Farm Production and Conservation Business Center.

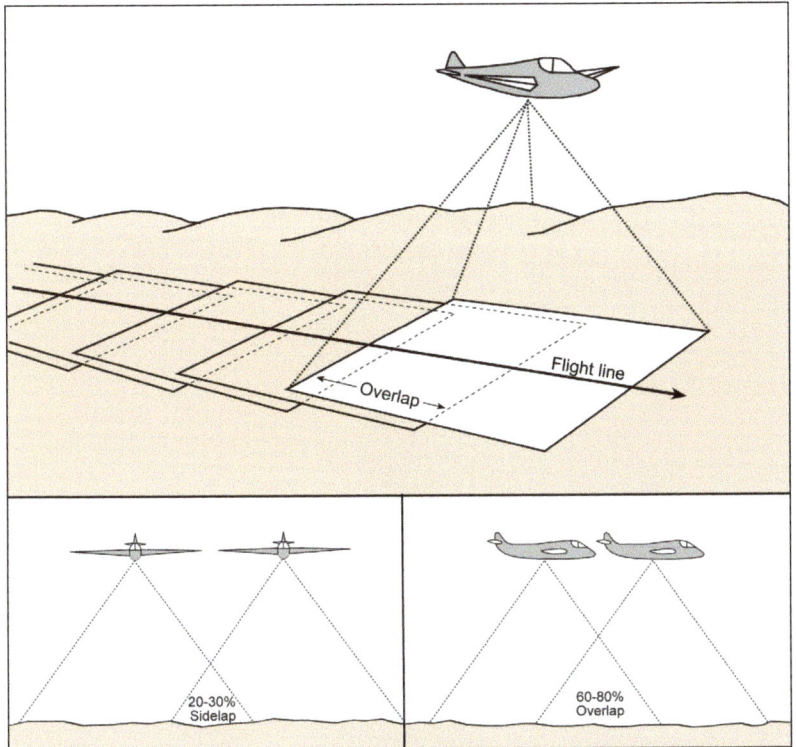

Figure 11.11. Adjacent aerial images are taken to ensure 60 to 80 percent overlap along the flight line and 20 to 30 percent sidelap between adjacent flight lines to ensure overlapping ground coverage.

Drone imagery

A **drone** is a type of aircraft able to fly remotely without a human pilot onboard. Drones can be operated by a human with a flight controller or autonomously using GNSS, avoidance sensors, and a preprogrammed flight path. For mapping, drones are used to capture high-resolution imagery and other spatial data for various applications.

To navigate autonomously, drones and other **uncrewed aerial vehicles** (**UAV**) rely on GNSS receivers, inertial navigation systems (INS), inertial measurement units (IMUs), compasses, and digital cameras. Drones can also be equipped with lidar and ultrasonic sensors. The drone is one component of an **uncrewed aircraft system** (**UAS**), which includes not only the UAV or drone but also the person on the ground controlling the flight and the system in place that supports them. Professional drones, like the quadcopter drone in figure 11.1, are used to collect digital images for maps.

To capture images for mapping, the drone follows a flight line like the one we illustrated in figure 11.11, taking overlapping digital images along each flight line. However, drone imagery requires more overlap than aerial imagery—a rule of thumb is 70 percent forward overlap and 65 percent sidelap. The mapping applications of drone imagery seem limitless, but we focus on a project in which drones were used to construct a digital twin for an area around Red Rocks Amphitheatre, just west of Denver, Colorado.

Overlapping planimetric drone images were captured at each of the locations indicated by the green dots in figure 11.12A. Above the amphitheatre, additional oblique images were acquired to capture the surrounding rock faces required to create the 3D model (figure 11.12B). The drone also stored its x,y,z location above the ground in the metadata for each image so that each image could be georeferenced. Site Scan for ArcGIS and Reality Engine were used to photogrammetrically stitch the overlapping images together and derive a high-resolution **3D mesh** (a 3D textured model of the project area in which the ground and above-ground feature facades are densely and accurately reconstructed). Other derived products included a DTM, DSM, lidar point cloud, and true ortho image (see chapter 10 for more on lidar, DTMs, and DSMs). The 3D mesh and drone images were then visually rendered with textures, lighting, shadows, and reflections to enhance the realism of the digital twin (figure 11.12C).

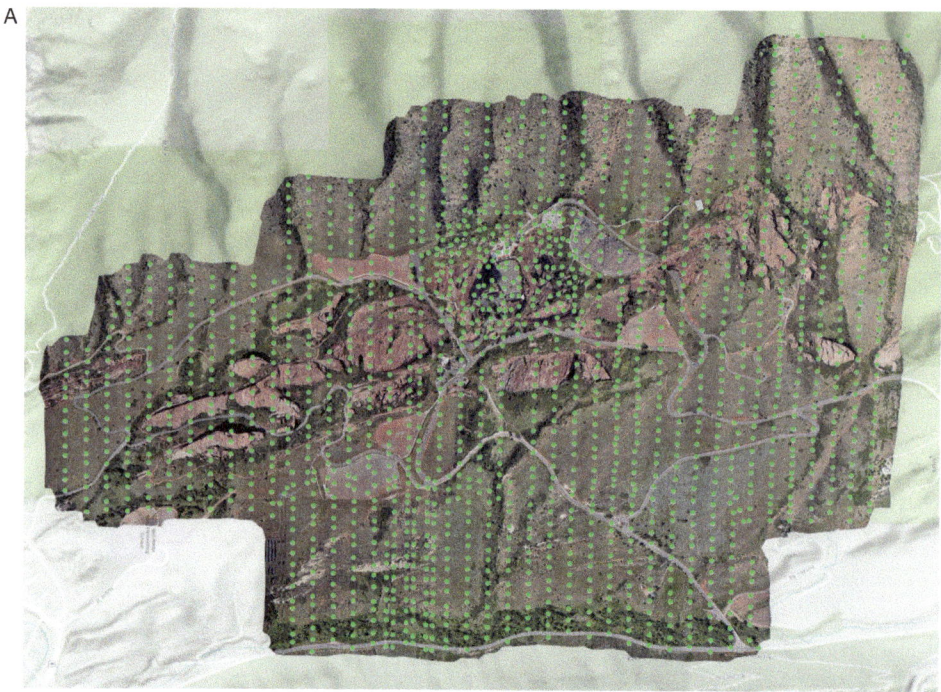

Figure 11.12. Digital imagery from a quadcopter drone was used in a project to create a digital twin of Red Rocks Amphitheatre near Denver, Colorado. Planimetric (*A*) and oblique (*B*) drone images were captured at the locations of the green dots in the first two images. Georeferenced drone imagery was photogrammetrically stitched together to create the 3D reality mesh, orthorectified imagery, and elevation models (*C*).

Orthophotos and orthophotomaps

Orthophotos

An **orthophoto** (**orthophotograph** or **orthoimage**) is an aerial photograph or satellite image that has been geometrically corrected, a process called **orthorectification**, so that the image scale is uniform. Rather than look radially outward from the nadir point to each feature on an aerial image, with an orthophoto, you look directly down on the landscape. Thus, all features on an orthophoto appear in their true planimetric position.

Creating an orthophoto from aerial images that cover an area is a mathematically demanding computer process. In most cases, some ground control (see chapter 1) is required to mathematically determine the camera location and orientation required to generate an orthophoto. The source for the ground control varies greatly and can include newer imagery.

Correction of scale difference and relief displacement in each image (figure 11.13A) is performed with the aid of DEM data covering the area (see chapter 10). Orthorectification software uses camera orientation information and the DEM to compute an image without ground displacement by repositioning each pixel on the digital image to remove both relief displacement and scale variation (figure 11.13B). Color images are also corrected to minimize the tonal and color differences.

If there is sufficient image overlap, more computationally intensive methods can correct these displacements and blend the images together into what are commonly referred to as **true orthos**. True orthos use digital aerial images to produce an incredibly detailed 3D point cloud of pixels. Specialized algorithms, such as those used in ArcGIS Pro, decide which pixel from which image needs to be selected for a 3D location. The orthophoto is built by making this decision for each pixel until the entire orthophoto has been built. This process allows for more accurate mapping of 3D objects, such as buildings, trees, and other structures. True orthos, as in figure 11.14, are currently being generated in metropolitan areas throughout the United States at about 30-centimeter spatial resolution for a range of applications, including surveying, land management, and infrastructure planning.

Orthophotomaps

Digital orthophotoquads

Orthophotos are such an important component of mapping technology that federal agencies in the United States have cooperated with state and local governments, as well as the private sector, to create **orthophoto quadrangles**, or **orthophotoquads**, for the nation. These images are available in the format of a **digital orthophotoquad (DOQ)**, which is an aerial or satellite image that has been corrected so that its pixels align with longitude and latitude lines on a map. The corrected images are mosaicked to cover the geographic extent of all or one-quarter of a USGS topographic quad (figure 11.15).

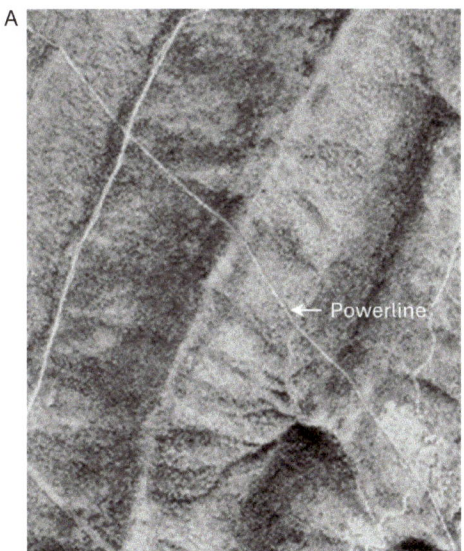
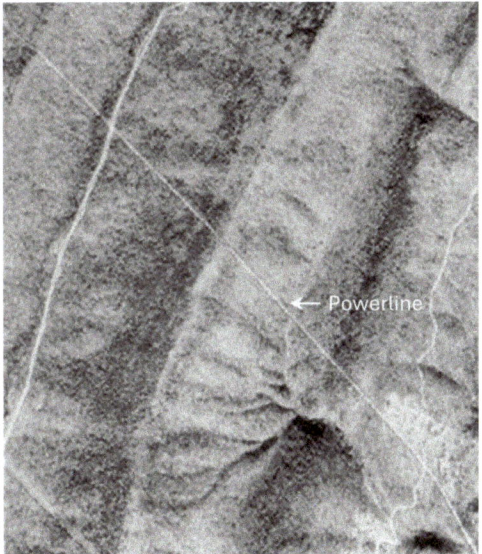

Figure 11.13. Relief displacement and scale variation in the aerial image of Tenth Legion, Virginia (*A*), are removed in the orthophoto of the same area (*B*). On the orthophoto, the planimetrically corrected power line running over the hills appears crooked in the aerial image because of relief displacement. Courtesy of the US Geological Survey.

Figure 11.14. A true ortho covering the northeastern part of San Francisco, California.

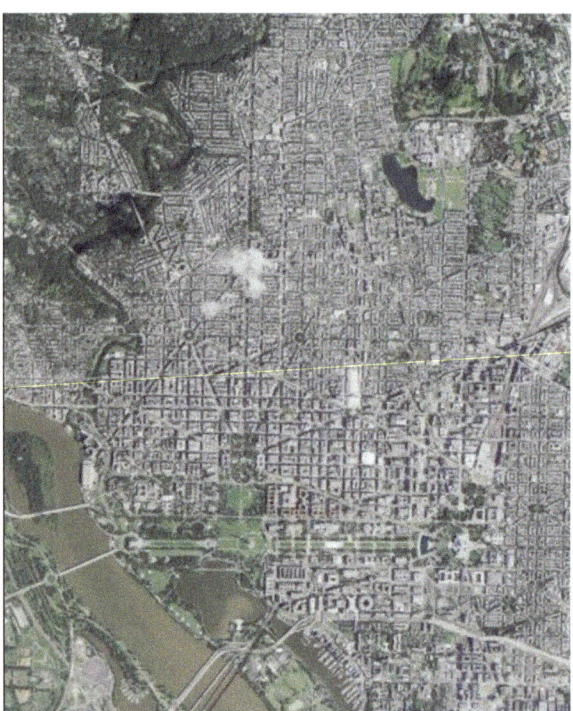

Figure 11.15. A full natural-color quarter-quad DOQ of central Washington, DC. Courtesy of the National Agriculture Imagery Program.

Annotated orthophotomaps

An orthophoto provides environmental detail that cannot be portrayed with conventional map symbols for point, line, and area features because you can see all the features in the landscape that are closest to the sensor (tops of trees, buildings, and so on). This level of detail contrasts with the advantages of a conventional map, on which the mapmaker uses symbols, text, selection, classification, generalization, and other techniques of cartographic abstraction (see chapter 6) to highlight important features, patterns, and themes.

However, mapmakers can overlay conventional map symbols on an orthophoto to create an **annotated orthophotomap** that emphasizes selected geographic information about the area. **Annotated orthophotomaps** are produced for a variety of mapping projects, as the two examples in figure 11.16 illustrate.

The annotated orthophotomap in figure 11.16A deals with monitoring and responding to April 2022 floods in the area surrounding Durban, South Africa. ArcGIS software was used to blend the orthophoto of the area with

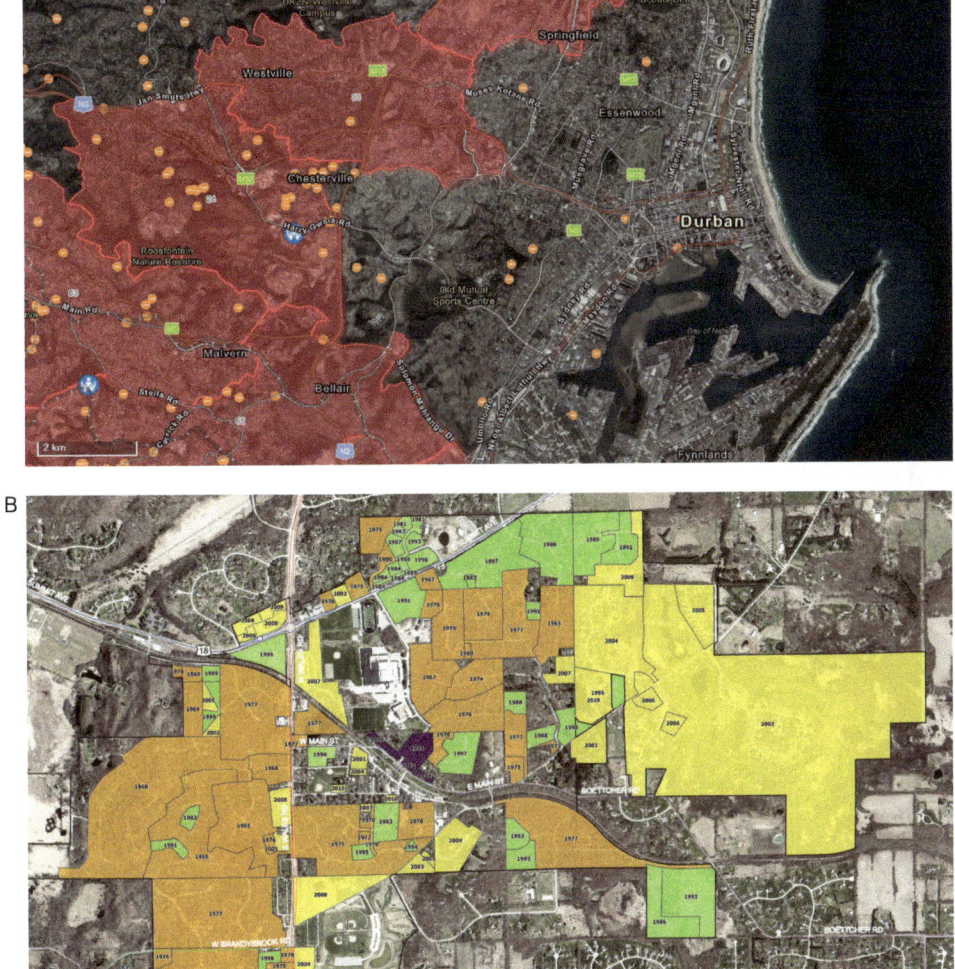

Figure 11.16. Annotated orthophotomaps of Durban, South Africa (*A*), and the village of Wales, Wisconsin (*B*). Courtesy of eThekwini Metropolitan Municipality, Durban, South Africa (*A*), and County of Waukesha, Wisconsin (*B*).

traditional map symbols and labels for city and town names, road names and numbers, points where roads were damaged by flooding, and a light-red tint overlying areas of high potential for flooding. The annotated orthophotomap in figure 11.16B was created for display at the centennial celebration of the village of Wales, Wisconsin, illustrating the expanding pattern of village development over time. The mapmaker used ArcGIS Pro to overlay orange, yellow, and green areas depicting subdivision development over time. The date of each subdivision plat was labeled in the center of its area, with other map annotation limited to street names and highway identifiers.

Satellite imagery

We mentioned previously that the higher the altitude of the sensor platform, the less the relief displacement and scale variation on the images collected. Images obtained by satellites orbiting the earth are at high enough altitudes that relief displacement of tall features, such as mountains, is minimal, and the only scale variation on the image is because of the earth's curvature. This scale variation can be corrected mathematically through the same computer processing software used to create orthophotos from digital aerial images, and the resulting satellite image can be made on any map projection (see chapter 3). Conventional map symbols can then be overlaid on the satellite image to create a **satellite image map**. Numerous satellites collect images of the earth and other planetary bodies, but we focus on two major types—weather satellites and earth resources satellites.

Weather satellites

If you are in the United States, your local internet, television, or newspaper weather report probably uses one or more weather satellite images from a **Geostationary Operational Environmental Satellite** (**GOES**), operated by NOAA to support weather forecasting, severe storm tracking, and meteorological research. GOES and similar weather satellites operated by other nations are placed in geostationary orbits.

A satellite in **geostationary orbit** is always in the same position with respect to the rotating earth. This orbit is also called the **Clarke orbit** because the science fiction author Arthur C. Clarke first proposed the idea of using a geostationary orbit for communications satellites in 1945. Because a geostationary orbit must be in the same plane as the earth's rotation, it is always in the **equatorial plane** (figure 11.17). By orbiting in the same easterly direction as the earth, at an altitude of 22,236 miles (35,786 kilometers) with a 24-hour **orbital period** that matches the earth's rotation, the satellite appears stationary (**synchronous**) to ground receiving stations.

The most frequently obtained imagery from space you will see in the United States is from **GOES-West** and **GOES-East**. Positioned in geostationary orbits above the equator at fixed longitudes of 137° W and 75° W, respectively, GOES-West and GOES-East obtain images of the western

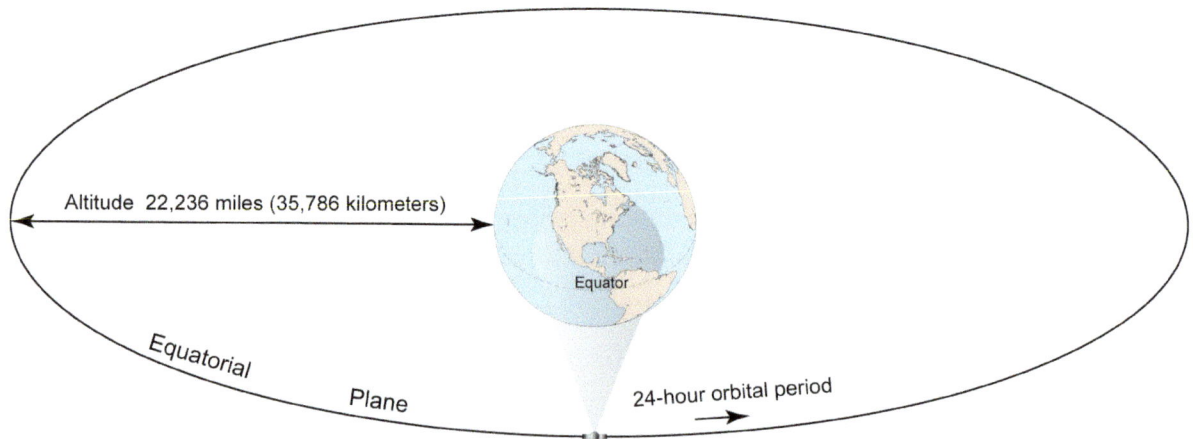

Figure 11.17. A satellite in a geostationary orbit moves easterly in the equatorial plane at an altitude of 22,236 miles (35,786 kilometers), with a 24-hour orbital period that matches the earth's 24-hour rotation. A large portion of the earth's surface can be imaged from this high altitude.

and eastern United States that are seen in the media and on mobile devices (figure 11.18). These two satellites record images of the western and eastern United States every 15 minutes and the full earth disk (nearly a hemisphere) every three hours.

A single GOES image covers millions of square miles of ground surface, making it possible to see the cloud patterns over half the United States simultaneously. By studying images taken every 15 minutes, you can easily see the detailed movement of clouds. This information is especially useful for monitoring severe rainstorms and snowstorms, as well as hurricanes and typhoons. This imagery is used to produce weather system videos and clips of natural disasters, such as volcanic eruptions and large wildfires (figure 11.19). The resolution of ground detail on weather satellite imagery is particularly coarse, however, because of its extremely high altitude. Consequently, the use of the imagery, often draped over a terrain model, is essentially restricted to broadscale atmospheric phenomena.

Weather satellite image maps for the conterminous

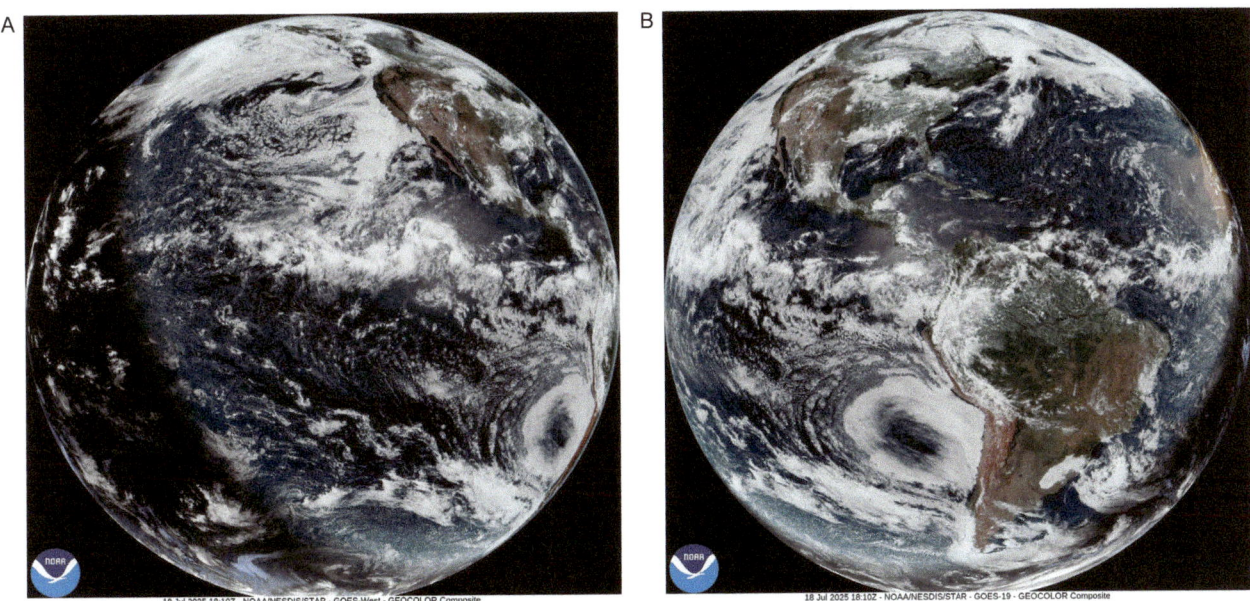

Figure 11.18. Examples of full NOAA GOES-West (*A*) and GOES-East (*B*) satellite images. Courtesy of the Cooperative Institute for Research in the Atmosphere and the National Oceanic and Atmospheric Administration.

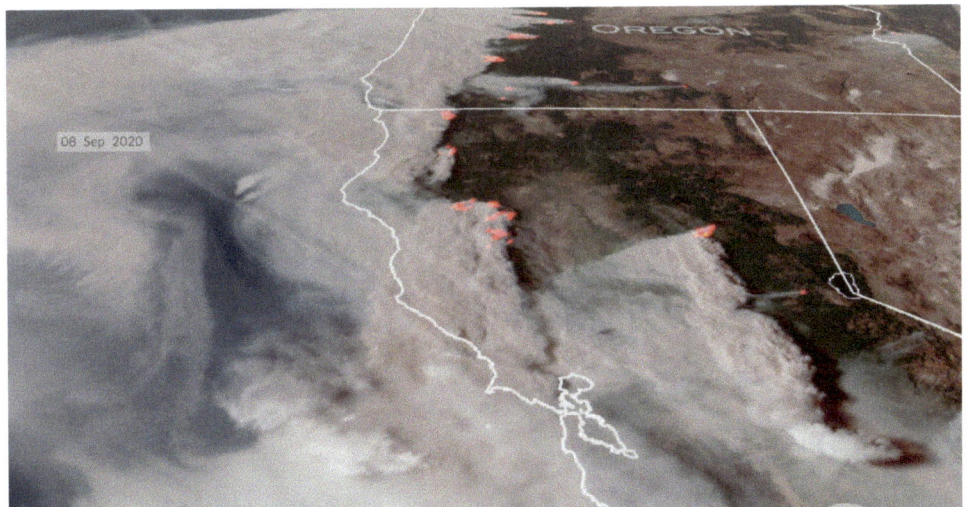

Figure 11.19. A GOES-West GeoColor still frame from a video clip showing hot spots and thick smoke plumes from multiple wildfires burning in Oregon and Northern California in September 2020. Courtesy of the Cooperative Institute for Research in the Atmosphere and the National Oceanic and Atmospheric Administration.

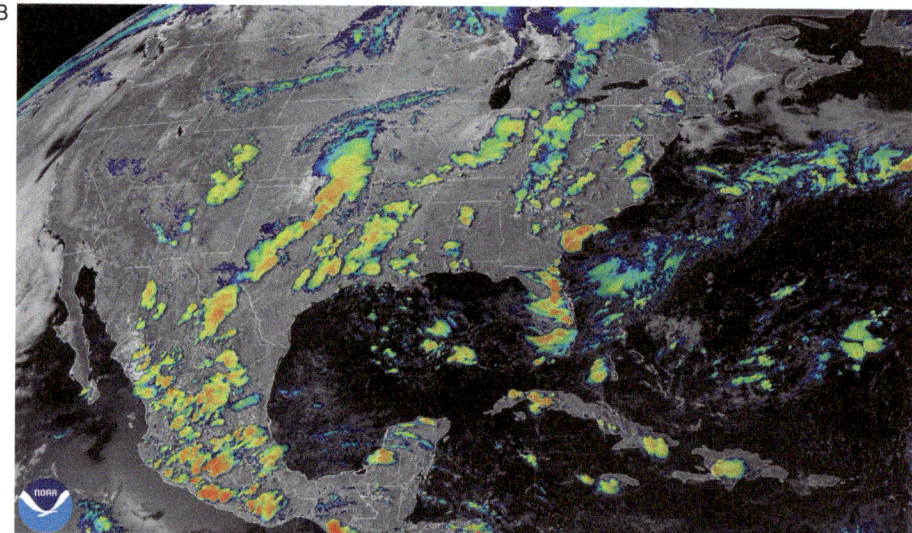

Figure 11.20. Composite GOES-East weather satellite GeoColor (*A*), sandwich (*B*), and dust (*C*) image maps of the conterminous United States allow you to see a close approximation of daytime natural-color imagery, cloud top formation and temperature, and airborne dust in the atmosphere. Courtesy of the National Oceanic and Atmospheric Administration.

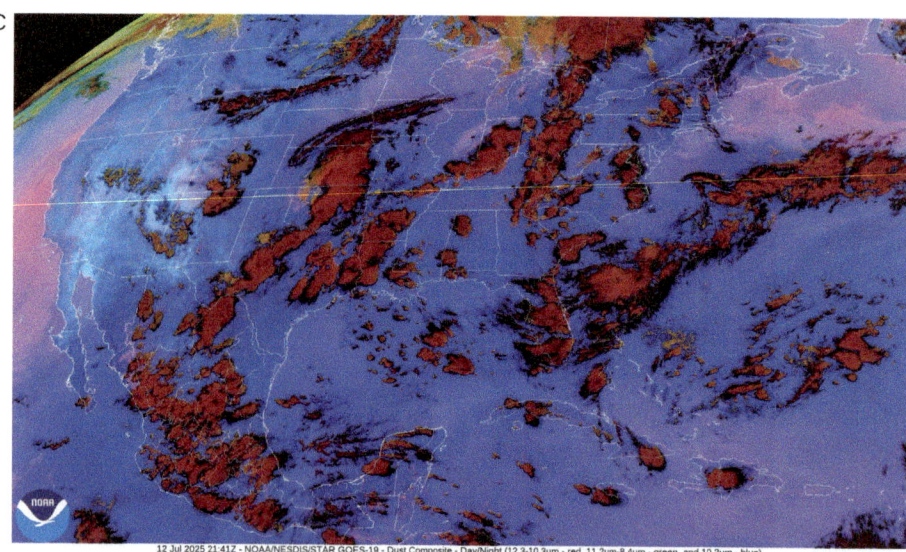

United States are made from GOES-West and GOES-East images collected in the visible and thermal-IR regions of the electromagnetic spectrum and in a mid-IR spectral region that is tuned to water vapor. Minimal geographic information (graticule lines, state boundaries, coastlines, and large lakes) is added to help you orient the map and see the geographic location of clouds and other atmospheric features.

The GOES-East GeoColor image map in figure 11.20A, closely approximates daytime natural-color imagery and thus gives an intuitive interpretation of meteorological and ground features. This composite image is also used to identify areas covered by smoke, blowing dust, smog, and anything else that has a unique color property. The "sandwich" composite image map (figure 11.20B) provides information on both cloud top formation and temperature, with the blue through orange color progression on the image map showing warm to cold cloud tops indicating more vertically developed cumulonimbus (thunderstorm) clouds. On the dust composite image map (figure 11.20C), dust appears pink or magenta (as over Texas and surrounding areas), dark brown indicates cold, thick clouds, tan shows mid-level thick clouds, and blue is associated with very thin clouds over warm surfaces (such as water). Dust plumes are easily distinguished from surrounding clouds, land surfaces (such as deserts), and oceans and other large water bodies.

Earth resources satellites

Multispectral imagery involves capturing a ground scene in different spectral bands of the electromagnetic spectrum so that the resulting images are geometrically identical. Different bands can be combined to produce images that emphasize certain characteristics of the environment (see figure 11.23). **Earth resources satellites** have become important multispectral sensing platforms. Earth resources satellites are launched with the primary mission of providing systematic, repetitive environmental image data, such as surface reflectance, wave height, surface temperature, or land elevation. Example images from many of the earth resources satellites described in this chapter can be viewed in ArcGIS Living Atlas of the World.

If a coarse spatial resolution sensor is used on the satellite, images are acquired from several hundred miles above the earth's surface and cover a relatively large ground area. If a high-resolution multispectral imaging device is used, a small ground area is recorded at a high spatial resolution.

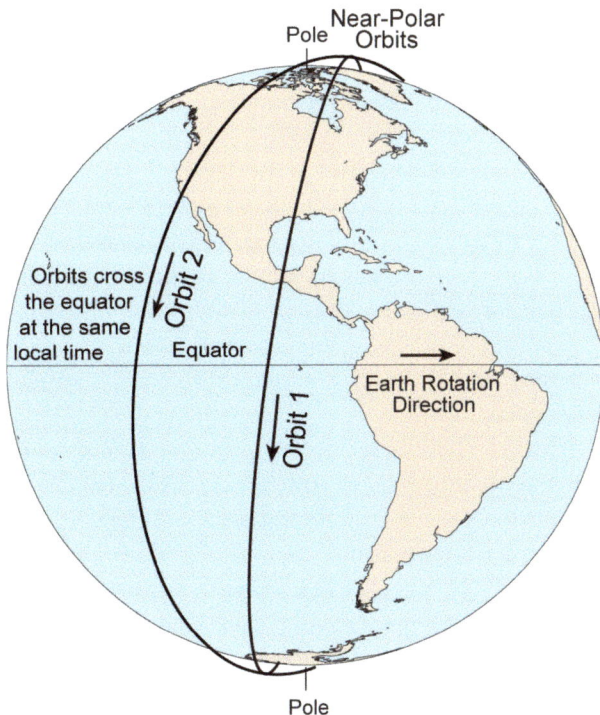

Figure 11.21. A sun-synchronous orbit crosses over the equator at approximately the same local time every day. This orbit allows consistent imaging with the angle between the sun and the earth's surface remaining relatively constant. This illustration shows two orbits of a sun-synchronous satellite with the same equatorial crossing time.

Most earth resources satellites are in **near-polar orbits**, circling the earth at a near-polar **inclination** (the angle between the equatorial plane and the satellite orbital plane—a true polar orbit has an inclination of 90 degrees). A **sun-synchronous orbit** is a special type of near-polar orbit in which the satellite passes over the same part of the earth at roughly the same local time each day (figure 11.21). This regularity enables data collection at consistent times as well as long-term image comparisons. Near-polar orbiting satellites provide a nearly complete view of the earth built up over many orbits because they orbit over all but the latitudes closest to the poles.

We examine imagery obtained from remote sensing devices on three earth resources satellites commonly used for image maps: Landsat, Sentinel-2, and MODIS. We then look at imagery from newer submeter-resolution sensors.

Landsat

In 1965, NASA and the USGS began the **Landsat** project to explore the potential for monitoring the earth from space. The first satellite, Landsat 1, was put into orbit in 1972. Landsat 8 (launched in 2013) and 9 (launched in 2021) are currently fully operational, with onboard remote sensing instruments collecting hundreds of images of the earth every day. These images provide an excellent orthoimage base for satellite image maps that are made worldwide for applications in agriculture, geology, forestry, regional planning, global change research, nautical charting, and other fields.

Landsat satellites are in sun-synchronous, near-polar orbits so that near-global coverage is possible because of the earth's rotation beneath the satellite (figure 11.22). Landsat 8 and 9 orbit the earth every 99 minutes from an altitude of 440 miles (709 kilometers), circling the earth roughly 14 times per day and passing over the same spot on earth at the same local time of day every 16 days. On each **descending node** (the daylight pass from north to south), remote sensing instruments record electromagnetic radiation from the ground over a strip that is 115 miles (185 kilometers) wide. This data is used to create images covering a ground area that is 115 miles wide by 105 miles high (185 by 170 kilometers).

Many image maps today are made from images acquired by the **Operational Land Imager (OLI)** sensor on Landsat 8 and 9. The more recent and widely used OLI sensor provides 30-meter resolution images from the visible (VIS) blue (bands 1 and 2), green (band 3), and red (band 4) portions of the electromagnetic spectrum, as well as the near-IR, or NIR, (band 5), and shortwave-IR, or SWIR, (bands 6, 7, and 9) portions of the spectrum. Two 100-meter resolution thermal-IR bands (bands 10 and 11) from the **Thermal Infrared Sensor (TIRS)** and a 15-meter resolution panchromatic band (band 8) are also provided. This is illustrated in figure 11.23. Pixel values for the more than 7,000 rows and 7,000 columns that make up an image in each band are transmitted to ground receiving facilities, where the data is processed into geometrically corrected images. There are two levels of Landsat data—Level 1 imagery is suitable for general use, while Level 2 imagery is preferred for applications requiring high accuracy and detailed analysis. You can download these images in the UTM map projection from the USGS EarthExplorer website.

Landsat OLI bands are often combined into **color-composite images**. **Band combination** is the process of merging different bands into a single image to enhance visual interpretation and analysis. By combining specific bands, you can more effectively highlight or analyze different features on and characteristics of the earth's surface. The combination you use depends on what you want to see in the image.

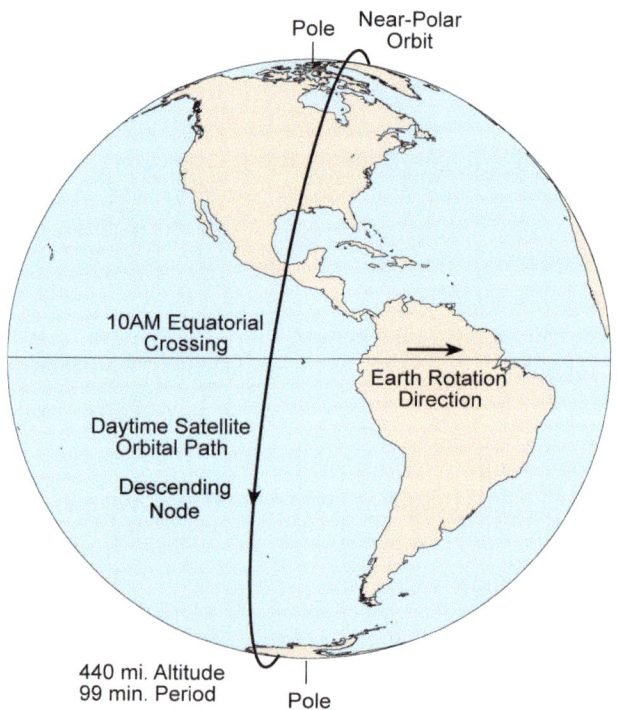

Figure 11.22. The sun-synchronous, near-polar orbits of Landsat 8 and 9 make it possible to provide nearly complete coverage of the earth's surface in 115-mile-wide (185-kilometer) swaths every 16 days.

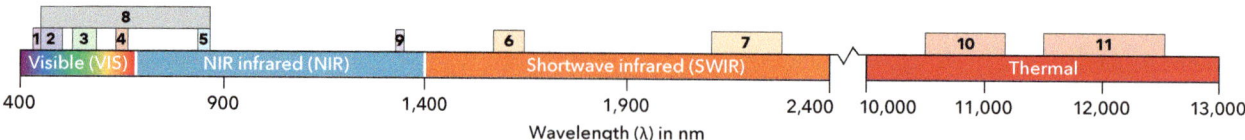

Figure 11.23. The bands for the Landsat OLI sensor. The Landsat 8 and 9 spectral and spatial parameters are nearly identical.

To create a composite image, different bands are assigned to the red, green, and blue channels of an RGB image. For example, a **true-color image** is produced by assigning the red, green, and blue bands (4, 3, and 2) to the RGB channels in that order. The result is a display of the earth in colors similar to what we might see with our own eyes. Any band combination other than red, green, and blue or any change in the order of their assignments to the RGB channels creates a **false color image**, so the possible number of false color images is nearly endless. Some band combinations are commonly used because of the useful images they produce. Appendix table A.7 describes and illustrates some of the common band combinations.

The Landsat 8 composite image in figure 11.24A, centered on Mount Vernon, Washington, and the surrounding Skagit River delta, is a true-color combination of the red, green, and blue bands (4, 3, and 2), whereas the color infrared composite image (*B*) is a combination of the NIR, red, and green bands (5, 4, and 3). The shortwave-IR image (*C*) is a combination of the SWIR, NIR, and red bands (7, 5, and 4). The fourth image (*D*) was created using the combination of SWIR, red, and blue bands (7, 4,

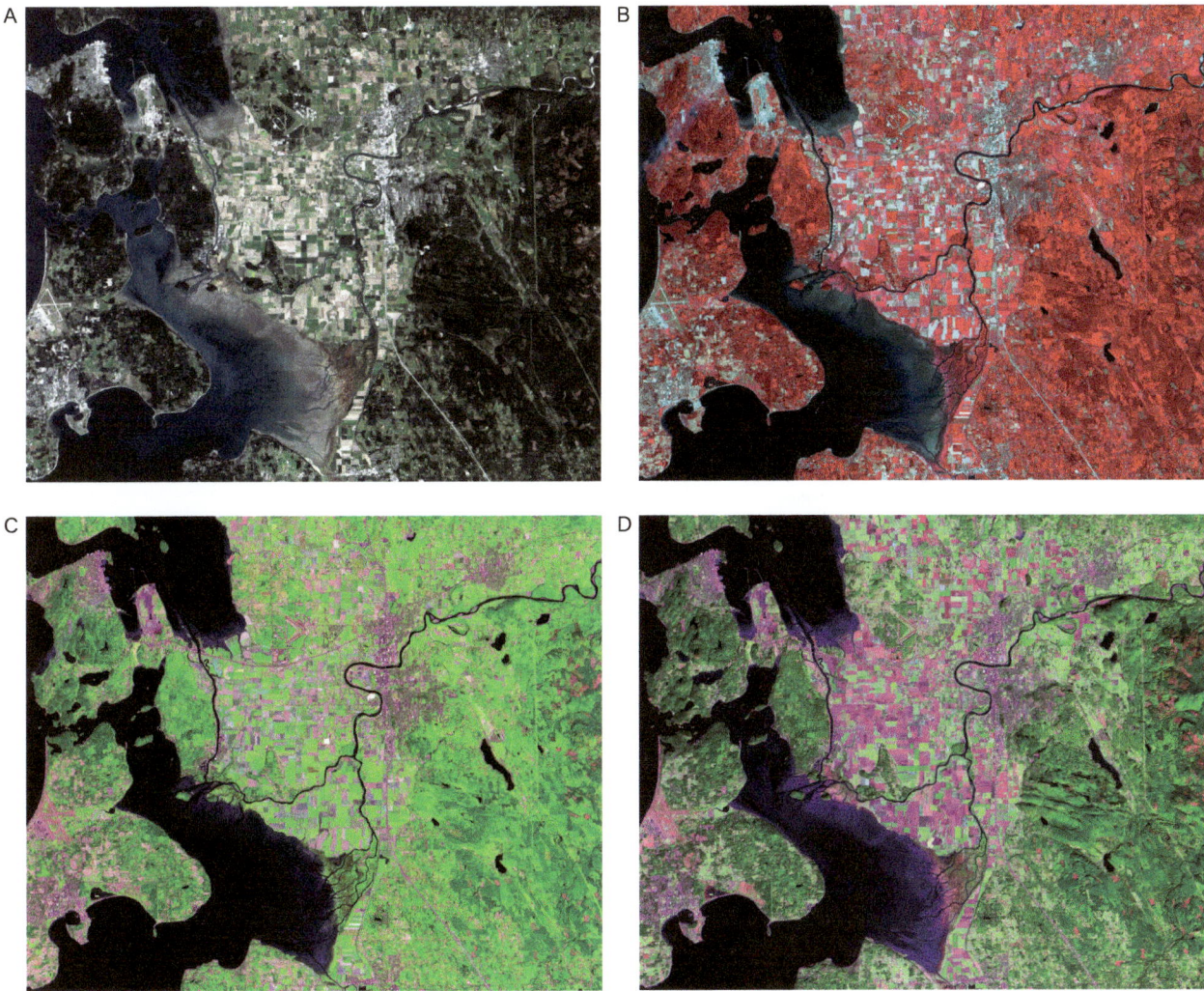

Figure 11.24. Landsat OLI bands 4, 3, and 2 are used to create a true-color composite image (*A*), and bands 5, 4, and 3 are used to create a color IR composite image (*B*) centered on Mount Vernon, Washington (north of Seattle). Bands 7, 5, and 4 are used to create the shortwave IR (*C*) composite image. The combination of bands 7, 4, and 2 (*D*) were used to create a NASA global Landsat mosaic. Courtesy of the Earth Resources Observation and Science (EROS) Center.

and 2)—the same as those used by the USGS to create a Landsat global mosaic image for NASA.

Sentinel-2

Sentinel-2 is an earth observation mission from the Copernicus Programme that acquires optical imagery at high spatial resolution (10 m to 60 m) over land and coastal waters. This system differs from other earth resources systems in that images are obtained from two sun-synchronous satellites that provide coverage for the continental land masses every five to seven days providing better coverage than the Landsat 16-day global coverage time. The two satellites are spaced one-half orbit apart, so they consequently cross the equator 180 degrees apart in longitude (figure 11.25). The satellite in daylight crosses the equator at about 10:30 a.m. local time on its descending node. Each satellite collects 13 bands of multispectral image data in the visible, near-IR, and shortwave-IR spectral regions, at a spatial resolution ranging from 30 to 180 feet (10 to 60 meters) with a swath width of 180 miles (290 kilometers).

Sentinel-2 imagery can be used to show different land characteristics over time. The example in figure 11.26 focuses on vegetation health and soil moisture. ArcGIS Image Server created the four images of agricultural fields along the Mississippi River, south of New Orleans, by using different sets of three bands from the 13 bands collected by the sensor. For example, a natural-color image was created using the red, green, and blue bands. The natural color image in figure 11.26A displays the optical wavelengths that our eyes naturally detect, the color infrared image (figure 11.26B) shows healthy vegetation in bright red, the short-wave infrared vegetation image (figure 11.26C) shows the most vigorous vegetation in bright green, and a water moisture index (figure 11.26D) image displays the highest moisture levels in blue.

MODIS

The **Moderate Resolution Imaging Spectroradiometer** (**MODIS**) sensor on the Terra and Aqua satellites, launched in 1999 and 2002, respectively, provides a complete view of the earth's surface every one to two days from a 438-mile (705-kilometer) altitude. The MODIS sensor collects data from 36 spectral bands, ranging from 0.4 µm (blue) to 14.4 µm (thermal-IR). Bands 1 and 2 (red and near-IR) have a 250-meter spatial resolution at the nadir, bands 3 through 7 (blue through mid-IR) have a 500-meter resolution, and

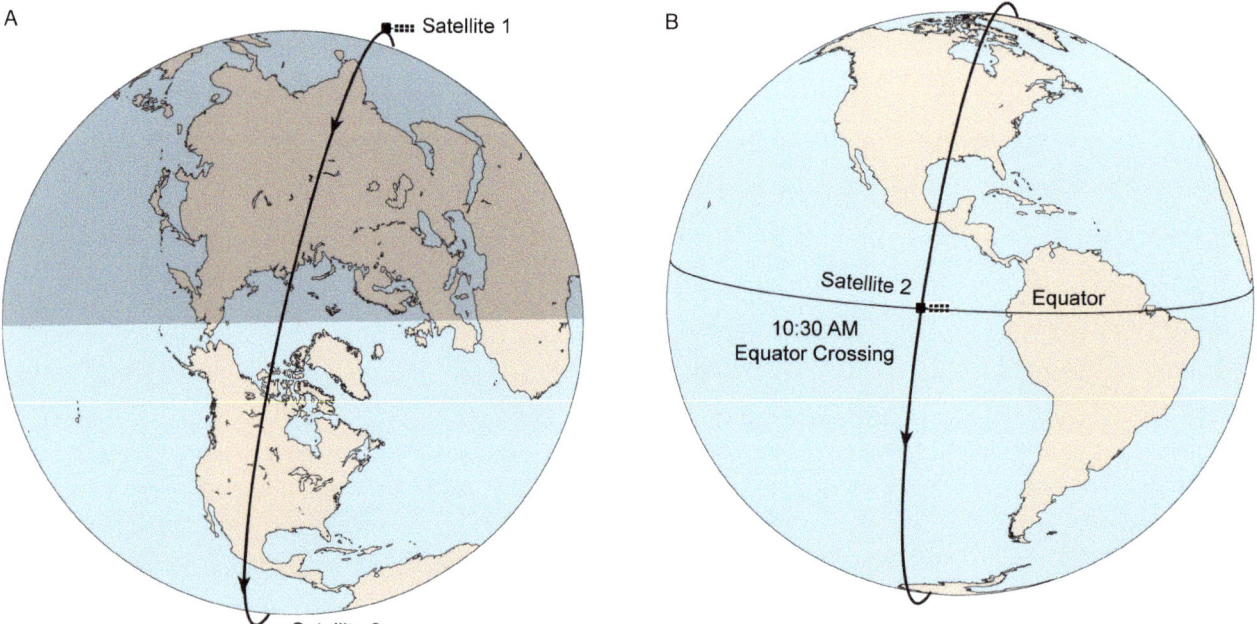

Figure 11.25. Two Sentinel-2 satellites are spaced one-half orbit apart, so they consequently cross the equator 180 degrees apart in longitude (*A*). The daytime satellite crosses the equator at about 10:30 a.m. local time on its descending node (*B*).

Figure 11.26. Agricultural fields in the Mississippi River delta downstream from New Orleans, Louisiana, are shown with four Sentinel-2 band combinations. Natural-color images (*A*) display the optical wavelengths that our eyes naturally detect. Color-infrared images (*B*) show healthy vegetation in bright red. Shortwave-infrared images (*C*) show the most vigorous vegetation in bright green. A normalized moisture difference index (*D*) is used to show the areas with the highest moisture levels in darker blue. Courtesy of Esri, European Commission, European Space Agency, and Amazon Web Services.

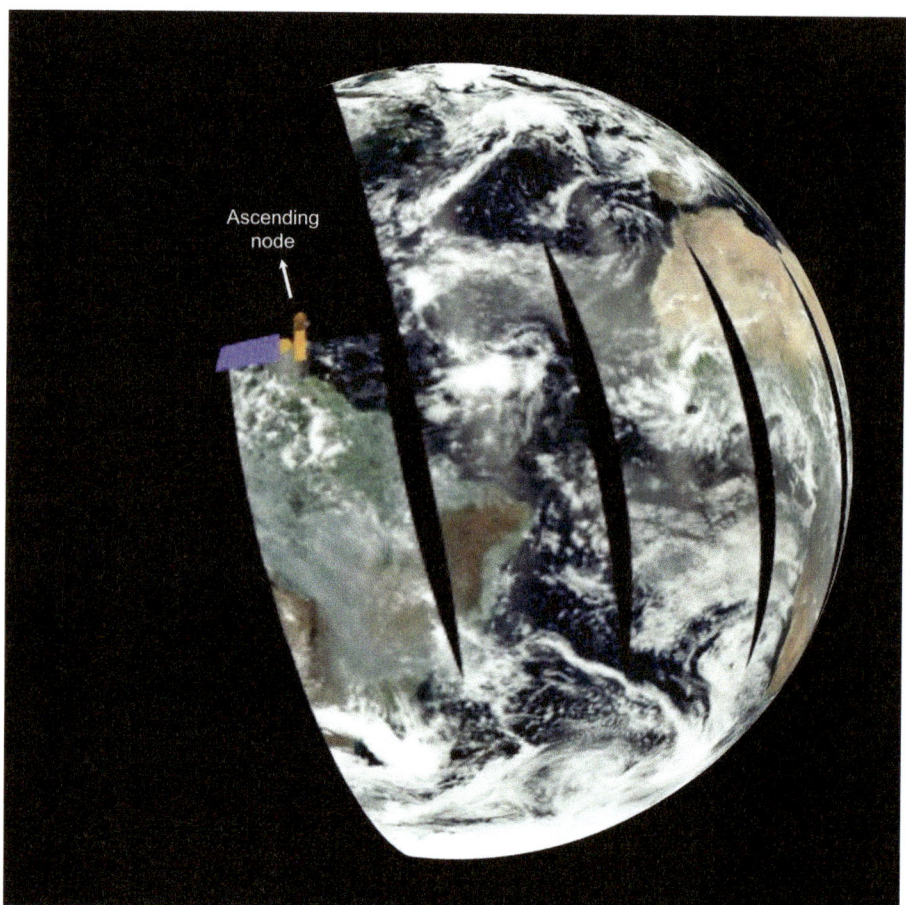

Figure 11.27. The MODIS sensor collects image data in 2,330-kilometer-wide swaths. Courtesy of the National Oceanic and Atmospheric Administration.

Figure 11.28. This natural-color image of an area around the Hawaiian Islands was created by mosaicking one-kilometer resolution MODIS imagery rendered with the blue, green, and red bands. Courtesy of Jacques Descloitres, MODIS Land Rapid Response Team at NASA Goddard Space Flight Center.

bands 8 through 36 (blue through thermal-IR) have a one-kilometer resolution. The 2,330-kilometer-wide image swaths shown in figure 11.27 allow a large portion of the earth to be imaged on each south to north orbit (daylight ascending node) at a coarser spatial resolution than other earth resources satellite sensors.

Images such as that of the Hawaiian Islands in figure 11.28 give a more general view of land and water features and earth phenomena. Scientists use these images to quantify and map land surface characteristics, such as land-cover type and extent, snow-cover extent, surface temperature, and forest fire occurrence. The one- to two-day coverage period for the entire earth means that changes in these surface characteristics can be studied and mapped day to day, week to week, or at monthly, seasonal, or annual intervals.

Submeter-resolution systems

Since 2014, the commercial satellite systems **WorldView-3** and **Pléiades NEO** have provided digital images of small areas from space for paying customers at the same or better spatial resolution as low-altitude aerial imagery. WorldView-3 and Pléiades NEO sensors are used aboard satellites in sun-synchronous, near-polar orbits at approximately 620-kilometer (385-mile) altitudes with an orbital period of around 100 minutes. These are submeter-resolution systems because both have a blue-to-near-IR wavelength panchromatic sensor with a 30-centimeter spatial resolution at the image nadir to coarser resolution away from the nadir.

Both WorldView-3 and Pléiades NEO have blue, green, red, and near-IR band multispectral data collection at a 1.25-meter spatial resolution, so that natural-color, CIR, and other images can be created from different band combinations and special image-processing procedures. For example, the WorldView-3 image of the US Capitol in Washington, DC, in figure 11.29 was created at 30-centimeter spatial resolution from the panchromatic and multispectral bands using pan sharpening. **Pan-sharpening** is a process to merge higher-resolution panchromatic and lower-resolution multispectral imagery to create a single high-resolution color image. The goal of pansharpening is to achieve the highest level of visual clarity and detail for an image.

Both systems have pointable sensors, giving them the ability to collect images away from vertical (off-nadir). **Pointable sensors** allow virtually all locations on earth to be imaged every few days, in swaths that at the nadir

Figure 11.29. A pan-sharpened 30-centimeter WorldView-3 natural-color image of the US Capitol building in Washington, DC. Courtesy of Esri, Vantor, Earthstar Geographics, and the GIS User Community.

vary from 14 kilometers (8.7 miles) for the Pléiades NEO satellite to 13 kilometers (8.1 miles) for the WorldView-3 satellite. Small portions of the earth can be imaged each day, totaling from 680,000 square kilometers (260,000 square miles, roughly the combined area of California and Oregon) for Worldview-3 to one million square kilometers (385,000 square miles, slightly less than the combined area of Texas and California) for the Pléiades NEO system. The Pléiades maximum daily coverage is significantly greater than WorldView-3 because it consists of two identical satellites in an orbit configuration that is similar to the Sentinel-2 system.

Most commercial clients request images of far smaller areas for specialized projects. These projects range from environmental emergencies to urban planning, security and crisis management, maritime monitoring, national defense issues, and agricultural field studies.

Hyperspectral systems

Hyperspectral imagery is similar to multispectral imagery in that digital image data related to energy reflected from or emitted by ground features is collected in different spectral bands. The difference is that hundreds of narrow bands are imaged so that the spectrum of visible through thermal-IR energy for each pixel in the scene is recorded. Features have unique **spectral signatures**, which are the differences in their reflectance or emittance characteristics with respect to wavelengths in the visible through thermal-IR spectrum. Digital image processing is used to identify features in the image by matching the values in different bands for each pixel with known spectral signatures for different types of features.

You can visualize the hundreds of bands ranging from blue to near-IR that compose a hyperspectral image as a cube of images, with each row as an image of one band, stacked top to bottom from shortest to longest wavelength. The cube in figure 11.30 graphically represents 224 spectral bands from an airborne hyperspectral sensor called AVIRIS. **AVIRIS** stands for the **Airborne Visible/Infrared Imaging Spectrometer**, a specialized instrument that captures high-resolution hyperspectral images of the earth's surface from an aircraft to provide detailed information about the composition and characteristics of the terrain, vegetation, and other features on the ground. The top of the hyperspectral cube is not a spectral band but rather a false-color image of Moffett Field Airport and vicinity in the San Francisco Bay Area.

AVIRIS is not a satellite system, but it has been flown

Figure 11.30. A hyperspectral cube shows 224 spectral bands from the AVIRIS sensor. Bands are shown with shortest (*top*) to longest (*bottom*) wavelength along the cube edges. Courtesy of the NASA Jet Propulsion Laboratory.

on four low- to high-altitude aircraft platforms at altitudes of approximately four kilometers and 20 kilometers. AVIRIS images are mainly used to identify, measure, and monitor physical features on the earth's surface and in the atmosphere based on how they absorb certain molecules and scatter electromagnetic radiation in the visible to near-IR range. Such studies focus on understanding physical processes related to the global environment and climate change. The AVIRIS natural-color image example in figure 11.31 is taken from a forestry study on the Washington coast where different stages in the forest harvest process were the prime concern. Recently clear-cut areas stand out in light pink, in contrast with the older uncut forest in dark green.

Map examples

Now we look at examples of satellite image maps that are produced from images collected by the earth resources satellites we have discussed. Like image maps made from aerial imagery, conventional map line work, area coloring, and lettering are commonly overlaid on geometrically corrected earth resources satellite images that serve as a locational basemap for the qualitative or quantitative information added to make the satellite image map. Although we cannot possibly catalog all the types of annotated satellite image maps, we can take a closer look at a representative example for New York from the Esri World Imagery Wayback app. We focus on the four examples in figure 11.32, which give you an idea of how satellite image maps are produced today. These examples are multiscale maps (see chapter 2), which is evident because the image resolution and level of detail in the overlaid road network and the number of feature labels increase as you zoom in on the map.

A further enhancement to satellite image maps is achieved by draping the images over the terrain surface (see chapter 10) to create an oblique-perspective satellite image. Conventional map symbols for geographic features and thematic data can also be overlaid on this base. The example in figure 11.33 shows a satellite image map of Hurricane Arthur, east of Florida, with precipitation extruded

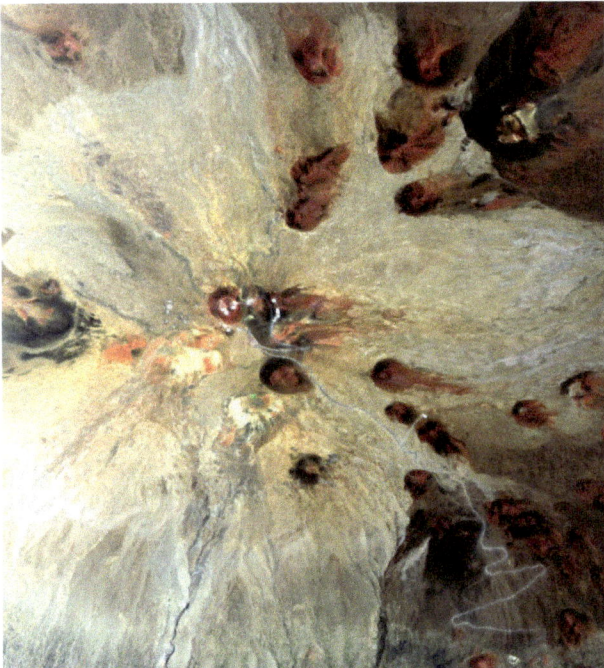

Figure 11.31. A natural-color AVIRIS image of a small section of the Washington coast created from three of the 224 spectral bands in the hyperspectral cube in figure 11.30. Courtesy of the NASA Jet Propulsion Laboratory.

Figure 11.32. These four satellite image maps of the New York metropolitan area from the Esri World Imagery Wayback archive illustrate how mapmakers overlay conventional map symbols on satellite images to create annotated image maps. The first two maps (on the top left) are based on TerraColor NextGen images by Earthstar Geographics, whereas the other maps use Vantor WorldView-3 images as the base. Courtesy of Esri, Vantor, Earthstar Geographics, and the GIS User Community.

above the surface to show the amount of rainfall. By showing the precipitation in 3D surface over the satellite image, meteorologists were able to better visualize the predicted path and 3D structure of the storm as it neared land.

Dynamic maps

Dynamic maps go beyond traditional static displays to include animated sequences of images that you can view using interactive navigation tools. In addition, these maps often allow you to interact with map symbols that you see in the animation that are linked to text, pictures, or video clips. You have probably used a dynamic map that gives you the feeling of flying over and around a part of the earth. You are likely to see fly-throughs (see chapter 10) in visual presentations, such as news programs, training simulators, and video games.

The 3D perspective view map in figure 11.35 is one scene from a fly-through of Alexander von Humboldt's 6,000-mile journey through six countries in South

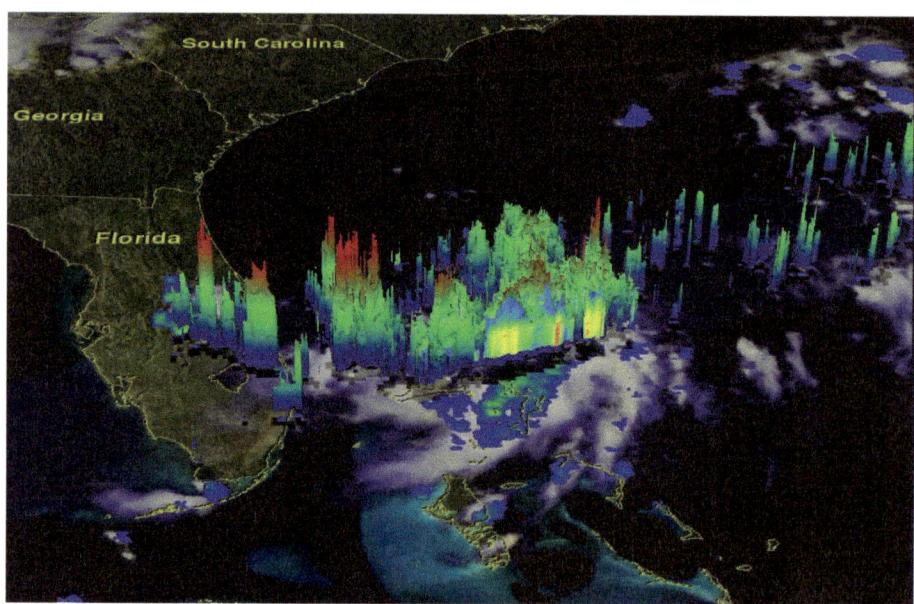

Figure 11.33. This oblique-perspective satellite image map of an area around the Bahamas is overlaid with Hurricane Arthur data that shows the horizontal distribution and vertical height of precipitation on June 30, 2015. Courtesy of the National Oceanic and Atmospheric Administration.

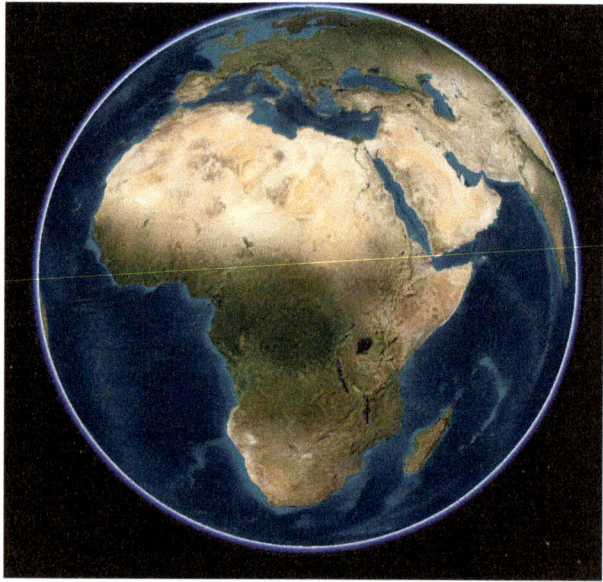

Figure 11.34. ArcGIS Earth image map.

Figure 11.35. One scene from a fly-through of Alexander von Humboldt's journey through South America from 1799 to 1804.

America in the early 1800s. Many find these maps compelling because the 3D terrain surface can be overlaid with satellite imagery to create a realistic basemap that can be overlaid with symbols and labels. The 3D terrain can also be overlaid with scanned maps, GIS data, or other georeferenced layers.

The possibilities for dynamic maps with imagery seem limitless, as remotely sensed data about the earth is being collected at an accelerating rate. This data can be visualized with spherical display systems such as OmniGlobe (figure 11.36) because any digital image map made on an equirectangular map projection, or plate carrée projection

Figure 11.36. This 47-inch (120-centimeter) diameter OmniGlobe in a "floating" design has an internal projection system that can display any global map or image, or a sequence of maps or images, on a spherical screen so that the earth's geometry is correctly shown everywhere. Image courtesy of Globoccess, © 2025.

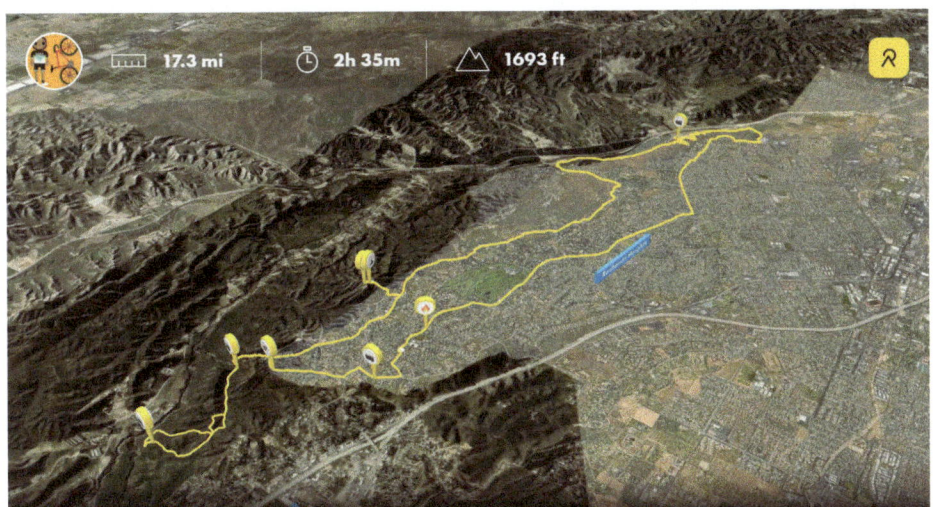

Figure 11.37. The route of a cyclist through Redlands, California, is displayed in yellow over a terrain-enhanced high-resolution satellite image. Photographs taken en route are linked to their acquisition spots, shown with yellow pin symbols. Courtesy of Relive.

(see chapter 3), can be displayed on the globe. Such 3D map displays may revolutionize how you use and benefit from traditional and image maps.

The future of imagery and maps looks bright and interesting. We have only discussed the aerial and satellite imagery that is most commonly used on maps today, but many other meteorological, earth resources, and military reconnaissance satellites are, and will continue to be, placed in orbit. Both static and dynamic maps will continue to be made from imagery that is collected by the diverse and ever-expanding suite of remote-sensing devices. For example, Relive is a fitness app that allows users to track their activities, such as running and cycling, using GNSS. The 3D web map in figure 11.37 depicts an actual bicycle ride displayed as a route draped over a satellite image combined with elevation-exaggerated terrain created from a DEM of the area.

Selected readings

Barrett, E. C., and L. F. Curtis. 1999. *Introduction to Environmental Remote Sensing*, 4th ed. Chapman and Hall.

Blom, J. D. 2010. "Unmanned Aerial Systems: A Historical Perspective." *Occasional Paper 37*. Combat Studies Institute Press.

Brown, C., and C. Harder. 2016. The ArcGIS Imagery Book: New View. New Vision. Esri Press.

Campbell, J. B., R. H. Wynne, and V. Thomas. 2023. *Introduction to Remote Sensing*, 6th ed. Guilford.

Chaminé, H. I., A. J. Pereira, A. C. Teodoro, et al. 2021. "Remote Sensing and GIS Applications in Earth and Environmental Systems Sciences." *Springer Nature Applied Sciences* 3 (870). https://doi.org/10.1007/s42452-021-04855-3.

Fox III, L. 2015. *Essential Earth Imaging for GIS*. Esri Press.

Kimball, J. S. 2011. "Earth Observation of Global Change: The Role of Satellite Remote Sensing in Monitoring the Global Environment." *EOS*. https://agupubs.onlinelibrary.wiley.com/doi/10.1029/2008EO320009.

Kramer, H. J. 2002. *Observation of the Earth and Its Environment: Survey of Missions and Sensors*. Springer-Verlag.

Lillesand, T. M., R. W. Kiefer, and J. W. Chipman. 2015. *Remote Sensing and Image Interpretation*, 7th ed. John Wiley & Sons.

Mesev, V., and L. R. Newcome. 2007. *Unmanned Aviation: A Brief History of Unmanned Aerial Vehicles*. Pen and Sword.

Paine, D. P., and J. D. Kiser. 2012. *Aerial Photography and Image Interpretation*. John Wiley & Sons.

Skidmore, A. (Ed.). 2002. *Environmental Modeling with GIS and Remote Sensing*, 1st ed. CRC Press. https://doi.org/10.4324/9780203302217.

CHAPTER 12
Maps and reality

In the 1927 murder mystery *The Man in the Queue*, Josephine Tey writes, "The very best map-reader has to suffer some severe shocks when he comes face to face with reality." This observation about map use, as disquieting as it may be, is as important as anything else you have learned about maps. Your aim, after all, is not to understand maps. Maps are just one means to your real goal—understanding the world.

If you do not look beyond map symbols to the reality they represent, you may defeat your purpose and end up with a warped view of your surroundings. The gravest accusation we can make about maps is that they taint our judgments about the environment when we allow them to. In this chapter, we look at the hazards of putting too much faith in maps, of not realizing their limitations, and of forgetting to look beyond the symbols of the map to the real world beyond.

The map as reality

We began work on this book with the reflection that maps mirror the world. Therefore, we suggested, the better your knowledge of maps, the wiser your spatial behavior and understanding. If we carry this analogy further, we see that some cautions are built in. A mirror is a useful tool, but it shows only a piece of reality. No one confuses its reflection with the real thing. Yet a surprising number of people treat maps' reflection of the world as if it were reality. Treating maps as reality may be partly due to the role that maps play in structuring our understanding of the world, as the Nobel Prize–winning author Abdulrazak Gurnah writes:

> I speak to maps. And sometimes they say something back to me. This is not as strange as it sounds, nor is it an unheard of thing. Before maps, the world was limitless. It was maps that gave it shape and made it seem like territory, like something that could be possessed, not just laid waste and plundered. Maps made places on the edges of the imagination seem graspable and placeable. And later when it became necessary, geography became biology in order to construct a hierarchy in which to place the people who lived in their inaccessibility and primitiveness in other places on the map. (Gurnah 2001)

You probably have explored new places using ArcGIS Earth or Google Maps, flying over what appears to be the real earth, complete with 3D terrain. What looks like the real earth, however, is composed of satellite images taken on a certain date, with a 3D surface created by draping the imagery over a terrain model with a deliberate vertical exaggeration to make the landscape appear real. With the rapid growth of **virtual reality** (immersive multimedia or computer-simulated reality), increasingly the challenge is to distinguish fact from fiction. With virtual reality, you are immersed in a computer-simulated environment that uses sensory stimuli to give the appearance—and help you feel the supposed effect—of your physical presence in different locations on earth or in imaginary worlds. You may play virtual-reality games with strikingly realistic landscapes, but they are not our real world—just as maps are not.

These "realistic" views are a reasonable **simulacrum** (likeness or representation) of reality—they attempt to create a "mirror world" by carefully copying the original. The image is intentionally distorted in an attempt to make the copy appear more "correct." But ignoring the reliable input of your senses and resorting to the map, which is only a representation of reality, you can arrive at a distorted understanding of reality. As Michael Agger, the culture editor of *The New Yorker*, wrote:

> As a simulacrum of the earth, Google Earth provides a safe space for unlimited voyeurism. You have instant access to

forbidden or dangerous places—North Korea, Mecca, the Kremlin, the favelas of Rio, the top of Everest. But mostly it's fun to hop around. Freed from physical constraints, the Google Earther perceives the planet as small, manageable, knowable, and interconnected. This bonhomie can be exhilarating. (Agger 2006)

Although they may attempt to be a mirror world, virtual worlds are deeply rooted in the practices and conventions of cartography.

Maps are a language for communicating aspects of the real world, but literally substituting maps for reality encourages a mechanical or technological rather than a humane view. Like words, maps can be used to obfuscate or misrepresent, and sometimes do both. A statement such as "the tornado wiped the town off the map" ignores the fact that a real town suffered. Such fuzzy thinking can confuse cause and effect. How often do you hear people speak of "the changing map" when they refer to a changing world? They see maps as actors rather than as reflectors of the world. This viewpoint suggests that if we can keep maps from changing, we can keep the world from changing—or that by changing a map, we can change the world.

The extremes to which such faulty reasoning can carry us are portrayed in the 1961 Joseph Heller novel, *Catch-22*. The men in a bomb squadron are ordered to bomb Bologna the next day. They stare pleadingly at the bomb line on the map, as if that mapped line is itself to blame.

> "I really can't believe it," Clevinger exclaimed to Yossarian in a voice rising and falling in protest and wonder.
>
> "It's a complete reversion to primitive superstition. They're confusing cause and effect. It makes as much sense as knocking on wood or crossing your fingers. They really believe that we wouldn't have to fly that mission tomorrow if someone would only tiptoe up to the map in the middle of the night and move the bomb line over Bologna. Can you imagine? You and I must be the only rational ones left."
>
> In the middle of the night Yossarian knocked on wood, crossed his fingers, and tiptoed out of his tent to move the bomb line up over Bologna. (Heller 1961)

The next morning the men consult the map. Sure enough, it shows that Bologna is captured. The mission is canceled, and everyone is happy.

The outcome here is a pleasant one. Usually, though, you get in trouble when you substitute maps for reality. Sometimes the result is merely inconvenience and narrow-mindedness. But the consequences can be more serious, as you will see later in this chapter.

Cartographic artifacts

Something you must realize is that much of what you see on a map does not exist in reality. Many map features are pure cartographic fiction—they are cartographic artifacts, the result of the mapping process. The first time you realize this, you may react as indignantly as Huck Finn did on a balloon trip with Tom Sawyer (in Mark Twain's *Tom Sawyer Abroad*). When Huck commented that they were flying over Illinois, Tom asked how he could be sure. Huck replied:

> "I know by the color. We're right over Illinois yet. And you can see for yourself that Indiana ain't in sight."
>
> "I wonder what's the matter with you, Huck. You know by the color?"
>
> "Yes, of course I do."
>
> "What's the color got to do with it?"
>
> "It's got everything to do with it. Illinois is green, Indiana is pink. You show me any pink down there, if you can. No sir; it's green."
>
> "Indiana pink? Why, what a lie!"
>
> "It ain't no lie; I've seen it on the map, and it's pink."
>
> You never see a person so aggravated and disgusted. He says: "Well, if I was such a numskull as you, Huck Finn, I would jump over. Seen it on the map! Huck Finn, did you reckon the states was the same color out-of-doors as they are on the map?"
>
> "Tom Sawyer, what's a map for? Ain't it to learn you facts?"
>
> "Of course."
>
> "Well, then, how's it going to do that if it tells lies?"
>
> "That's what I want to know." (Twain 1894)

Huck's question is a good one: How can a map give you facts if it tells lies? The answer is central to an understanding of maps. To tell the truth, a map must lie. As you saw in chapter 6, cartographic abstraction is required to transform reality into a map. This map transformation is done through the cartographic processes of selection, generalization, classification, and symbolization. These processes are helpful ways to organize and reduce the detail in the environment. But the abstraction process adds cartographic artifacts that bear little or no relation to reality. The danger is that the artifacts so dominate the map that you mistake them for real aspects of the environment.

Most map readers are not as naive as Huck Finn. Few of us expect a map's colors to be the same on the ground, a road on a map to look the same as a real road, or a tree's symbol to look like its leafy real-life counterpart. It is obvious that colors and other symbols are artifacts of the mapping process and have nothing to do with the environment. Yet even sophisticated map users can mistake cartographic artifacts for real geographic features, as the following example shows.

Image maps are created from aerial photographs and satellite images, yet many people do not think of them as maps at all. Photos, they reason, cannot contain misleading artifacts because they show the environment exactly as it is. This perception is one of the biggest map misconceptions. Image maps, like all maps, contain distortions and artifacts from the methods used to make them.

For example, atmospheric and light conditions can alter tones and textures on images. This problem especially applies to image mosaics, which are made by fitting several images together. Often, a mosaic of satellite images shows the mottled effect of different tones and textures. Some edges at which the images join remain unnoticeable, whereas others stand out clearly. Without an awareness of how the image map is made, you can confuse these artifacts of mosaicking with real environmental patterns. Like Huck, you may see as reality what is merely the result of the mapmaker's technique.

Even on a single air photo or satellite image, features can vary dramatically in tone. A lake in one area of an image might appear as a light tone, for example, whereas a similar lake in another area appears dark. You might jump to the conclusion that these variations indicate different water depths, turbidity levels, or algal growth. But if the other lakes in the image also show a variation in tones, you can assume that light reflection, not water quality, provides the explanation.

You cannot anticipate every possible artifact. The best defense is to learn to recognize the effects of various mapping methods and then envision the effects of these methods on the map's appearance. This recognition will keep you from interpreting something as reality that, in fact, is only a cartographic artifact.

The missing essence

The fact that many things on maps do not exist in reality is only one side of the coin. Even more serious is that many things that do exist in reality are missing from maps. Indeed, it seems that the most beautiful and humanizing parts of our lives are absent from maps. In his journal (November 9, 1860, entry), Henry David Thoreau laments the inadequacy of a map:

> How little there is on an ordinary map! How little, I mean, that concerns the walker and the lover of nature. Between the lines indicating roads is a plain blank space in the form of a square or triangle or polygon or segment of a circle, and there is naught to distinguish this from another area of similar size and form. Yet the one may be covered, in fact, with a primitive oak wood, like that of Boxboro, waving and creaking in the wind, such as may make the reputation of a county, while the other is a stretching plain with scarcely a tree on it. The wauling woods, the dells and glades and green banks and smiling fields, the huge boulders, etc., etc., are not on the map.

By condensing geographic phenomena into symbols, we make features in the environment easy to locate but sterile. In *Travels with Charley: In Search of America*, John Steinbeck ([1962] 1971) records which highways he took on his cross-country trip and adds:

> I can report this because I have a map before me, but what I remember has no reference to the numbers and colored lines and squiggles.

How can a map possibly capture the totality of life's vivid and meaningful experiences—the touches, sounds, smells, tastes, and the inner connection between people and their surroundings? These traits are the lost parts in translating "reality" to a map, and we must reconstitute them through the process of map interpretation or risk forgetting the true relationship between an object and its map symbol.

Imaginative map use

To gain the most from maps, you must be willing to stretch your imagination and allow symbols to conjure up their full meaning in your mind. Map symbols are meaningless in themselves. They are meant to direct your thoughts beyond the map to the environment. Here, (from *Cable-Car*, by June Drummond), for example, is what an imaginative map reader saw in one line symbol:

> Finally, there was the fact that this was a frontier region. That meant far more than a line on a map. It meant all the treachery, the corruption, the bravado, and the watchfulness of such a confrontation. (Drummond 1967)

Every symbol holds hidden meanings, but you must look for them. It is up to you to step from map to real world, and a giant step it is. Getting at realities beneath appearances is rarely easy. Consider the Oregon Natural Disasters map in figure 12.1. At a glance, you would never know it portrays misfortunes over 70 years; it essentially uses county outlines like cells in a spreadsheet to enumerate devastating floods, forest fires, severe storms, and other disasters. The matter-of-fact shorthand does not begin to show the decimated forests and fields, shattered towns, homeless people, and ruined lives caused by these calamities. It takes a vast, visionary leap to look at such symbols and picture reality.

Poorly designed maps aside, given a reasonable chance, almost anyone can use imagination to look for the meanings behind map symbols. The "armchair traveler" has perfected the technique. Before travel was as widespread as it is today, writers often remarked that those who stayed at home could still have exciting journeys—by map. In *Don Quixote*, Miguel de Cervantes ([1605] 1902) writes that one can "journey all over the universe in a map, without the expense and fatigue of traveling, without suffering the inconveniences of heat, cold, hunger, and thirst."

A map inspires the imaginative person to enter the reality it depicts. In *Heart of Darkness*, Joseph Conrad ([1902] 1960) notes the imaginative appeal of maps:

> Now when I was a little chap I had a passion for maps. I would look for hours at South America, or Africa, or Australia, and lose myself in all the glories of exploration. At that time there were many blank spaces on the earth, and when I saw one that looked particularly inviting on a map (but they all look like that) I would put my finger on it and say, "When I grow up I will go there."

That even blank spaces generate excitement underscores the imagination's power to give a map depth. As Aldo Leopold (1949) puts it in *A Sand County Almanac*: "To those devoid of imagination, a blank place on the map is a useless waste; to others, the most valuable part."

In *Deliverance*, the author James Dickey (1970) also suggests the map's potential to evoke more than the composite of its symbols. In the novel, some friends plan a trip down a river that will soon be obliterated by a new dam. For the narrator Ed, the map has power and vitality—almost a life of its own.

> "When they take another survey and rework this map," Lewis said, "all this in here will be blue."

> I leaned forward and concentrated down into the invisible shape he had drawn, trying to see the changes that would come, the nighttime rising of dammed water bringing a new lake up with its choice lots, its marinas and beer cans, and also trying to visualize the land as Lewis said it was at that moment, unvisited and free. … I looked around the bar

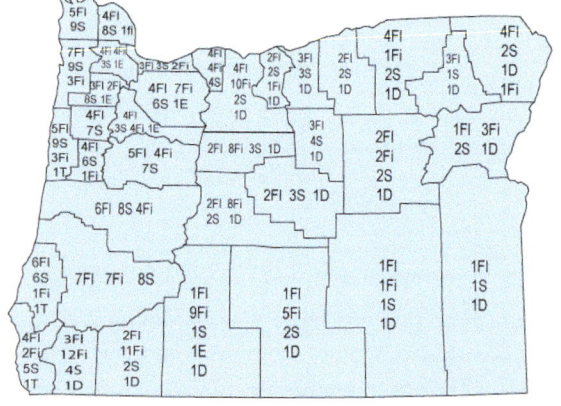

Figure 12.1. This map shows Oregon counties that were designated disaster areas because of six types of natural disasters that occurred from 1953 to 2023. Because of the poor graphic design, it is hard to achieve a meaningful visual impression of the geographic distribution and frequency of each type of disaster.

and then back into the map, picking up the river where we would enter it.

The author's wording gives life to the map; we can almost see the river running. Using such phrases as "looked … into the map," Dickey suggests that a map is not just something at which you glance but something into which you can see. For Ed, the map holds an unknown dimension, wonder, and excitement. He sees a crooked blue line on paper and, at the same time, a raging river. As he studies the map, he begins to want to see even more:

> [The map] was certainly not much from the standpoint of design. The high ground, in tan and an even paler tone of brown, meandered in and out of various shades and shapes of green, and there was nothing to call you or stop you on one place or the other. Yet the eye could not leave the whole; there was a harmony of some kind. Maybe, I thought, it's because this tries to show what exists. And also because it represents something that is going to change, for good. There, near my left hand, a new color, a blue, would seep upward into the paper, and I tried to move my mind there and nowhere else and imagine a single detail that, if I didn't see it that weekend, I never would; tried to make out a deer's eye in the leaves, tried to pick up a single stone. (Dickey 1970)

William Least Heat-Moon is a nonfiction author of the same ilk. His first book, *Blue Highways: A Journey into America* (1982), is a travel memoir documenting a 13,000-mile road trip in his van following separation from a teaching job and his spouse. He planned the trip to follow "blue" highways, the color with which secondary roads were symbolized in old road atlases (figure 12.2). By eschewing highways and big cities, he stated his cartographic plan in this manner:

> With a nearly desperate sense of isolation and a growing suspicion that I lived in an alien land, I took to the road in search of places where change did not mean ruin and where time and men and deeds connected.

Figure 12.2. A map of William Least Heat-Moon's 1982 circular route around the United States, during which he traveled only "blue" highways—the out-of-the-way roads drawn in blue on the *Rand McNally Road Atlases* of the time. Data courtesy of Michael L. Hess.

Moving from a singular focus on one color symbol on maps encompassing the contiguous United States, Least Heat-Moon turned his attention to studying the map blankness for one rural location, Chase County, Kansas. He integrated features he observed on historical maps with traveling throughout the area for the better part of a decade to explore the county and talk with its citizens. His subsequent book, *PrairyErth* (1992), popularized the idea of a "deep map," a narrative of the geography, history, ecology, and all other components of a locale whose maps might be described at first glance as sparse. This would naturally be impossible to capture on one traditional map and took the author 624 pages to describe!

This ability to let your imagination carry you from map to reality is the sign of skillful map use. Once you have this knack, maps will be truly beautiful and useful to you, evoking images and emotions beyond the printed paper or computer screen. But if you are unwilling to let your imagination move from a map's abstract image to the complex world, you will never realize the map's full value.

Map misuses

There is a big difference between using a map to help you understand reality and using a map as a substitute for reality. You give a map life through your imagination, but do not imagine that a map will return the favor if you give it too much credence over your own perceptions. Accepting a map as a surrogate for reality, you make it all too easy to view the environment in impersonal, machinelike terms. What's the trade-off? Nothing great. Environmental features are dwarfed on maps. People are invisible. The cultural environment (farms, cities, roads) looks like a child's board game. The world is reduced to an unchanging background against which you play out your life. So, remind yourself how limited maps are, and never lose sight of the missing essence that is found in reality and not on a map.

In *I Was a Savage* by Prince Modupe (1957), a Nigerian educated in the United States, Modupe describes his eagerness to return to his people and share the wonders of maps. To his surprise, his father was less than enthusiastic:

> Maps are liars, he told me briefly. ... The things that hurt do not show on the map. The truth of a place is in the joy and the hurt that come from it. I had best not put my trust in anything as inadequate as a map, he counseled.

Many skeptics share this distrust of maps, and with good reason. Maps separate you from the environment. If you see yourself as part of your environment, you know you cannot harm it without harming yourself. If you reduce your environment to symbols on a map, however, you are more likely to misuse it. You can misuse maps offhandedly just as you can deface a person's photo, whereas you would not think of altering the appearance of the real person. The important difference is, when you paint a mustache on a political poster, the politician does not really suffer. But when you distance yourself through map misuse, others bear the consequences.

When environmental decision-makers treat a map as mere symbols and disregard the real-life places, people, or things it represents, the results can be disastrous. City planners, for example, can be tempted to build a new freeway along the straight line that looks most convenient on the map. But the route might separate people based on race or socioeconomic status or disrupt the ecology of an area in which an endangered species lives. The map way is not always the best way. Might it not be better to take a more roundabout path, along which there are considerations of building community and preserving habitat?

It cannot be overemphasized that the reason for studying maps is to understand the real world, not the map world. People are drawn to applications that allow them to take account of the multiple, interrelated realities that influence one another—the ecology as a whole. Geographic information systems, for example, allow you to look at mapped information consistently, layer upon layer, drilling deeper for information. But even with such sophisticated tools, the key is still in the interpretation, and we must never forget that the results of our analyses can change lives. When you make plans based on maps, it is not the map but reality that stands to gain or lose. This fact may seem obvious. Yet time and again, people draw map boundaries for projects without regard for the people and region affected. Subdivisions are constructed, dams built, neighborhoods renewed—no doubt, it looks good on some map, yet it is sometimes without any thought for the consequences.

This blindsided allegiance to maps is also one of the tragedies of war. People die when the way that looked simplest on the map turns out to be not the wisest move on the field. A commander might study a map, decide that a certain spot makes a good observation point, and give the order to march. While peering at symbols on a map, it is all too convenient to ignore such things as terrain, climate, and morale. Yet these factors influence the soldiers obeying orders. Although it takes only a second to put that *X* on

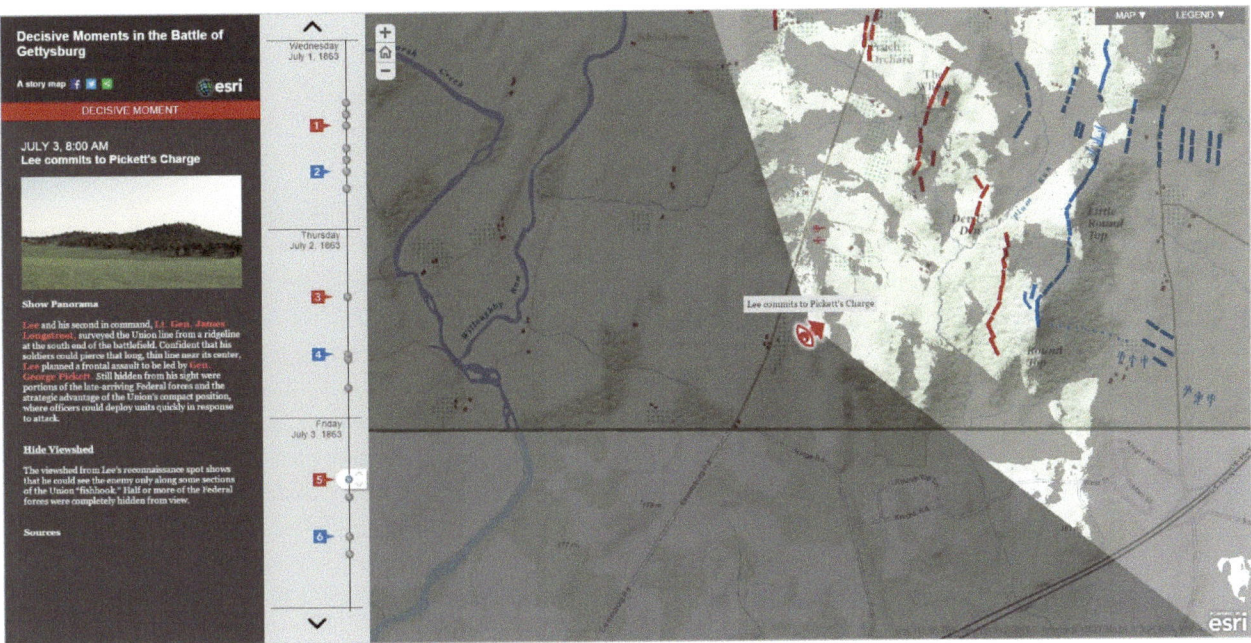

Figure 12.3. A viewshed analysis performed with ArcGIS shows that half or more of the Union forces were hidden by the terrain from Confederate General Robert E. Lee's view when he decided to launch his doomed assault on the last day of the Battle of Gettysburg, thus ending the three-day battle and Lee's campaign into Pennsylvania. Courtesy of Anne Knowles and Esri.

the map, it might take a huge amount of money, weeks of time, and a terrible toll on lives to carry out the mission. "Decisive Moments in the Battle of Gettysburg" is an online story map (figure 12.3) that uses GIS to examine the viewpoints of Union and Confederate commanders at key points in the Battle of Gettysburg. Panoramic landscapes help you understand that what the commanders could—or could not—see had a significant effect on the decisions they made. For example, what General Robert E. Lee was unable to see when he decided to launch his doomed assault, Pickett's Charge, on July 3, 1863, was that half or more of the Union forces were completely hidden from view by the terrain.

People who surround themselves with maps are liable to arrive at cold strategies, untempered by feeling, if they forget who or what will bear the consequences of their decisions. Again, figures in literature often remind us of this truth. In *The Naked and the Dead*, Norman Mailer (1948) describes such an unimaginative map reader. Major Dalleson, making strategy decisions with maps during World War II, "had a picture for a moment of the troops moving sullenly along a jungle trail, swearing at the heat, but he couldn't connect that to the figures on the map. An insect crawled sluggishly over his desk and he flicked it off." As casually, as thoughtlessly as Dalleson brushes an insect from his desk, decisions are made and people's fates decided.

Major General Cummings, another character in Mailer's novel, wants to reduce reality to mere "figures on the map." He is frustrated that soldiers' emotions keep his plans from being as effective in real life as in theory. At one point, Mailer writes, Cummings threw down his pencil and "stared with febrile loathing at the map board by his cot. By now, it was a taunt to him." He can control the map world, but not the real world.

War encourages military officers to substitute maps for reality. Miles from the battlefield, surrounded by "complete" map intelligence, they find the consequences of their actions easier to handle. To make a few lines on a map is painless. To visualize soldiers and their loved ones, landscapes and their ecosystems, all whose doom is sealed by those mapped lines, is another matter. How different it would be if the decision-makers were there when their decisions were carried out!

Even the soldiers who carry out these decisions are not always in touch with reality, as newspaper accounts of the Vietnam War during the 1970s remind us. Calling it the "Impersonal War," journalists' reports of the time lend some sense of how complicit map use was in depersonalizing the act of bombing from the air. Almost no place names were printed on the maps the crews used. Only intersecting coordinates marked the targets, which were given code numbers in place of names.

Enabling warfare by remote control, technology further distances decision-making from the realities of fighting and killing. In the 1990s, the media referred to the Persian Gulf War as the "Nintendo War" and praised "smart" bombs for their "surgical strikes." Fortified with high-tech gadgetry and digital maps, those waging war can increasingly take the option to avoid looking into the face of destruction by not linking cause and effect with the use of Predator and similar bomb-equipped drones remotely piloted from distant command centers. In a 2015 interview with NBC News, drone operator Michael Hass stated:

> We were very callous about any real collateral damage. Whenever that possibility came up, most of the time it was a "guilt by association" or sometimes we didn't even consider other people that were on screen.

Of course, map misuse by using maps as a way to distance yourself from reality or from the consequences of your decisions is by no means solely responsible for war's atrocities. Yet we can trace many difficulties, military and otherwise, to the map's almost innate insensitivity to crucial human and environmental features, as well as to a map user's failure to take this missing essence into account.

All maps are not equally insensitive, of course. Because some types of maps represent reality better than others, decisions based on one map will not be the same as those based on another. Image maps, for example, show much more information than conventional line maps. The consequences of military actions are also more evident on an image map, providing extra information for subsequent decision-making. Likewise, layered GIS maps intend to enable wise decision-making by showing extra information that can be toggled on and off, thereby providing the user with various stages and facets of assessment.

A degraded environmental image

Another result of treating maps as reality may be to hinder us from experiencing a rich, full life, as Steinbeck ([1962] 1971) reminds us in *Travels with Charley*:

> For weeks I had studied maps, large scale and small, but maps are not reality at all—they can be tyrants. I know people who are so immersed in road maps that they never see the countryside they pass through, and others who, having traced a route, are held to it as though held by flanged wheels to rails.

Imagine what Steinbeck might have written about the tyranny of GNSS navigation devices. As he notes, a map represents only a few of many truths. Its purpose is to help us see more, but if you rely too much on it, you may instead see less. He continues:

> There are map people whose joy is to lavish more attention on the sheets of colored paper than on the colored land rolling by. I have listened to accounts by such travelers in which every road number was remembered, every mileage recalled, and every little countryside discovered. Another kind of traveler requires to know in terms of maps exactly where he is pin-pointed every moment, as though there were some kind of safety in black and red lines, in dotted indications and squirming blue of lakes and the shadings that indicate mountains.

By realizing the map's limitations, Steinbeck found his cross-country trip far more satisfying than his friends who were beholden to maps. For him, a map was useful as a framework for his memories. Maps enhanced, rather than detracted from, his enjoyment because he was wisely aware that "maps are not reality at all."

Abstract decision-making

Although there are dangers in treating maps as a stand-in for reality, such an approach can bring advantages, in consciously employing the map's abstract nature as a buffer and for balance. Sometimes, decision-makers who deal with emotional issues (and pressure groups) find it easier to do their jobs by working with symbols. Our values and feelings are less aroused when working with abstractions than confronting ugly or distasteful issues directly. An objective removed position can also help keep short-term satisfaction from compromising the long-range good.

In *Night Flight*, Antoine de Saint-Exupéry (1931) describes a decision-maker who learns that only by treating maps as reality can he do his duty. The character Riviere has the job of tracing the flights of pilots who carry the mail at night. In the pioneering days of air mail delivery depicted in this novel, such flights involved great hazard, and Riviere knows that each time he sends a plane into the darkness, he risks a pilot's life. As he stares at the map with the airlines traced in red, he muses:

"On the face of it, a pretty scheme enough—but it's ruthless. When one thinks of all the lives, young fellows' lives, it has cost us! It's a fine, solid thing and we must bow to its authority, of course; but what a host of problems it presents!"

With Riviere, however, nothing mattered save the end in view.

For a moment, Riviere allows his imagination to make the leap from the cold, solid "face" of the map to the reality it symbolizes. To accomplish "the end in view," however, he must veer abruptly from this reality and think of the map as an abstract thing. Otherwise, he cannot keep sending pilots into such danger. He would be better off, perhaps would suffer less, if he were a more unimaginative map reader.

In a way, map users are like doctors, who must put emotional distance between themselves and their patients. Otherwise, tragic cases may impair their judgment. Some doctors succeed so well that patients complain that they are treated as diseases or broken bones, not as the frightened and hurting people they are. We hope for doctors who strike a balance between emotion and rationality. When using maps to make decisions that affect other people, we must achieve just such a balance. Overly emotional decisions are as undesirable as overly insensitive ones. When we must transcend our human limitations, maps offer a useful counterbalancing perspective.

Reality as a map

We have seen how dangerous it is to treat maps as reality. Just as hazardous is to treat reality as if it were a map—to regard the world as something laid out in maplike form. How many people look down at the earth from an airplane and say, "It looks just like a map"? Such a statement is comparable to gazing at the Grand Canyon only to remark, "It looks just like a postcard." This attitude strips an incredible sight of its grandeur. Viewing the world as if we had seen it all before, as if it were laid out neatly, as in a map, wherever we are, becomes a predictable, limited thing—fragmented, split, weakened, and drained.

Regrettably, confusing maps with reality is a common occurrence. We all know people who claim to have visited nearly every state in the nation—when pressed, they admit that all they saw of at least some states was an airport or a freeway. Even business executives who travel the world may be familiar with little more than the views from airports, restaurants, or hotels. Although the map says that they are in a distant state or foreign country, in their reality they may as well be at an airport, restaurant, or hotel in their hometown.

Marlowe, Conrad's narrator in *Heart of Darkness*, describes the map-minded way we put a buffer between ourselves and an unfamiliar environment:

Next day I left that station at last, with a caravan of sixty men, for a two-hundred-mile tramp.

No use telling you much about that. Paths, paths, everywhere; a stamped-in network of paths spreading over the empty land, through long grass, through burnt grass, through thickets, down and up chilly ravines, up and down stony hills ablaze with heat; and a solitude, a solitude, nobody, not a hut. ([1902] 1960)

To maintain his sanity in the African jungle, Marlowe reduces the environment to a series of connected pathways, like lines on a map. The abundance of plant and animal life between those paths he dismisses as "empty land." Many people simplify the world in this way when thrust into an unfamiliar setting. But it is a bad habit to develop. With this maplike conception of reality, losing your orientation is easy. Whatever is unknown becomes, as it did for Marlowe, a wilderness to be avoided.

Wilderness does not have to be the backwoods kind, of course—for some people, a trip to the central city is as adventuresome, or unsettling, as a journey to the Yukon. We fear what we do not understand. If we stop infusing our world with the structure of a map and explore its true variety, we might find many of our fears unfounded. Then, the ones we find that do have foundation, we can confront and overcome or find a way around—either way, breaking new ground as we do.

The error of treating reality as a map can shape your whole perception of the world. It is like looking at the world through another's eyes instead of your own. There is a lesson about image maps that also applies to all maps: Maps and imagery pervade our world, structure our reality, and teach us to look at the world in a maplike way, as if our perception were trained by what we are given to perceive. One limitation in this view of reality is that the map is valued almost as much as the thing itself.

Our perceptions lead us to how we decide to act or react, and human beings are gifted with many and various ways of perceiving and assessing our surroundings and experience. Your perception need not be limited by how mapmakers or anybody else structure the "accepted" images of reality. Some may think that a maplike view of reality is more objective, but it is not. Maps are no more objective than photographs; they reflect the way one person or another views the world—and what they mean comes to life through your interpretation of it.

Survival

The deficiency in either treating the "map as reality" or "reality as a map" is that both views separate you from—and set you up to ignore—your environment. Neither stance positions you to perceive the interrelationships on which your quality of life, and potentially your survival, depend.

Maps are a metaphor for a limited kind of experience. Mapped features may have as little relation to the world as a telephone number to its subscriber or an ID number to its feature. You must place maps in perspective as a limited communication device—as only one of many. To put too much emphasis on maps is to screen out much of what is crucial to environmental behavior.

Rather than see the environment as mapmakers do, biased by the tools of their profession, try to take an omniscient viewpoint. Only when you see the whole environment as one system, as events that are separate but united in time and space, will you appreciate all that maps have to offer. While viewing a map, you can then unite yourself with the environment and look into, rather than at, map symbols.

This task of visualizing mapped features in all their richness and complexity is not simple. It takes effort and practice. The job is hardest when you are sitting in an environmentally controlled room scrutinizing artificial map symbols, far removed from the phenomena you are studying and the procedures used to create the map. In addition, your job is complicated by an obvious but often overlooked consideration: We all are human.

We all are human

Mapping as a communication process is influenced by human shortcomings throughout. Both as mapmaker and map user, we are fallible. We crave simplicity; we are strong in some areas, weak in others; we are prejudiced sometimes, uninformed other times; and our integrity, honesty, and insight are never beyond question. We are human, and our judgments, faults, and biases creep into the mapmaking and the map interpreting process.

Every map reflects myriad decisions made by the mapmaker. It is not a copy but a semblance of reality that is filtered through the mapmaker's motives and perceptions. The map is partly a representation of reality and partly a product of its maker. The mapmaker's individual knowledge, skill, and integrity all enter into the map's design. If 10 mapmakers are given the same mapping task, some maps will be effective and beautiful, and others will fail to communicate as well or look as pleasing.

Results vary widely, but the blame for poor map communication does not rest entirely with the mapmaker. As the map user, you, too, approach maps with the baggage of your experience, motives, and skill. Your biggest danger is that you will see what you want to see or expect to see. You can guard against this tendency by keeping an open mind and applying what you have learned about map reading.

Sometimes the penalty for being a human link in map use is disaster. A pilot misreads a map, and an airplane crashes. An accident occurs when a driver misses the freeway exit and tries to correct. Map coordinates are misunderstood, and soldiers are bombed by friendly forces.

Still, the price of misusing maps is more often frustration than disaster, and even this frustration can be moderated by understanding the nature of maps. There is no way to eliminate human failings or map limitations, but by knowing they exist, you can learn to live with them.

Living with map limitations

You have seen how the abstract, generalized nature of maps introduces the potential for error and exploitation. Yet these same qualities are what make maps so valuable for showing the big picture. If a map's strengths are also its weaknesses, the opposite is true as well. A map is remarkably useful as long as you do not ask it to do things for which it is not designed. You must not, for instance, ask a map to be the same as reality—if it was, it would lose its unique clarifying function.

One problem with maps is that they are rarely tailored to the requirements of the individual user. Therefore, they are seldom perfect for specific needs. Most mapmakers have only a vague idea of who will use their maps. But you can learn to live with this map weakness, too. Maps are available in an infinite variety. You can save yourself grief by finding the best map available for each situation, rather than using one map for all purposes. Increasingly, with the aid of digital mapping applications, you can tailor the map you see to be the map you need.

If you are aware of map limitations, it is usually easy

to make up for them. It makes sense to bring as much experience and information as possible to bear on map interpretation. The best navigators are those who augment map information with all the direct "ground truth" they can. Pilots do not land their planes and captains do not dock their ships with maps—they rely on direct visual or instrumental contact during these final, crucial moments. Part of using maps shrewdly involves knowing when to go beyond them and rely instead on your senses, intuition, or world knowledge.

Every map represents just one way of looking at reality. If one map does not serve your needs, do not become disenchanted with maps. Perhaps you can find another map that better fits your requirements. Or maybe you must supplement your map or turn to another source of information. But if you give up on maps entirely, you cheat yourself, for at another time they may be exactly what you need. Strike a balance between too much faith and not enough faith in maps.

In this chapter, we stress how maps are incomplete, limited, and often faulty, so in contrast, Alvin Toffler ([1970] 1971) points out in *Future Shock* that maps need not be perfect to be useful:

> Even error has its uses. The maps of the world drawn by the medieval cartographers were so hopelessly inaccurate, so filled with factual error, that they elicit condescending smiles today when almost the entire surface of the earth has been charted. Yet the great explorers could never have discovered the New World without them. Nor could the better, more accurate maps of today have been drawn until men, working with the limited evidence available to them, set down on paper their bold conceptions of worlds they had never seen.

The path ahead

Our maps of today, though pitifully lacking in many respects, as you have seen, serve an indispensable function as well, by moving us to the next step toward something better. We are in a wonderful new age in map use, in which online maps, many produced using GIS, play a major role in synthesizing and communicating increasing types of information.

When you think of map use in the future, you must think big. Through the Internet of Things (IoT), anything that can be connected will be connected. Devices, vehicles, buildings, and people will collect data through electronics, software, and sensors and share it over digital networks. The impact on map use is interesting to think about. Imagine that you are in your vehicle one morning on the highway on your way to work. The car in front of you brakes suddenly, but you avoid a collision because your car senses the danger, sees that the passing lane is clear, and drives itself around the other vehicle. Your car senses a traffic jam ahead and finds a new route around the delay. On the spur of the moment, you decide to stop for a latte. Your car finds the nearest coffee shop, your smartphone calls in your order, and your bank pays for your drink in advance. Ten minutes later, you are back on your way. Your calendar notices that you will now be late for your first appointment, so your phone sends a text to the person you are supposed to meet with. When you get to work, your car finds the closest available parking space, and the person you are meeting receives a message that you are on your way into the office. Devices that know where you are and where things are around you will make all this possible. In some ways, the maps of the future know more about the real world than you do.

The ubiquity of maps is hardly noticeable. Your reliance on maps may increase without your knowing it. It might seem as if you will not have to put as much into interpreting the environment around you. But the ability of maps to help you navigate, explore, and understand your environment will not change. Maps will still be able to give you new insight, help you make better decisions, and enable you to learn interesting and useful things. In the end, you are still on a long path toward understanding the world, and every new map carries you one step closer to your goal.

Selected readings

Agger, M. 2006. "The Trainspotters of Google Earth: What Happens When the Internet and Geography Collide." *Hyperlink* blog. Posted August 26.

Cervantes, Miguel de. (1605) 1902. *Don Quixote of La Mancha*. John Grant.

Conrad, Joseph. (1902) 1960. *Heart of Darkness and The Secret Sharer*. The New American Library.

Dickey, James. 1970. *Deliverance*. Houghton Mifflin.

Drummond, June. 1967. *Cable-Car*. Holt, Rinehart & Winston.

Gurnah, Abdulrazak. 2001. *By the Sea*. The New Press.

Heller, Joseph. 1961. *Catch-22*. Simon & Schuster.

Henrickson, A. K. 1975. "The Map as an 'Idea': The Role of Cartographic Imagery During the Second World

War." *The American Cartographer* 2, no. 1 (April): 19–53.

Huxtable, Ada. 1973. "How We See, or Think We See, the City: Architecture." *The New York Times* (November 25), section 2:26.

Keates, J. S. 1972. "Symbols and Meaning in Topographic Maps." *International Yearbook of Cartography* 12: 168–80.

Kingsbury, P., and J. P. Jones. 2009. "The 'View from Nowhere'? Spatial Politics and Cultural Significance of High-Resolution Satellite Imagery." *Geoforum* 40 (July): 4. www.sciencedirect.com/science/article/pii/S0016718508001735.

Least Heat-Moon, W. 1982. *Blue Highways: A Journey into America*. Little, Brown.

Least Heat-Moon, W. 1992. *PrairyErth: (A Deep Map)*. Houghton Mifflin.

Leopold, A. 1949. *A Sand County Almanac*. Oxford University Press.

Mailer, Norman. 1948. *The Naked and the Dead*. Holt, Rinehart & Winston.

Monmonier, M. 1991. *How to Lie with Maps*. University of Chicago Press.

Muehrcke, P. C. 1973. "Beyond Abstract Map Symbols." *Journal of Geography* 73 (8): 35–52.

Muehrcke, P. C. 1974. "Map Reading and Abuse." *Journal of Geography* 73 (5): 11–23.

Muehrcke, P. C., and J. O. Muehrcke. 1974. "Maps in Literature." *Geographical Review* 64 (3): 317–38.

NBC News. 2015. "Former Drone Pilots Denounce 'Morally Outrageous' Program." December 7.

Prince Modupe. 1957. *I Was A Savage*. Frederick A. Praeger.

Quam, L. O. 1943. "The Use of Maps in Propaganda." *Journal of Geography* 42 (1): 21–32.

Ristow, W. W. 1957. "Journalistic Cartography." *Surveying and Mapping* 17 (4): 369–90.

Robinson, A. H., and B. Bartz-Petchenik. 1976. *The Nature of Maps: Essays Toward Understanding Maps and Mapping*. University of Chicago Press.

Saint-Exupéry, Antoine de. (1931) 1961. *Night Flight*. The New American Library.

Steinbeck, John. (1962) 1971. *Travels with Charley: In Search of America*. Bantam Books.

Tey, Josephine. 1927. *The Man in the Queue*. Peter Davies.

Thoreau, Henry D. 1860. *Journal Manuscript 32*, November 9, 1860, entry.

Toffler, Alvin. (1970) 1971. *Future Shock*. Bantam Books.

Twain, Mark. 1894. *Tom Sawyer Abroad*. Charles L. Webster.

Vernon, J. 1973. *The Garden and the Map*. University of Illinois Press.

Wood, D., and J. Fels. 1986. "Designs on Signs: Myth and Meaning in Maps." *Cartographica* 23 (3): 54–103.

Wright, J. K. 1942. "Map Makers Are Human: Comments on the Subjective in Maps." *Geographical Review* 32, no. 4 (October): 527–44.

Appendix

This appendix has seven tables that contain useful geographic information referenced in chapters throughout the book. Basic geographic information such as the longitude of prime meridians used previously on maps published by different countries (table A.1) may be necessary for determining distance on historical maps. Metric and Imperial unit distance and area equivalents (table A.2) are important because the book is written for an international audience that uses different measurement units. Knowing the exact spacing of parallels and meridians, as well as the areas of quadrilaterals, at different latitudes (tables A.3, A.4, and A.6) is important for precise distance and area measurements on the ellipsoid. Map scales associated with the zoom levels for web maps that use the web Mercator projection are given in table A.5. Common Landsat 8 and 9 band combinations are described and illustrated in table A.7.

Table A.1. Prime meridians used previously on maps published in different countries and their longitudinal distances from the Greenwich meridian (in degrees, minutes, seconds)

Prime meridian	Distance from Greenwich meridian
Amsterdam, Netherlands	4°53′01″ E
Athens, Greece	23°42′59″ E
Beijing, China	116°28′10″ E
Berlin, Germany	13°23′55″ E
Bern, Switzerland	7°26′22″ E
Brussels, Belgium	4°22′06″ E
Copenhagen, Denmark	12°34′40″ E
Ferro, Canary Islands	17°40′00″ W
Helsinki, Finland	24°57′17″ E
Istanbul, Türkiye	28°58′50″ E
Jakarta, Indonesia	106°48′28″ E
Lisbon, Portugal	9°07′55″ W
Madrid, Spain	3°41′15″ W
Moscow, Russia	37°34′15″ E
Oslo, Norway	10°43′23″ E
Paris, France	2°20′14″ E
Rio de Janeiro, Brazil	43°10′21″ W
Rome (Monte Mario), Italy	12°27′08″ E
Stockholm, Sweden	18°03′30″ E
St. Petersburg, Russia	30°18′59″ E
Tokyo, Japan	139°44′41″ E
Washington, DC, USA	77°02′14″ W

Table A.2. Metric and Imperial distance and area equivalents

Converting from Imperial units		
Multiply	**By**	**To obtain**
acres	0.4046856	hectares
acres	43,560.0*	square feet
acres	4,046.856	square meters
acres	0.0015625*	square miles
acres	4,840.0*	square yards
feet	30.48*	centimeters
feet	0.0003048*	kilometers
feet	0.3048*	meters
feet	0.00018939394	miles
square feet	0.000022956	acres
square feet	929.0304*	square centimeters
square feet	0.09290304*	square meters
square feet	0.00000003587	square miles
inches	2.54*	centimeters
inches	0.0254*	meters
inches	0.000015782	miles
inches	0.027777778	yards
square inches	6.4516*	square centimeters
square inches	0.00064516*	square meters
square inches	645.16*	square millimeters
miles	160,934.4*	centimeters
miles	5,280.0*	feet
miles	63,360.0*	inches
miles	1.609344*	kilometers
miles	1,609.344*	meters
miles	1,760.0*	yards
miles	0.86897624	nautical miles
square miles	640.0*	acres
square miles	27,878,400.0*	square feet
square miles	2.589988110647	square kilometers
yards	91.44*	centimeters
yards	0.0009144*	kilometers
yards	0.9144*	meters
yards	0.000568182	miles
square yards	0.000206611	acres
square yards	0.83612736*	square meters
square yards	0.0000003228305	square miles

Converting from metric units		
Multiply	**By**	**To obtain**
centimeters	0.03280839895	feet
centimeters	0.3937007874	inches
centimeters	0.00001*	kilometers
centimeters	0.01*	meters
centimeters	0.000006213711922	miles
centimeters	0.01093613298	yards
square centimeters	0.001076391042	square feet
square centimeters	0.15500031	square inches
hectares	2.471054073	acres
hectares	107,639.1042	square feet
kilometers	100,000.0*	centimeters
kilometers	1,000.0*	meters
kilometers	1,093.613298	yards
kilometers	3,280.839895	feet
kilometers	39,370.07874	inches
kilometers	0.6213711922	miles
square kilometers	247.1054	acres
square kilometers	10,763,910.42	square feet
square kilometers	0.386102158496	square miles
meters	100.0*	centimeters
meters	3.280839895	feet
meters	39.37007874	inches
meters	0.001*	kilometers
meters	0.0006213711922	miles
meters	1.093613298	yards
square meters	0.0002471054	acres
square meters	10.76391042	square feet
square meters	0.0000003861003	square miles
square millimeters	0.00001076391042	square feet

Converting geographic angular units		
Multiply	**By**	**To obtain**
degrees	0.017453292	radians
radians	57.2958	degrees

* Constants are exact.

Table A.3. Variation in the length of a degree of latitude

Latitude	Meters	Statute miles	Latitude	Meters	Statute miles	Latitude	Meters	Statute miles
0°–1°	110 567.3	68.703	30°–31°	110 857.0	68.883	60°–61°	111 423.1	69.235
1°–2°	110 568.0	68.704	31°–32°	110 874.4	68.894	61°–62°	111 439.9	69.246
2°–3°	110 569.4	68.705	32°–33°	110 892.1	68.905	62°–63°	111 456.4	69.256
3°–4°	110 571.4	68.706	33°–34°	110 910.1	68.916	63°–64°	111 472.4	69.266
4°–5°	110 574.1	68.708	34°–35°	110 928.3	68.928	64°–65°	111 488.1	69.275
5°–6°	110 577.6	68.710	35°–36°	110 946.9	68.939	65°–66°	111 503.3	69.285
6°–7°	110 581.6	68.712	36°–37°	110 965.6	68.951	66°–67°	111 518.0	69.294
7°–8°	110 586.4	68.715	37°–38°	110 984.5	68.962	67°–68°	111 532.3	69.303
8°–9°	110 591.8	68.718	38°–39°	111 003.7	68.974	68°–69°	111 546.2	69.311
9°–10°	110 597.8	68.722	39°–40°	111 023.0	68.986	69°–70°	111 559.5	69.320
10°–11°	110 604.5	68.726	40°–41°	111 042.4	68.998	70°–71°	111 572.2	69.328
11°–12°	110 611.9	68.731	41°–42°	111 061.9	69.011	71°–72°	111 584.5	69.335
12°–13°	110 619.8	68.736	42°–43°	111 081.6	69.023	72°–73°	111 596.2	69.343
13°–14°	110 628.4	68.741	43°–44°	111 101.3	69.035	73°–74°	111 607.3	69.349
14°–15°	110 637.6	68.747	44°–45°	111 121.0	69.047	74°–75°	111 617.9	69.356
15°–16°	110 647.5	68.753	45°–46°	111 140.8	69.060	75°–76°	111 627.8	69.362
16°–17°	110 657.8	68.759	46°–47°	111 160.5	69.072	76°–77°	111 637.1	69.368
17°–18°	110 668.8	68.766	47°–48°	111 180.2	69.084	77°–78°	111 645.9	69.373
18°–19°	110 680.4	68.773	48°–49°	111 199.9	69.096	78°–79°	111 653.9	69.378
19°–20°	110 692.4	68.781	49°–50°	111 219.5	69.108	79°–80°	111 661.4	69.383
20°–21°	110 705.1	68.789	50°–51°	111 239.0	69.121	80°–81°	111 668.2	69.387
21°–22°	110 718.2	68.797	51°–52°	111 258.3	69.133	81°–82°	111 674.4	69.391
22°–23°	110 731.8	68.805	52°–53°	111 277.6	69.145	82°–83°	111 679.9	69.395
23°–24°	110 746.0	68.814	53°–54°	111 296.6	69.156	83°–84°	111 684.7	69.398
24°–25°	110 760.6	68.823	54°–55°	111 315.4	69.168	84°–85°	111 688.9	69.400
25°–26°	110 775.6	68.833	55°–56°	111 334.0	69.180	85°–86°	111 692.3	69.402
26°–27°	110 791.1	68.842	56°–57°	111 352.4	69.191	86°–87°	111 695.1	69.404
27°–28°	110 807.0	68.852	57°–58°	111 370.5	69.202	87°–88°	111 697.2	69.405
28°–29°	111 823.3	68.862	58°–59°	111 388.4	69.213	88°–89°	111 698.6	69.406
29°–30°	111 840.0	68.873	59°–60°	111 405.9	69.224	89°–90°	111 699.3	69.407

Note: Degrees of latitude are measured along the meridian on the WGS84 ellipsoid.

Table A.4. Areas of quadrilaterals of 1° extent

Lower latitude	Area in square kilometers	Area in square miles						
0°	12,308.09	4,752.16	30°	10,642.47	4,109.06	60°	6,123.64	2,364.34
1°	12,304.44	4,750.75	31°	10,533.66	4,067.05	61°	5,935.01	2,291.51
2°	12,297.14	4,747.93	32°	10,421.62	4,023.79	62°	5,744.46	2,217.94
3°	12,286.21	4,743.71	33°	10,306.39	3,979.30	63°	5,552.08	2,143.66
4°	12,271.63	4,738.08	34°	10,188.00	3,933.59	64°	5,357.88	2,068.68
5°	12,253.39	4,731.04	35°	10,066.48	3,886.67	65°	5,161.97	1,993.04
6°	12,231.56	4,722.61	36°	9,941.87	3,838.56	66°	4,964.38	1,916.75
7°	12,206.05	4,712.76	37°	9,814.18	3,789.26	67°	4,765.19	1,839.84
8°	12,176.94	4,701.52	38°	9,683.49	3,738.80	68°	4,564.44	1,762.33
9°	12,144.23	4,688.89	39°	9,549.80	3,687.18	69°	4,362.18	1,684.24
10°	12,107.89	4,674.86	40°	9,413.15	3,634.42	70°	4,158.56	1,605.62
11°	12,067.92	4,659.43	41°	9,273.60	3,580.54	71°	3,953.53	1,526.46
12°	12,024.41	4,642.63	42°	9,131.15	3,525.54	72°	3,747.24	1,446.81
13°	11,977.30	4,624.44	43°	8,985.85	3,469.44	73°	3,539.73	1,366.69
14°	11,926.61	4,604.87	44°	8,837.75	3,412.26	74°	3,331.05	1,286.12
15°	11,872.35	4,583.92	45°	8,686.89	3,354.01	75°	3,121.29	1,205.13
16°	11,814.57	4,561.61	46°	8,533.30	3,294.71	76°	2,910.51	1,123.75
17°	11,753.24	4,537.93	47°	8,377.07	3,234.39	77°	2,698.75	1,041.99
18°	11,688.41	4,512.90	48°	8,218.17	3,173.04	78°	2,486.14	959.90
19°	11,620.06	4,486.51	49°	8,056.69	3,110.69	79°	2,272.70	877.49
20°	11,548.24	4,458.78	50°	7,892.69	3,047.37	80°	2,058.51	794.79
21°	11,472.95	4,429.71	51°	7,726.18	2,983.08	81°	1,843.64	711.83
22°	11,394.19	4,399.30	52°	7,557.23	2,917.85	82°	1,628.17	628.64
23°	11,312.01	4,367.57	53°	7,385.85	2,851.68	83°	1,412.17	545.24
24°	11,278.21	4,334.52	54°	7,212.17	2,784.62	84°	1,195.70	461.66
25°	11,137.44	4,300.17	55°	7,036.18	2,716.67	85°	978.84	377.93
26°	11,045.08	4,264.51	56°	6,857.93	2,647.85	86°	761.67	294.08
27°	10,949.38	4,227.56	57°	6,677.51	2,578.19	87°	544.21	210.12
28°	10,850.36	4,189.33	58°	6,494.94	2,507.70	88°	326.60	126.10
29°	10,748.06	4,149.83	59°	6,310.33	2,436.42	89°	108.88	42.04

Table A.5. Web map scales (at the equator) associated with zoom levels

Level of detail	Map scale for image tile layers	Map scale for vector tile layers	Usage suggestion for vector tile layers
0	591,657,527.59	295,828,763.80	Hemisphere
1	295,828,763.80	147,914,381.90	Continent
2	147,914,381.90	73,957,190.95	Country
3	73,957,190.95	36,978,595.47	Countries or States
4	36,978,595.47	18,489,297.74	Country or State

5	18,489,297.74 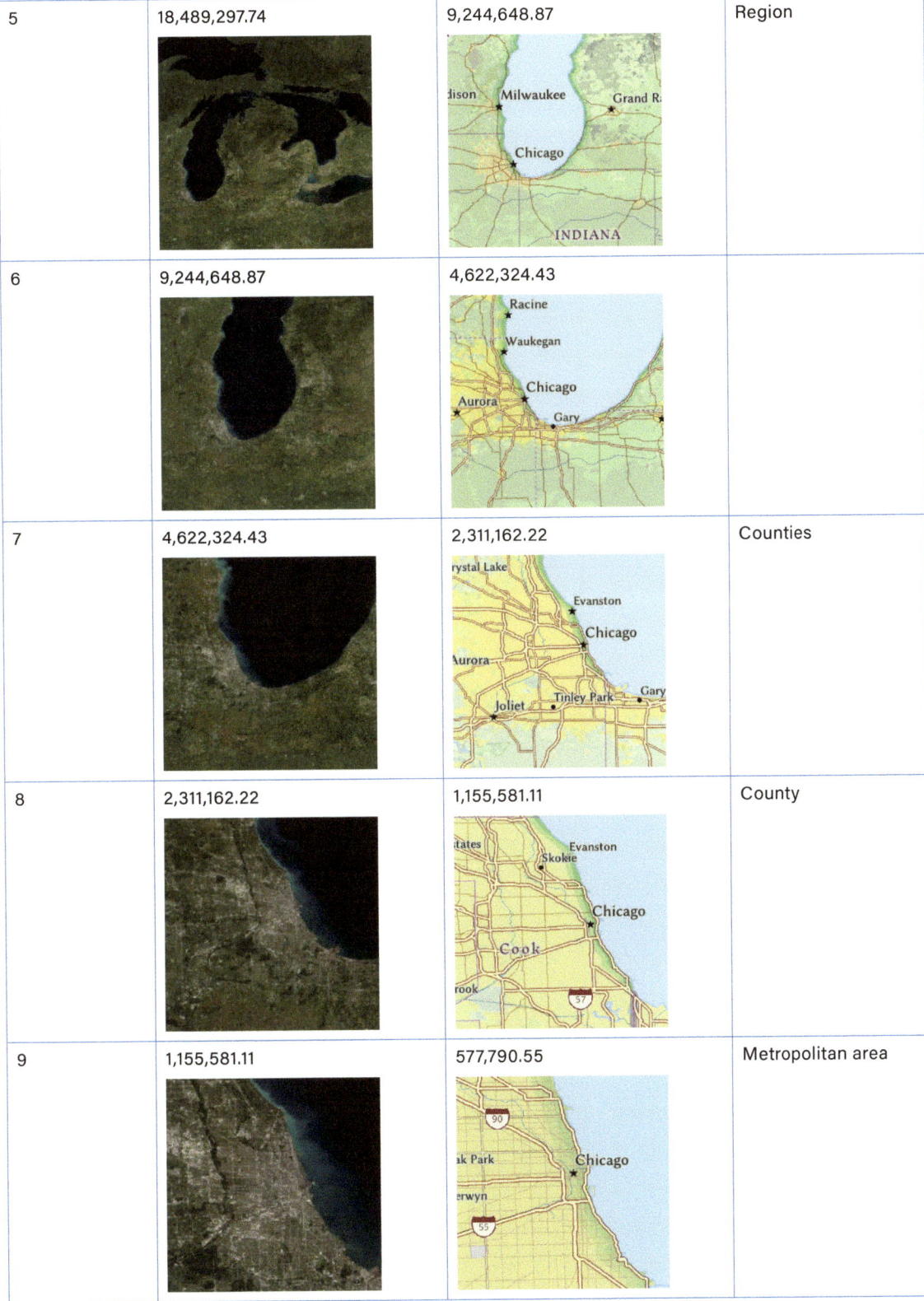	9,244,648.87	Region
6	9,244,648.87	4,622,324.43	
7	4,622,324.43	2,311,162.22	Counties
8	2,311,162.22	1,155,581.11	County
9	1,155,581.11	577,790.55	Metropolitan area

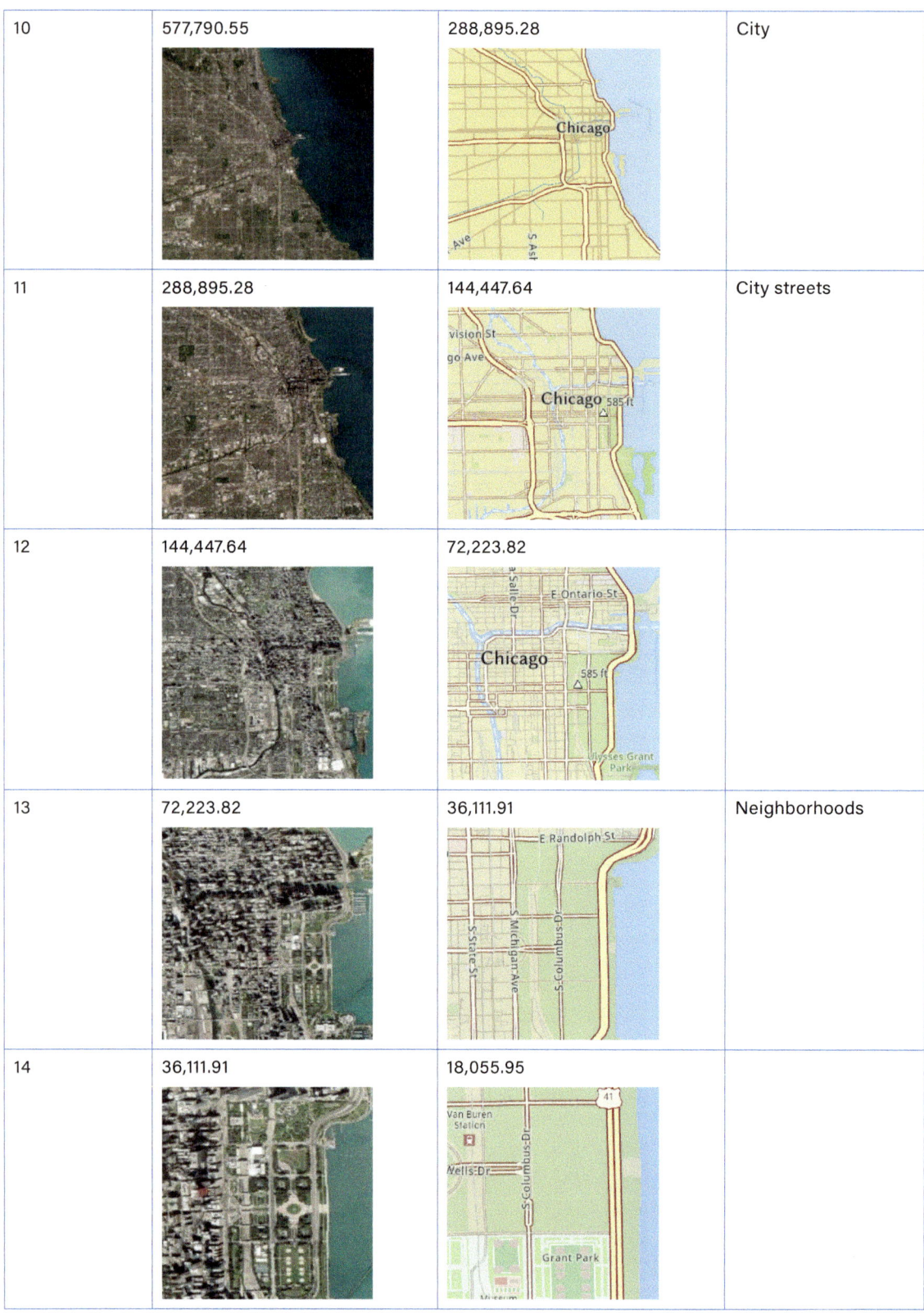

Appendix 307

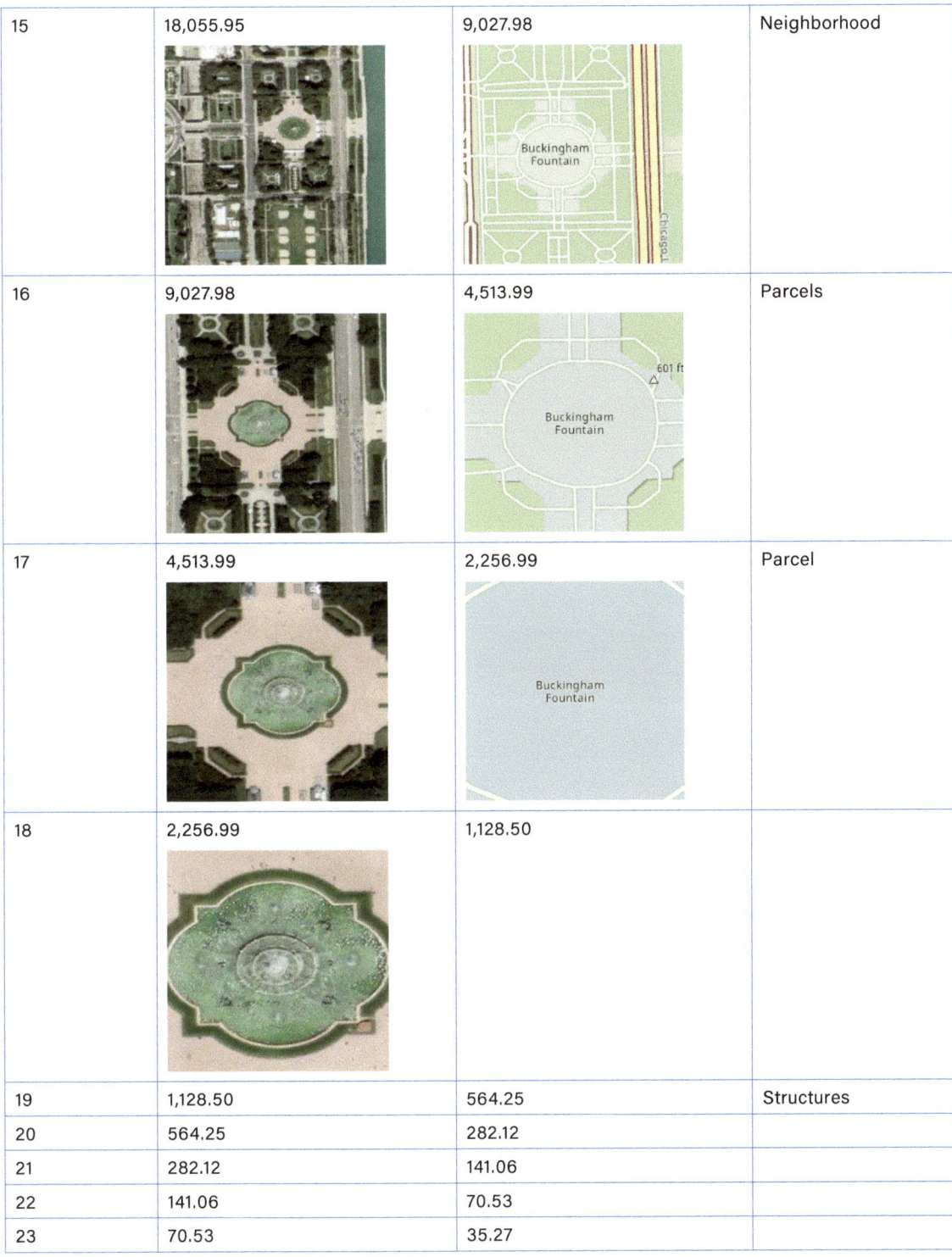

15	18,055.95	9,027.98	Neighborhood
16	9,027.98	4,513.99	Parcels
17	4,513.99	2,256.99	Parcel
18	2,256.99	1,128.50	
19	1,128.50	564.25	Structures
20	564.25	282.12	
21	282.12	141.06	
22	141.06	70.53	
23	70.53	35.27	

Esri World Imagery Map courtesy of Esri, Vantor, Earthstar Geographics, and the GIS User Community. National Geographic Style Map courtesy of the ArcGIS Living Atlas of the World team.

Table A.6. Variation in the length of a degree of longitude at different latitudes

Latitude	Meters	Statute miles
0°	111 321	69.172
1°	111 304	69.162
2°	111 253	69.130
3°	111 169	69.078
4°	111 051	69.005
5°	110 900	68.911
6°	110 715	68.795
7°	110 497	68.660
8°	110 245	68.504
9°	109 959	68.326
10°	109 641	68.129
11°	109 289	67.910
12°	108 904	67.670
13°	108 486	67.410
14°	108 036	67.131
15°	107 553	66.830
16°	107 036	66.510
17°	106 487	66.169
18°	105 906	65.808
19°	105 294	65.427
20°	104 649	65.026
21°	103 972	64.606
22°	103 264	64.166
23°	102 524	63.706
24°	101 754	63.228
25°	100 952	62.729
26°	100 119	62.212
27°	99 257	61.676
28°	98 364	61.122
29°	97 441	60.548
30°	96 488	59.956
31°	95 506	59.345
32°	94 495	58.716
33°	93 455	58.071
34°	92 387	57.407
35°	91 290	56.725
36°	90 166	56.027
37°	89 014	55.311
38°	87 835	54.579
39°	86 629	53.829
40°	85 396	53.063
41°	84 137	52.281
42°	82 853	51.483
43°	81 543	50.669
44°	80 208	49.840
45°	78 849	48.995
46°	77 466	48.136
47°	76 058	47.261
48°	74 628	46.372
49°	73 174	45.469
50°	71 698	44.552
51°	70 200	43.621
52°	68 680	42.676
53°	67 140	41.719
54°	65 578	40.749
55°	63 996	39.766
56°	62 395	38.771
57°	60 774	37.764
58°	59 135	36.745
59°	57 478	35.716
60°	55 802	34.674
61°	54 110	33.623
62°	52 400	32.560
63°	50 675	31.488
64°	48 934	30.406
65°	47 177	29.315
66°	45 407	28.218
67°	43 622	27.106
68°	41 823	25.988
69°	40.012	24.862
70°	38 188	23.729
71°	36 353	22.589
72°	34 506	21.441
73°	32 648	20.287
74°	30 781	19.127
75°	28 903	17.960
76°	27 017	16.788
77°	25 123	15.611
78°	23 220	14.428
79°	21 311	13.242
80°	19 394	12.051
81°	17 472	10.857
82°	15 545	9.659
83°	13 612	8.458
84°	11 675	7.255
85°	9 735	6.049
86°	7 792	4.842
87°	5 846	3.632
88°	3 898	2.422
89°	1 949	1.211
90°	0	0

Note: Degrees of longitude are measured along the parallel on the WGS84 ellipsoid.

Table A.7. Common Landsat 8 and 9 band combinations

Application	Bands	Uses	Example
Natural color	4 3 2	Excellent for visualizing and interpreting Earth's surface features as they appear to the human eye.	
Color infrared	5 4 3	Healthy vegetation is bright red, and stressed vegetation is dull red.	
Shortwave infrared	7 5 4	Useful for studying vegetation health and stress, change detection, disturbed soils, soil type, and camouflage detection.	
Agriculture	6 5 2	Vigorous vegetation is bright green, stressed vegetation is dull green, and bare areas are brown.	

Geology	7 6 2	Enables easier visualization and extraction of delineation of major structural features such as thrust faults and folds, detection of textural characteristics of igneous and sedimentary rocks, and mapping of lithological and geological features such as hydrothermal altered rocks.	
Bathymetry	4 3 1	Useful for bathymetric mapping water applications.	
Normalized difference vegetation index (NDVI)	(NIR − R) / (NIR + R)	Thick vigorous vegetation is dark green, sparse or dry vegetation and bare ground ranges from brown to tan, and water is white.	
Normalized difference moisture index (NDMI)	(NIR − SWIR1) / (NIR + SWIR1)	Higher moisture areas range from dark to light blue, and drier areas range from tan to white.	

Imagery courtesy of Esri, NASA, US Geological Survey, Microsoft, and Amazon Web Services.

Index

#

3D cadastre, 131
3D maps, 13–14
3D mesh, 267
3D surface maps, 225
7.5-minute topographic maps, 33, 50
100,000-meter square cells, 110
100,000-meter square identifier, 110

A

absolute relief maps
    benchmarks, 237
    contours, 239–41
    hypsometric tints, 241–43
    isobaths, 241
    soundings, 237–39
    spot elevations, 237
accessibility, 17
accessible maps, 17
acres, 117
actual scale, 63
aerial imagery
    aerial photos, 4
    capture, 266
    color-infrared imagery, 266
    drone imagery, 267–68
    natural-color imagery, 264–66
    orthophotomaps, 269–72
    orthophotos, 269
    panchromatic imagery, 264
aeronautical charts, 10, 33, 69
Agger, Michael, 287–88
agricultural census, 186
Airborne Visible/Infrared Imaging Spectrometer (AVIRIS), 282–83
air navigation, 10
air quality maps, 7
air rights, 131
Albers equal-area conic projection, 79–80

amounts, 203
anaglyphs, 254
analog image, 259
angular deformation, 80
animated maps, 14
annotated orthophotomaps, 269
antipodal meridian, 32
appearance compensation, 210
ArcGIS, 271, 272
ArcGIS Earth, 60, 287
ArcGIS Image Server, 278
ArcGIS Living Atlas of the World, 275
ArcGIS Pro, 269
ArcGIS Site Scan, 267
area, 66
area feature locations, 195–97
area feature maps
    qualitative thematic maps, 191–93
    quantitative thematic maps, 213–20
area features
    qualitative thematic maps, 186–87
    quantitative thematic maps, 206–8
area point symbol maps, 213
arpent, 117
arpent sections, 117
arroyos, 119
aspect, 70–72
aspect ratio, 32, 85
Astronomical Almanac, 30
attribute-based classification, 143
attributes, 197–98
authalic sphere, 31
automated vehicle location (AVL) technology, 112
automatic traffic recorders (ATRs), 206

azimuth, 66
azimuthal equidistant projections, 74–75
azimuthal projections, 66, 67

B

Babylonian sexagesimal system, 25
band combination, 276
bar scale, 45–46
baseline, 122
basemap, 6
basemaps, 180
bathymetric contour, 241
bathymetric maps, 241
bathymetry, 38, 62, 233
bearing, 76
benchmarks, 35, 237
Bertin, Jacques, 153
binned maps, 220
bivariate choropleth map, 154
block diagram, 251
boundary-walking, 121
bounds, 116
building information modeling (BIM), 131
building setback lines, 128

C

cadastral maps, 129
cadastre, 129
cardinal directions, 2
Cartesian coordinate system, 91
cartograms, 215–20
cartographic abstraction
    classification, 141–50
    contextual generalization, 139
    definition, 137
    generalization, 139–41
    selection, 137–39
cartographic artifacts, 288
cartographic maps, 1, 3

cartometric analysis, 10
cartometric maps, 10
cartometrics, 10
case, 70
categorical map, 191
celestial globes, 59
census, 186
census data, 186–87
Census of Population and Housing, 188
Centaurus, 30
central conic projections, 69
central cylindrical projections, 68
central meridian, 72–73, 95
central parallel, 73
centrifugal forces, 27
Cervantes, Miguel de, 290
chains, 116
change maps
    qualitative thematic maps, 195–98
    quantitative thematic maps, 227–30
charts, 6, 9–10, 33, 69, 74, 239
choropleth maps, 144, 215
ChromaDepth maps, 254
chronometer, 30
Clarke, Arthur C., 272
Clarke 1866 ellipsoid, 28, 35, 95
Clarke orbit, 272
classification
    attribute-based classification, 143
    definition, 141
    proximity-based classification, 143
    qualitative data classification, 143
    quantitative data classification, 144–50
class intervals, 144
closed traverse, 128
clustering, 215
CMY color model, 160–61
cognitive maps, 1
color
    additive color mixing, 159
    additive primaries, 159
    catagorical schemes, 161
    CMYK color mixing, 159
    continuous, 161
    diverging schemes, 162
    models, 159–61
    offset printing, 159
    progressions, 161
    properties, 158
    qualitative schemes, 161
    ramps, 161
    schemes, 161–62
    sequential schemes, 161–62
    subtractive color mixing, 158
    subtractive primaries/process colors, 158–59
    systems, 158–59
color-composite images, 276
color-infrared (CIR) film, 266
color-infrared imagery, 266
color vision deficiency, 17
community grants, 119
compass rose, 175
complete coverage multizone layer, 98
completeness, 63
composite index, 227
composite indicators, 227
composite variable maps, 227
conformal conic projections, 69
conformality, 65–66
conformal maps, 66
conformal projections, 65–66
congressional townships, 123
conic projections, 69–70, 79–80
Conrad, Joseph, 290, 295
conterminous United States, 46
contiguity, 215
contiguous cartograms, 218
continuity, 64
continuous hypsometric tints, 241
continuously operating reference stations (CORS), 40
continuous surface maps, 220–25
continuous surfaces, 209–10
contour interval, 240
contour lines
contours, 239–41
control points, 35
converging meridians, 32
correspondence, 63

Cossin, Jean, 81
counts, 203
course, 116
cross-blended hypsometric tints, 243
crowdsourcing, 205
Crux Australis (Southern Cross), 30
cyclical phenomena, 230
cylindrical equal area projections, 68
cylindrical projections, 68–69, 76–78
cylindrical projections family, 68

D

dasymetric maps, 215
data collection unit, 186
data normalization, 210
datum, 35
datum contour, 239
decimal degrees, 25–26
degrees (°), 25
democratization of cartography, 133
depression contours, 241
depressions, 241
depth contour, 241
depth curve, 241
depth sounders, 238
Descartes, René, 91
developable surface, 60, 67
Dickey, James, 290–91
differential leveling, 235
digital elevation model (DEM)
    data collection, 235–36
    definition, 233
    hypsometric tints, 241–43
digital image, 259
digital orthophotoquad (DOQ), 269
digital surface model (DSM), 233, 267
digital terrain model (DTM), 233, 267
digital twins, 181–82
digitization, 184
diorama map, 252
discrete hypsometric tints, 241
discrete layer tints, 241
distance, 64–65
distortion, 99–104
distribution, 179

distributive flow maps, 212
divine proportion, 176
DMS (degrees/minutes/seconds), 25
Dominion Land Survey, 128
Donation Land Claim Act of 1850, 119
donation land claims (DLCs), 119–20
donation land grants, 119–21
dot density maps, 223–25
dot maps, 215
dot size, 224
dot value, 224
drone imagery, 267–68
dynamic maps
    definition, 14
    imagery, 284–86
    qualitative thematic maps, 198–201
    quantitative thematic maps, 230–31
    relief portrayal, 254–56

E

earth
    as geoid, 37–39
    as oblate ellipsoid, 27–28
    as sphere, 23–27
earth resources satellites, 275–381
easements, 128
easting, 93, 95
ecoregion map, 191
elevation profile, 240
elevation tints, 241
ellipsoid distance, 99
ellipsoids, 27–28, 62, 101–4
enumerate, 186
enumeration area, 186
enumeration unit, 186
enumerators, 188
Epsilon, 30
equal area, 66
equal-area projections, 67, 68
equal-area world maps, 66
equator, 24
equatorial aspect, 70
equatorial plane, 272
equidistance, 64–65

equidistant cylindrical projections, 76
equidistant projections, 65, 68
equirectangular projection, 76
equivalence, 66
Eratosthenes, 23–24
Euclidean geometry, 2

F

false-color images, 266, 277
false easting, 93, 95
false northing, 93, 95
fathom, 239
field observation, 184
field sketch, 184
Flamsteed, John, 81
Flannery, James, 210
flight lines, 266
flow lines, 212–13
flow maps, 212–13
fly-through, 254, 284
focal length, 264
fonts, 163–64
foreshortening, 85
formal balance, 179
forward overlap, 266
French arpent land grants, 117
French long lots, 117–19
front overlap, 266
functional divisions, 124

G

Gauss, Carl, 78
Gauss-Krüger projection, 78
General Land Office (GLO), 120
general-purpose basemaps, 180
general purpose maps, 6
generating globe, 63, 66
geocentric coordinates, 62
geocentric latitude, 24–25, 29, 33
geocentric longitude, 25–26, 29
geocentric view, 1
geodetic latitude
    appearance on maps, 33
    definition, 29
    determining, 29–31
    on large-scale maps, 35–37
    map projections, 62

geodetic longitude
    definition, 29
    determining, 29–31
    on large-scale maps, 35–37
    map projections, 62
Geodetic Reference System of 1980 (GRS80), 36, 95, 103, 104
geographer's line, 122
geographic coordinate, 25
geographic information systems (GIS)
    apps, 60, 76, 86, 87, 184, 267, 269, 271, 272, 275, 278, 285, 287, 293, 294, 297
    binned maps, 220
    color models, 159–61
    dasymetric maps, 215
    data analysis, 203, 227
    data normalization, 210
    land information systems, 109
    low-distortion projections, 101–2
    map compilation, 135
    map transformation process, 5
    modeling line features, 186
    tactile maps, 17
    use in state/local government, 104–5
geoid, 37–39, 62
geoidal undulation, 62
geoid height, 38–39, 103
geologic maps, 7–8
geometric distortion, 64
geometric reference framework, 2
geomorphology, 247
geospatial technologies, 5
Geostationary Operational Environmental Satellite (GOES), 272
geostationary orbit, 272
global geoid, 38
Global Navigation Satellite System (GNSS), 29–31, 36–37, 40, 41, 94, 96, 97, 104, 112, 184
Global Positioning System (GPS), 5, 29, 40
globes, 59–60
glyphs, 227
gnomonic projections, 68, 74

GOES-East, 272–75
GOES-West, 272–75
golden mean, 176
golden ratio, 176
golden rectangle, 176
golden section, 176
golden spiral, 176
Goode, J. Paul, 81, 82
Goode homolosine projection, 81–83
Google Earth, 287–88
Google Maps, 77, 287
government lots, 125
graduated symbol maps, 210–12
graduated symbols, 210, 215
graticule
    appearance on maps, 33–34, 72
    characteristics, 24–27
    equal-area world maps, 66
    properties, 31–33
gravity, 28
great circle, 32
Greenwich Mean Time (GMT), 30
Greenwich meridian, 26
grid coordinate systems
    Cartesian coordinate system, 91
    conformal conic projections, 69
    definition, 91
    determination on maps, 105–9
    grid cell location systems, 109–13, 109–14
    grid convergence, 106
    grid coordinate measurement, 107–9
    grid coordinates, 92–105
    grid orientation, 106
    grid zone designation, 110
grid distance, 99, 101
ground distance, 99, 101
ground features, 53
ground resolution, 56–57
ground sample distance (GSD), 262
Gurnah, Abdulrazak, 287

H
hachures, 251
half sections, 124
Harrison, John, 30
heat maps, 222
Heat Moon, William Least, 291–92
hectares, 117
height above the ellipsoid (HAE/ellipsoidal height), 38–39
Heller, Joseph, 288
hemisphere, 60
hexmap, 220
hillshading, 246
hillsigns, 251
Hipparchus, 24, 73
Historical Topographic Map Collection (HTMC), 46–48
Homestead Act of 1862, 120
homogeneity, 188
horizontal datums/horizontal reference frames, 35
HSV color model, 161
human-made features, 10
Humboldt, Alexander von, 284
hydrologic units, 208
hyperspectral systems, 282–83
hypsometric tints, 241–43
hypsometry, 233

I
image map, 259
image quality
    relief displacement, 263–64
    sensor height, 262–63
    spatial resolution, 262
    spectral sensitivity, 261–62
imagery
    aerial imagery, 264–68
    dynamic maps, 284–86
    image quality, 261–64
    orthophotomaps, 269–72
    orthophotos, 269
    remote sensing, 259–86
    satellite imagery, 272–84
image sensors, 259
image tile layers, 50
Imperial measurement system, 31
Imperial units, 31
inclination, 275
inclusive maps, 17
index contours, 240–41
informal balance, 179
initial point, 121
inset maps, 175
interactive maps, 14, 15
intermediate contours, 241
International Meridian Conference, 26
international reference meridian, 36
international reference pole, 36
International Terrestrial Reference Frame (ITRF), 37
Internet of Things (IoT), 205, 297
interpolation, 209
interrupted Goode homolosine projection, 82
interrupted homolosine projection, 82
interrupted projection, 82
isobaths, 241
isochrons, 220
isohyets, 220
isoline maps, 220–22
isolines, 209

J
Jenks, George, 149

K
Krüger, Johann, 78

L
Lambert, Johann Heinrich, 75
Lambert azimuthal equal-area planar projections, 75
Lambert azimuthal equal-area planar projections, 68
Lambert conformal conic projection, 79, 95
land information systems, 109, 129
land miles, 31
land navigation, 10
Land Ordinance of 1785, 121
land partitioning
    definition, 115
    irregular systems, 115–21
    land records, 128–31
    regular systems, 121–28
land records, 128–31
land record system, 128

Landsat, 276–78
landscape drawings, 252
large-scale equidistant projections, 87
large-scale maps
    characteristics, 33–35
        geodetic latitude/longitude, 35–37
        map scale, 46–50
latitude of center, 73
latitude of origin, 73
layered zones, 98
layer tints, 241
lead line, 237–38
legend, 174
legend patches, 178
leveling, 37
levels of detail (LODs), 50
lidar return, 236
light detection and ranging (lidar), 235–36
linear distortion, 99, 101
linear interpolation, 209
line feature locations, 195
line-feature maps, 191
line feature maps
    qualitative thematic maps, 191
    quantitative thematic maps, 212–13
line features
    qualitative thematic maps, 185–86
    quantitative thematic maps, 206
line of tangency, 68
links, 116
lobe, 82
local geoid, 38
location, 195–97
location-based service (LBS), 206
locator maps, 175–76
longitude of center, 73
longitude of origin, 73
longitudinal distance, 55–56
long lots, 117
low-altitude images, 262–63
low-distortion projections, 69, 101–4

low-distortion projections (LDPs), 101–2

M

Mailer, Norman, 293
map analysis, 19
map audience, 133
map design
    cartographic abstraction, 137–50
    color, 158–62
    compilation, 135
    considerations, 133–35
    definition, 19, 133
    map projections, 135–37
    map scale, 135
    symbols, 150–58
    type, 162–64
map interpretation, 19
map layouts
    golden ratio, 176–78
    harmony, 179
    layout design, 176–79
    map elements, 174–76
    proportion, 178–79
    web maps, 179–82
MapMaker web app, 14
map marginalia, 174
map organization
    basic design concepts, 167–73
    closed form, 167
    color contrast, 169–71
    drop shadow, 167
    figure-ground organization, 167
    illumination, 167
    legibility, 171–73
    screening, 167
    vignette, 167
    visual angle, 172
    visual contrast, 169–71
    visual hierarchy, 167–69
map perspective, 237
map projections
    commonly used, 73–86
    families, 59, 64–70
    globes versus flat maps, 59–61
    map design, 135–37
    parameters, 70–73
    process, 61–62

    properties, 62–64
    selection, 87
    on sphere/ellipsoid, 86–87
map purpose, 133
map reading, 19
maps
    as art, 17–18
    cartographic artifacts, 288–89
    characteristics, 3–4
    as communication, 296
    credibility, 5
    definition, 1
    geodetic latitude/longitude on large-scale maps, 35–37
    graticule appearance, 33–35
    grid coordinate determination, 105–9
    imagery, 259–86
    limitations, 289, 294, 296–97
    as metaphor, 296
    misuses, 292–94
    PLSS appearance, 127–29
    popularity, 4–5, 297
    as positional reference system, 23
    as reality, 287–97
    types, 5–18
map scale
    conversions, 51–52
    determining, 52–57, 87
    expressing, 43–45
    large-scale maps, 33–37, 46–50
    map design, 135
    multiscale maps, 50–51
    small-scale maps, 46–50
    topographic maps, 46–50
map surrounds, 174
map transformation process, 5
map use, 19, 289–92
Marinus of Tyre, 76
mean center of population, 183
mean high water (MHW), 239
mean lower low water (MLW), 238–39
mean low water (MLW), 238
means, 144
mean sea level (MSL), 37–38
medium-scale maps, 46
mental maps, 1–2

Mercator, Gerhardus, 76
Mercator projection, 46, 51, 68–69, 76–79, 87
meridians, 26–27, 30, 32, 33–35, 36, 55–56, 66, 71, 74, 95, 102, 122
metadata, 175
meter, 31
metes, 115–17
Metonic cycle, 37
metric system, 31
metric units, 31
Mexican land grants, 119
mid-infrared (mid-R), 261
Military Grid Reference System (MGRS), 109–12
mimetic symbols, 153
minutes ('), 25
Moderate Resolution Imaging Spectroradiometer (MODIS), 278–81
Mollweide, Carl B., 80
Mollweide projection, 67, 80, 87
monument, 35
multidimensional data, 230
multidirectional relief shading, 246–47
multilayer zone layers, 98
multipurpose cadastre, 129–31
multispectral imagery, 275
multivariate areas, 195
multivariate lines, 193–95
multivariate maps
   qualitative thematic maps, 193–95
   quantitative thematic maps, 225–31
multivariate points, 193
multivariate symbols, 225–27
multizone complete zone layer, 98
multizone partial zone layer, 98

N

National Agriculture Imagery Program (NAIP), 264
National Geodetic Survey (NGS), 39, 97, 98, 101, 102
National Geodetic Vertical Datum of 1929 (NGVD29), 38
National Geographic Society, 85
national mapping agencies (NMAs), 46
National Oceanic and Atmospheric Administration (NOAA), 39, 174, 178, 272
National Spatial Reference System (NSRS), 39–41, 97
National Spatial Reference System (NSRS) modernization, 40–41
natural-color imagery, 264–66
Natural Earth projection, 84
natural features, 10
nautical charts, 9–10, 33
nautical miles, 31
navigational maps, 32
near-infrared (near-R), 261
near-polar orbits, 275
neat line, 174
network flow maps, 212
Newton, Isaac, 27, 28
noncontiguous cartograms, 218–19
nonintersecting map projection surface, 101
normal aspect, 70–71
normal distribution, 144
normal orientation, 175
North American Datum of 1927 (NAD27), 35–36, 95
North American Datum of 1983 (NAD83), 35–36, 39, 40, 95, 96, 105
North American Datum of 1983 (NAD83) reference frame, 40
North American Treaty Organization (NATO), 109
North American Vertical Datum of 1988 (NAVD88), 38, 39, 40–41
northing, 93
North Polar Zone (NPZ), 94–95
Northwest Territory, 121

O

oblate ellipsoid, 27–28
oblique aspect, 70
oblique-perspective maps, 86, 237, 250–52
oblique-perspective projections, 85–86
observations, 144
open traverse, 128
Operational Land Imager (OLI), 276
operational layer, 180
orbital period, 272
Oregon coordinate reference system (OCRS), 101, 103–4
Oregon Lambert system, 101, 105
orientation indicator, 175
orthographic projections, 68, 73
orthometric height, 38–39
orthophoto, 4
orthophotomap, 4
orthophoto quadrangles, 269
orthorectification, 269
outliers, 144
overlap, 266

P

panchromatic imagery, 264
pan-sharpening, 281
parallax, 252
parallels, 23, 31, 32–33, 54–55, 66, 70, 74, 87, 95, 102
Paris meridian, 27
partial-coverage multizone layer, 98
Paterson, Tom, 84
people with disabilities, 17
pictographic symbols, 153
pixel density, 56–57
planar projection family, 68
planar projections, 67–68, 73–75
plane surveying, 128
plan perspective maps, 237, 246–50
plate carrée ("flat square"), 76
platform, 259
plats, 115
platted lots, 125
platted subdivision, 115
Pléiades NEO, 281
pointable sensors, 281–83
point clouds, 236
point feature locations, 195

point feature maps
    qualitative thematic maps, 189–91
    quantitative thematic maps, 210–13
point features
    qualitative thematic maps, 183–85
    quantitative thematic maps, 205–6
point of beginning (POB), 116, 128
point of tangency, 67–68
point-to-point correspondence, 63
polar aspect, 70
polar-aspect azimuthal equidistant projections, 87
polar axis, 27
Polaris (North Star), 29–30
population census, 186
population count, 203
population dependence, 210
positional reference system, 23
prime meridian, 26–27, 30
principal meridian, 122
principal scale, 63
prism maps, 220
private grants, 119
projection axis, 101
projection surface, 67
property abstracts, 124
proportional symbol maps, 210–12
proportional symbols, 210, 215
proximity, 215
proximity-based classification, 143
pseudocontiguous cartogram, 219
pseudocylindrical map projections, 80–85
Public Land Survey System (PLSS), 115, 120, 121–28, 129
Pueblo grants, 119

Q

quadrangle, 48, 50
quadrants, 91
quadrats, 204
quadrilateral maps, 33
quadrilaterals, 32, 50, 66, 102

quads, 50
qualitative data, 143
qualitative data classification
    definition, 143
    qualitative visual variables, 153–54
qualitative thematic maps
    area features, 186–87
    data accuracy, 187–88
    dynamic maps, 198–201
    homogeneity, 188
    line features, 185–86
    mapping change, 195–98
    multivariate maps, 193–95
    qualitative information, 183–85
    single-theme maps, 189–93
qualitative visual variables
    area patterns, 153
    arrangement, 153–54
    hue, 153
    line patterns, 153
    mimetic symbols, 153
    orientation, 153
    pattern arrangement, 153
    pattern orientation, 153
    pictographic symbols, 153
    shape, 153
quantitative data classification
    class breaks, 144, 150
    classing methods, 144, 147–50
    critical break, 150
    critical value, 150
    equal frequency intervals, 148
    equal intervals/equal-range intervals, 148
    Jenks optimization/Jenks natural breaks, 149
    natural breaks, 149–50
    number of classes, 144–47
    quantile (equal frequency) intervals, 148
    quartiles, 148
    quintiles, 148
quantitative thematic maps
    data normalization, 210
    dynamic maps, 230–31
    mapping change, 227–30

    multivariate maps, 225–31
    quantitative information, 203–10
    single-theme maps, 210–25
quantitative visual variables
    definition, 154
    pattern texture, 154
    quantitative data, 154
    saturation, 154
    size, 154
    texture, 154
    value, 154
quarter-quarter sections, 124
quarter sections, 124

R

radial flow maps, 212
radius of curvature, 28
raised-relief globe, 244–45
raised-relief maps, 245
random samples, 204
range lines, 123
real property, 129
real-time data, 14
real-time maps, 14–15
reference grids, 113–14
reference layer, 180
reference maps, 5–6, 7
reference material, 54
relative relief, 243
relative relief maps, 243–45
relief, 233
relief portrayal
    absolute relief maps, 237
    anaglyphs, 254
    DEM data collection, 235–36
    elevation data, 233
    map perspective, 237
    oblique-perspective maps, 250–52
    plan perspective maps, 246–50
    relative relief maps, 243–52
    spatial resolution, 233–35
    True 3D, 252–54
    variable perspective maps, 243–45
    vertical exaggeration, 236
relief shading
    contours, 250

definition, 246
  hypsometric tints, 249–50
  shadowing, 247–49
remote sensing, 259
representation, 3, 5, 19
representative fraction (RF), 43, 46, 51–52
responsive design, 181
returns, 236
RGB color model, 159–60, 161
rhumb lines, 76
Robinson, Arthur, 83
Robinson projection, 83
route line, 195
Royal Observatory, Greenwich, England, 26

S

Saint-Exupéry, Antoine de, 294
sample point, 204
Sanson, Nicholas, 81
Sanson-Flamsteed projection, 81
sans serif fonts, 164
satellite image map, 272
satellite imagery
  definition, 259
  earth resources satellites, 275–381
  hyperspectral systems, 282–83
  maps, 283–84
  submeter-resolution systems, 281–82
  weather satellites, 272–75
scale, 62–63, 139–41
scale bar, 45–46
scale bar expression, 45
scale conversions, 51–52
scale factor (SF), 63, 66, 70, 72, 101
scale indicator, 175
scale variation, 263
scientific maps, 8–9
secant-case cylindrical projections, 70, 72
secant-case planar projections, 70, 72
seconds ("), 25
sections, 124
seigneuries, 117
seismometers, 203

semi-major axes, 27, 28
semi-minor axes, 27, 28
sensors, 259
Sentinel-2, 278
serial imagery, 259
sextant, 29
shaded relief, 246
shape, 65–66
ship traffic maps, 7–8
sidelap, 266
side overlap, 266
simulacrum, 287
single-theme maps
  qualitative thematic maps, 189–93
  quantitative thematic maps, 210–25
sinusoidal projections, 81
"slippy" web maps, 50–51
small circle, 32
small multiples, 197–98
small-scale maps, 33, 46–50
smooth terrain models, 245
sounding pole, 238
soundings, 237–39
source scale, 139
South Polar Zone (SPZ), 94–95
Spanish land grants, 119
sparklines, 228
spatial resolution, 233–35
spatial samples, 203–5
spatiotemporal data, 228
SPCS 83, 97
special purpose maps, 6
spectral bands, 262
spectral signatures, 282
spherical geometry, 2
spot elevations, 237
stacked scale bar, 45
stadion, 24
standard parallels, 70
standard symbol, 190
star charts, 74
state grids, 104–5
state plane coordinate system (SPCS), 95–104
State Plane Coordinate System of 1927 (SPCS 27), 95–98

State Plane Coordinate System of 1983 (SPCS 83), 95–98
State Plane Coordinate System of 2022 (SPCS2022), 97–104
statewide zone layer, 98
static maps, 14
statute miles, 31
Steinbeck, John, 289, 294
stepped terrain models, 245
stereographic projections, 68, 73
stereoscopic pairs, 252
stereoscopic views, 252–54
story maps, 15–17
strata, 204
stratified sample, 204
stream gauges, 203
stream reach, 206
subdivision, 115
subdivision plats, 128
submeter-resolution systems, 281–82
subsurface rights, 131
sun-synchronous orbit, 275
survey, 121
surveying, 116
survey irregularities, 125–26
survey mark, 35
survey townships, 123
symbolization, 150
symbols
  bivariate symbols, 154–58
  categorical data, 151
  classed data, 151
  definition, 150
  graphic marks/graphic primitives, 150
  interval data, 151–53
  level of measurement, 150–53
  multivariate symbols, 154–58
  nominal data, 150–51
  ordinal data, 151
  pixels, 150
  ratio data, 153
  redundant symbols, 154
  visual variables, 153–54
  voxels, 150
synchronous, 272
systematic samples, 204

T

tactile maps, 17
tangent-case planar projections, 70
tangent-case projections, 70
terrain models, 245
terrain profile, 240
terrain surface, 233
Thales of Miletus, 74
thematic maps, 6, 7–9
thermal-infrared (thermal-R), 261
Thermal Infrared Sensor (TIRS), 276
Thoreau, Henry David, 289
time series maps, 228
Toffler, Alvin, 297
topographic contours, 240
topographic maps, 4, 10, 33, 46–50, 87, 105
topographic surface, 97, 99, 101
topography, 10, 38, 62
topo maps, 46
topos, 10
Township and Range System, 121
township lines, 123
townships, 123–24
transect lines, 204
transverse aspect, 70
transverse aspect cylindrical projection, 72
transverse Mercator projection, 78, 95
traverse, 128
Tropic of Cancer, 24
Tropic of Capricorn, 24
True 3D, 252–54
true-color images, 265, 277
true-direction projections, 66, 68
true perspective conic projections, 69
true perspective cylindrical projections, 68
true perspective projections, 67, 68
Tufte, Edward, 228
Twain, Mark, 288–89
type
    definition, 162–63
    properties, 163–64
typeface, 163

U

unclassed maps, 145
unclosed traverse, 128
uncrewed aerial vehicles (UAV), 267
uncrewed aircraft system (UAS), 267
uninterrupted Goode homolosine projection, 81
unity, 63
universal polar stereographic (UPS) system, 94–95, 109
universal transverse Mercator (UTM) system, 92–94, 104, 105, 106–10
US Bureau of Land Management (BLM), 120, 127
US Census Bureau, 186
US Department of Homeland Security (DHS), 112–13
US Forest Service (USFS), 127
US Geological Survey (USGS), 33, 46–50, 103, 105, 119, 127, 208
US Geological Survey (USGS) TopoBuilder web app, 48
US National Grid (USNG), 106, 112–13
US National Grid for Spatial Addressing, 112
US National Ocean Service, 239
US National Park Service (NPS), 190
US Topo (UST), 48

V

variable contour density, 240
variable perspective maps, 243–45
variable scale bar, 45–46
varied perspective maps, 237
vector flow maps, 213
vector tile layers, 50
vertical exaggeration, 236
vertical images, 263
vertical perspective projections, 85
vertical reference datum (vertical datum), 37
virtual reality, 287
visible light, 261
visual center, 179
visual equilibrium, 178

W

water/marine navigation, 10
watersheds, 208
weather satellites, 272–75
web apps, 179
web maps, 6, 14, 50–51, 85, 135, 164, 176, 179–82
web Mercator projection, 77–78
white space, 179
Winkel, Oswald, 84–85
Winkel tripel projection, 84–85
Wisconsin transverse Mercator (WTM) system, 104
word scale, 43–45, 51–52
World Geodetic System of 1984 (WGS84), 28, 31, 35, 36–37, 38, 53, 86–87
WorldView-3, 281

X

x-axis, 91, 95

Y

y-axis, 91

Z

zero contour, 239
zone, 98
zone layer, 98
zoom level, 50
z-value, 13

About Esri Press

Esri Press is an American book publisher and part of Esri, the global leader in geographic information system (GIS) software, location intelligence, and mapping. Since 1969, Esri has supported customers with geographic science and geospatial analytics, what we call The Science of Where. We take a geographic approach to problem-solving, brought to life by modern GIS technology, and are committed to using science and technology to build a sustainable world.

At Esri Press, our mission is to inform, inspire, and teach professionals, students, educators, and the public about GIS by developing print and digital publications. Our goal is to increase the adoption of ArcGIS and to support the vision and brand of Esri. We strive to be the leader in publishing great GIS books, and we are dedicated to improving the work and lives of our global community of users, authors, and colleagues.

Acquisitions
Stacy Krieg
Alycia Tornetta
Jenefer Shute

Product Engineering
Craig Carpenter
Maryam Mafuri

Editorial
Carolyn Schatz
Mark Henry
David Oberman

Production
Monica McGregor
Victoria Roberts

Sales & Marketing
Eric Kettunen
Sasha Gallardo
Beth Bauler

Contributors
Christian Harder
Matt Artz

Business
Catherine Ortiz
Jon Carter
Jason Childs

Related titles

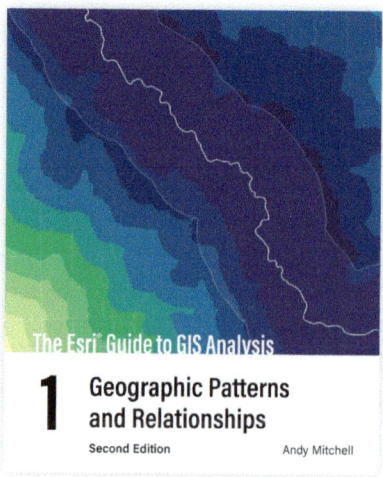

The Esri Guide to GIS Analysis: Volume 1, Geographic Patterns and Relationships, second edition
Andy Mitchell
9781589485792

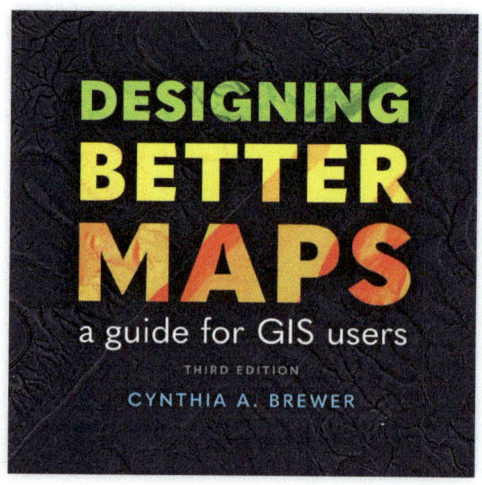

Designing Better Maps: A Guide for GIS Users, third edition
Cynthia A. Brewer
9781589487826

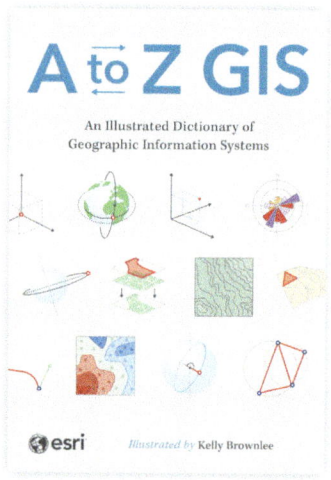

A to Z GIS: An Illustrated Dictionary of Geographic Information Systems, third edition
Esri, illustrated by Kelly Brownlee
9781589488113

Cartography.
Kenneth Field
9781589484399

For more information about Esri Press books and resources, or to sign up for our newsletter, visit

esripress.com.